Second Edition

ANIMAL BEHAVIOR:
CONCEPTS, PROCESSES, AND METHODS

LEE C. DRICKAMER
Williams College

STEPHEN H. VESSEY
Bowling Green State University

WADSWORTH PUBLISHING COMPANY
Belmont, California
A Division of Wadsworth, Inc.

Library of Congress Cataloging in Publication Data

Drickamer, Lee C.
  Animal behavior.

  Includes bibliographies and index.
  1. Animal behavior. I. Vessey, Stephen H.
II. Title.
QL751.D72 1986   591.5'1   85-28222

ISBN 0-87150-919-9

Sponsoring Editor: Jean-François Vilain
Production Coordinator: Robine Storm van Leeuwen
Editorial Assistant: Traci Sobocinski
Production: East of the Sun
Cover Designer: Robine Storm van Leeuwen
Cover Photo: Visuals Unlimited / A. J. Copley © 1985
Typesetting: Progressive Typographers, Inc.
Cover Printing: Lehigh Press Lithographers
Printing and Binding: The Maple-Vail Manufacturing Group

Printed in the United States of America
86 87 88 89 90 — 10 9 8 7 6 5 4 3 2

For John A. King, on the occasion of his retirement
—LCD

For my wife, Kristin, and for Dr. David E. Davis
—SHV

PREFACE

The study of animals and their behavior has been approached from several perspectives: ethology, comparative psychology, behavioral ecology, and, more recently, evolution. In this text, although we place some emphasis on evolution as a foundation for the study of behavior, we expose the student to a number of approaches and viewpoints and show that these different approaches are complementary, not mutually exclusive.

This text is suited for undergraduates taking their first course in animal behavior and assumes only a limited knowledge of general biology. Our approach and the overall themes regarding concepts, processes, and methods also make the book useful for more advanced undergraduate and graduate students. The nineteen chapters cover all the topics encompassed by modern animal behavior. Because each chapter is self-contained, instructors can select the coverage that suits their personal approach or the needs of their particular course.

Our presentation centers on the questions researchers ask about animal behavior and on the methods they use to answer these questions. We first define and elaborate on the concepts and processes that underlie behavior; and then, in the context of appropriate research examples, we present the methods animal behaviorists use. In this manner the student is introduced to a variety of viewpoints that have contributed to the richness of the discipline.

The text is divided into five major parts. Part I covers the background for the study of animal behavior, including history, methods, approaches, genetics, and evolution of animal behavior. In Part II the mechanisms and processes that control behavior are dealt with. Part III examines the behaviors of individuals and groups from proximate and ultimate perspectives. Part IV treats topics in behavioral ecology; and Part V is devoted to the evolution of behavior and of social behavior. The first half of the book emphasizes *how* (proximate) questions, the latter half of the text places stronger emphasis on *why* (ultimate) questions. We feel that this organization of the study of animal behavior—from questions pertaining to events inside the animal to discussions of complex environmental interactions—gives a more complete picture of the discipline of animal behavior today than would have been afforded by a functional scheme of classification (for example, maintenance activities and coping with the environment) or by an organization based on Niko Tinbergen's scheme of four questions about behavior (causation, development, evolution, and function).

CHANGES IN THE SECOND EDITION

The preceding paragraphs introduced the first edition of our book. Reactions to that edition encouraged us to keep the philosophy and approach that made it successful. This revision allowed us to fine-tune the material and to make changes that will improve its usefulness. Some of the following changes were suggested by colleagues who used the book:

- All chapters have been updated to reflect the latest research. The end-of-chapter references now include articles published in 1985.

- All chapters have been revised to make the presentation and flow of material even clearer and more accessible.

- The discussions of genetics (Chapter 5) and of the nervous system (Chapter 6) have been carefully fine-tuned and made more accessible.

- Chapters 8 and 9, entitled "Biological Timekeeping" and "Development of Behavior," respectively, have been thoroughly revised, reorganized and expanded.

- Chapter 11, "Sexual Behavior and Reproduction," and Chapter 19, "Evolution of Societies," have also been extensively revised and reflect the most current understanding of these topics.

- "Habitat Selection," (Chapter 15) now introduces Part IV and is followed by "Feeding Relationships." This reversal of order from the first edition makes for a more logical flow of presentation.

- Finally, new drawings and photographs make this edition more pedagogically effective and attractive.

    As in the first edition, discussion questions, extensive references, and a list of annotated suggested readings follow each chapter and provide points of departure for further exploration. A subject index and an author index facilitate access to the material. Finally, we have paid careful attention to the illustrations, which are meant to support and enhance the textual material and to convey a sense of the excitement of the field; we hope we have succeeded in this attempt.

ACKNOWLEDGMENTS

We would like to extend our sincere thanks to a number of individuals who have helped us in many ways during the several years that it has taken to complete this book. Special thanks go to our colleagues who read all or part of the manuscript at different stages and provided helpful criticism. To the following who reviewed the first edition: Terry Christenson — Tulane University; Victor DeGhett — SUNY College at Potsdam; Kenneth Able — SUNY at Albany; Robert Mathews — University of Georgia; Ronald Barfield — Rutgers – The State University; Irwin Bernstein — University of Georgia; David Barash — University of Washington; H. B. Graves — The Pennsylvania State University; and Gordon Gallup, Jr. — SUNY at Albany; to Donald Dewsbury, University of Florida, and George Waring, Southern Illinois University, who shared their and their students' experience with the first edition. Their comments helped us shape the revision. And thanks to the following, who read all or parts of the second edition manuscript at different stages and provided helpful criticism: Patricia DeCoursey — University of South Carolina – Columbia; Maureen Powers — Vanderbilt University; George Waring — Southern Illinois University; Arthur Coquelin — University of California – Los Angeles; Terry Christenson — Tulane University; Carl Bassi — Vanderbilt University.

Jerry Lyons encouraged us to write the book, many years ago; Jean-François Vilain deserves special mention for his help and patience through the good times and bad of these two editions. Final manuscript editing and preparations were handled in a competent and able manner for both editions by Robine Storm van Leeuwen; her touch is evident throughout the book. Nancy Murphy and Jeanne Bolden of East of the Sun helped greatly with the production of this edition. To these people we are most grateful. We would also like to thank Kristin Vessey for her editorial advice and typing and Karen Drickamer for her typing. Both our wives deserve a special thanks for their enduring patience as we labored to put the manuscript together.

Lee C. Drickamer
Williamstown

Stephen H. Vessey
Bowling Green

CONTENTS

PART ONE   STUDY OF ANIMAL BEHAVIOR   2

1   INTRODUCTION   5

Why We Study Animal Behavior   5

How We Study Animal Behavior: Four Processes   6

Physiological processes 6;   Social processes 8;   Behavior-
ecological processes 8;   Evolutionary processes 9

Evolutionary Perspective   10

Comparative Studies   10

2   HISTORY OF ANIMAL BEHAVIOR   13

Interest in Animal Behavior   13

Early humans 13;   Classical world 14

Foundations of Animal Behavior Studies   16

Theory of evolution by natural selection 16;   Comparative
method 18;   Theories of genetics and inheritance 19

Four Experimental Approaches   20

Studies of function and evolution 20;   Studies of
mechanisms 22;   Behavioral ecology 25;   Sociobiology 26

3   APPROACHES AND METHODS   32

Ethology   33

Comparative Psychology   37

Design Features in Animal Behavior Studies   39

Definitions and records 39;   Design of the
experiment 41;   Variation and variance 46;   Field
methods 47;   Observational methods 48;   Methodological
pitfalls 50

4   GENETICS AND EVOLUTIONARY PROCESSES   57

Genetics   58

Species and gene pools 58;   Basic principles 59

Evolutionary Processes   62

Relative genetic fitness 62;   Survival 62;   Darwin and biological
evolution 63

Population Genetics   65

Mutation and recombination 65;   Causes of evolution 67

Selection, Adaptation, and Speciation   69

Natural selection 69;   Adaptation 71;   Speciation 74

Genetic Variation, Kinship, and Units of Selection   77

Polymorphism 77;   Measurement of genetic
relationship 78;   Units of selection 78

Evolutionary Stable Strategies   80

Problems with Evolutionary Explanations   82

PART TWO   CONTROL OF BEHAVIOR   86

5   BEHAVIOR GENETICS   89

Questions in Behavior Genetics   90

Methods of Assessing Genetic Determination   *90*

Inbreeding 91;   Strain differences 92;   Selective
breeding 93;   Cross-fostering 97;   Mutations 99

Modes of Inheritance   *100*

Twin Studies   *101*

Genes to Behavior   *102*

Genetics and Evolution   *103*

Gene frequency 103;   Adaptations 105

6   THE NERVOUS SYSTEM AND BEHAVIOR   *111*

The Nervous System   *111*

Communication of impulses 112;   Properties of nerve cells 114

Sensation and Perception   *115*

Sensory equipment 115;   Perception systems 117;   Sensory
filters 120

Evolution of the Nervous System   *122*

Simple systems 122;   Radially symmetrical nervous
system 122;   Bilaterally symmetrical nervous
system 122;   Generalized vertebrate nervous system 126

Methods of Investigating the Nervous System   *129*

Recording neural activity 129;   Transection 132;   Neural
stimulation 133;   Lesions and split-brain
techniques 136;   Functional neuroanatomy 139;
Psychopharmacology and neurotransmitters 141;
Cannulation 142;   Transplantation 143

Potential Problems   *143*

7   HORMONES AND BEHAVIOR   *151*

Comparison of the Endocrine and Nervous Systems   *152*

Invertebrate Endocrine Systems   *153*

Vertebrate Endocrine Systems   *154*

Endocrine gland secretions 156;   Endocrine gland interactions 158

Experimental Methods   *160*

Activational Effects   *161*

Secondary sexual characteristics 161; Aggression and sexual
behavior patterns 161; Sexual attraction 163;
Eclosion 163; Life stages 164; Molting 164; Color change 165

Organizational Effects   165

Endocrine-Environment-Behavior Interactions   169

Reproductive sequence in ring doves 170; Parturition and maternal
behavior in rats 174; Reproduction in lizards 176

Hormone-Brain Relationships   178

8   BIOLOGICAL TIMEKEEPING   187

Biological Rhythms   188

Types of rhythms 189; Endogenous pacemaker 191;
Zeitgebers 195; Model 196; Location and physiology of the
pacemaker 196

Significance of Biological Timekeeping   202

Ecological adaptations 202; Diurnality 203;
Hibernation 205; Migration 206

PART THREE   BEHAVIORS OF INDIVIDUALS IN GROUPS   214

9   DEVELOPMENT OF BEHAVIOR   217

Methods in Behavior Development   217

Aspects of experimental design 218; Testing procedures 219

Embryology of Behavior   219

Nervous system development 220; Sensory and motor
stimulation 221; Prenatal maternal experience affects offspring
behavior 224

Early Postnatal Events   225

Imprinting 225; Development of feeding and food preferences 229

Juvenile Events: Birds and Mammals   232

Deprivation experiments 233; Bird song 236; Puberty in female
house mice 238

Juvenile Events: Insects and Fish   240

Insects 241; Fish 244

Play Behavior 246

Canids 247;   Keas 249;   Human children 249

Development into Adult Life 250

Nature/Nurture/Epigenesis 252

Genetic influences 252;   Environmental
influences 256;   Epigenesis 257

10   LEARNING AND MOTIVATION 268

Types of Learning 268

Habituation 269;   Classical and operant
conditioning 269;   Classical and operant conditioning
compared 272;   Other aspects of learning 274

Phylogeny of Learning 278

Protozoa and Coelenterata 279;   Platyhelminthes and
Annelida 280;   Mollusca 281;   Arthropoda 283;   Vertebrata
284;   Comparative learning 286

Constraints on Learning 287

Preparedness 287;   Methodological constraints 288

Memory 290

Theories of memory 290;   Storage mechanisms 290

Motivation 291

Lorenz's model 292;   Deutsch's model 293

11   SEXUAL BEHAVIOR AND REPRODUCTION 301

Ultimate Factors in Reproductive Behavior 302

Diversity of reproduction 302;   Costs and benefits of sex 302

Sex Ratios 304

Sexual selection 305;   Intersexual selection 306;   Intrasexual
selection 308;   Mating systems 310;   Monogamy 311;
Polygyny 312;   Polyandry 313

Ecology and Mating Systems 314

Parents and Offspring 315

Parental investment 315;   Parental care and ecological
factors 321;   Parent-offspring conflict 322

Proximate Factors in Reproductive Behavior   323

Climatic-hormonal interaction 323; Social-hormonal
interaction 325

Reproduction in a Primate — The Rhesus Monkey   332

Mating behavior 332; Birth and infancy 336; Behavior
development 338

12   AGGRESSION   344

Agonistic Behavior   344

Types of competition 345; Extreme forms of aggression:
Cannibalism 346

Social Use of Space   347

Definitions of the use of space 348; Size and boundaries of
territory 351

Dominance   353

Dominance hierarchies 354; Dominance hierarchy in the rhesus
monkey 356; Advantages of dominance 358

Internal Factors in Aggression   359

Limbic system 359; Hormones 360; Genetics 361

External Factors in Aggression   361

Learning and experience 361; Pain and frustration 362

Social Factors   363

Restraint of Aggression   364

Displays 364; Games theory model 365; Relevance for
humans 366; Social control and social disorganization 367

13   COMMUNICATION   374

How Signals Convey Information   375

Discrete and graded signals 375; Distance and
duration 375; Composite signals, syntax, and
context 376; Metacommunication 377

Measurement of Communication   378

Observation 378; Quantification 378

Functions of Communication   381

Group spacing and coordination 382; Recognition 382;

Reproduction 384;   Agonism and social status 385;
Alarm 386;   Hunting for food 387;   Giving and soliciting
care 388;   Soliciting play 389;   Synchronization of hatching 389

Channels of Communication   389

Odor 389;   Sound 391;   Touch 394;   Surface
vibration 394;   Electric field 395;   Vision 396

Language   398

Food location in honeybees 398;   Language in chimpanzees 401

Origins of Display   403

Honesty in Communication   404

14   MIGRATION, ORIENTATION, AND NAVIGATION   411

Migration   411

Birds 411;   Mammals 417;   Invertebrates 418;   Others 420

Orientation and Navigation   420

Early work 420;   Birds 420;   Mammals 427;
Invertebrates 431;   Fish 431;   Amphibians and reptiles 432

*PART FOUR   BEHAVIORAL ECOLOGY
AND ANIMAL POPULATIONS*   442

15   HABITAT SELECTION   445

Factors Restricting Habitat Use   446

Dispersal 447;   Behavior 450;   Other species 450;   Physical and
chemical factors 452

Choice of Breeding Sites   452

Dispersal or philopatry 452;   Microhabitat choice and reproductive
success 455;   Environmental cues 455

Determinants of Habitat Preference   458

Heredity 458;   Early experience 458;   Tradition 462

16   FEEDING RELATIONSHIPS   467

Ecosystems and Trophic Levels   468

Foraging Strategies   470

Movement rules within a patch 471; Choice of food
items 472; Moving to a new patch 476

Feeding Techniques 477

Trophic levels 478; Resource partitioning 478; Modifying food
supply 478; Trapping and detecting 481; Aggressive
mimicry 483; Using tools 484

Feeding and Social Behavior 485

Defending a territory 485; Group feeding 486; Social
carnivores 489

Defense against Predators 495

Individual strategies 495; Social strategies 499

17 BEHAVIOR AND POPULATION REGULATION 508

Limiting Factors 508

Density-independent and density-dependent
factors 510; Predators 512; Intraspecific competition 512

Population Self-Regulation 512

Behavioral mechanisms 514; Behavioral-physiological
mechanisms 521; Behavioral-genetic mechanisms 523

Evolution of Population Self-Regulation 524

PART FIVE BEHAVIOR AND EVOLUTION 532

18 EVOLUTION OF BEHAVIOR PATTERNS 535

Microevolution 536

Evidence for the Evolution of Behavior 537

Phylogeny 537; Study of adaptation 547; Behavioral isolating
mechanisms in speciation 549

Tradition 551

19 EVOLUTION OF SOCIETIES 557

Examples of Complex Social Systems 558

Colonial invertebrates 558; Social insects 559; Vertebrates 561

Evolutionary Advantages and Disadvantages of Living in Groups   568

Cooperation through group advantage 569

Cooperation through Selfishness   572

The selfish herd 572;   Kin selection 573;   Reciprocal
altruism 576;   Parental manipulation of offspring 581;   Ecological
factors in cooperation 582

Implications for Understanding Human Social Behavior   583

Human mating systems 584;   Kin selection 585

Author Index   592

Subject Index   599

ANIMAL BEHAVIOR:
CONCEPTS, PROCESSES, AND METHODS

PART ONE

STUDY OF ANIMAL BEHAVIOR

1 Introduction

2 History of Animal Behavior

3 Approaches and Methods

4 Genetics and Evolutionary Processes

1

INTRODUCTION

We humans, like all animals, are bound in complex and vital relationships with members of our own species, with members of other species, and with the environment. Our survival, like that of all animals, depends on our ability to procure food and shelter, to find mates and produce offspring, and to protect ourselves from the elements and, at least in the early stages of human history, from predators. Since our species has always been involved in a competitive and dependent relationship with animals, we should not be surprised to find that people have always been interested in animal behavior. The relationship of the Pleistocene big-game hunters with their prey was no more intimate or important than our relationship with the insects and rodents that compete for our crops or spread disease.

WHY WE STUDY ANIMAL BEHAVIOR

Unlike other animals, however, we have demonstrated a desire for knowledge about the world around us that transcends our survival needs. Thus, of the several factors that stimulate our interest in the concepts and processes of animal behavior, curiosity about the living world is perhaps the strongest. By observing and experimenting with animals, we can learn a great deal about the relationships between animals and their environments and about the internal processes that govern their behavior. Others of us, interested in establishing general principles common to all behavior, undertake comparative studies of diverse species and formulate models that explain the phenomena we observe. Still others are concerned with trying to understand our own species — the brain mechanisms and the behavioral biology and evolution of humans. Those who desire to maintain and preserve the environment investigate the physiology and behavioral processes of animals in order to save endangered species. We can also use our knowledge to control economically costly environmental pests.

HOW WE STUDY ANIMAL BEHAVIOR: FOUR PROCESSES

The return of migratory red-winged blackbirds *(Agelaius phoeniceus)* from winter haunts in the warmer southern latitudes is a certain sign of the arrival of spring in the north-central United States (Figure 1–1). It is also part of an annual cycle of events, which has physiological, genetic, ecological, and evolutionary aspects. We can ask a multitude of complex, interrelated questions about this cycle, and we can test these questions experimentally. How do we go about it?

Part One of this book provides background for examining the approaches and methods used to formulate and test questions in animal behavior. We first examine the historical roots of the four major approaches to the study of behavior: comparative psychology, ethology, behavioral ecology, and sociobiology. We then look at the methodological problems involved in designing experiments to test hypotheses about behavioral processes. Finally, we discuss some basic theoretical concepts of evolution and genetics in order to understand the dynamics of evolutionary processes and the ways in which they have shaped behavior.

In using this book we have two major goals: first, to learn about four key processes of animal behavior — physiology and genetics, social interactions, ecology, and evolution — and their significance; and second, to learn how to formulate questions about behavior and to explore some of the methods used to investigate animal behavior. In the rest of this chapter, we make some observations about the behavior of red-winged blackbirds, we use these observations to question how and why the behavior is generated, and we see how our twin goals of learning about processes and learning how to ask questions will be realized.

PHYSIOLOGICAL PROCESSES

From mid-March to early April, male redwings arrive in small flocks. They establish individual territories in suitable habitats, usually marshy areas or wetlands.

FIGURE 1-1 Annual cycle of behavioral changes related to reproduction in red-winged blackbirds

The redwings' annual cycle provides us with many opportunities for studying the physiological, social, behavior-ecological, and evolutionary processes of animal behavior.

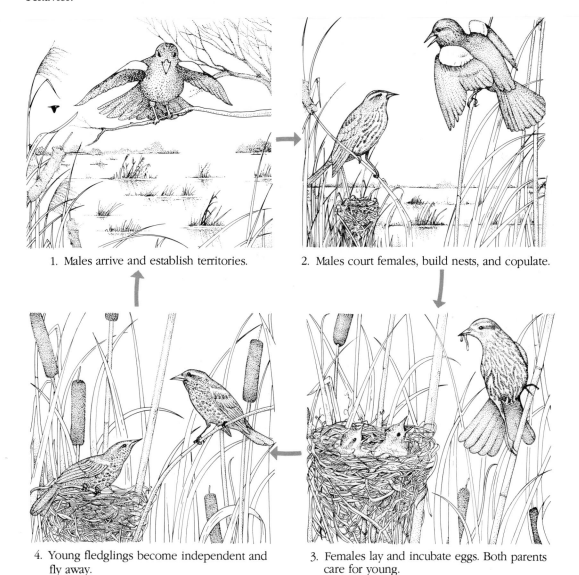

1. Males arrive and establish territories.

2. Males court females, build nests, and copulate.

4. Young fledglings become independent and fly away.

3. Females lay and incubate eggs. Both parents care for young.

- The social system of male redwings during migration is characterized by gregarious flocking, but during the breeding season the social system is characterized by individually defended territories; what internal changes within individual males are involved in this behavioral shift?

- How do the genetic inheritance and past experiences of each individual male redwing interact, resulting in the observed patterns of territory defense?

The **physiological processes** occur in the biochemistry and interrelated functions of cells, tissues and organs. In Part Two we examine the mechanisms that directly control behavior: genetics, the nervous system, the hormonal system, and biological clocks.

SOCIAL PROCESSES

Female redwings arrive in mid-April, and each finds a mate. Sometimes more than one female shares the same territory with a male mate. The female initiates nest-building activity. Courtship and copulation take place by early May. After completion of a nest, each female lays two to five eggs; the young birds hatch after an incubation period of several weeks.

- What behavior patterns are involved in male-female interactions after the females arrive and during the process of rearing the young?
- What forms of communication are used by the redwings during courtship?
- Why do some females share a male's territory with other females?

The study of **social processes** includes examining the performance of individual animals and patterns of interaction among families and social groups of the same species. In Part Three we explore the behaviors of individuals and groups. We examine the factors in an organism's past experience and those in its present situation that shape observed behavior: development, learning, and motivation. We then look at specific aspects of social behavior: sexual behavior and reproduction; aggression; communication; and migration, orientation, and navigation.

BEHAVIOR-ECOLOGICAL PROCESSES

Young redwings are kept warm and are fed by the parents for about six weeks while they grow, become fully feathered, and eventually learn to fly. They achieve full independence from their parents by early August. Throughout this period the male continues to defend the territory.

- Why do redwings prefer to make their nests in certain types of habitats?

- What foods do redwings consume and what behavior patterns do they use to obtain these foods?

- What factors may account for the observed division of labor between males and females?

Behavioral ecology is the study of functional aspects of behavior. Behavioral-ecological processes arise from the interaction of animals with the living and non-living environment, such as populations, ecological communities, and ecosystems. In Part Four we consider functional problems that relate behavior to the ecological context: feeding relationships, habitat selection, and population regulation.

EVOLUTIONARY PROCESSES

Adult red-winged blackbirds form mixed-sex flocks and fly south in mid-September, followed several weeks later by flocks of younger birds.

- How has this particular species-specific social system developed over many hundreds of generations?

- Why has this particular social system developed for redwings instead of some alternate one, such as large nesting colonies?

Evolutionary processes produce behavioral patterns that develop in a particular species through natural selection. Evolution produces changes in morphology, physiology, and behavior over long periods of time. Our concern in Part Five is specifically with evolutionary processes that have affected the historical development of individual behavior patterns, social behavior, and social organization.

While in this book the study of animal behavior is divided along the lines of four processes, the division is only a convenient way of organizing our study. The four processes are not separate entities, but are actually highly interrelated.

We might, for example, ask what behavior patterns are important for the interaction among male redwings during the establishment and maintenance of territories? This question is primarily concerned with social processes; in answering it, we would focus on the redwings' general social organization, the types of behavior patterns they exhibit, the sequence in which the behaviors occur, where in the territory they occur, and how they relate to the acquisition and defense of territory. A complete answer to our question involves all four processes. For example, there is feedback between social processes and the physiological changes within individual birds; internal changes in hormone levels may produce shifts in behavior patterns. We must find out why redwings prefer certain types of habitat and what factors determine territory size. Finally, we will want to know not only how the behavior patterns involved in territoriality have evolved, but also how and why territoriality itself has developed over many hundreds of generations.

EVOLUTIONARY PERSPECTIVE

Until recently, many animal behaviorists focused on questions of how behavior is directly produced and controlled. We call these structural and mechanical influences **proximate factors.** The development of sociobiology has given a strong impetus to the investigation of "why" questions about the ultimate, indirect evolutionary factors that produce behavior. This approach has both advantages and disadvantages. Because the evolutionary perspective requires behaviorists to study animals in their natural contexts, it generates important questions about the functions of behavior.

However, the evolutionary viewpoint has been criticized for overusing the concept of adaptation as an explanatory device and for relying on "just-so" explanations of the evolution of behavior. The fossil record provides bones and teeth and environmental evidence, but behavior itself is not preserved and must be inferred often from sketchy and incomplete evidence. We should approach the study of animal behavior from the dual perspectives of proximate and ultimate factors; understanding how immediate factors affect behavior is just as important as explaining why behavior is functional.

COMPARATIVE STUDIES

Comparative studies are often used to explore the ecology and evolution of behavior. Several brief examples will illustrate the importance of the comparative method.

Dewsbury (1972) examined descriptions of the patterns of copulatory behavior for a variety of mammal species. By comparing these patterns — of males as they mate with females — we note significant differences among mammal species (see Figure 11–13). The males varied with respect to four major characteristics of the mating pattern: multiple ejaculations, multiple intromissions, thrusting, and the presence of a copulatory lock at ejaculation. Dewsbury suggests that these differences may be significant in terms of maintaining the reproductive isolation of a species. In Chapter 4 we further examine the underlying evolutionary processes leading to reproductive isolation, and in Chapter 11 we look at the significance of pattern differences with regard to reproduction.

A second example concerns the ecology and social behavior of three species of marmots (Barash 1973a, b). Woodchucks *(Marmota monax)* inhabit fields at lower elevations, reproduce each year, are usually solitary and aggressive toward one another, and have juveniles that disperse at one year of age. Olympic marmots *(M.*

olympus) inhabit higher elevation meadows, reproduce every other year, are colonial and highly tolerant of conspecifics, and have juveniles that disperse at three years of age (Figure 1–2). A third species, yellow-bellied marmots *(M. flaviventris),* live at intermediate elevations, generally reproduce annually but sometimes skip a year, are colonial but moderately aggressive, and have juveniles that disperse at two years of age.

These differences among three species of marmots can be correlated with the length of the annual growing seasons at the various elevations. Short growing seasons at high elevations result in less food available to individuals living in those habitats, whereas the longer growing seasons at lower elevations result in larger food supplies. Barash (1974) relates the behavior and ecology of these species by noting that aggressiveness is a key factor in the dispersal of juveniles. Where less food is available, as with the Olympic marmot, the young grow and mature more slowly; adults are more tolerant, and a colonial social system is possible, probably because of the need for delaying dispersal of the young until they are capable of independent existence. Where food is more plentiful, as with the woodchucks' environment, juveniles can attain sufficient size to become independent of their parents after only one year, and adult aggressiveness is correspondingly more prevalent.

Animal behaviorists also use comparative studies to draw analogies between behaviors of diverse types of animals, including humans. We should be aware that analogies can be instructive, but they can also be misleading. We should rely on comparative studies only to form tentative conclusions and to suggest hypotheses to be tested further. We would be incorrect to conclude, for example, that because red-winged blackbirds compete for territorial space, humans also compete for territories in the same way or for the same reasons.

FIGURE 1–2   A family grouping of Olympic marmots, a species that lives at higher elevations
Since the growing season is shorter, less food is available, juveniles grow more slowly and do not disperse until they are three years of age. The result is a colonial type of social system.
Source: Photo by David P. Barash.

Our studies of red-winged blackbirds may suggest some of the functions and processes of territoriality — but only for redwings. We may then apply some of this knowlege to comparative studies of spatial relationships and their functions in humans. We may use our knowledge about other species to generate questions about human patterns of spacing, but we can answer questions about human behavior only by studying people directly.

References

Barash, D. P. 1973a. The social biology of the Olympic marmot. *Anim. Behav. Monogr.* 6:171–249.

———. 1973b. Social variety in the yellow-bellied marmot *(Marmota flaviventris). Anim. Behav.* 21:579–584.

———. 1974. The evolution of marmot societies: A general theory. *Science* 185:415–420.

Dewsbury, D. A. 1972. Patterns of copulatory behavior in male mammals. *Quart. Rev. Biol.* 47:1–33.

2

HISTORY OF ANIMAL BEHAVIOR

Long before people conducted scientific studies to learn why and how animals have come to look and behave as they do, humans and their prehuman ancestors left evidence—both deduced by us and drawn and written by them—of their interest in the natural world. Some of this interest originated in need. Because animals were a primary source of food, clothing, and materials for tools and shelter, knowledge of their behavior was necessary for successful hunting. Interest in animal behavior also stemmed from human curiosity about the natural world. In this chapter we examine how and why people have studied animal behavior, from the early days of human existence, through the emergence of animal behavior as a scientific discipline in the nineteenth century, to the experimental and theoretical approaches of the present.

INTEREST IN ANIMAL BEHAVIOR

EARLY HUMANS

For many thousands of years humans and their ancestors have been hunters and eaters of meat. The early hominids and the first *Homo erectus* practiced a crude variety of hunting. Peking man, a form of *Homo erectus* of 400,000 years ago, was an accomplished hunter and user of fire and made tools from animal bones.

L. S. B. Leakey (1903–1972), an anthropologist known best for his discoveries of early hominid remains in Tanzania, proposed and tested a hunting strategy that was based on knowledge of animal behavior—a strategy that early hunters may have used to capture a rabbit or other small animal. Leakey suggested that, upon sighting the prey at about fifteen meters distance, the hunter should dash directly toward the animal; a small animal initially freezes in such a situation. Within two or three meters of the prey, the hunter should turn sharply either left or right, because the typical escape behavior of the prey is to make a sudden dash in one direction or the other. If both prey and hunter go to the left, the hunter is upon the animal and can grab it bare-handed (as Leakey demonstrated), or the hunter may use a club or stone to strike the prey. If the hunter guesses incorrectly, he should stop, turn, and wait for the animal to stop so the process can be repeated, perhaps resulting in a successful capture.

Early *Homo sapiens* must have been close observers of animal habits and characteristics. They needed to be familiar with the behavior of animals not only to know where and how to hunt their prey, but also to protect themselves from potential predators. Hunters of the Upper Paleolithic (35,000 to 10,000 years ago) probably used fire to drive animals over cliffs or into culs-de-sac where they could be slaughtered with rocks or clubs. A ravine with at least 100 mammoth carcasses has been located in Czechoslovakia, and the remains of thousands of horses that had been stampeded over a cliff have been discovered in France.

Prehistoric cave paintings in France and Spain have revealed other aspects of humankind's relationship to animals. These paintings realistically depict game animals of all sorts in a way that suggests close observation of the animals at various times in their life cycles; in addition, some of the drawings are symbolic representations of actual hunting scenes (Figure 2–1). However, while early people were of necessity aware of the animals in their environment, their knowledge of animal behavior was limited to strictly practical concerns.

CLASSICAL WORLD

Interest in animal behavior in the classical world stemmed from curiosity about natural phenomena and a desire to record and categorize observations. For example, Aristotle (384–322 B.C.) wrote ten volumes on the natural history of animals, in which we note the first extensive use of the observational method. The following brief excerpts, translated from the original Greek, give us a flavor of what Aristotle's observations were like (the first two passages are true, the last is false) (Ley 1968, pp. 36–37):

They say that the cuckoos in Hellice, when they are going to lay eggs, do not make a nest, but lay them in the nests of doves or pigeons, and do not sit, nor hatch, nor bring

FIGURE 2-1   Cave painting depicting man aiming arrow at deer
Cave paintings, such as this one found in Alpera, Spain, indicate that even in the early stages of human history people knew something about animal behavior, particularly about those habits and traits of animals that were important either as potential food sources or as possible predators on humans themselves.
Source: The Bettmann Archive.

up their young; but when the young bird is born and has grown big, it casts out of the nest those with whom it has so far lived.

In Egypt they say there are some sandpipers that fly into the mouths of crocodiles and peck their teeth, picking out the small pieces of flesh that adhere to their teeth; the crocodiles like this and do them no harm.

The goats in Cephallaria apparently do not drink like other quadrupeds, but every other day turn their faces to the sea, open their mouths and inhale the air.

The Roman naturalist Pliny (A.D. 23–79) also made extensive observations of the natural world. A quote from his *Natural History* provides some insight into the anthropomorphism that characterized Roman perceptions of animal behavior (Nordenskiöld 1928, p. 55):

Amongst land animals the elephant is the largest and the one whose intelligence comes nearest that of man, for he understands the language of his country, obeys commands, has a memory for training, takes delight in love and honour, and also possesses a rare thing even amongst men — honesty, self-control and a sense of justice; he also worships stars and venerates the sun and moon.

FOUNDATIONS OF ANIMAL BEHAVIOR STUDIES

The study of animal behavior did not begin to emerge as a scientific discipline until the latter part of the nineteenth century. Three major developments contributed significantly to the study of behavior: (1) publication of the theory of evolution by natural selection, (2) development of a systematic comparative method, and (3) studies in genetics and inheritance.

THEORY OF EVOLUTION BY NATURAL SELECTION

For several centuries European ships made voyages of exploration and discovery to all parts of the globe. Often scientists were officially attached to the voyages, as Charles Darwin (1809–1882) himself was. These scientists and other crew members made observations of exotic fauna and flora and brought back live and preserved specimens to zoos and laboratories in Europe, where scholars could observe, record, and speculate about the anatomy, behavior, and interrelationships of these newly discovered species. The following passage from Darwin's account (Figure 2–2) of the Galápagos Islands lizard illustrates the kind of observations he made in the natural setting (Darwin 1845, p. 336):

They inhabit burrows, which they sometimes make between fragments of lava, but more generally on level patches of the soft sandstone-like tuff. The holes do not appear to be very deep, and they enter the ground at a small angle; so that when walking over these lizard warrens, the soil is constantly giving way, much to the annoyance of the tired walker. This animal, when making its burrows, works alternatively the opposite sides of its body. One front leg for a short time scratches up the soil, and throws it towards the hind foot, which is well placed so as to heave it beyond the mouth of the hole. That side of the body being tired, the other side takes up the task, and so on alternatively.

Like all major scientific paradigms, the theory of evolution drew on contributions by and suggestions from the work of other scientists. In 1798 Thomas Malthus (1766–1834), in his *Essay on the Principle of Population,* hypothesized that humans have the reproductive potential to rapidly overpopulate the world and outstrip the available food supply; that is, population increases geometrically, while the food supply increases arithmetically. The inevitable result is disease, famine, and war. Malthus's theory was an important influence on Darwin's thinking about the competition for survival among members of a species. Geologist Sir Charles Lyell (1797–1875), a friend of Darwin's, made observations of rock strata and successions of fossils which gave evidence of a process of continuous change in living material through time, an idea that was at odds with the biblical suggestion of simultaneous

FIGURE 2–2   Charles Darwin testing the speed of an elephant tortoise on the Galápagos Islands
Darwin wrote extensive descriptions of animal and plant forms and was co-author with A. R. Wallace, of the paper presented before the Royal Society in London in 1859 that set forth the theory of evolution by natural selection.
Source: The Bettmann Archive.

creation of all living things. This evidence of geological change led others to the idea that species themselves are not fixed entities. The artifical selective breeding of domesticated stocks by English farmers provided additional support for the thinking of both Darwin and A. R. Wallace (1823–1913).

Wallace's voyage to the Malay archipelago and Darwin's travels on the *Beagle* to South America and the South Pacific, combined with their other studies and the intellectual influences of the time, led each man independently to formulate the theory of evolution by natural selection. The theory states that while each animal species has a high reproductive potential, the number of animals of a species remains relatively constant over time. Thus there is competition for survival. Variation in traits exists within animals of one species. Because some traits are more advantageous than others in the competition for survival, the operational process of natural selection occurs. Those species members that are able to survive to produce offspring contribute their characteristics to subsequent generations through their young.

The following passage from *The Origin of Species* illustrates that Darwin clearly recognized the central role of animal behavior in determining the outcome of competition (Darwin 1859, p. 94):

Amongst birds, the contest is often of a more peaceful character. All those who have attended to the subject, believe that there is the severest rivalry between the males of many species to attract, by singing, the females. The rock-thrush of Guiana, Birds of Paradise, and some others, congregate; and successive males display with the most elaborate care, and show off in the best manner, their gorgeous plumage [Figure 2–3]; they likewise perform strange antics before the females, which, standing as spectators at last choose the most attractive partner.

Darwin concluded that species were not fixed entities. The theory of evolution by natural selection was able to account for changes, through time, within a species and also for the gradual appearance of new species. Recent developments in other biological fields, genetics in particular, have modifed the theory of evolution by natural selection proposed by Darwin and Wallace. Today, some evolutionary biologists believe that evidence from the fossil record and genetic mechanisms supports the claim that rates of evolution vary through time (Stanley 1981). Change through evolution and, in particular, the appearance of new species may occur more rapidly during some time periods than at other times.

COMPARATIVE METHOD

George John Romanes (1848–1894) is generally credited with formalizing the use of the **comparative method** in studying animal behavior. For Romanes the comparative method involved studying nonhuman animals to gain insights into the behavior of humans. Romanes sought to support Darwin's theory of natural selection through his proposal that mental processes evolve from lower to higher forms and that there is a continuity of mental processes from one species to another. He argued that while people could really know only their own thoughts, they could infer the mental

FIGURE 2-3 Male bower bird in display
As in the passage quoted from Darwin, male birds of a variety of species display to attract females. The male bower bird builds a bower and adorns it with brightly colored objects.
Source: Photo by Gerald Borgia.

processes of animals, including other humans, from knowledge of their own. For Romanes the similarities between the behavior of humans and that of other animals implied similar mental states and reasoning processes in humans and in nonhuman species. He suggested that a sequence could be constructed for the evolution of various emotional states in animals. Worms, which exhibit only surprise and fear, were placed lowest on this scale; insects were said to be capable of various social feelings and curiosity; fish showed play, jealousy, and anger; reptiles displayed affection; birds exhibited pride and terror; and finally, various mammals were credited with hate, cruelty, and shame.

Romanes's theory relied largely on inferences rather than on recorded facts; he made substantial use of anecdotes. A movement led by another Englishman, C. Lloyd Morgan (1852–1936), sought to counteract these faults by using the **observational method.** Morgan's basic tenet was that only data gathered by direct experiment and observation could be used to make generalizations and to develop theories.

Morgan is probably best known for his **law of parsimony,** which is now axiomatic in animal behavior studies, "In no case may we interpret an action as the outcome of the exercise of a higher psychical faculty if it can be interpreted as the outcome of the exercise of one which stands lower in the psychological scale (Morgan 1896, p. 53). This statement, also called **Morgan's canon,** has been interpreted to mean that in the analysis of behavior we must seek out the simplest explanations for observed facts. Where possible, complex hypotheses should be reduced to their simplest terms to facilitate the clearest understanding of the mechanisms that control behavior.

THEORIES OF GENETICS AND INHERITANCE

The third development that greatly influenced research in animal behavior was the birth of the science of genetics and the development of modern theories of inheritance. In the 1860s Gregor Mendel (1822–1884) reported on his findings from a series of breeding experiments on garden peas. These studies established key principles of the laws of inheritance of biological characteristics. Present-day behavioral biology is based on the combination of evolutionary theory, which provides an explanation of the processes by which traits can change through time, and genetics, which explains how traits are passed from one generation to another.

Like the morphological and physiological traits, an animal's behavior also has a genetic component. Thus, behavior may change as a species evolves. This means that, as scientists, we can explore the genetic variation underlying various behaviors, just as others have investigated the role of genetic inheritance affecting morphology and physiology. Behavior-genetic analysis had its beginnings in these early studies of inheritance and was then greatly expanded in the 1930s by the work of R. A. Fisher and others. Modern behavior-genetic analysis (see Hirsch 1967; Fuller and

Thompson 1978) is a powerful tool used by many animal behaviorists; of this we shall learn more in Chapter 5.

FOUR EXPERIMENTAL APPROACHES

The ideas, methods, and theories established during the latter half of the nineteenth century are the foundation of today's four experimental approaches to the study of animal behavior. (1) Ethology is concerned primarily with the functional significance and evolution of behavior patterns. (2) Comparative psychology and physiology seek to determine the underlying causes of behavior, the mechanisms that control observed animal behavior. (3) Behavioral ecology, an approach that has emerged in recent decades, explores the ways in which animals interact with their living and nonliving environments. (4) Sociobiology applies the principles of evolutionary biology to the study of social behavior in animals. We now examine briefly the historical development of each approach.

STUDIES OF FUNCTION AND EVOLUTION

ETHOLOGY.   The systematic study of the function and evolution of behavior, called **ethology,** is now a little over a century old. One of its most important principles is that behavioral traits, like anatomical and physiological traits, can be studied from the evolutionary viewpoint. C. O. Whitman (1842–1910) made extensive observations of display patterns, which he termed **instincts,** in various species of pigeons. Whitman found that he could use **displays**—patterns of behavior exhibited by animals that function as communications signals—to classify animals according to similarities and differences in behaviors.

The **ethogram,** a complete inventory of the behaviors performed by a species, is a starting point for many ethological studies. The ethologist can then formulate specific questions about the adaptiveness and function of particular behavior patterns. A student of Whitman's, Wallace Craig (1876–1954), defined two key categories of behavior patterns from his work with doves and pigeons. The first category includes the variable actions of an animal, such as its searching behavior to find food, a nest site, or a mate, and are called **appetitive behaviors.** The second category includes stereotyped behaviors—behaviors that are repeated without variation—such as the act of mating or the killing of prey; these are called **consummatory behaviors.**

Another major area of inquiry in the ethological approach is determining the way in which key stimuli trigger specific behavior patterns. J. von Uexküll demonstrated that animals perceive only limited portions of the total environment with their sense organs and central nervous systems. This sensory-perceptual world was termed the *Umwelt* by von Uexküll. Among the stimuli recorded by the sense organs, certain specific cues, which ethologists call **sign stimuli,** trigger particular stereo- typed responses. For example, the enlarged belly of the female three-spined stickle- back fish triggers courtship behavior by male sticklebacks (Figure 2–4).

Credit for the synthesis of these early findings and the further development of modern ethology belongs largely to two men, Konrad Lorenz (1903–) and Niko Tinbergen (1907–). Lorenz is noted for his studies of genetically programmed behavior and for his work on the importance of specific types of stimulation for young animals during critical periods of early development. Modern ethology's concern with four areas of inquiry — causation, development, evolution, and func- tion of behavior — is developed from a scheme proposed by Tinbergen. (As psychol- ogist Thomas McGill noted, the first six letters of the alphabet can be used to remeber these questions: *Animal Behavior Causation Development Evolution Function.*) The culmination of the development of animal behavior studies came in the fall of 1973, when the Nobel prize for physiology and medicine was awarded to three ethologists: Konrad Lorenz, Niko Tinbergen, and Karl von Frisch (1886–1982), Von Frisch had pioneered studies of bee behavior and communication.

Modern ethology is characterized by varied types of investigations that range from traditional observational studies in natural environments (Eibl-Eibesfeldt 1972; Geist 1971) to experiments on the physiological bases of behavior; (Bentley and Hoy 1974). The latter studies are indistinguishable from those conducted by

---

FIGURE 2-4   Courtship of male and female three-spined stickleback
The enlarged belly of the female three-spined stickleback fish (top) is a sign stimulus for the male of the species (bottom) to court and to entice the female to enter the nest he has built.

physiological psychologists. Some ethologists work primarily with behavior genetics and the evolution of behavior (Manning 1971; Gerhardt 1979) or explore the relationships between hormones and behavior (Hinde 1965; Truman, Fallon, and Wyatt 1976) or the nervous system and behavior (Nottebohm 1981). Others work on research problems in the field or in a laboratory setting that resembles the natural habitat. By employing experimental manipulation to test specific hypotheses, Kummer (1971) investigated the effects of transplantation of individuals on social behavior of baboons, Wickler (1972) studied the significance of color patterns in fish, and recruitment in honey bees (Gould 1975).

STUDIES OF MECHANISMS

PERCEPTUAL PSYCHOLOGY.   Several distinct approaches to discovering the mechanisms underlying behavior emerged during the mid-nineteenth century. Researchers who were concerned with the mind/body dichotomy studied the relationships between physical and mental processes. Gustav Fechner (1801–1887), one of the founders of perceptual psychology and psychophysics, was most interested in separating the processes of sensation (body) and perception (mind). His goal was the objective measurement of sensation — the reception of environmental stimuli through the senses such as those of sight or hearing — and the comparison of these direct measurements to subjective perceptual interpretations of the sensations.

PHYSIOLOGICAL PSYCHOLOGY.   Modern physiological psychology developed from early attempts to relate physiological properties and events within an organism to external behaviors. For example, Pierre Fluorens (1794–1867) surgically removed portions of the brains of pigeons and recorded the resulting changes in the birds' behavior. Hermann von Helmholtz (1821–1894) studied the speed of conduction of nerve impulses and, later, the physiology of vision. He measured the speed of nerve conduction in an ingenious manner by experimenting on the motor neurons of frogs, which trigger muscle contractions. Helmholtz stimulated one point on a nerve near the muscle and then a second point on the same nerve further away from the muscle. The difference in amount of time elapsed between stimulus and muscular contraction in the two measurements is the conduction time for the distance between the two stimulus points, from which the speed of conduction can be calculated.

FUNCTIONALISM.   By the late 1800s and early 1900s, Europe was no longer the exclusive center of behavioral studies, and individuals were conducting research investigations in comparative psychology at a number of laboratories in the United States. Two major new theoretical and experimental points of view arose during this period: functionalism and behaviorism. The **functionalists,** headed by John Dewey (1859–1952), studied the functions of the mind and how the mind operates, in contrast to how the mind is structured. Functionalists attempted to answer three major questions: (1) How does mental activity occur? (2) What does mental activity accomplish? and (3) Why does mental activity take place? Functionalism employed objective observations rather than introspection as its primary method.

An important outcome of the functionalist approach was the introduction into psychology of the notion prevalent in biology of **adaptive behavior**—the notion that behaviors have particular functions in terms of survival value for the animal in its natural habitat. To these early psychologists, the concept of adaptive behavior implied that the response to a stimulus changes the sensory situation in a way that alters the original conditions that produced the response. For example, pain disappears when a sharp splinter is removed from the hand, and the original condition—the existence of a splinter—is also altered.

BEHAVIORISM. John B. Watson (1878–1958) was the principal founder of a new approach to the study of behavior, **behaviorism.** The basic tenet of behaviorists is that behavior consists of an animal's responses, reactions, or adjustments to stimuli or complexes of stimuli. Thus, most behaviors of an organism are products of its past experiences; the mind is a *tabula rasa* (a blank slate) at birth and cumulates all subsequent experiences. Psychology then becomes the study of the ways in which past events affect behavior. To what degree can we predict and control behavior based on a knowledge of an animal's previous experiences? The methods utilized by Watson and his followers, for example B. F. Skinner (1904–   ), were strictly objective. Reports of subjective feelings or emotions were, by definition, not acceptable as scientific data. It is interesting to note that this restriction forced the behaviorists to study human behavior in much the same way they studied the behavior of any other animal, without benefiting from their subjects' verbal judgments or reports of feelings and perceptions.

ANIMAL PSYCHOLOGY. Concurrent with the development of these theoretical viewpoints was the emphasis by Edward L. Thorndike (1874–1949) on the need for systematic, replicable experiments in comparative animal psychology. Thorndike used the puzzle box (Figure 2–5) to perform a series of experiments on learning tasks, with cats as test subjects. A cat was placed in the box, which was fastened shut; the cat could open the door by manipulating a shuttle-lever system to obtain a reward placed outside the box. From these experiments Thorndike concluded that much of animal learning takes place by trial and error and that rewards are a critical component of learning processes.

Animal psychology today is a diverse mixture of subdisciplines, both new and old. The study of comparative learning and learning theory is still quite important, as the work of Bitterman (1975), Seligman (1970), and Zolman (1982) exemplifies. The development of behavior is also a subject of continued investigation by individuals like Gottlieb (1970) and Oppenheim (1982) and by groups like the one developed at the Wisconsin Regional Primate Laboratory by Harlow and colleagues (Suomi and Harlow 1971, 1977). The study of the physiological processes underlying behavior has also diverged into several pathways, exploring hormones and

FIGURE 2–5   Thorndike puzzle box
A cat placed inside the cage can clearly see the reward, in this case a fish, placed outside. In order to obtain the reward the cat must learn to manipulate a shuttle-lever system that raises the door of the puzzle box.

behavior (Lehrman 1965; McEwen et al. 1979; Crews 1980); neural correlates of behavior (Hubel and Wiesel 1965; Wayner and Carey 1973) and brain chemistry; and psychopharmacology and behavior (Kelly et al. 1979). The cross-fertilization of genetics and behavior also produced a new subdiscipline called behavior genetics, which is concerned with the hereditary bases of behavior and how the interactions of genetics and environment affect behavior (Hirsch 1967; McGill and Haynes 1973; Oliverio 1983).

In 1950 Frank Beach, in his presidential address to the American Psychological Association, stressed that the discipline of comparative psychology was devoting too much attention to the white rat as a subject, while ignoring many other types of available vertebrate organisms. In more recent years others (Lockard 1971; Hodos and Campbell 1969) again called attention to the lack of an evolutionary perspective in comparative psychology and to the incorrect use of the rat as a model for other organisms, especially humans. These critiques have stimulated more truly comparative investigations, as exemplified by the work of Dewsbury (1972, 1975) on reproductive behavior in rodents. More attention has also been given to the natural context and actual field investigation of animals (Lockard 1971; Barash 1973, 1974).

COMPARISONS.   Discerning whether a particular study has been conducted by an ethologist or comparative psychologist may be difficult at first. If we understand the historical differences between these two approaches, we can better appreciate the synthetic approach that characterizes the behavior studies of the past several decades.

Ethology was developed, largely in Europe, by researchers trained in biology. Ethologists traditionally observed a wide variety of animals in nature and conducted experiments under conditions that mirrored the natural setting as closely as possible.

They concentrated their efforts on exploring questions of ultimate causation — the "why" questions of the evolution and function of behavior.

Comparative psychology originated primarily in America. Until recently, psychologists worked most often under controlled laboratory conditions. Much of their research was carried out on small rodents, particularly the domesticated rat. Comparative psychologists placed primary emphasis on proximate issues — the "how" questions of the physiological and developmental mechanisms underlying observed behavior patterns. Dewsbury (1984) has recently published a volume on the history of comparative psychology that provides many details regarding the development of concepts and theories in this field and many insights into the persons responsible for the experimental and theoretical work. Dewsbury defines comparative psychology as the attempt to make comparisons across species in order to develop principles of generality regarding animal behavior. He examines the course of development that has characterized this field since 1900 and notes the many myths that have come to be associated with what scientists and nonscientists alike have come to believe a comparative psychologist is and does.

Since the mid-1950s the distinctions between ethology and comparative psychology have been slowly disappearing. Several events have opened communication between scientists of the two approaches, including the biennial meetings of the International Ethological Conference, many cross-visitations between researchers in Europe and America, and the publication of a number of international animal behavior journals. A common approach, starting with the notion of species-typical behaviors, is beginning to emerge from this exchange of information. **Species-typical behaviors** are actions and displays that are broadly characteristic of a species and that are performed in a similar manner by all its members. Recently ethologists have begun to ask more questions about the mechanisms of causation of behavior, and, conversely, comparative psychologists have become increasingly concerned with questions about the evolution and adaptive functions of behaviors.

At least one major, long-standing controversy in animal behavior has been largely resolved during recent years. Ethologists believed that much of an animal's behavior was instinctive or preprogrammed and was not affected to any great extent by experience. Psychologists claimed that learning and experience were the major determinants of behavior. Today, most animal behaviorists accept that neither of these viewpoints is entirely correct. Instead, as we shall see in Chapter 9, the current focus is on the interactions of genotype, physiology, and experience as determinants of behavior and on how the degree of genetic and experimental determination of observed behavior patterns differs among animal species.

BEHAVIORAL ECOLOGY

In the past five decades a third approach to the study of animal behavior has emerged. **Behavioral ecology,** with origins in zoology, examines the ways in which animals interact with their living and nonliving environments. Environment as used here includes **conspecifics**—animals of the same species—other living animals within the same ecological community, plants, and inorganic features of the habitat.

Behavioral ecologists are concerned with both ultimate and proximate questions about behavior. Consider, for example, the behavior and habitat selection of a rabbit living on the edge of a large field (Figure 2–6). Ultimate constraints operating in this instance include the physiological tolerances of the rabbit. Variables like temperature and moisture level restrict the rabbit to certain environments, and its digestive system can break down only certain types of foods, primarily vegetation. Proximate factors include the rabbit's past experience with types of habitat and foods, which may lead to a particular preference for home site and diet. Ultimate factors establish the limits, and proximate factors affect the behavior of an animal within those limits.

Behavioral ecologists, trained primarily in zoology, ecology, and related fields, are also greatly influenced by the thinking and methods of comparative animal psychology. Behavioral ecologists begin investigating a research problem in the field and define various questions, regarding, for example, population regulation or predator-prey relations. (For instance, does the predator maximize its net rate of energy intake by utilizing some form of optimal foraging strategy?) Certain aspects of the overall problem under investigation—for instance, determination of what features of the prey are most important for predator recognition and detection—may require more systematic experiments than can be done in the field setting. The behavioral ecologist often brings specific, testable hypotheses into the laboratory, where controlled experiments are conducted, and then attempts to relate laboratory findings back to the natural field setting.

Among the investigations in behavioral ecology, those of Emlen (1952a, b) on bird behavior and energy budgets, Davis (1951) on population biology, and King (1955) on social behavior in relation to habitat were notable for the way in which they established topical areas for research work within the developing discipline. In more recent years, topics that have received particular attention by investigators include foraging strategy (Krebs 1978), the ecology of sex and strategies of reproduction (Askenmo 1984; Clutton-Brock et al. 1979) and social systems in relation to ecology (Christenson 1984).

SOCIOBIOLOGY

The most recent approach to animal behavior reached maturity in 1975 with the publication of *Sociobiology: The New Synthesis* by E. O. Wilson. **Sociobiology** applies the principles of evolutionary biology to the study of social behavior in animals. It is, in effect, a hybrid between behavioral biology from the ethological perspective, with an emphasis on ultimate questions, and population biology, with an ecological perspective. As with the other approaches to animal behavior, sociobiology relies heavily on the comparative method. Similarities and differences in social systems

FIGURE 2-6   Factors affecting the habitat selection of a rabbit
Limitations and constraints that have come about through evolution determine the
range of habitats and diets for the rabbit. Developmental and experiential factors
influence the immediate choices of the animal.

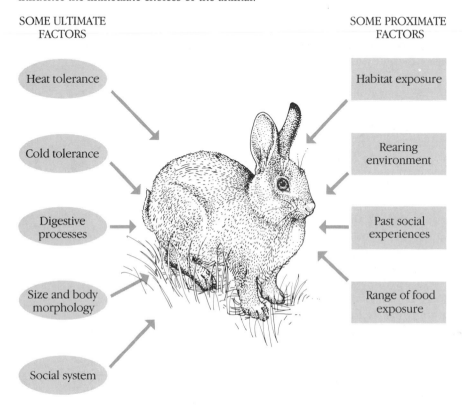

SOME ULTIMATE
FACTORS

SOME PROXIMATE
FACTORS

Heat tolerance

Cold tolerance

Digestive
processes

Size and body
morphology

Social system

Habitat exposure

Rearing
environment

Past social
experiences

Range of food
exposure

are examined for diverse groups of animals living in a wide variety of habitats and
situations to ascertain what, if any, general patterns or rules emerge to explain the
social behavior of a species.

The various theories and corollaries that constitute sociobiology have their
roots in many earlier works. Among the most significant are the writings of Williams
(1966) on natural selection and the concept of adaptation, Trivers (1971, 1972) on
evolutionary aspects of altruism and parental behavior, and Hamilton (1964, 1971)
on the genetic theory underlying the evolution of social behavior. Studies conducted
under the general heading of sociobiology range from those on altruism (Sherman
1977) in ground squirrels, to research on strategies for reproduction in damselflies
(Waage 1979) and parental investment in wasps (Werren 1980).

Since the mid-1970s sociobiology has pervasively influenced research on animal behavior. Faced with the challenge of devising new research questions and methods to test aspects of sociobiological theory, investigators have reexamined older data in light of new predictions. One area of prediction and hypothesis, and the source of considerable controversy, is the application of sociobiology to *Homo sapiens.* Sociobiologists argue that the same principles used to investigate the social behavior of animals can be applied to investigate the social behavior of humans. Other individuals argue that sociobiology is merely a form of biological determinism. A complete resolution of this controversy is probably not possible.

SUMMARY

Archaeological evidence indicates that early humans had a practical, working knowledge of the behavior of animals, particularly of those that were potential food sources or predators. By Greek and Roman times, writers like Aristotle and Pliny recorded extensive observations about and deductions from natural phenomena.

Three developments of the last half of the nineteenth century contributed to the emergence of animal behavior as a scientific discipline. First, Darwin and Wallace, each working from his own data and from ideas of previous investigators, independently put forth the theory of evolution by natural selection. Second, Romanes pioneered the development of the comparative method and used it initially to study mental evolution. Third, with Mendel's work on inheritance and the rediscovery and development of his findings at the turn of the century, modern theories of genetics and evolution emerged.

Ethology encompasses studies of the functional significance and evolution of behavior. Behavioral traits, like physical or physiological traits, are viewed in evolutionary terms and are thus subject to natural selection pressures. Traditionally, ethologists have made many of their research observations in a natural setting. The research objectives of modern ethology range from observational studies and field experiments, conducted to assess the function of behavior patterns, to investigations of the physiological bases of behavior.

Derived from these diverse beginnings, four major approaches characterize the current study of animal behavior. Studies of the mechanisms controlling behavior have historically been conducted primarily by comparative animal psychologists and physiologists. Although much of the early psychological research relied heavily on introspection and inference, these methods were later replaced by systematic, objective observations and replicable experiments. Modern animal psychologists explore such areas as physiological control of behavior, sensation and perception, learning processes, and behavior genetics.

Behavioral ecology generates studies that examine the ecological relationship between an organism and its living and nonliving environment. The questions are asked from an ecological and sometimes evolutionary viewpoint, and investigations conducted in both field and laboratory settings utilize systematic, controlled experimentation.

Sociobiology has emerged as an approach to the study of animal behavior in the past several decades. It applies principles of evolutionary biology to the study of social behavior in animals. Sociobiological research concentrates on field observations of social groups of animals in their natural environments.

Discussion Questions

1. Select a bird or mammal that is easily visible during daylight hours (even a domestic cat or cow) and watch the animal for several hours. Record as many as you can of the different behaviors and their frequencies and patterns of occurrence.

2. Consider yourself a prehistoric *Homo sapiens*. What facts, characteristics, and habits would you want to know of the other animals in your environment?

3. You have been asked to study the behavioral biology of the zinger *(Zingus zingu)*, a snake living in the deserts of the American Southwest. Formulate several questions you would wish to answer concerning the zinger from each of the four different perspectives discussed in this chapter.

Suggested Readings

Klopfer, P. H. 1974. *Introduction to Animal Behavior: Ethology's First Century,* revised ed. Englewood Cliffs, N.J.: Prentice-Hall.
Written originally as a textbook and revised once, this short
book gives a solid introduction to the ethological perspective on animal behavior.

Gould, J. L. 1981. *Ethology: The Mechanisms and Evolution of Behavior.* New York: Norton.
Chapters 1–5 provide an expanded synopsis of the history
of animal behavior, though the coverage of some aspects of
the history of comparative psychology are underrepresented.

Mortenson, F. J. 1975. *Animal Behavior: Theory and Research.* Monterey, Calif.: Brooks/Cole.
A compact treatment of the origins and development of the
various themes that together make up much of modern
animal behavior.

References

Askenmo, C. E. H. 1984. Polygyny and nest selection in the pied flycatcher. *Anim. Behav.* 32:972–980.

Barash, D. P. 1973. The social biology of the Olympic marmot. *Anim. Behav. Monogr.* 6:171–249.

———. 1974. Social behavior of the hoary marmot *(Marmota caligata) Anim. Behav.* 22:257–262.

Beach, F. A. 1950. The snark was a boojum. *Amer. Psychol.* 5:115–124.

Bentley, D., and R. R. Hoy. 1974. The neurobiology of cricket song. *Amer. Scientist* 23:34–44.

Bitterman, M. E. 1975. The comparative analysis of learning. *Science* 188:699–709.

Crews, D. 1980. Interrelationships among ecological, behavioral and neuroendocrine processes in the reproductive cycle of *Anolis carolinensis* and other reptiles. *Adv. Study Behav.* 11:1–75.

Christenson, T. E. 1984. Behaviour of colonial and solitary spiders of the Theridid species *Anelosimus eximus. Anim. Behav.* 32:725–734.

Clutton-Brock, T. H., et al. 1979. The logical stag: Adaptive aspects of fighting in red deer. *Anim. Behav.* 27:211–225.

Darwin, C. R. 1845. *The Voyage of the* Beagle. London: Dent.

———. 1859. *The Origin of Species.* London: Dent.

Davis, D. E. 1951. The relation between level of population and pregnancy of Norway rats. *Ecology* 32:459–461.

Dewsbury, D. A. 1972. Patterns of copulatory behavior in male mammals. *Quart. Rev. Biol.* 47:1–33.

———. 1975. Diversity and adaptation in rodent copulatory behavior. *Science* 190:947–954.

———. 1984. *Comparative Psychology in the Twentieth Century.* Stroudsburg, Pa.: Hutchinson Ross.

Eibl-Eibesfeldt, I. 1972. Similarities and differences between cultures in expressive movements. In J. M. Argyle and R. Hinde (eds.), *Nonverbal Communication.* Cambridge: Cambridge University Press.

Emlen, J. T. 1952a. Social behavior in nesting cliff swallows. *Condor* 54:177–199.

———. 1952b. Flocking behavior in birds. *Auk* 69:160–170.

Fuller, J. L., and W. R. Thompson. 1978. *Foundations of Behavior Genetics.* St. Louis, Mo: Mosby.

Geist, V. 1971. *Mountain Sheep: A Study in Behavior and Evolution.* Chicago: University of Chicago Press.

Gerhardt, C. 1979. Vocalizations of some hybrid treefrogs: Acoustic and behavioral analyses. *Behaviour* 49:130–151.

Gottlieb, G. 1970. *Development of Species Identification in Birds.* Chicago: University of Chicago Press.

Gould, J. L. 1975. Honey bee recruitment. *Science* 189:685–693.

Hamilton, W. D. 1964. The genetical theory of social behaviour. I and II. *J. Theoret. Biol.* 7:1–52.

———. 1971. Geometry for the selfish herd. *J. Theoret. Biol.* 31:295–311.

Harlow, H. F., M. K. Harlow, and S. J. Suomi. 1971. From thought to therapy: Lessons from a primate laboratory. *Amer. Scientist* 59:538–549.

Hinde, R. A. 1965. Interaction of internal and external factors in integration of canary reproduction. In F. A. Beach (ed.), *Sex and Behavior.* New York: Wiley.

Hirsch, J. 1967. Behavior-genetic analysis. In J. Hirsch (ed.), *Behavior-genetic Analysis.* New York: McGraw-Hill.

Hodos, W., and C. B. G. Campbell. 1969. Scala naturae: Why is there no theory in comparative psychology? *Psychol. Rev.* 76:337–350.

Hubel, D. H., and N. Wiesel. 1965. Receptive fields and functional architecture in two nonstriate visual areas of the cat. *J. Neurophysiol.* 28:229–289.

Kelly, R., et al. 1979. Biochemistry of neurotransmitter release. *Ann. Rev. Neurosci.* 2:399–446.

King, J. A. 1955. Social behavior, social organization, and population dynamics in a black-tailed prairie dog town in the Black Hills of South Dakota. *Contrib. Lab. Vert. Biol. Univ. Mich.* 67:1–123.

Krebs, J. R. 1978. Optimal foraging: Decision rules for predators. In J. R. Krebs and N. B. Davies (eds.), *Behavioral Ecology.* Sunderland, Mass. Sinauer.

Kummer, H. 1971. *Primate Societies: Group Techniques of Ecological Adaptation.* Chicago: Aldine-Atherton.

Lehrman, D. S. 1965. Interaction between internal and external environments in the regulation of the reproductive cycle of the ring dove. In F. A. Beach (ed.), *Sex and Behavior*. New York: Wiley.

Ley, W. 1968. *Dawn of Zoology*. Englewood Cliffs, N.J.: Prentice-Hall.

Lockard, R. B. 1971. Reflections on the fall of comparative psychology: Is there a message for us all? *Amer. Psychol.* 26:168–179.

McEwen, B. S., et al. 1979. The brain as a target for steroid hormone action. *Ann. Rev. Neurosci.* 2:65–112.

McGill, T. E., and C. M. Haynes. 1973. Heterozygosity and retention of ejaculatory reflex after castration in male mice. *J. Comp. Physiol. Psychol.* 84:423–429.

Manning, A. 1971. Evolution of behavior. In J. L. McGaugh (ed.), *Psychology: Behavior from a Biological Perspective*. New York: Academic Press.

Morgan, C. L. 1896. *Introduction to Comparative Psychology*. London: Scott.

Nordenskiöld, E. 1928. *History of Biology*. New York: Knopf.

Nottebohm, F. 1981. A brain for all seasons: Cyclical anatomical changes in song control nuclei in the canary brain. *Science* 214:1368–1370.

Oliverio, A. 1983. Genes and behavior: An evolutionary perspective. *Adv. Study of Behav.* 13:191–219.

Oppenheim, R. W. 1982. Preformation and epigensis in the origins of the nervous system and behavior: Issues, concepts and their history. *Perspectives in Ethology* 5:1–100.

Seligman, M. E. P. 1970. On the generality of the laws of learning. *Psychol. Rev.* 77:406–418.

Sherman, P. 1977. Nepotism and the evolution of alarm cells. *Science* 197:1246–1253.

Stanley, S. M. 1981. *The New Evolutionary Timetable*. New York: Basic Books.

Suomi, S. J., and H. F. Harlow. 1977. Early separation and behavioral maturation. In A. Oliverio (ed.), *Genetics, Environment and Intelligence*. Elsevier: North-Holland.

Trivers, R. L. 1971. The evolution of reciprocal altruism. *Quart. Rev. Biol.* 46:35–57.

———. 1972. Parental investment and sexual selection. In B. Campbell (ed.), *Sexual Selection and the Descent of Man*. Chicago: Aldine.

Truman, J. W., A. M. Fallon, and G. R. Wyatt. 1976. Hormonal release of programmed behavior in silk moths. *Science* 194:1432–1433.

Waage, J. 1979. Dual function of the damselfly penis: Sperm removal and transfer. *Science* 203:916–918.

Wayner, M. J., and R. J. Carey. 1973. Basic drives. *Ann. Rev. Psychol.* 24:53–80.

Werren, J. 1980. Sex ratio adaptations to local mate competition in a parasitic wasp. *Science* 208:1157–1159.

Wickler, W. 1972. *The Sexual Code: The Social Behavior of Animals and Men*. Garden City, N.Y.: Doubleday.

Williams, G. C. 1966. *Adaptation and Natural Selection*. Princeton: Princeton University Press.

Wilson, E. O. 1975. *Sociobiology: The New Synthesis*. Cambridge: Harvard University Press.

Zolman, J. F. 1982. Ontogeny of learning. *Perspectives in Ethology* 5:275–324.

3

APPROACHES AND METHODS

For any observations of animal behavior—for example, the annual reproductive cycle of red-winged blackbirds *(Agelais phoeniceus)* outlined in Figure 1–1 (see page 7)—we need to develop a framework to formulate testable hypotheses. We can ask many questions about, for example, the redwings' territoriality, social system, and habitat preference or about the internal events that take place in males and females during the changing seasons. Which questions are the most significant? In the first section of this chapter we examine in detail how the two historically predominant approaches to the study of behavior—ethology and comparative psychology—have posed and answered questions, and we consider the advantages and disadvantages of each approach.

Variation in how individual animals behave creates methodological problems for the investigator. For example, some red-winged blackbirds perform territorial boundary displays repeatedly in the face of an intruder (Figure 3–1), while others display much less frequently under the same circumstances. Nest sites used by redwings may vary in their height above ground and the types of plant material used in construction. How can experimental design take account of this variation? We examine some general principles of experimental design for problem-oriented behavioral research in the second part of the chapter.

Finally, we explore some problems and pitfalls of experimental research. An awareness of the problems and limitations of the various methods and techniques will help us evaluate the results of behavioral investigations and formulate better questions about behaviors we observe.

FIGURE 3-1   Territorial display by male red-winged blackbird Redwings display when faced with an intruder attempting to enter the territory. The wingspread display is often associated with the conkaree call. *Source:* Photo by Sarah Lenington.

ETHOLOGY

Ethology can be defined as the biology of behavior, or the exploration of functional and evolutionary questions relating to *why* an animal exhibits certain behavior patterns under particular sets of social and environmental circumstances. The first step in studying any animal is to compile an **ethogram,** a complete inventory of the behaviors performed by animals of the species. Behaviors can be divided either into broad descriptive categories, such as courtship, nesting, sleeping, and feeding, or into more restricted units, such as specific patterns shown during various phases of courtship (Figure 3–2). The ethogram serves as a basis for posing questions about the adaptive value and ecological importance of the various behavior patterns.

AN ETHOLOGICAL STUDY.   By examining an ethological study in detail, we can see how ethologists use both observations from nature and field experiments to formulate and test questions. Niko Tinbergen studied the behavior of gulls for many years. Among the interesting problems he investigated was a curious phenomenon he first noted in his early observations of gulls (Tinbergen 1963; Tinbergen et al. 1962). After a young black-headed gull chick *(Larus ridibundus)* hatches and frees itself from the shell, the parents carefully remove the remnants of the shell from the nest area by picking up pieces of shell in their bills and flying off to some distant location to drop

FIGURE 3-2   Ethogram showing courtship pattern in orange chromide *(Etroplus maculatus)*

An ethogram can be compiled for all behaviors of a species or for only selected aspects of behavior.

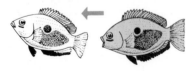

a. *Charging:* an accelerated swim of one fish toward another

b. *Tail Beating:* an emphatic beating of the tail toward another fish

c. *Quivering:* a rapid, lateral, shivering movement that starts at the head and dies out as it passes posteriorly through the body

d. *Nipping:* an O-shaped mouth action that cleans out the (presumptive) spawning site

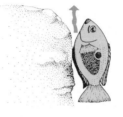

e. *Skimming:* the actual spawning movement whereby the fish places its ventral surface against the spawning site and meanders along it for a few seconds

them (Figure 3–3), a behavior also seen in many other bird species. Why do parent birds remove the eggshells? What selection pressures could have led to the evolution of such a behavior?

Tinbergen and his colleagues considered several explanations of the behavior. First, the sharp edges of the shells might injure the chicks, or the shells might clutter the nest and hamper attempts by the parents to brood and feed chicks. These explanations seemed unlikely because the eggshells of black-headed gulls are thin and easily crushed. The investigators also noticed that the eggshells are mottled on the outside and white on the inside. Although the exterior may camouflage the eggs during incubation, the bright, white interior may attract potential predators, such as crows *(Corvus corax)* or herring gulls *(Larus argentatus)*, which rely on visual cues for detecting prey. Thus, a more attractive hypothesis is that the parents remove the

FIGURE 3–3   Young gull chick beside its shell fragments
The chick's parents remove the shell fragments soon after hatching. Tinbergen and his colleagues showed that the white color of the inside of the shell, which normally shows when the shell is broken, attracts more predators than the mottled outside of the shell.

white eggshells to protect their black-headed young. The behavior of a related bird species, the kittiwake *(Rissa tridactyla)*, provides additional support for this explanation. Kittiwakes do not remove the shells after chicks hatch; their chicks are white and thus not camouflaged at all. However, kittiwakes nest on cliffs in regions where there are few predators.

TESTABLE HYPOTHESES.   One testable hypothesis that arose from these observations was the following: An eggshell left in a nest exposes the brood to a higher rate of predation. To test this hypothesis Tinbergen and his co-workers performed field experiments in a large gullery in England. They scattered a number of black-headed gull eggs over the dunes just outside the gullery; some were the natural, mottled coloration, and others were painted white. For a number of days they watched the dune site from a blind and recorded the rates of predation on the two types of eggs by crows and herring gulls. The results, shown in Table 3–1, clearly indicated that the white eggs were taken in greater numbers by both species of predators.

TABLE 3–1   Predation of black-headed gull eggs

	Eggs Painted White	Eggs with Natural Mottled Coloration
Taken by crow	14	8
Taken by herring gull	19	1
Taken by others	10	4
Total taken	43	13
Total not taken	26	55

Source: Data from Tinbergen (1963).

TABLE 3-2   Number of eggs taken and
distance between whole gull eggs and
broken shells

	15 cm	100 cm	200 cm
Eggs taken	63	48	32
Eggs not taken	87	102	118

Source: Data from Tinbergen (1963).

Tinbergen and his colleagues designed a second experiment as a further test of the main hypothesis. They laid eggs out in the dune valley, and placed half-broken eggshells at varying distances from the whole eggs. The results revealed that the greater the distance between the whole egg and the half-broken eggshell, the lower the probability of detection of the whole egg (Table 3–2). These and several additional experiments substantiated the idea that it is advantageous for adult gulls to remove the shells when the chicks hatch. Selection should favor those gulls who remove the shells.

ADVANTAGES AND DISADVANTAGES.   Ethologists proceed from observations, usually made in a natural setting, to the formulation of hypothetical explanations of the functions and evolutionary development of behaviors. The best explanations are those that lend themselves to experimental testing, such as the reason for eggshell removal by black-headed gulls. One major advantage of the ethological approach is that it is based on an evolutionary perspective; the functions and adaptive significance of behaviors can be discerned only under field conditions and through an understanding of how evolution affects morphology, physiology, and behavior.

Two of the disadvantages of the ethological approach are of particular importance. First, working in the natural setting, ethologists cannot control the environment and lack knowledge of the observed animals' past history. Since factors like weather, habitat differences, or seasonal changes cannot be manipulated by the investigator, the functional significance of observed variations in behavior may be difficult to ascertain.

Second, providing a hypothetical evolutionary explanation for a behavior is often easy, but devising and conducting an experiment that will thoroughly and convincingly test the hypothesis can be difficult. Not all behaviors lend themselves to observation and analysis; this may be one reason why we tend to look at reproductive behaviors when we are interested in questions about adaptation. For example, we can hypothesize that the cooperative hunting behavior seen in some African wild dogs (*Lycaon pictus*) that prey on various ungulates (hoofed herd animals) evolved as a means by which a group of dogs could succeed in isolating, running down, and killing a larger prey. A single dog could probably achieve this feat rarely, if at all. Testing such a hypothesis directly, however, is simply not possible because we cannot retrace the steps in the evolutionary sequence that led to the present situation. Given the right set of observational data, we might be able to conclude that single dogs or pairs of dogs were less successful in their hunting efforts than larger groups, but these data would be only suggestive and would not provide a direct test of the original hypothesis about the evolution of group hunting.

COMPARATIVE PSYCHOLOGY

COMPARATIVE ANALYSES.   **Comparative psychology** can be defined as the discipline devoted to comparative studies of behavior in nonhuman animals and, in some instances, in humans. The behaviors most often studied are those having to do with learning and development. The primary emphasis is on "how" questions about the mechanisms that underlie observed behavior patterns. Research in comparative psychology may begin with the identification and characterization of classes of behavior patterns in two or more species. Comparative analyses lead to the discovery of relationships between various types of behaviors and among species. A second approach is to select a species that is most appropriate for investigating a particular problem. Here, as several investigators (notably, Beach 1950; Hodos and Campbell 1969; Lockard 1971; Dewsbury 1984) have noted, the frequent use of the domesticated rat has produced criticism of comparative psychology both from within and from outside the discipline.

A PSYCHOLOGICAL STUDY.   By examining a specific research project, we can get a clearer look at the thinking and methods of comparative psychology and note some of the advantages and disadvantages of this approach. Turpin (1977) studied the effects of early social experience on the environmental preferences of rats *(Rattus norvegicus)* by testing two groups, one in which each animal had been raised in isolation from the time of weaning and another in which the animals had been raised in groups. Half the isolation cages and half the group cages were painted with a pattern of vertical black and white stripes, and the remaining cages were painted with horizontal black and white stripes. Young rats were exposed to these conditions for fifteen days after weaning and were then tested. Testing consisted of presenting each subject with a choice of two chambers, one painted with vertical stripes and the other with horizontal stripes. Preferences were measured according to the amount of time the subject spent in each chamber.

The results, shown in Figure 3–4, indicate that while the isolation-raised rats preferred the familiar environment—that is, the stripe pattern to which they had been exposed for fifteen days—the group-raised rats showed a distinct preference for the unfamiliar chamber. These findings suggest that the presence or absence of conspecifics during some types of early experiences may be critical in determining how that experience affects a young animal's subsequent reactions to new stimuli. This particular investigation, then, illustrates an attempt to elucidate some underly-

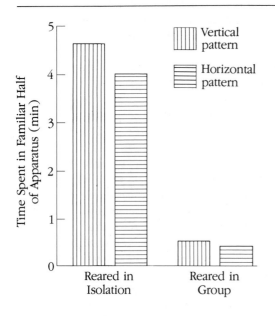

FIGURE 3-4   Effects of early social experience on environmental preference in rats
Rats reared in isolation spent more time in the familiar half of the apparatus, whereas rats reared in groups spent more time in the unfamiliar half. The pattern of lines — vertical or horizontal — does not affect the rats' choices.

ing developmental mechanism to explain observed behavior — the "how," or proximate, questions about animal behavior.

ADVANTAGES AND DISADVANTAGES.   Many features of this study characterize investigations in comparative psychology and illustrate some of the advantages and disadvantages of this approach. First, the experience and environment of the subject animals can be controlled, and the investigator can draw exact or nearly exact conclusions about the effects of treatments on behavior. Second, one or two variables can be manipulated in a systematic and replicable experimental design to ascertain their effects without the possibly confounding influences of other, uncontrolled variables. Last, the investigation is concerned primarily with the development of a preference for novel or familiar environments, and the proximate mechanisms underlying this preference.

Certain drawbacks are inherent in the comparative psychology approach. First, the animals studied are often, as in the previous example, a laboratory stock, housed in an environment that is relatively simple, compared with the natural habitat. How does domestication of rats or other laboratory animals affect behavior, and what are the effects of housing them in unnatural conditions? Comparisons of wild and laboratory rats suggest that some behaviors are affected but others are not (Price 1972, 1984; Price and Huck 1976; Price and Belanger 1977; Boice 1972). For example, wild rats are more active than their domestic counterparts, and laboratory strains exhibit different patterns of learning than do those captured in the wild. Also, aspects of maternal behavior, including pup retrieval and nest construction, differ in the two types of rat. In general, it appears that during the process of domestication, selection may have produced a greater flexibility in behavior. Investigators must be constantly aware that laboratory confinement or housing in unnatural conditions may have changed the behavior of their subject animals.

The approach used by comparative psychologists has also made them prone to the error of failing to view their experimental methods and results in a complete context involving ecology and evolution. We should be concerned with both physiological and evolutionary explanations for observed patterns of behavior in a species. For example, knowing something about the natural environment of a species and its recent phylogenetic history is necessary in studies of learning. A species that has a narrow range of diet preferences may not perform well in a laboratory learning situation in which various types of foods are used as reinforcement. Nor might such a species be expected to exhibit much flexibility in diet even when its food experiences are varied under laboratory conditions. In the study just discussed we might, for example, want to question the ecological relevance of vertical and horizontal stripes in the world of the rat. As animal behaviorists we must constantly be concerned about the *Umwelt*—the sensory-perceptual world of the species we are investigating.

Both ethology and psychology have a great deal to contribute to the investigation and understanding of animal behavior. Because both "how" and "why" questions must be tested, an integrated approach that combines the methods of each discipline can provide the best total analysis of behavior.

DESIGN FEATURES IN ANIMAL BEHAVIOR STUDIES

We now examine some principles and problems associated with conducting experiments in animal behavior. To provide a basis for examining these principles, we can consider the design of an investigation of how the hormone testosterone affects aggressive behavior in male Mongolian gerbils *(Meriones unguiculatus)* (Figure 3–5). Our test animals will be normal intact male gerbils, castrated males, and castrated males given testosterone replacement treatments.

DEFINITIONS AND RECORDS

Our first task is to agree on which behaviors, of the many exhibited by gerbils, we will consider to be aggressive. We begin by observing several pairs of gerbils interacting. By watching and discussing the various behaviors exhibited we may be able to decide which behaviors we will consider aggressive.

FIGURE 3–5   Mongolian gerbil
(Meriones unguiculatus)
These animals, native to the Gobi Desert and nearby regions in Mongolia and China, have become widely used test subjects in laboratory investigations of behavior. Here two unfamiliar males meet, and one sniffs the other, which freezes.
Source: Photo by Lee C. Drickamer.

All observers must record similar behavior patterns in exactly the same way. (For an excellent discussion of the problems involved in describing behavior, read Drummond, 1981.) To ensure that record keeping remains consistent, we need a permanent record of the behaviors. Each investigator can refer to this record when making observations so that definitions of behaviors will not change during the course of many weeks or months of study. In addition, other scientists are able to use the visual or written definitions of aggressive behaviors to repeat the experiments or to make comparisons between gerbils and other rodents.

Films, photographs, drawings, and written definitions are the standard types of records that can be made. **Films** can be taken of the various interactions to provide an unequivocal record of the patterns recorded as aggressive behaviors. However, films have two disadvantages: they are difficult to refer to during daily observations, and other scientists cannot use the definitions of aggressive behavior patterns without having copies of the film.

Photographs or **drawings** are easily referred to during experiments and can be published as part of the permanent record of the research. Figure 3–6 illustrates aggressive behavior patterns commonly observed in rodents.

In addition to films, photographs, or drawings, animal behaviorists often provide **written definitions** of behavior patterns. Thus, for example, in the study of gerbils we can record three patterns as aggressive behavior:

- Attack: One gerbil bites, kicks, claws, or pushes the other, possibly inflicting physical harm.
- Chase: One gerbil vigorously pursues another in order to catch and attack it.
- Aggressive Groom: One gerbil mouths the neck, flanks, or ano-genital region of the other.

As rigorous as these definitions may be, we still need to check the agreement between observers about the behavior patterns. Thus we must conduct a series of **interobserver reliability** tests. Two observers simultaneously watch several pairs of male gerbils interacting, and each observer maintains a separate record of the frequencies of the three kinds of aggressive behavior. Then we compute the correlation between the two sets of data; a high positive relationship between the two data records indicates a good agreement between what the two observers have seen and recorded for the behavior patterns.

When we report the results of the experiment, we must be careful to specify all aspects of the methods used in conducting the investigation. In this way another scientist, working at another time and in another place, can independently **replicate** our experiment and verify, contradict, or modify our conclusions.

DESIGN OF THE EXPERIMENT

HYPOTHESIS FORMULATION.   Once agreement is reached on what we will consider to be aggressive behavior, our next step is to design the experiment. First, we must formulate specific hypotheses, or questions for testing. Two types of hypotheses are involved in behavioral research: experimental hypotheses and null hypotheses.

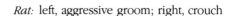

FIGURE 3-6   Aggressive behavior patterns in small rodents
Classification of aggressive behavior patterns of small rodents enables scientists to observe and record similar behavior patterns in exactly the same way.
Source: Data from E. C. Grant and J. H. Mackintosh, ''A Comparison of the Social Postures of Some Common Laboratory Rodents,'' *Behaviour* 21 (1963):246–259.

Hamster: nose–nose   *Rat:* left, aggressive groom; right, crouch

Gerbil: left, aggressive; right, submissive   *Mouse:* left attacking right

Ideas that we develop by synthesizing the reported investigations of other scientists with our own ideas about a research topic are termed **experimental hypotheses.** In our present example we may want to know: (1) is aggression in adult male gerbils dependent upon the hormone testosterone? and (2) do male gerbils begin to fight before or after puberty? These hypotheses lead us to a series of predictions. For example, if testosterone is critical for adult male gerbil aggression, then eliminating testosterone by castration should lessen or eliminate aggression in castrated males. Further, based on hypothesis (2) we can predict that male gerbils will not fight before the age of puberty, when production of testosterone increases.

The **null hypothesis** proposes that the behaviors of animals being given different experimental treatments do not differ significantly. The null hypothesis cannot be proved; it can only be accepted or rejected to provide a basis for interpreting answers to the experimental hypothesis.

It is important to understand that animal behavior research is probabilistic. This is one of the basic principles governing investigations in psychobiology. If two animals engage in a fight one day, animal A may defeat animal B. But on another occasion, B may defeat A. The reasons for this reversal are not readily apparent or easily explained. In another example, consider the problem of measuring how fast a springbok can run 100 meters when it flees from a cheetah. Some antelope will run faster than others. To generalize about the speed of all antelope of a particular species, we need an average measure of speed for many animals and a statement about how much variation in speed to expect among members of the test group. Clearly, an exact determination of fighting ability or running speed cannot be made, so animal behaviorists must deal with probabilities and predictions of behavior. Although we will not deal directly with statistics in this text, knowledge of statistics and experimental design is necessary for advanced work in animal behavior. For now, a few basic principles of experimental design can provide us with a background for understanding and critiquing behavioral research.

Our second task in the gerbil study, then, is to formulate a statistical hypothesis — a null hypothesis. Our null hypothesis might be: There are *no* differences in the levels of aggression of normal, intact male gerbils, castrated male gerbils, or castrated male gerbils given daily injections of synthetic testosterone to replace the hormone normally secreted by the testes.

From the null hypothesis we derive some alternative outcomes for the experimental test and can examine their interpretation. If indeed there were no significant statistical differences among all the treatment groups, we would accept our null hypothesis. We would then return to experimental hypothesis (1) above and its predictions. Our major prediction would be incorrect: no basis exists from the test data to claim that testosterone plays a critical role in male gerbil aggression. However, if the intact males and those receiving daily injections of synthetic testosterone exhibited statistically significantly higher levels of aggression than castrated males, we would reject the null hypothesis. Returning once again to the experimental hypothesis, we could conclude that the test revealed an important role for testosterone affecting aggression in male gerbils. Regardless of whether we accept or reject the null hypothesis, there may be important biological conclusions to be drawn from our experimental findings. (For expansion of these ideas and a discussion of the distinctions between hypotheses and deductions see Moore, 1984.)

DESIGNATION OF VARIABLES. The next step in a research project is to establish the exact parameters to be investigated — that is, the independent and dependent variables. **Independent variables** are the factors the investigator manipulated to define the treatment groups and conditions of the experiment. To test the first question regarding testosterone in male gerbils, at least three treatment groups are necessary: (1) normal, intact adult male gerbils; (2) castrated adult male gerbils; and (3) castrated adult male gerbils given daily injections of synthetic testosterone. The outline of the experiment might look like that shown in Table 3–3. In addition to controlling the treatment groups, the experimenter must systematically control many other aspects of the experiment, including the ages and previous rearing experiences of the gerbils, the length of the observation period, and the size of the arena in which the males interact. The experimenter should hold these factors constant for all experimental test groups so that the *only* manipulated factor in the design is the presence or absence of the hormone testosterone. In the present study we will pair each gerbil with another male from the same treatment group in an open circular arena two meters in diameter for a ten-minute observation period.

The measures of the behaviors that are observed and recorded are the **dependent variables.** In this experiment the frequencies of the three aggressive behaviors will serve as dependent variables.

CONTROL GROUPS. In addition to the design features discussed earlier, at least three other considerations arise in the course of setting up an experiment. Many experiments require control groups in order to provide a baseline for comparison with other experimental groups. A **control group** is an observed, unmanipulated set of

TABLE 3–3 Outline of experiment to test effects of testosterone on aggression in adult male gerbils
Groups 4 and 5 serve as controls for manipulations performed on gerbils in groups 2 and 3 to ascertain that the treatment, not the manipulation, affects aggressive behavior.

	Behavior Patterns (Frequencies)		
Treatment Groups	*Attack*	*Chase*	*Aggressive Groom*
1. Normal, intact males			
2. Castrated males			
3. Castrated males given testosterone injections			
4. Sham-operated intact males			
5. Castrated males injected with oil			

subjects, so the investigator can see what subjects do *without* the experimental condition. In the gerbil experiment, the intact males can provide data on "normal" aggression levels in gerbils. The frequency of aggressive behavior by the pairs of castrated males and castrated males given testosterone injections can be compared with that of the normal, intact males of the control group.

In other instances control groups serve to demonstrate that some portion of the experimental treatment procedure has not produced an unwanted effect that confounds the results. The surgery performed on some test subjects, including the process of anesthetizing them and making an incision in the scrotum to remove the testes, may have altered their aggressive tendencies. Castrated males may exhibit higher or lower levels of aggression because of changes in blood testosterone levels due to testes removal, or the shifts in the levels of aggression may result solely from the surgery. Thus we must add a fourth treatment group—sham-operated males (Table 3–3, group 4). All surgical processes except actual removal of the testes are carried out on these males. If the surgical processes have no effect on aggression, then we would expect sham-operated males to exhibit the same amounts of aggressive behavior as do normal, intact males. Unless we have a control group, we will not be able to distinguish whether any observed behavioral effects in the treatment group receiving testosterone-replacement therapy are the result of the hormone or of the injection. Thus we need to add a fifth treatment group of gerbils that are injected with only the peanut oil used to dissolve the testosterone for injection. For each manipulation in an experimental design, we need to determine what, if any, control groups are appropriate to the investigation.

Control groups are unnecessary in some behavioral studies. For example, in research that simply describes the behavior patterns of various animal species, no control groups are needed. Nor would control groups be necessary in an assessment of a particular behavior pattern among groups of one species living in different habitats—for example, grooming behavior in groups of squirrel monkeys *(Saimiri spp.)*. In this situation we would simply make comparisons between the monkeys in various habitats and draw conclusions about the relationship of differences in grooming patterns to differences in habitat features.

INDEPENDENT DATA POINTS.   The need to gather numerical pieces of information that are independent of each other is probably the most difficult principle for behavioral biologists to comprehend and is based on both biological and statistical arguments. We can consider the issue of behavioral independence in our gerbil experiment in several ways. For any pair of gerbils, the behavior of one male will definitely influence the behavior of the other male. The principle of independent data points dictates that data from both males cannot be used as two distinct points in the same analysis. There are at least two mathematical solutions to the problem. We can combine the records of aggressive behavior for the two gerbils to produce one score for each pair. Or, data for "winners" and "losers" could be analyzed separately, but we would then need preestablished criteria for defining winners and losers.

In the gerbil investigation, we will use each animal only once because the aggressive performance of the male the second time it is tested is partly a function of its experiences in the first interaction. After being severely attacked in the first encounter, some males become submissive during all subsequent interactions, and erroneously low estimates of aggression frequencies would result. In contrast, another male might become superaggressive after winning its first encounter; reusing this male would result in erroneously high estimates of aggression frequencies.

Another aspect of the problem of independent data points is related to animals living in groups. For example, in an investigation of the feeding rate in adult female geese living in flocks, the total number of females and males living in the flock could influence the feeding rate of each individual female (Figure 3–7). The independent units for data collection and analysis in this example are flocks; an average feeding rate for all females in a flock would provide data for making comparisons between flocks of different sizes or resident in different geographical locations or habitats. Assuming that each of the females in a single flock represents a

FIGURE 3–7   Canada geese feeding on ground beside pond
In a flock of geese the behavior of each bird may be influenced by actions of the other birds. Information gathered on the behavior of an individual bird cannot generally be considered as independent of similar data on another bird in the same flock.
Source: Photo by Hugo H. Schroder from Frank W. Lane.

separate data point would be incorrect; the behavior of each is not necessarily independent. If we wish to design and conduct the experiment properly, we must pay considerable attention to the difficult but necessary requirement of keeping each data point independent.

The principle of independent data points also derives from one of the critical requirements of most statistical tests utilized by animal behaviorists — that is, that each piece of information used in the analysis must be independent of every other piece of information. Special statistical tests have been developed to handle data analysis in experiments in which the same tests have been repeated on the same animals over time. These tests do not require complete independence of data points for information gathered on the same animal.

SAMPLE SIZE.   A final factor in designing an experiment is the sample size — the number of animals that must be used in the investigation. Each experimental treatment group must contain enough animals to provide a complete and accurate assessment of the behavior. If the experiment is concerned with only qualitative descriptions of behavior, then the sample size should be large enough to permit an observer to record the behavior fully, including providing sound estimates of average values for the dependent variables and of the amount of variation around the averages. In addition, most statistical tests require a certain minimum sample size for a correct analysis of the data. Deciding on the correct sample size to use for a particular experiment requires some practical experience with both the animal being studied and the type of statistical test to be used in analyzing the results.

VARIATION AND VARIANCE

For most morphological, physiological, and behavioral traits animals exhibit variation in their appearance or performance. For example, we noted that red-winged blackbirds exhibit variation in frequency of displays and in selection of nest sites.

The variation may be either discrete or continuous. **Discrete** variation involves those measures that can take on only certain values. For example, clutch size (number of eggs from a pair in a season) in birds is a discrete measure; a species may lay 1, 2, 3, or up to 6 or more eggs per clutch, but they do not lay 1.5 or 3.6 eggs. Other measures are **continuous;** they can take on any value between some lower and upper limit. The nesting territories of red-winged blackbirds may vary from a few square meters to thousands of square meters (Figure 3–8). Values assigned for different birds can be anywhere within this range, limited only by how refined we wish to make our measurement.

To assess the amount of variation that occurs for a particular behavior or trait in a sample, we use two principal measures, the mean ($\overline{X}$) and the variance (s^2). We can calculate the mean and variance for any group of values for a trait regardless of whether the variable is discrete or continuous. The **mean,** or central tendency, is

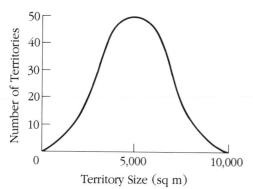

FIGURE 3-8   Territory size in red-winged blackbirds
Many measurable behavioral traits show some form of continuous variation. Territory size in redwings provides such an example. This graph illustrates territory size for a large sample of over 500 birds.

the arithmetic average. Thus, from a sample of four clutches of eggs from yellow-shafted flickers we might get values of 2, 4, 6, and 8 — with a mean of 5 eggs per clutch. In a second sample of four clutches the values of 4, 5, 5, and 6 eggs would also produce a mean of 5 eggs per clutch.

The **variance** is an estimate of the amount of deviation of the values in a sample around the mean of that sample; it is, by definition, the average squared deviations of the values from the mean of the sample. We calculate the variance as

$$s^2 = \frac{(X_i - \overline{X})^2}{N - 1},$$

where X_i represents the individual values in the sample, $\overline{X}$ is the mean, and N is the total number of values in the sample. So, for our first sample of clutches the variance is 6.7, whereas for the second sample, the variance would be 0.7. The means for the two samples are identical, but the variances are quite different; there is much more variation around the mean in the first sample.

We shall refer to the notions of mean and variance in both general and specific terms at a number of junctures throughout the book. Two other measures of the variation around the mean, the **standard deviation** and the **standard error of the mean,** are also often encountered in animal behavior literature.

FIELD METHODS

The same requirements for design features of laboratory experiments apply to field situations. One of the early pioneers in conducting animal behavior research in laboratory and field settings, T. C. Schnierla (1950) used an experimental strategy that combined field investigation with appropriate laboratory tests of specific points. He was also a strong proponent of the comparative method and carried out behavioral observations under varied field conditions.

In recent years, field studies of many animal species, particularly of primates, have focused considerable attention on research strategies and methods of collecting data under field conditions. Topics like focal-animal sampling, instantaneous sampling, and so forth are beyond the scope of this book. (See J. Altmann 1974; Dunbar 1976; Hinde 1973; and Lehner 1979 for excellent treatment of the subject of observational methods in the study of behavior.)

We should note two important points. First, despite the absence of rigid laboratory-type control, field studies should not be considered less accurate than laboratory investigations. Second, the same constraints and methodological considerations apply whether we are working in the field or laboratory.

OBSERVATIONAL METHODS

In addition to the topics covered elsewhere in this chapter and at other appropriate points in the book, several selected issues should be discussed briefly in conjunction with the use of observational methods for studying animal behavior. Most of these issues have particular relevance to investigations conducted in the field, but many points also pertain to laboratory conditions. (For a thorough treatment of these and related topics see Lehner, 1979.)

OBSERVING VERSUS WATCHING.   Most of us watch animals on a daily basis, whether they are pets, wildlife, or simply other humans. We may notice when something about the behavior of a dog is different than usual or when particular bird species flock to the feeder in greater numbers. These visual and mental notes constitute **watching. Observing** animal behavior involves systematic recording (writing, tape recording, or filming) of the activities of particular animals, usually with specific test questions governing the nature of the data that are recorded.

Making thorough, proper observations requires that the observer maintain an awareness of some critical considerations. Two of these — the problem of equal observability of all of the subjects under study and the establishment of interobserver reliability — are discussed elsewhere in this chapter. The latter is important both in terms of the agreement between two or more people recording data on the same animals and may act as a check against possible sex or age biases. One observer may be recording a disproportionate number of some actions for a particular age or sex class.

Another key consideration for conducting observational research is scheduling the sessions to coincide with the time of day (and year) when the behaviors we want to study are likely to occur with reasonable frequency, but not such that a distorted picture is obtained. For example, it may be erroneous to select *only* the peak periods for the occurrence of a particular class of behaviors, such as aggression, as this would give a biased impression, but it would be equally unwise to study the same animals primarily during their resting period.

Proper observational research requires certain equipment. Generally, binoculars or spotting scopes are needed to identify animals and to accurately see some aspects of their behaviors. For some species it may be necessary to construct a blind or other unobtrusive vantage point in order to record the animals' activities without disrupting their normal patterns. We must always be aware of the fact that the mere presence of an observer can influence the subjects' activities, in both field and laboratory settings.

Last, some form of reliable, accurate scheme of data collection must be devised. Construction of proper data sheets and ways of encoding behaviors and keeping track of the identities of subjects requires some practical experience. Lehner (1979) devotes a section of his book to this subject.

IDENTIFYING SUBJECTS.   A variety of schemes for identifying and marking individual animals for observational experiments have been developed. Among these are using ear tags, colored leg bands, numbered tattoos, and fur dyes; notching the ears; clipping the fur in various patterns; and/or using natural markings and behavioral traits to identify subjects. At least two major problems should be mentioned in conjunction with marking animal subjects. First, some of the techniques just mentioned may alter the behavior of the subject, or they may influence the appearance or actions of the subject, resulting in changes in the reactions of other animals to the subject. This could indirectly influence changes in the subject's behavior. Second, some investigators have adopted the practice by which animals are identified and recorded in their data by names, such as "Swifty," "the Old One," and the like. This practice should be avoided. When we name an animal based on its physical appearance or some behavioral trait, we attribute to it certain qualities that will almost certainly result in subsequent bias in our observations. Even the practice of randomly assigning human names (such as "Ralph" and "Sue") can result in the association of particular traits with some names; this can also bias behavioral observations.

NATURAL VARIATION VERSUS MANIPULATION.   We often begin a behavioral study by making general observations of the subjects. When we have decided on the hypotheses to be tested and the general approach, we arrive at several critical questions: Can the hypotheses be investigated without any manipulation — that is, can we rely on natural variation? Do we need to consider the possibilities for manipulating either the subject(s) or the environment? For some studies it will be necessary only to devise a usable scheme for recording the data. For instance, if we were interested in comparing the frequencies of foraging behavior for a particular bird species in deciduous versus in conifer forests, no manipulations would be necessary, as the natural variation in the two types of habitats provides us with the basis for a testable hypothesis.

If, however, the hypotheses being tested involves the effects of some social variable — for example, flock composition on foraging behavior — it may be necessary to manipulate the number and sexes of birds in the foraging flocks by netting or trapping to remove some birds. Alternatively, we might be interested in whether provisioning the habitats with supplementary food would alter the foraging rates of the birds in either habitat. In this instance we would be manipulating the environment.

METHODOLOGICAL PITFALLS

Several potential pitfalls may entrap the unwary investigator; we become aware of most of them through experience. Three areas deserve special attention: (1) perceptual worlds, (2) correlation versus causation, and (3) differential observability of animals.

PERCEPTUAL WORLDS.   One important task of animal behaviorists is to gain a thorough understanding of the perceptual world—the *Umwelt,* of each animal. Every animal lives in a different perceptual world. All too frequently animal behaviorists assume that the animals they are studying live in the world of our human perceptions; this situation is only rarely true. For example, bats and many rodents emit sounds at very high frequencies, well beyond the range of human hearing (Griffin 1958; Noirot 1972). These high-frequency sounds are important in bats for navigation and catching prey (Figure 3–9 and Figure 3–10) and in mice for communication

FIGURE 3-9   Bat in flight using large ears for echolocation
Animals live in perceptual worlds that differ from that of humans. To properly understand behaviors of animals we must also investigate their varying perceptual worlds.
Source: New York Zoological Society Photo.

—sonar waves

FIGURE 3-10   Bat's system of echolocation
Bats navigate by echolocation whereby they emit high energy sounds that bounce off objects in the environment and return to their sensitive hearing systems as echoes.

echo of sonar waves

between the young pups and the mother. To understand the behavior patterns of animals and to test various behavioral phenomena properly, we must know something about animals' sensory systems and how organisms interpret various stimuli in their environments.

CORRELATION VERSUS CAUSATION.   When explaining why a particular behavior occurs, we must be aware of a second problem area — confusing *correlation* with *causation.* For example, if we were investigating reproduction in song birds and found that the birds always mate at the same time of the year and that mating commences soon after the beginning of heavy spring rains, we would probably find a good correlation between the onset of rains and subsequent mating behavior. However, it would be incorrect to infer that rainfall itself is a direct cause of mating behavior. Increased rainfall may lead to changes in both the quality and quantity of vegetation available to the birds for food. Changes in diet may in turn produce hormonal changes, which may then lead to behavioral changes. In addition, other factors may affect the mating behavior of the birds, including external factors such as number of hours of daylight, availability of nesting materials, or social relations with other birds, and internal factors such as biological clocks (see Chapter 8) or previous mating experience.

Thus, explaining the timing of the onset of mating behavior in song birds involves considerably more than simple correlation between environmental and behavioral changes. Correlation is an important tool for structuring our thinking, and for formulating hypotheses about the causes of behavior, but we must not confuse correlation with causation. Any of several possible sequences can explain a particular behavior. Each step in any hypothetical sequence must be analyzed experimentally to confirm the causal connections between events in the sequence.

DIFFERENTIAL OBSERVABILITY OF ANIMALS.   A third potential pitfall for investigators conducting studies of animal behavior centers on how long and for what period of time they observe different animals within the group. If investigators observe juveniles, for example, more frequently than other group members, the conclusions of the investigation could be significantly biased. A good illustration is the relationship between social dominance status and mating frequency of adult male rats at a garbage dump (Table 3–4). To test this relationship we need the following information: (1) the social rank of each adult male rat, (2) the observed frequency of mating by each male, and (3) the number of times and length of time each male was observed.

We can determine social rank by watching the males at the dump for several weeks. The top-ranking male (B) is the rat that can defeat all other six adult male rats living at the dump. The next highest male (D) can defeat the remaining five, and so forth down to the lowest-ranking rat (M), who loses all his fights with the other males. We must also record how many times we see each male copulating with an adult female (column 3). On the basis of the frequency of copulation and rank of the male rats, we may conclude that higher social status among males is associated with a greater amount of mating activity. However, this may not be true; what if the lower-ranking male rats were not seen as often? The lower-ranking males may be seen less frequently and seen to mate less frequently but they could be engaging in frequent mating behavior in areas not visible to the observer.

To complete the analysis we need to know the number of times and length of time each male rat is seen in a standard time period — say, one month. If all the males are seen about the same number of times and for the same total amount of time, then the correlation between higher social status and more mating activity would be

TABLE 3-4   Social dominance and mating in adult male rats at a garbage dump

Male	Social Rank	Observed Matings/ Month	Times Observed/ Month (hr)	(Corrected) Number of Matings/Month/ Times Observed (hr)
B	1	18	85	0.21
D	2	15	80	0.19
K	3	14	68	0.21
G	4	10	41	0.24
C	5	7	30	0.23
J	6	5	26	0.19
M	7	4	22	0.18

valid. If, however, as Table 3–4 shows, males are seen more frequently or less frequently depending on social rank, then we must make a correction for *observability*. The data in column 5 of Table 3–4 have been corrected for the observability of each male and are expressed as the number of matings per month per time observed. In this case there are no differences in the mating rates of males of different ranks. While this finding does not mean that the original conclusion is invalid, the corrected results do indicate that further study will be necessary to determine what the lower-ranking, less-visible males are doing when the observer cannot see them.

SUMMARY

An ethological study of animal behavior begins with an ethogram—a complete inventory of an animal's behaviors—and seeks to answer "why" questions by determining the functions and evolutionary developments of behavior, as is exemplified by Tinbergen's studies of the functional significance of eggshell removal in gulls. The advantages of the ethological approach are that animals are studied in their natural habitat, where functional relationships are more readily discerned, and that this approach deals with problems from an evolutionary perspective. However, the ethological approach sacrifices control over environmental parameters and over knowledge of an animal's history. In addition, devising a good test of the evolutionary significance of a behavior pattern is difficult.

A study of animal behavior from the approach of comparative psychology begins with classification and comparison of the behaviors of different species in order to discover relationships between them. This approach is exemplified by a comparative study on the environmental preferences of isolation-raised and group-raised rats. In the comparative psychology approach, experimenters have a high degree of control over both test subjects and environmental conditions; they can manipulate one or more variables while they hold all others constant. However, the breadth and strength of the conclusions of these studies are limited by the almost exclusive use of laboratory stock for test subjects, by the influence of "domestication" on laboratory animals, and by an overemphasis on methodology to the detriment of a more complete view of the context and functional importance of a behavior pattern.

Because animal behavior deals with probabilistic phenomena, correct design of experiments is of great importance. The investigator must define the behavior patterns being investigated; must state the experimental hypotheses, or questions about observed biological phenomena, and the null hypotheses, or statements about differences between experimental treatment groups; and must designate the independent variables, or the factors that are controlled and manipulated by the investigator, and the dependent variables, or the behavior patterns that are measured as outcomes of the experiment.

In addition, the investigator must determine if control groups are necessary, either to provide a foundation for comparison or to demonstrate that the factors being manipulated, and not the manipulative processes themselves, are responsible for the observed results; must insure the independence of data points; must use sufficient numbers of animals in the experiment to provide proper analytical results; and must keep accurate and complete records of experimental procedures so that experiments can be replicated. Field investigations of animal behavior operate under the same design constraints and methodological requirements as do laboratory experiments.

Behavioral scientists must be aware of the need to understand the perceptual world of the animal being investigated and of the problem of differences in observability in animals of different ages, sex, or social status. Correlations between environmental factors and behaviors should not be interpreted as causes of behaviors.

Discussion Questions

1. Animal behaviorists ask questions concerning causation, development, evolution, and function. Some species of gulls pick up shellfish, fly up a short distance in the air, and drop the shellfish on rocks to break them open. Describe the specific series of questions you might ask in each of the four areas about this behavior and present the methods you would use to answer these questions. What specific hypotheses would you test? What experimental designs would you use to test these hypotheses?

2. A study was conducted on the effects of early feeding experiences on later food preferences in guinea pigs. The animals were grouped five per cage, and each of the three groups was fed only one type of food—A, B, or C—for three weeks. Then the guinea pigs were tested individually to determine their food preferences in a "cafeteria" situation, with all three food types present. The results are shown in the table that follows. What can you say about the relative effects of early feeding experiences in guinea pigs? What principles of experimental design and methodology have been violated in conducting this experiment?

Food Eaten in Cafeteria Test (g/24 hr)

		Type A	Type B	Type C
	Type A	9.6	2.8	5.7
Rearing	Type B	3.8	6.2	7.8
Food	Type C	1.4	.6	16.2

3. To conduct sound experiments in animal behavior, an understanding of proper methodology and experimental design is necessary. What other reasons can you cite for acquiring a firm knowledge of these principles for work in animal behavior?

Suggested Readings

Beach, F. A. 1960. Experimental investigations of species-specific behavior. *Amer. Psychol.* 15:1–18.
Defines species-specific behavior and discusses a total
framework for questions in comparative psychology. Beach
has been one of the pioneers in developing this field.

Colgan, P. W., ed. 1978. *Quantitative Ethology.* New York: Wiley.
An edited volume of articles dealing with the design and
analysis of behavior investigations. Probably more for the
advanced student than for the beginner.

Dewsbury, D. A. 1973. Comparative psychologists and their quest for uniformity. *Ann. N.Y. Acad. Sci.* 223:147–167.
A sound criticism of the approaches and methods of
comparative psychology. Presents and discusses specific
problems and suggests alternative patterns of thinking.

Lehner, P. N. 1979. *Handbook of Ethological Methods.* New York: Garland STPM Press.
Excellent introduction to the variety of methods and
approaches to the study of animal behavior.

Tinbergen, N. 1963. On aims and methods of ethology. *Zeit. Tierpsychol.* 20:410–433.
Written by a Nobel laureate and founding father of modern
ethology. Presents the four-question framework used for
the study of behavior and comments on the methods used
by ethologists.

References

Altmann, J. 1974. Observational study of behavior: Sampling methods. *Behaviour* 49:227–267.

Beach, F. A. 1950. The snark was a boojum. *Amer. Psychol.* 5:115–124.

Boice, R. 1972. Some behavioral tests of domestication in Norway rats. *Behaviour* 42:198–231.

Dewsbury, D. A. 1984. *Comparative Psychology in the Twentieth Century.* Stroudsburg, Pa.: Hutchinson Ross.

Drummond, H. 1981. The nature and description of behavior patterns. In P. P. G. Bateson and P. H. Klopfer (eds.), *Perspectives in Ethology,* vol. 4. New York: Plenum.

Dunbar, R. I. M. 1976. Some aspects of research design and their implications in the observational study of behavior. *Behaviour* 58:78–98.

Grant, E. C., and J. H. Mackintosh. 1963. A comparison of the social postures of some common laboratory rodents. *Behaviour* 21:246–259.

Griffin, D. R. 1958. *Listening in the Dark.* New Haven: Yale University Press.

Hinde, R. A. 1973. On the design of check-sheets. *Primates* 14:393–406.

Hodos, W., and C. B. G. Campbell. 1969. *Scala naturae:* Why there is no theory in comparative psychology. *Psychol. Rev.* 76:337–350.

Lehner, P. N. 1979. *Handbook of Ethological Methods.* New York: Garland STPM Press.

Lockard, R. B. 1971. Reflections on the fall of comparative psychology. *Amer. Psychol.* 26:168–179.

Moore, J. A. 1984. Science as a way of knowing—evolutionary biology. *Amer. Zool.* 24:467–534.

Noirot, E. 1972. Ultrasounds and maternal behavior in small mammals. *Dev. Psychobiol.* 5:371–387.

Price, E. O. 1972. Domestication and early experience effects on escape conditioning in the Norway rat. *J. Comp. Physiol. Psychol.* 79:51–55.

———. 1984. Behavioral aspects of animal domestication. *Quart. Rev. Biol.* 59:1–32.

Price, E. O., and P. L. Belanger. 1977. Maternal behavior of wild and domestic Norway rats. *Behav. Biol.* 20:60–69.

Price, E. O., and U. W. Huck. 1976. Open-field behavior of wild and domestic Norway rats. *Anim. Learn. Behav.* 4:125–130.

Schnierla, T. C. 1950. The relationship between observation and experimentation in the field study of behavior. *Ann. N.Y. Acad. Sci.* 51:1022–1044.

Smith, D. G. 1976. An experimental analysis of the function of red-winged blackbird song. *Behaviour* 56:136–156.

Tinbergen, N. 1960. *The Herring Gull's World.* New York: Basic Books.

———. 1963. The shell menace. *Nat. Hist.* 72:28–35.

Tinbergen, N., et al. 1962. Egg shell removal by the black-headed gull *Larus ridibundus* L.: A behavior component of camouflage. *Behaviour* 19:74–118.

Turpin, B. 1977. Early social experience and environmental preference in rats. *J. Comp. Physiol. Psychol.* 91:29–32.

4

GENETICS AND EVOLUTIONARY PROCESSES

To understand and interpret animal behavior, we must be familiar with the theory and processes of genetics and evolution. Like their morphology and physiology, the behaviors of animals have undergone changes, both large and small, that have allowed animals to adapt to their various habitats and have ensured their reproduction success. We can use our knowledge of evolutionary processes to formulate tentative explanations of the ultimate causation of observed behavior patterns and to design experiments to test the functional significance of these behaviors.

Since modern evolutionary theory is based on the principles of genetics, we examine basic genetics in the initial sections of the chapter. We then take an in-depth look at evolutionary theory and investigate the relationship between genetics and evolution under the topic of population genetics. To find out how the process of evolution works, we explore the ways in which evolutionary processes result in adaptive changes within species as well as in the production of new species. Finally, we examine genetic variation, kinship, and units of selection in the evolutionary process and conclude with some cautionary comments on the use of evolutionary explanations for behavioral phenomena.

GENETICS

SPECIES AND GENE POOLS

Although the concept of species has been defined in many different ways, there is no universally accepted definition. In biological terms a **species** consists of animals that share similar characteristics and habits and that interbreed freely with one another (Mayr 1970) but that are reproductively isolated from other species in nature — that is, individuals of one species cannot or do not breed with members of another species. Biological species are dynamic entities, changing over time, and thus are one of the primary units of study for evolutionary biologists.

Each individual possesses a number of **genes,** or basic units of heredity; the number is characteristic for a species. The genes, contained on chromosomes, establish the fundamental framework for the development of anatomical, physiological, and behavioral traits. The complete set of genes for any organism is called its **genotype;** the actual physical, physiological, and behavioral traits that can be observed are called the **phenotype.** The **gene pool** of a species consists of a hypothetical collection of all of the genes of all reproductive individuals.

The concept of distinct species of animals or plants is also employed with a somewhat different meaning to categorize living organisms. In the standard classification scheme utilized by zoologists, the species is the smallest recognized unit. Although some species, like dogs or horses, are subdivided into subspecies, races, or breeds, generally all members of these subunits can interbreed (Mayr 1963). Each succeeding higher order of classification (Table 4–1) is designated on the basis of the degree of evolutionary and genetic relatedness of the species in question, so that animals of different species that share many characteristics and a recent common ancestry are classified within the same genus, or group of related species. The more

TABLE 4–1   Classification for some animals
The basic unit of animal classification is the species. Each species may be further classified in a hierarchy of categories based on characteristics shared with other animals and on patterns of evolutionary ancestry.

Common Name	Kingdom	Phylum	Class	Order	Family	Genus	Species
rhesus monkey	Animalia	Chordata	Mammalia	Primates	Cercopithecidae	*Macaca*	*mulatta*
Norway rat	Animalia	Chordata	Mammalia	Rodentia	Muridae	*Rattus*	*norvegicus*
red-winged blackbird	Animalia	Chordata	Aves	Passeriformes	Icteridae	*Agelaius*	*phoeniceus*
three-spined stickleback	Animalia	Chordata	Osteichthyes	Gasterosteiformes	Gasterosteidae	*Gasterosteus*	*aculeatus*
migratory locust	Animalia	Arthropoda	Insecta	Orthoptera	Acrididae	*Schistocera*	*gregoria*

closely two animal species are classified to one another, the greater will be the percentage of overlap between the gene pools of the two species.

BASIC PRINCIPLES

A **karotype** is a diagrammatic representation of the contents of a cell nucleus, including the chromosome number and their size and form. Each somatic cell of most organisms is **diploid,** and contains $2n$ chromosomes, arranged in n homologous pairs (Figure 4–1). Since each sexually produced offspring must have $2n$ chromosomes, the **gametes** (ovum and sperm) are **haploid;** they each have a single set of

---

FIGURE 4-1   Human karyotype
There are 23 pairs of homologous chromosomes in each body cell of the human ($2n = 46$).
Source: Photo by J. S. Yoon, Bowling Green State University.

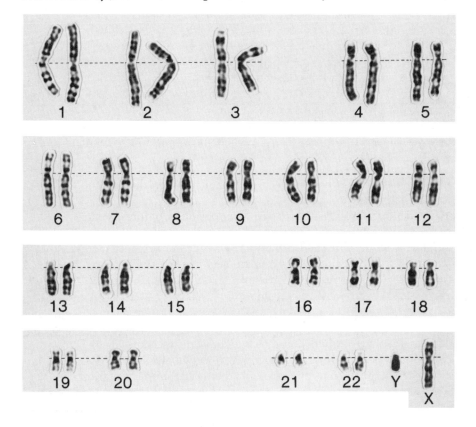

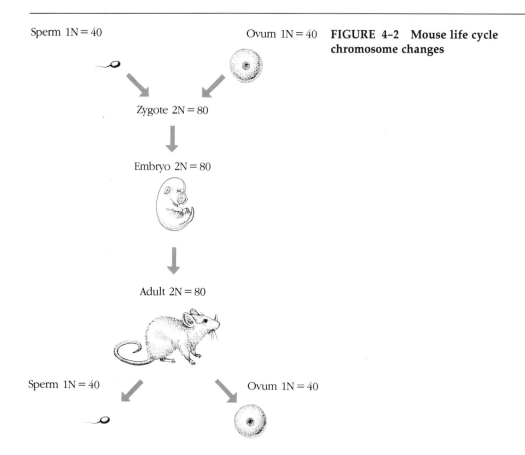

Sperm 1N = 40

Ovum 1N = 40   **FIGURE 4-2   Mouse life cycle chromosome changes**

Zygote 2N = 80

Embryo 2N = 80

Adult 2N = 80

Sperm 1N = 40

Ovum 1N = 40

unpaired chromosomes, a total of $1n$. Changes in chromosome numbers during the life cycle of a mouse are diagramed in Figure 4–2.

Genes are arranged along chromosomes at particular positions, or **loci.** Alternative forms of a gene, called **alleles,** may be present at a given locus. Although many alternative forms of a single gene can exist, an individual can have only two possible alleles of any gene at a particular locus, one on each chromosome. Since literally thousands of pairs of alleles are involved in a normal fertilization, the task of analyzing the genetic bases of behavior is quite complex.

Some alleles at a particular locus are **dominant** and express an effect when present on only one of the paired chromosomes. In contrast, other alleles are **recessive;** their effect is expressed only when both chromosomes have the recessive allele. Another possible condition is **partial dominance,** in which two alleles interact to produce a genetic expression in the phenotype somewhere between the effects of the two alleles.

An animal is **homozygous** for a given locus when both alleles at that locus are identical — for example vv or $++$ alleles in the mouse. When the alleles at the same locus are different — for example, $+v$ — a **heterozygous** condition results. Animal populations, such as inbred strains of mice, that are homozygous for all or most of their loci, are referred to as genetically **homogeneous;** all members of such a population have essentially the same set of genes, or genotype. In contrast, when the individuals are heterozygous — there is variation in the alleles found at many loci — the population is likely to be **heterogeneous.**

To gain a clearer picture of the dominant or recessive effects of genes, we can consider the following example. Suppose we are dealing with just one locus, which controls whether a mouse is normal or a waltzer — that is, one that shows aberrant circling and erratic movement patterns. Also, there are only two alternative alleles, $+$ and v, for the locus in question (see Table 4–2). If a normal male mouse of genotype $+v$ mates with a normal female bearing $+v$ alleles, then one-fourth of the offspring will be $++$, one-fourth will be vv, both homozygous conditions, and one-half of the progeny will be heterozygous, $+v$. As these genes are expressed in the phenotype, three-fourths of the progeny will be normal and one-fourth will be waltzers, because wild-type $+$ alleles are dominant to recessive v alleles.

How do genes affect the biochemical and physiological processes within the cells and organs of an animal? Modern genetic research indicates that specific gene sequences of DNA code for the structure and synthesis of particular proteins. Proteins are found in cell walls, in the structures of organelles within cells, and in the liquid medium within cells, where they are critical features of many ongoing chemical reactions. Proteins are also a basic unit of construction for enzymes, which are catalysts in biochemical processes within the cell.

TABLE 4–2   Possible combinations of two alleles, $+$ and v, at one particular locus in the mouse, when a $+v$ female mates with a $+v$ male
The $+$ allele is dominant to the v allele.

		Female Parent	
		$+$	v
Male Parent	$+$	$++$	$+v$
	v	$+v$	vv

Genotype	*Phenotype*
1 $++$	= normal
2 $+v$	= normal
1 vv	= waltzer

EVOLUTIONARY PROCESSES

Today, the most widely accepted evolutionary theory, based on the principle of reproductive success, is that of the synthetic school (see Mayr 1963; Stebbins 1966; Huxley 1974). The outcome of reproductive activity by an organism can be considered in terms of the passage of genes to the next generation; the more successful members of a species are those that pass on their genes, and thus their characteristics, to the next generation.

RELATIVE GENETIC FITNESS

Relative genetic fitness is a critical concept in the synthetic theory of evolution by natural selection; it refers to the probability that a particular genotype will be present in the offspring of the species being examined. Fitness values range from 0 to 1. When this number is multiplied by the frequency of a genotype in the present producing generation, the resulting product indicates the expected frequency of that genotype in the next generation. For example, a fitness value of 0.0 for a particular genotype would mean that no organisms in the subsequent generation would carry that genotype. A fitness value of 1.0 would mean that the genotype would be represented in the subsequent generation at the same rate that it is present in the producing generation.

We should note here that many concepts in genetics have underlying assumptions, not all of which always hold true. The concept of fitness as just presented is based on the assumptions that the population is of sufficient size and that there is no differential movement into or out of the population of genotypes. However, fitness, even with its limitations, is an extremely useful concept since it describes the relative ability of a particular genotype to continue in succeeding generations.

SURVIVAL

To pass on its genes, an animal must first survive by finding food and shelter and by avoiding predation and disease. Then the animal must successfully mate and the offspring must survive to reproduce. Much of animal behavior study is concerned with how animals play the two parts of the evolutionary "game"—survival and reproduction. Those physiological and behavioral traits having a genetic basis that are favorable for survival and reproduction are more likely to be passed on to future generations. Traits that are less favorable or that put the animal at a disadvantage will slowly disappear from the population because the genotype fails to continue in succeeding generations.

Phenotypes, or characteristics, are the "pieces" with which organisms play the evolutionary game. The phenotype is the physical manifestation of the genetic blueprint. A series of developmental stages occurs, involving interaction of the genetic information with the environment, to result in the phenotypic manifestation. The phenotype is the end result of **epigenesis.** Epigenesis is viewed as the complex unfolding of an organism's morphology, physiology, and behavior through the organism's intrinsic genetic program and through extrinsic stimulation and experience (see Chapter 9).

DARWIN AND BIOLOGICAL EVOLUTION

Biological evolution through natural selection can be defined as a change in the frequencies of genes within a population, thereby altering the gene pool from one generation to the next. Thus, biological evolution encompasses both small changes that result in minor modifications **(adaptation)** and longer-term changes that result in new forms **(speciation).** Although Darwin knew little about genes, he was able to assemble the important elements of the process of biological evolution. The following account, summarized in Figure 4–3, attempts to reconstruct the stages by which Darwin arrived at the theory of natural selection (from Mayr 1977). Darwin derived the initial facts from Malthus's 1798 essay on population.

Fact 1: All species are capable of overproducing. For example, one pair of houseflies could produce more than 6 trillion progeny in one year, if unchecked. Even elephants, which are not particularly fast breeders, could, in 750 years, produce 19 million descendants from one pair.

Fact 2: Populations of species tend to remain relatively stable over time. The death rate tends to equal the birth rate, and most fluctuations are temporary or are repeated cyclically. For example, close monitoring of a population of white-footed mice *(Peromyscus leucopus)* over a number of years shows that their number fluctuates annually, decreasing to between 5 and 20 in a two-hectare plot in winter and increasing to between 30 and 100 in the summer, after spring breeding. Although these may seem like wide fluctuations, they are in fact small, given the range of possibilities.

Fact 3: Limitation of resources. Populations do not go on increasing indefinitely because resources are limited; eventually they run out of something. We usually think of food as most important, but other requirements, such as shelter or nesting space, may limit the population. Many species, particularly invertebrates, are limited by local weather conditions involving, for example, storms or seasonal shifts in temperature, which can eliminate many members of a population.

Inference 1: Struggle for existence among individuals. Exponential population growth combined with a fixed supply of resources results in a struggle for existence among individuals. Although this idea was not new with Darwin—others had suggested that such a struggle is a beneficial feedback device to maintain the balance of nature—he emphasized that much of the competition is among members of the

FIGURE 4–3   Components of natural selection theory
Source: E. Mayr, "Darwin and Natural Selection," *American Scientist* 65 (1977): 321–327. Reprinted with permission.

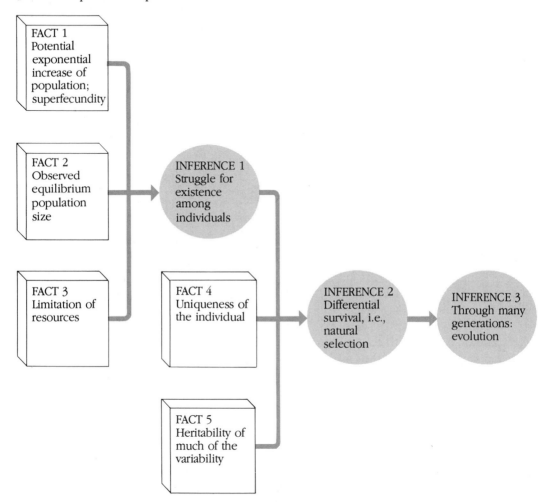

same species. Competition refers not only to predator-prey interactions but also to a struggle among animals that need the same limited resources. As we shall see, most competition is subtle, and more aggressive animals do not necessarily leave more offspring. In the mouse populations mentioned above, only one or two pups from a litter of four or five survive to the weaning age of three to four weeks. Should they survive to weaning age, their life expectancy is just ten weeks, even though the laboratory animals of this species can live one to two years.

Fact 4: Uniqueness of the individual. From his studies of animal breeding, Darwin confirmed what others had observed: Every animal in a herd is different from every other, and extreme care must be used to select the sires and dams from which to breed the next generation.

Fact 5: Heritability of much of individual variation. Although Darwin's understanding of genetics was meager, he assumed correctly that the potential for variation is always present and that it is renewed in every generation.

Inference 2: Differential survival, or natural selection. Excessive fertility and individual variation come together to set the stage for evolution. Some individuals have traits that enable them to outcompete their conspecifics, and these individuals are more likely to survive and reproduce.

Inference 3: Through many generations—evolution. The natural selection of individuals with heritable qualities, continued over many generations, leads to gradual change in the appearance of individuals—representing a change in gene frequency in the population.

POPULATION GENETICS

The subdiscipline of population genetics deals quantitatively with two aspects of evolution: the sources and the dynamics of variations. More specifically, population genetics inquires into (1) the basic units of genetic variation at the level of the gene and chromosome and (2) the causes of change in the frequency with which these units occur in populations.

MUTATION AND RECOMBINATION

The two main sources of genetic variation are mutation and recombination through sexual reproduction. **Mutation** is a heritable change in the genetic material in the form of an alteration in the sequence of nucleotide pairs in the DNA molecule. Most spontaneous mutations are copying errors that occur in the course of DNA replication. Mutations can occur from minute chemical changes that result in the substitution of one nucleotide for another in the DNA molecule (point mutations) or as major structural changes in the chromosome (chromosomal aberrations). The latter can occur in a number of ways, such as loss of a piece of chromosome (deletion); breakage of a piece, which then flips end over end and reattaches (inversion); or transfer of a piece from one chromosome to another (translocation) (see Figure 4–4).

Recombination occurs in sexually reproducing species when homologous chromosome pairs cross over, so that part of the maternal chromosome exchanges with part of the paternal chromosome. Further variability is introduced at each

FIGURE 4-4   Chromosome alteration by recombination
Chromosomes can be altered by deletion, inversion, translocation, and recombination. The process of recombination, shown here, occurs during the formation of gametes (ovum or sperm). Portions of homologous chromosomes exchange equivalent sections that produce new patterns of genes that are represented in some of the gametes produced.

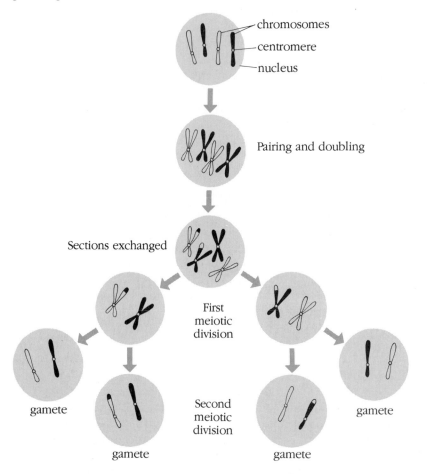

generation because the chromosomes sort randomly and independently during cell division, producing new combinations of genetic material. The haploid gametes then contain some maternal and some paternal chromosomes, plus some that are mixtures (due to crossing over).

It has been estimated that 10,000 gene loci, many bearing multiple alleles, exist in humans. The total number of possible diploid genotypes is therefore astronomical.

CAUSES OF EVOLUTION

HARDY-WEINBERG FORMULA.   Does the scrambling of genes through segregation and recombination cause evolution? In other words, does one allele increase in the following generation at the expense of another? The **Hardy-Weinberg formula** allows us to sample a population and determine whether a set of alleles is at equilibrium or whether the relative frequencies change as a result of sexual reproduction.

In order to apply the Hardy-Weinberg formula, we must assume that the population is **panmictic**—that is, each individual has an equal opportunity to mate with any member of the opposite sex, and the genes in the population are mixed randomly. (We should note that completely random mating probably does not exist in any species in nature.) In addition, we assume that gene frequencies do not change and that evolution is not occurring at that locus. To use the Hardy-Weinberg formula, we first assume that gene frequencies don't change. If the observed frequencies don't fit the values expected from Hardy-Weinberg, then we reject that assumption.

Suppose that there are only two alleles, A and a, at a given locus; each gamete carries either A or a. If p equals the relative frequency of A, and q equals the relative frequency of a, then $p + q = 1$. Will sexual reproduction itself alter p and q? What will be the frequency of the diploid genotypes AA and aa in the next generation?

If p is the frequency of the A allele, it is also the probability that a gamete will contain A; similarly, q is the probability that a gamete will contain a. Because the probability of the union of independent events is equal to the product of their separate probabilities, the probability that the two A gametes will come together is $p \times p$, or p^2. Likewise, the frequency of Aa is 2pq, and that of aa is q^2; the frequencies must total 1 (see the matrix of all possible combinations in Figure 4–5). The equation $p^2 + 2pq + q^2 = 1$ is known as the Hardy-Weinberg formula.

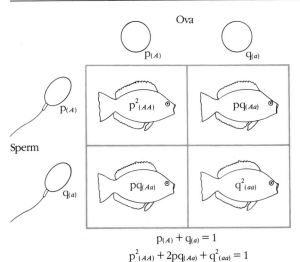

FIGURE 4-5   Hardy-Weinberg equilibrium
When all alleles have an equal chance of being transmitted from generation to generation, their relative frequencies can be described by the Hardy-Weinberg formula. A two-by-two matrix illustrates the possible combinations of offspring from sperm and ova. Under these conditions alleles A and a are at a particular locus with frequency $p(A)$ and $q(a)$.

$$p_{(A)} + q_{(a)} = 1$$
$$p^2_{(AA)} + 2pq_{(Aa)} + q^2_{(aa)} = 1$$

Let us use some real numbers to see how the Hardy-Weinberg predicts gene frequencies and tests whether the population is at equilibrium for the alleles. Suppose we mix 6,000 *AA* individuals with 4,000 *aa* individuals. The initial diploid frequencies are *AA* = 0.60, *Aa* = 0.00, and *aa* = 0.40. The value of *p* is 0.6 and that of *q* is 0.4. In the next generation and every generation thereafter, the frequency of *AA* is $AA(p^2) = 0.36$; *Aa* $(2pq) = 0.48$; and $aa(q^2) = 0.16$. The frequency of the *A* allele is thus $0.36(p^2) + \frac{1}{2}(0.48)(2pq) = 0.36 + 0.24 = 0.60$, the starting frequency. In other words, the gene frequencies do not change. Thus segregation and recombination do not cause evolution.

Use of the Hardy-Weinberg formula enables us to sample a population at one point in time and determine whether or not it is at equilibrium for a set of alleles, without having to take repeated samples across several generations, as the following example illustrates. The Quinault Indians, a Northwest Coast group, were studied by Hulse (1971) for the incidence of the *MN* blood group. Of a sample of 201 individuals, 77 were homozygous for the *M* allele *(MM)*, 101 were heterozygous *(MN)*, and 23 were homozygous for the *N* allele *(NN)*. Is the *MN* trait under selective pressure?

To find the expected frequencies in the next generation, we first calculate *p* and *q* and then substitute those numbers into the Hardy-Weinberg equation to get the proportion of each phenotype. Finally, we calculate the expected numbers and compare them with the original sample.

In this case *p* = 0.63 and *q* = 0.37. The proportions in the next generation are 0.397 *MM*, 0.466 *MN*, and 0.137 *NN*. Converting these proportions to individuals (based on the original sample of 201), we get 80 *MM*, 94 *MN*, and 28 *NN* — close to the existing frequencies. For each set of values of *p* and *q*, there is only one expected combination of homo- and heterozygotes, which should persist indefinitely if the conditions of Hardy-Weinberg equilibrium are met. We have therefore demonstrated that the population is panmictic and that no selection for the *MN* trait is occurring.

FORCES FOR CHANGE.   If sexual reproduction does not cause evolution, what does? One force for genetic change is **mutation pressure.** Our discussion of mutations has indicated that if, for example, allele *A* mutates to allele *a* in an individual, then *p* and *q* will change in the population. The problem is that mutations are relatively rare, on the order of 10^{-5} per gene per replication, or less. In fact, Wilson and Bossert (1971) have calculated that it would take about 10,000 generations to reduce the frequency of a particular allele to one-third its original value.

A second, more rapid way to change gene frequencies is by **gene flow.** Animals that are moving from one population to another have gene frequencies that are different from those in the population where these animals are moving (see Figure 4–6). A third force for change is **natural selection,** which may be defined as the differential change in relative frequency of genotypes due to differences in the ability of their phenotypes to obtain representation in the next generation.

A fourth force to consider is **genetic drift** — the alteration of gene frequencies

FIGURE 4-6 Evolution by gene flow
Population B, with a low frequency of the white allele, gains members from
population A, which has a high frequency of the white allele. The frequencies of the
white allele would be expected to increase in future generations of population B.

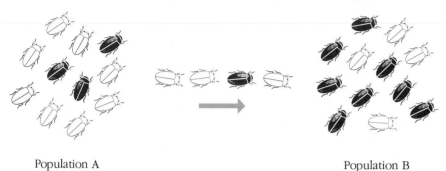

Population A   Population B

through sampling error. Genetic drift operates to some degree in all populations but
is significant only when populations are small—from 10 to 100 individuals. In a
small population, gene frequencies can change greatly because each chance mating
combination has a big impact on the total population. In the same way, the batting
average of a baseball player who hits once for three times at bat is .333; if he gets a hit
on his next at bat, his average will jump to .500. If a player's average of .333 is based
on 300 at bats, his next hit will increase his average to only .336.

   The ultimate fate of any allele is that it is either lost ($p = 0$) or fixed ($p = 1$) in
the population. Thus heterozygosity tends to be lost as a result of genetic drift. A
special case of drift is the **founder effect,** a situation in which new populations are
started by small numbers of individuals—for example, one in which a pair of birds
colonizes an island. These individuals can carry only a fraction of the genetic variability of the parental population; thus the new population is likely to differ from the
old. In a second example, in an isolated woodlot the starting population of white-
footed mice before spring breeding is sometimes only one or two pairs. This event
forms a genetic bottleneck that results in widely fluctuating gene frequencies.

SELECTION, ADAPTATION, AND SPECIATION

NATURAL SELECTION

We have defined natural selection as a change in the relative frequency of genotypes
that derives from the differential ability of their phenotypes in gaining representation in the subsequent generation. Three major types of selection can occur with

respect to genotype/phenotype variations: (1) stabilizing selection, (2) directional selection, and (3) disruptive selection (Figure 4–7).

STABILIZING SELECTION.   If selection is stabilizing, the average or normal individuals have an advantage, in terms of reproduction success, over the extreme variants. **Stabilizing selection** occurs when the environmental conditions that are most important for survival of a particular population remain relatively constant over a long period of time (Figure 4–7a). This type of selection does not favor a specific optimum genotype; rather, the norm is represented by a number of variable genotypes that are well adapted to the existing set of environmental conditions.

DIRECTIONAL SELECTION.   In **directional selection** a regular shift in certain of the adaptive characteristics of a species takes place. When there is a progressive shift in some aspect of the environment, or when a population migrates into a new area with altered environmental conditions, directional selection can occur (Figure 4–7b). The Darwin's finches of the Galápagos Islands provide an example of this type of selection. The ground finches and cactus-feeding finches *(Geospiza* spp.) have evolved from the ancestral fringillid type which colonized the islands. Among the species in this group there are types which have large stout beaks and which consume a diet of large, hard seeds. Another group in the islands has more refined, pointed beaks useful for obtaining nectar from flowers. Peter Grant and his colleagues (Grant, 1981; Grant and Abbott, 1980; Grant and Grant, 1980) propose that these differences in beak morphology are the result of competition for food sources. This resulted in directional selection for specific beak characteristics adapted for different feeding strategies and food habits.

Artificial selection used in the domestication and breeding of animals, like cattle and dogs, and in the development of specific strains of laboratory animals, like mice and fruit flies, is also an example of directional selection. We shall say more about artificial selection in the discussion of behavior genetics in Chapter 5.

FIGURE 4–7   Stabilizing, directional, and disruptive selection
Each type of selection can occur during evolution. Each type is shown with the original population ("before selection"). The expected result ("after selection") is graphed in terms of the frequencies of phenotypes within a population.

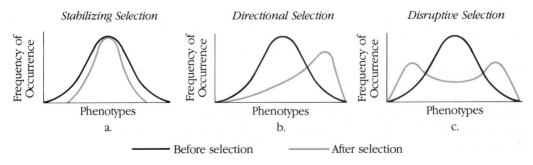

DISRUPTIVE SELECTION. When individuals of a species, formerly in a homogeneous environment, are later subjected to different selection pressures in various parts of the species' geographical range, **disruptive selection** may occur (Figure 4–7c). Different variants of a trait or behavior may be more adaptive for particular conditions at different locations within the geographical range of the species. Disruptive selection — or, as it is sometimes labeled, **diversifying selection** — is relatively rare in nature. One possible example of disruptive selection has been reported for an African butterfly species *(Papilio dardanus)*. The color pattern of this butterfly is characteristic for a particular locale and, in each case, it matches the pattern of another butterfly species in that locale that is distasteful to predators. This mimicry protects the *Papilio dardanus* from being preyed upon. Intermediate color patterns do not occur; if they did, they would not match those of any particular local species. Thus we can postulate that disruptive selection has produced a series of discrete types (Clarke and Sheppard 1960), each matching a local form, as a means of avoiding predation.

ADAPTATION

The term **adaptation** has two meanings. It can refer to all of the traits and characteristics that enable an organism to survive and reproduce. Thus we may speak of an animal as being adapted for life under a particular set of ecological conditions. For example, the bills of many bird species are adapted for obtaining a variety of food types, ranging from insects to seeds to fish (Figure 4–8).

Adaptation is also used to refer to the evolutionary selection that leads to specializations of anatomy, physiology, and behavior. In this sense, adaptations are temporary stopping points in long-term evolutionary actions; observing the actual processes is much more difficult.

When discussing adaptation as an evolutionary process we need to define several critical terms:

- The **ecological niche** of an organism consists of its location and functional role in the ecological system; niche refers to traits like nest sites, food habits, and seasonal and daily activity rhythms. The niche is an *n*-dimensional space that includes all of the animal's characteristics and its place in the ecological system (Hutchinson 1957).
- **Competition** occurs when some resources necessary for survival and reproduction are available in limited supply and are needed by animals of different species (interspecific competition) or by individuals of the same species (intraspecific competition). Here we are concerned primarily with interspecific competition.
- The **environment** of an organism consists of its living and nonliving surroundings. Thus living animals of the same and different species, vegetation, soil, geological formation, climate, and so on are all components of the environment of a terrestrial organism.

FIGURE 4–8   Bill adaptation in four species of birds
The bills of (a) the spoonbill, (b) the golden eagle, (c) the hawfinch, and
(d) the belted kingfisher show some of the variation in the many types of
bills that are adapted to each species' specialized techniques in obtaining and eating
food. The spoonbill feeds on aquatic detritus; the eagle is a carnivore and uses its
hooked bill to tear flesh; the hawfinch consumes seeds; and the kingfisher dives to
capture fish.
Source: Photos by (a) Cornell Laboratory of Ornithology, (b) Eric Hosking, (c) Eric
Hosking, and (d) G. Ronald Austing.

(a)

(b)

(c)

(d)

ENVIRONMENTAL ADAPTATIONS. Two major selective forces, environment and competition, influence the evolution of adaptations in animals. The effects of these two forces cannot be readily separated, but we can make some useful distinctions. General environmental adaptations refer to the traits and habits of an animal that probably have developed in response to climate or to various ecological features of the habitat. For example, females of several species of salmon that are native to the Pacific Northwest and neighboring parts of Canada and Alaska have spawning grounds on the rocky bottoms of glacial streams. Before she drops her eggs at a selected site, a female lies on her side and rapidly moves her body and tail to dislodge sand and smaller stones from among the rocks. The dropped eggs are fertilized by sperm from a male and then fall into the cracks and crevices that were created by the female's actions. After laying the eggs, the female moves slightly upstream and repeats the body movements. Dislodged sand and gravel settle over the eggs, which are more secure against currents and better concealed from potential predators.

COMPETITIVE ADAPTATIONS. Adaptations can also result from competition between individuals of different species. An axiom of ecology states that no two species can occupy exactly the same niche in the same geographical area. Animals that compete for the same resources (e.g., food, nest sites) have some degree of niche overlap; the severity of competition between individuals of two species varies with the degree of niche overlap. Two possible outcomes of severe competition are local extinction of the individuals of one species or the development of some form of coexistence through a change in the habits of the individuals of one or both species. To reduce competition and niche overlap, the individuals of one (or both) species may develop adaptations that alter patterns of resource utilization.

Two related salamander species, *Plethodon hoffmani* and *P. punctatus* (Figure 4–9), share the same moist environment in the Appalachian Mountains. The two

FIGURE 4–9 Salamander
Salamanders are widely distributed and occupy both aquatic and terrestrial habitats. Where two closely related species occupy similar habitats, behavioral changes can result to reduce the overlap of the niches they occupy. *Source:* Photo by Joyce Kilmer, courtesy Stephen Tilley.

species share several characteristics — for example, both are nocturnal — so we may ask whether they compete for certain critical resources that may be available in limited supply and, if so, to what extent. A study by Fraser (1976) has demonstrated that adults of these two species of salamander have dissimilar patterns of utilization of critical resources of food and habitat; the degree of niche overlap is lessened and competition reduced. The dissimilar patterns of resource utilization may be based, in part, on different adult body sizes and thus different resource requirements. There is, however, strong potential competition between adults of the smaller species *(P. hoffmani)* and juveniles and subadults of the larger species *(P. punctatus)*, which are similar in size; staggered feeding times and occupation of different habitats by the two species probably act to lessen the degree of niche overlap and minimize the extent of this competition. We can see from this example how adaptation may have evolved in two species that occupy the same area and that may, at some time in the past, have had a high degree of niche overlap.

Interspecific competition may also manifest itself in a phenomenon called **character displacement** (Brown and Wilson 1956), which sometimes occurs when two closely related species that have similar niche requirements overlap geographically. In the zone of overlap, the two species show sharper differentiation for some traits than each does in the nonoverlapping zone. For example, two species of cricket frogs *(Acris)* (Figure 4–10) have quite similar calls in eastern Texas and in southern Georgia, areas of nonoverlapping distribution (Blair 1958, 1974). However, the calls are sharply differentiated where the species overlap in central Mississippi. Because of this differentiation, each species can more easily identify members of its own species, where both species are present.

We should note that these examples are of static rather than dynamic situations. Whether two coexisting species evolved separately and have come to have overlapping distributions or whether they evolved different body sizes and patterns of resource utilization as a result of sharing the same range at some time in the past is difficult or impossible to determine.

In addition to competition, predation, disease, and mutualism can also affect the processes of adaptation through evolution by natural selection. Predation, involving adaptations of both the predator organism and its potential prey, will be examined in some detail in Chapter 16. The role of disease in affecting individuals and populations is discussed in Chapter 17.

SPECIATION

The formation of new animal forms through evolution is called **speciation.** In order for populations of a single species to diverge and eventually to evolve into separate species, two conditions must be met: (1) a barrier must form between populations of a species, and (2) a reproductive isolating mechanism must develop while the populations are separated so that when and if the barrier disappears, any separation in genotypes that has developed will not be eliminated by exchange of genes. Although

FIGURE 4-10   Character displacement in two species of cricket frogs

Acris gryllus and *A. crepitans* can be found in the southeastern areas of the United States. Where the distributions are nonoverlapping, calls from the two species are similar; but in zones of overlap, the calls are sharply differentiated.

Source: Adapted from L. S. Dillon, *Evolution: Concepts and Consequences,* 2nd ed. (St. Louis: C. V. Mosby Company, 1978). Reprinted with permission of the publisher and the author.

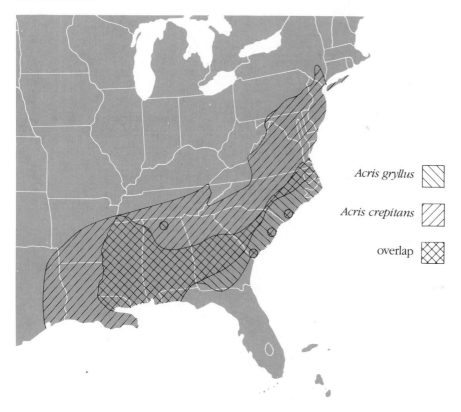

not observable directly, speciation of animal forms comes about in several ways (Mayr 1963; Huxley 1974; Futuyma 1979); we consider two of them here.

GEOGRAPHIC ISOLATION.   Speciation by **geographic isolation** is a gradual process involving several steps or stages (Figure 4–11). As the figure shows, a beetle species may initially be divided into several subspecies. If the subspecies *A* and *B* are then separated by a barrier—for example, a mountain range—then evolutionary changes in the now independent populations may develop so that individuals in subspecies *A* may emerge from pupae in early summer, while those in *B* may emerge

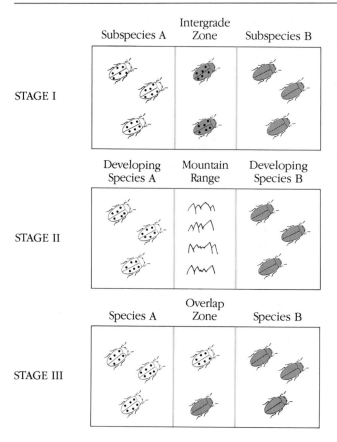

FIGURE 4-11   Speciation by geographic isolation in beetles
During a period of geographical isolation, subspecies A and B (Stage I) develop separate breeding seasons so that when they are again free to intermingle (Stage II) the two forms have become reproductively isolated and can not be considered separate species.

in late summer. When the mountain range wears down, perhaps over thousands of years, and the individuals from the two groups are again free to intermingle, there will no longer be any gene flow between the two groups because of the difference in their annual cycles. We can now say that separate species *A* and *B* have evolved. Geographic isolation is believed by most investigators to have been the most common cause of speciation among animals.

Not all evolutionary biologists agree about the existence of a second form of speciation, **sympatric speciation**. In this process two forms diverge from a single species into separate species even though they inhabit the same geographical region. In certain insect species, for example, all members live out their entire life cycles on one plant species. A mutation that causes some individuals to transfer to and remain permanently on another plant species could result in two distinct species that are isolated from each other by virtue of their exclusive use of different plant species. In this instance the selection of a particular plant species is a key factor in reproductive isolation.

REPRODUCTIVE ISOLATION. How do species remain distinct? What are the mechanisms that evolve to maintain the separateness of species? **Species-isolating mechanisms** are generally of two kinds: **prezygotic** and **postzygotic.** That is, they may occur before mating and zygote formation (prezygotic mechanisms), or during the development of the zygote or during the life of a viable hybrid produced by a mating between members of different species (postzygotic mechanisms). These isolating mechanisms are outcomes of the speciation process.

The importance of species-isolating mechanisms lies in the use of energy for and the efficiency of reproduction. When no viable progeny are produced, there is a high cost in genetic fitness for an animal to expend energy and gametes on nonproductive matings. Evolution by natural selection has produced mechanisms to ensure efficient mating activities.

There are several kinds of **prezygotic isolating mechanisms.** (1) In **habitat isolation,** members of related species do not interact or mate with one another because of preferences for different habitat types (see Chapter 17). (2) **Seasonal** or **temporal isolation** involves mating activity at different times of the year or different times of the day and eliminates the possibility of matings between members of related species (see Chapter 8). (3) **Behavioral isolation** occurs when members of different species are selecting mates; they choose only conspecific partners on the basis of patterns of coloration and sometimes quite complex courtship displays (see Chapter 11). (4) When members of two species have such different genital structures that mating is not possible, the effect is called **mechanical isolation.** (5) In some cases copulation may take place but, due to chemical and structural incompatibilities, no fertilization takes place and thus no zygote is formed.

Postzygotic mechanisms entail a genetic disharmony in the hybrids. The term **genetic disharmony** refers to a broad class of phenomena that may lead to inviability of the hybrids, sterility of any hybrids that are produced, or the production of weak or sterile second-generation hybrids. Also, some hybrids may fail to mate successfully because they exhibit courtship patterns that are different from either parent species, which leads to a kind of behavioral isolation.

GENETIC VARIATION, KINSHIP, AND UNITS OF SELECTION

POLYMORPHISM

One might expect that, through selection, alternative alleles at a locus would disappear. In fact, as many as 30 percent of the loci in fruit flies are **polymorphic**—that is, there are two or more alleles for that locus present in the gene pool for the population. What maintains this polymorphism? One of several possible explanations is **heterozygote superiority**—in some environments the heterozygote is more fit than either of the homozygotes. An example of this phenomenon occurs in humans:

heterozygosity for the sickle-cell trait provides protection from malaria. In parts of Africa and India and in some Mediterranean countries, the prevalence of malaria as a cause of mortality is sufficient to have produced a frequency of the sickle-cell gene in excess of 0.1, even though this gene in the homozygous state is usually lethal.

Another mechanism for producing balanced polymorphism is cyclical selection. The numbers of some species of animals, such as meadow voles *(Microtus)*, undergo drastic cyclical fluctuations. Krebs and his colleagues (1973) have argued that one allele is favored during earlier phases of population growth and another during the peak and decline phases of the population cycle. Others (Crow and Kimura 1970) have argued that this polymorphism is due to selectively neutral genes that are changing in frequency through genetic drift and are of no adaptive value to the individuals possessing them.

MEASUREMENT OF GENETIC RELATIONSHIP

In social groups of most species, the members are related and thus inbred to some extent. On the one hand, inbreeding reduces heterozygosity, thus diminishing the adaptability and benefits of heterozygote superiority, and it increases the possibility of the accumulation of deleterious recessive alleles. On the other hand, the closer the genetic relationships of group members, the more intricate are the social bonds and degree of cooperation, as we shall see later.

Several measures of relationship are used in population genetics; only one is discussed here. The **coefficient of relationship,** designated by r, is the proportion of genes in two individuals that are identical because of common descent. For example, in parent-offspring relationships, only one-half of the genes from one parent go to an offspring because the gametes are haploid; thus $r = \frac{1}{2}$. The r for full siblings is, on the average, also one-half because they share one-fourth of the genes from each parent. A general rule for computing r is to count the number of paths back to the common ancestor and to the related individual. The number of paths is the power to which $\frac{1}{2}$ is raised to give r. If there is more than one common ancestor, the paths must be traced separately and their probabilities added together. Figure 4–12 gives examples of the calculation of r.

UNITS OF SELECTION

Up to this point we have assumed that the unit of selection is the individual. It is possible, however, to argue that there are other units, all the way from genes themselves up to entire ecosystems. The notion that the gene is the unit of selection is relatively recent (Williams 1966) and is expounded by R. Dawkins (1976). The individual may be thought of as a unique and temporary vehicle for pieces of DNA, which replicate themselves and are potentially immortal.

FIGURE 4-12 Coefficients of relationship

To compute the proportion of genes that are identical in two individuals because of common descent, count the number of paths connecting those individuals by way of a common ancestor. This number is the power to which ½ is raised. Thus for half-siblings there are two paths, so ½ is raised to the second power. If there is more than one common ancestor, the process must be done separately for each common ancestor and the probabilities summed.

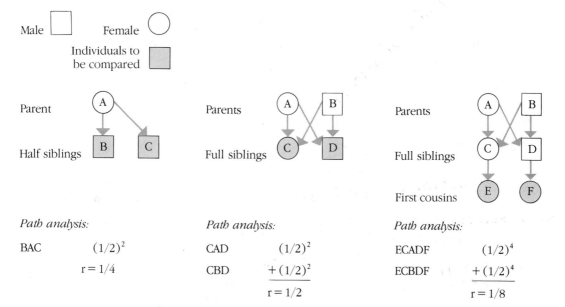

Path analysis:		*Path analysis:*		*Path analysis:*	
BAC	$(1/2)^2$	CAD	$(1/2)^2$	ECADF	$(1/2)^4$
	$r = 1/4$	CBD	$+ (1/2)^2$	ECBDF	$+ (1/2)^4$
			$r = 1/2$		$r = 1/8$

We can consider interdemic selection and kin selection to be at two ends of a continuum of relatedness above the level of the individual. By **interdemic selection** we mean selection operating at the level of the **deme,** which is a local population whose members breed randomly. **Kin selection** is defined as selection acting through close relatives. Although all members of a species are related to some extent, we assume that the deme is large enough and outbred enough so that the average coefficient of relationship is low. Instead of thinking that individuals or genes survive and reproduce or disappear, we can think of entire demes or populations that survive and colonize new areas or become extinct. In the latter case, of course, the entire gene pool is lost. The theory of **group selection** (Wynne-Edwards 1962) —that entire demes may be the units of selection—holds that a trait will evolve if reproductive restraint confers reproduction advantages on the entire group and if selfishness with regard to reproducing is disadvantageous for the group. Wynne-Edwards developed his theory to explain a situation in which an animal apparently behaves in a manner that is beneficial to the group but that might reduce that individual's own fitness. Evolutionary biologists have been reluctant to invoke group selection because it is likely to be much slower than individual selection. Wynne-Edwards's ideas, discussed further in Chapter 18, have thus been generally rejected.

Smaller units within the deme are frequently comprised of related individuals who share a certain amount of their genetic material through common descent. We can think of these kin units as extensions of individuals and apply the same rules of natural selection to them that we did earlier to individuals. Kin selection is thus an extension of the selfish gene idea. We would expect closely related individuals to cooperate among themselves and to compete with other units of close relatives.

Individuals may reduce their own fitness by acts that increase the fitness of others **(altruism).** Hamilton (1964) showed how such behavior could evolve via kin selection. He introduced the concept of **inclusive fitness,** which is the sum of direct and indirect fitness. **Direct fitness** is determined by the reproductive success of one's own offspring. **Indirect fitness** is determined by the reproductive success of relatives other than one's own offspring. For altruistic genes to spread by kin selection the ratio of benefit to the recipient (b) to cost to the altruist (c) must be greater than the reciprocal of the coefficient of relationship (r) (Hamilton 1964). That is,

$$\frac{b}{c} > \frac{1}{r}.$$

In brothers, who share one-half their genes, $r = \frac{1}{2}$ and therefore b/c must exceed 2 for altruistic genes to spread. In other words, if an individual more than doubles the fitness of his brother through an altruistic act that causes that individual to leave no offspring, genes promoting that behavior could spread through the population. The more distant the relative, the lower is r, and the higher the ratio of benefit to cost must be. These aspects of selection are discussed further in Chapter 19.

EVOLUTIONARY STABLE STRATEGIES

In recent years behavioral ecologists have applied new mathematical models to problems of behavior and evolution through natural selection. One such model is the **optimality theory,** based on the idea that individuals that are maximally efficient at certain activities — foraging or seeking mates, for example — should have an advantage in selection in order to ensure the greatest propagation of their genes in subsequent generations (MacArthur 1965; Maynard Smith 1976).

This approach presents at least one problem, however: What if there are several possible "best" solutions to a situation? Here we move into the realm of contingent probabilities, where the optimum behavior of an individual in a particular situation depends on the behavior of other individuals. In situations that involve multiple variables, behavioral ecologists and sociobiologists employ the concept of an **evolutionary stable strategy (ESS)**—a strategy that, if adopted by most members of a population, cannot be bettered by an alternative strategy. An example will clarify what an ESS is.

The common yellow dung fly *(Scatophaga stercoraria)* is found in pastures where cattle are grazing. The males and females meet at pats of fresh cow dung and mate; the female oviposits and the larvae develop in the cow pat. At fresh cow pats, the ratio of mature, reproducing males to females is 4 or 5 to 1. What is the best strategy (ESS) a male can follow to achieve a maximum frequency of successful matings? As we might expect, the strategy is multifaceted. Consider first the question of where the males search for females in relation to a fresh cow pat. The expected (based on the mathematics of ESS theory) and observed captures for mating of females by males in various zones around a fresh cow pat are graphed in Figure 4–13 (Parker 1974, 1978). Males may mate with females after capturing them or by taking over a female that has been captured by another male. The decision factors are the length of time it takes to mate, how much time it takes to dislodge a competitor, the age of the dung pat, and when to move on (emigrate) in search of another fresh cow pat.

From this example we can see how natural selection might operate on several related patterns simultaneously. The behaviors in this example can, theoretically at least, be described by an evolutionary stable strategy. The use of ESS theory is increasing rapidly among behavioral ecologists, both as a modeling tool and as a means of stimulating questions to be tested.

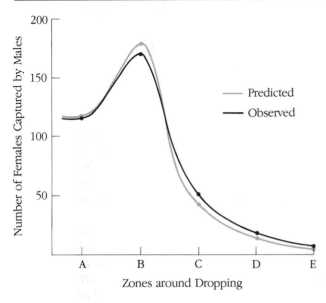

FIGURE 4–13 Captures of male *Scatophaga* at fresh cow pats Comparisons are shown between predicted and observed captures in zones on and around the cattle droppings. The predicted curve is based on an ESS (see text). Zone A = the surface of the cow dropping; Zone B = zone nearest the dropping on the grass to a distance of 20 cm; Zone C = a similar grassy band in the area from 20 to 40 cm from the edge of the cow pie; Zone D = 40 to 60 cm region; and Zone E = 60 to 80 cm from the dropping.
Source: G. A. Parker, "Reproductive Behavior and the Nature of Sexual Selection in *Scatophaga stercoraria* L." *Evolution* 28 (1974):93–108. Reprinted with permission.

PROBLEMS WITH EVOLUTIONARY EXPLANATIONS

Although evolutionary theory is a useful tool for developing explanations for observed behavior patterns, it should not be used indiscriminately. We should note the following cautionary comments about the evolutionary perspective.

First, we must take care that our reasoning is not circular. New species evolve in several ways. We may argue correctly that in the course of evolution, species-isolating mechanisms have developed to ensure efficient use of energy and gametes in mating. It would be incorrect, however, to claim that two species evolved where previously there was but a single species because the species developed a species-isolating mechanism. The isolating mechanism evolved during speciation and should not be presented as the cause of speciation.

Second, the recording of behavioral phenomena presents a "static" picture on the evolutionary time scale. To visualize the past events and dynamic processes that have influenced the present adaptations of a species is difficult. In discussing the evolution of behavior patterns, we must guard against overextending our hypothetical explanations in accounting for the functional significance and natural selection pressures that led to each aspect or detail of a complex behavioral sequence. We should provide evolutionary explanations only if we can support them with data or if we can use them to devise experiments to test a hypothesis. In short, we should not feel compelled to provide a complete evolutionary explanation for each existing behavior pattern.

Third, in discussing the evolution of behavior, we must bear in mind that the environment is a passive and not an active force. An environment itself does not impose a strategy for evolution on an organism. The adaptations that characterize an animal species represent temporary stopping points in evolution and are the results of the action of natural selection upon the phenotypes within that species. The environment provides the setting for evolution, and thus the parameters of the environment are critical determinants of the nature of the adaptations that evolve.

SUMMARY

The primary units of study for evolutionary biologists are species, or groups of animals that share similar characteristics and habits and that interbreed freely with one another but that are reproductively isolated from all other species. The classification of animal types is hierarchical and is based on commonality of ancestry and similarity of characteristics.

Each organism of a species possesses a characteristic number of genes, or basic units of heredity. The complete set of genes for any organism is its genotype; the actual physiological or behavioral traits that can be observed, the phenotype; and the hypothetical collection of genes within a species is the gene pool. For each locus on a chromosome, one of various alleles, or alternative forms of a gene, may be present. These alleles can be dominant or recessive or can exhibit partial dominance.

We measure biological evolution through natural selection in terms of fitness and changes in gene frequencies. Traits that have a genetic basis and that are favorable for survival and reproduction will be found in higher frequency in subsequent generations. The theory of evolution accounts for the difference in the rate of survival and reproduction of some animals in the face of limited resources on the basis of the heritability of variation.

As we look more closely at the process of evolution, we can glance briefly at a subdiscipline — population genetics — which provides us with the useful tool of the Hardy-Weinberg formula that enables us to sample at a point in time rather than over several generations in order to determine whether or not there is an equilibrium for certain sets of alleles. Among the forces for genetic changes are mutation pressure, gene flow, genetic drift, and natural selection — the latter subdivided into stabilizing selection, directional selection, and disruptive selection.

Another evolutionary process is adaptation, in which changes occur over time in an animal species so that members of that species are well suited for life in that particular environment. Adaptations come about as populations respond, through natural selection, to aspects of the environment (which we define as an organism's living and nonliving surroundings), to interspecific competition, and to niche differentiation.

Speciation, the evolutionary process by which new forms appear, can occur only when populations of a species are separated and isolating mechanisms develop during the period of separation. After the separating barrier is removed, the species maintain their distinct identities through pre- and postzygotic isolating mechanisms. Genetic variation, or polymorphism, is maintained through several mechanisms, including heterozygote superiority and balanced polymorphism.

Although we assume, when we examine evolutionary theory, that the unit of selection is the individual, several theories have proposed other units, ranging from whole ecosystems, or demes, at one end of the spectrum to genes (DNA) at the other. Evolutionary theorists also examine group selection and kin selection, both of which are based on altruistic acts, or acts performed by individuals that benefit other individuals or the group even though their own fitness is reduced.

The concept of evolutionary stable strategies (ESS) has been developed by behavioral ecologists to apply to situations in which the optimum behavior of an individual animal in a particular situation is a function of multiple variables.

Evolutionary theory is useful for explaining observed behavior. It has its limitations, however — namely, we are taking a static look at behavior when in fact evolution is a dynamic process.

Discussion Questions

1. Animal species must maintain a certain degree of flexibility to be able to change in response to environmental changes. At the same time, due to competition and other factors, these species must attain a certain degree of specialization. What are the dynamics of "walking the fine line" between flexibility and specialization?

2. Explain how random events can affect the stages of the evolutionary processes discussed in this chapter.

3. How might natural selection operate to facilitate the evolution of species-isolating mechanisms?

4. Outline the major steps in adaptation and speciation.

5. What is the role of environment in evolution?

6. Charles Darwin married his first cousin, Emma Wedgewood. What is the coefficient of relationship between the Darwins' children?

Suggested Readings

Barash, D. P. 1983. *Sociobiology and Behavior,* 2nd ed. New York: Elsevier North-Holland.
Condensed treatment of the new hypotheses and thinking of sociobiology. Very readable style, excellent introductory chapters, and solid use of examples throughout.

Dawkins, R. 1976. *The Selfish Gene.* New York: Oxford University Press.
Written for both scientists and nonscientists. Presents an extreme viewpoint on the role of genes and DNA with original and thought-provoking ideas.

Williams, G. C. 1966. *Adaptation and Natural Selection.* Princeton: Princeton University Press.
Readable, somewhat philosophical, treatise on aspects of modern evolutionary thought. Williams's ideas are now receiving much more attention than when they were first written.

Wilson, E. O., and W. H. Bossert. 1971. *A Primer of Population Biology.* Sunderland, Mass.: Sinauer Associates.
An excellent introduction to population genetics.

References

Blair, W. F. 1958. Mating call in the speciation of anuran amphibians. *Amer. Natur.* 92:27–51.

———. 1974. Character displacement in frogs. *Amer. Zool.* 14:1119–1125.

Brown, W. L., and E. O. Wilson. 1956. Character displacement. *Systematic Zoology* 5:49–64.

Clarke, C. A., and P. M. Sheppard. 1960. Supergenes and mimicry. *Heredity* 14:175–185.

Crow, J. F., and M. Kimura. 1970. *Introduction to Population Genetics Theory.* New York: Harper & Row.

Dawkins, R. 1976. *The Selfish Gene.* New York: Oxford University Press.

———. 1982. *The Extended Phenotype.* Oxford: Oxford University Press.

Fraser, D. F. 1976. Coexistence of salamanders in the genus *Plethodon:* A variation on the Santa Rosalia theme. *Ecology* 57:238–251.

Futuyma, D. J. 1979. *Evolutionary Biology.* Sunderland, Mass.: Sinauer.

Grant, P. R. 1981. Speciation and the adaptive radiation of Darwin's finches. *Amer. Sci.* 69:653–663.

Grant, P. R., and I. Abbott. 1980. Interspecific competition, null hypotheses and island biogeography. *Evolution* 34:332–341.

Grant, P. R., and B. R. Grant. 1980. The breeding and feeding characteristics of Darwin's finches on Isla Genovesa, Galapagos. *Ecol. Monogr.* 50:381–410.

Hamilton, W. D. 1964. The genetical theory of social behavior: I and II. *J. Theoret. Biol.* 7:1–52.

Hulse, F. 1971. *The Human Species.* New York: Random House.

Hutchinson, G. E. 1957. Concluding remarks. *Cold Spr. Harb. Symp. Quant. Biol.* 22:415–427.

Huxley, J. 1974. *Evolution: The Modern Synthesis.* London: Allen and Unwin.

Krebs, C. J., et al. 1973. Population cycles in small rodents. *Science* 179:35–41.

MacArthur, R. H. 1965. Patterns of species diversity. *Biol. Rev.* 40:410–533.

Maynard Smith, J. 1976. Evolution and the theory of games. *Amer. Sci.* 64:41–45.

Mayr, E. 1963. *Animal Species and Evolution.* Cambridge: Harvard University Press.

———. 1970. *Populations, Species and Evolution.* Cambridge: Harvard University Press.

———. 1977. Darwin and natural selection. *Amer. Sci.* 65:321–327.

Parker, G. A. 1974. The reproductive behavior and the nature of sexual selection in *Scatophaga stercoraria* L. IX. Spatial distribution of fertilization rates and evolution of male search strategy within the reproductive area. *Evolution* 28:93–108.

———. 1978. Searching for mates. In J. R. Krebs and N. B. Davies (eds.), *Behavioural Ecology: An Evolutionary Approach.* Oxford: Blackwell Scientific.

Stebbins, G. L. 1966. *Processes of Organic Evolution.* Englewood Cliffs, N.J.: Prentice-Hall.

van Tyne, J., and A. J. Berger. 1959. *Fundamentals of Ornithology.* New York: Wiley.

Williams, G. C. 1966. *Adaptation and Natural Selection.* Princeton: Princeton University Press.

Wilson, E. O., and W. H. Bossert. 1971. *A Primer of Population Biology.* Sunderland, Mass.: Sinauer.

Wynne-Edwards, V. C. 1962. *Animal Dispersion in Relation to Social Behavior.* London: Oliver and Boyd.

PART TWO

CONTROL OF BEHAVIOR

5   Behavior Genetics

6   The Nervous System and Behavior

7   Hormones and Behavior

8   Biological Timekeeping

5

BEHAVIOR GENETICS

Part Two provides a foundation for understanding the control of behavior as manifested in the processes of receiving information from internal sources and the external environment and producing observed behavior. Because critical information affecting the morphology, physiology, and behavior of an organism is contained in its genotype, we begin with this chapter on behavior genetics. Chapters 6 and 7 explore the roles of the nervous and hormonal systems in receiving external cues and generating behavioral responses. Chapter 8 examines the phenomenon of the biological clock — that is, mechanisms of behavior control evidenced in many organisms by patterns or cycles of activity — for example, daily rhythms.

   Both genetics and behavior followed separate paths of scientific development for several decades, but in the 1960s a synthesis of these two research areas led to modern behavior-genetic analysis. Beginning as early as the late nineteenth century and continuing into the 1930s, many psychologists and zoologists, notably Robert Tryon, Sewall Wright, and R. A. Fisher, were questioning the effects of inheritance on behavior; these early investigators provided the basic groundwork for techniques and theories that have been more fully expounded since the 1960s. In this chapter we examine several aspects of modern behavior genetics, beginning with the kinds of questions asked by behavior geneticists. We then discuss the methods used in behavior-genetic investigations and the significance of the findings of this research. We conclude with some comments on behavior genetics and evolution.

QUESTIONS IN BEHAVIOR GENETICS

Behavior genetics is an expanding field that is not yet fully defined. Although further development of the field may change the research objectives, certain basic questions will continue to be posed about the role of genes in influencing and controlling behavior (Hirsch 1967).

QUESTIONS OF CONTROL AND DEVELOPMENT.   This approach begins by asking whether individuals within a species differ regarding some particular behavioral trait and whether these intraspecific differences, or **variations,** have a possible or probable genetic basis. Another question is what the mechanism of inheritance of the behavioral trait is. In some cases we may be able to ask how many genes are responsible for influencing a particular behavior pattern, or how gene action affects the development of various behaviors. We could further ask what proportion of the variation is attributable to genetic influences and what proportion is attributable to environmental influences. (Gene-environment interactions during development, the process of epigenesis, will be treated more fully in Chapter 9.)

QUESTIONS OF POPULATION AND EVOLUTION.   Research in this area begins with the determination of the frequencies in a population of the genes known to influence particular behaviors. Is the behavior in question adapted to the species' present environment? Do gene frequencies shift in response to selection pressures associated with changes in the environment? Finally, do phylogenetic comparisons of related species reveal similarities or differences in genotype and behavior that may be used to determine the evolutionary history of certain behavior patterns?

METHODS OF ASSESSING GENETIC DETERMINATION

Behavior geneticists use two basic approaches in their research: they may hold environment constant in order to explore the effects of genetic variation, or they may hold genetics constant in order to study the effects of varying the environment.

The starting point for investigations is a simple formula:

$$V_T = V_G + V_E + V_I$$

where V_T represents the total, or phenotypic, variation observed in a population of animals for a particular trait; V_G represents the genotypic component of the total variation; V_E, the environmental component; and V_I, the variation attributable to interactions between genotypic and environmental factors ($V_I = V_G \times V_E$). In other words, the phenotypic variation is the sum of genotypic, environmental, and inter-active forces. (For a brief discussion of variation and the measurement of sample variance, refer to Chapter 3.)

In the basic formula $V_T = V_G + V_E + V_I$, behavior geneticists generally as-sume that $V_I = 0$, leaving only genotypic and environmental variation. The as-sumption that $V_I = 0$ is made to simplify the formula. V_I is not a readily determined quantity, since it represents the interaction of both the genotypic and environmental influences on behavior — the differential responsivity of animals of different geno-types to given environmental influences. We should note, however, that $V_I = 0$ when V_G or $V_E = 0$, or when there is no interaction between genotype and environ-ment (Plomin et al. 1977). Since behavior development and gene expression require interaction with the environment, in most instances, the assumption that $V_I = 0$ is incorrect. Moreover, behavior geneticists generally agree that the equations $V_G = 0$ or $V_E = 0$ are impossibilities in normal populations. Although we shall proceed here with the customary assumption that $V_I = 0$, we must question how accurate and how generalizable our conclusions are.

Investigators have used a variety of techniques to explore the relationships between genetics and behavior. We will now examine some of these techniques, including the underlying theory and actual research examples.

INBREEDING

Behavioral geneticists use homogeneous strains of animals to study the effects of environmental parameters on behavior while the genetic component is held con-stant. One method of achieving genetic homozygosity — that is, a population of homogeneous animals with the same genotype — is by inbreeding, or using brother-sister matings for many generations. For inbred strains $V_G = 0$, so $V_T = V_E$. In mice, after about thirty generations of inbreeding, some 98 to 100 percent of the allele pairs are homozygous. During the inbreeding process, many recessive alleles, which may be lethal or otherwise detrimental in terms of successful reproduction, attain the homozygous condition. Thus to obtain a viable inbred strain of mice, we must begin with many replicates of brother-sister matings because many lines will die out before a high degree of homozygosity is achieved.

If our experimental subjects have a common genetic background, we can manipulate various aspects of the environment to determine the relative importance of the external parameters that influence behavior. For example, Cooper and Zubek (1958) studied two inbred strains of laboratory rats *(Rattus norvegicus)*. One, desig-nated the bright strain, could perform a series of maze-learning tasks with very few errors. The other, labeled the dull strain, made numerous errors while performing the same tasks. At twenty-five days of age, rats from the two strains were reared for forty days in either (1) a restricted environment (wire cages, view of a gray wall, only food and water in the cage); (2) an enriched environment (objects such as mirrors,

TABLE 5–1   **Mean error scores for bright and dull rats reared in restricted, enriched, or normal environments and given a series of maze-learning tasks beginning at 65 days of age**

Strain	Restricted	Enriched	Normal
		Rearing Condition	
Bright	169.7	111.2	117.0
Dull	169.5	119.7	164.0

Source: Data from Cooper and Zubek (1958).

tunnels, and ramps in the cage, view of a highly decorated wall); or (3) the normal laboratory cage setting. At sixty-five days of age, rats from each of the six treatment groups (two strains, three rearing environments) were tested on the same series of maze problems. The results (Table 5–1) show that the enriched environment did not improve the performance of bright-strain rats; however, the restricted environment negatively affected their performance. The performance of dull-strain rats was not affected by rearing under restricted conditions, but those reared in an enriched environment had a lower error rate. As expected, bright-strain control rats, housed in the normal environment, made fewer errors than did dull-strain rats housed in the normal environment. The data demonstrate a clear interaction of inheritance and environment; hence we aren't always correct in assuming $V_I = 0$.

The clear-cut conclusions reached in this study are not possible in all behavior-genetic analyses. Often the performance of test animals falls in between the two extremes, thereby complicating the interpretation of results. We should note that the conclusions from studies using inbred strains are limited to the particular strain of animals and the specific variables measured in an investigation. Numerous additional factors, such as methodological differences between laboratories and the age, sex, and previous experience of test subjects may affect the interpretations and conclusions of such studies.

STRAIN DIFFERENCES

We can use systematic examination of two or more genetically homogeneous inbred strains of the same species, maintained under the same constant environmental conditions, to compare and assess the effect of heredity on behavior.

Nest-building behavior by small mammals may be important under natural conditions in terms of both temperature regulation and maternal behavior. Mice living in different climates or nesting in different habitats might be more likely to survive and successfully rear offspring if their nesting habits properly match the conditions imposed by the environment. For example, geographical differences in nest size have been documented for various species and subspecies of deermice *(Peromyscus)* living at different latitudes in North America (King, Maas, and Weisman 1964).

Lynch and Hegmann (1972) used an experimental design that held environment stable and varied inherited traits to assess how variations in genotype affect a particular behavior pattern ($V_E = 0$, so $V_T = V_G$). They compared the nest-building activity of house mice *(Mus musculus)* from five different inbred strains, which were reared in the same environment. Over five consecutive days, they measured the amount of cotton (in grams) that individually caged male and female mice of each strain used to make their nests. The combined results for both sexes indicated that each strain differed significantly in the average amount of material used in nest construction each day: strain 1 used 1.3 g; strain 2, 1.3 g; strain 3, 0.7 g; strain 4, 1.1 g; and strain 5, 0.7 g (Figure 5–1).

The results demonstrate genetic variance in nest-building behavior in house mice. The existence of such genetic variation is a necessary, but not sufficient, condition for a behavioral response to selection pressure. Studies of inherited differences in behavior between strains serve to point out which behavioral traits in different types of organisms are genetically labile and thus could be affected by selection pressure.

SELECTIVE BREEDING

Domesticated pets and livestock and many types of ornamental and agricultural plants are the end products of selective breeding programs. In modern behavior genetics research, artificial selection techniques are used, for example, to demonstrate that a trait is under some degree of genetic control. Thus, the trait could be

FIGURE 5–1 House mouse nests in laboratory
The difference in size of nests constructed by mice can have a genetic basis. These two nests were built by house mice selected for construction of large and small nests.
Source: Photo by Lee C. Drickamer.

Y—MAZE

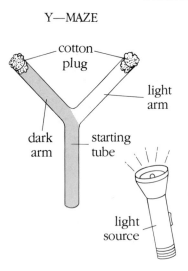

FIGURE 5-2   Y-maze used for selection experiments
This apparatus, which was dark except for the lighted arm, was used to test the light responses of fruit flies.

affected by selection pressures. Artificial selection is also used to estimate the degree to which genes are expressed in phenotypes.

ARTIFICIAL SELECTION TECHNIQUES.   Artificial selection is based on the assumption that some degree of genetic variability exists for the trait in question. Animals that exhibit the extremes of the distribution in a large sample population are then selected for breeding over several generations, and the strength of expression is tested.

For example, Hirsch and Boudreau (1958) tested whether reactions of fruit flies *(Drosophila melanogaster)* to light were under genetic control. They used a Y-maze with one lighted arm and one dark arm (Figure 5–2) to determine the responses of large groups of flies in terms of the percentage of the group that approached the light source in each of a series of ten trials (Figure 5–3a). Flies that made the greatest number of approaches (high line) and those that made the fewest approaches (low line) were then bred separately. Progeny of matings of high-line flies were tested in the apparatus, and individuals that made the greatest number of approaches to light were bred to produce the next generation. The same procedure was followed for the low line — that is, flies making the fewest approaches to light were selected to serve as parents for each succeeding generation. For both high and low lines, the procedure was carried out for a total of twenty-nine generations. Figure 5–3b and c show the frequency distributions of responses to light for the high and low lines after two and after twenty-nine generations. By the last generation, little overlap remained between the responses of flies from the two lines. Hirsch and Boudreau concluded that the photic responses of this species of fruit fly are under some degree of genetic control. This conclusion leads directly to the question of how much genetic control is involved.

Another example of artificial selection combines this technique with some

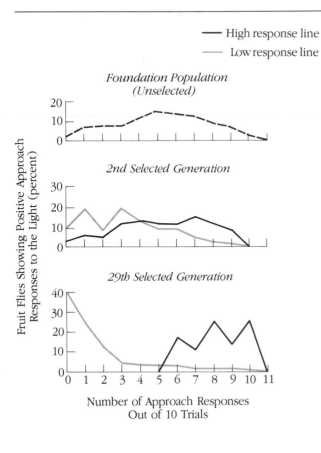

— High response line
— Low response line

Fruit Flies Showing Positive Approach Responses to the Light (percent)

Foundation Population (Unselected)

2nd Selected Generation

29th Selected Generation

Number of Approach Responses
Out of 10 Trials

FIGURE 5-3   Distributions of light approach scores of fruit flies
Graph (a) measures the responses of the initial population of flies, and graphs (b) and (c) measure the high and low response lines after 2 and 29 generations of artificial selection. *Source:* J. Hirsch and J. C. Boudreau, "Studies in Experimental Behavior Genetics. I. The Heritability of Phototaxis in a Population of *Drosophila melanogaster*," *Journal of Comparative and Physiological Psychology* 51 (1958):647–651. Copyright 1958 by the American Psychological Association. Adapted by permission of the publisher and the author.

degree of inbreeding, i.e., mating cousins or nieces. Pfaffenberger and colleagues (1976) used this combined technique to investigate the production of guide dogs for the blind. The experiments involved several lines of German shepherds. Among the favorable traits achieved or enhanced by artificial selection and inbreeding were a reduced walking speed, improved overall temperament, and greater uniformity of response to various training procedures.

MEASUREMENT OF HERITABILITY.   **Heritability,** (h_b^2) used in a broad sense, is an estimate of the **degree of genetic determination** ($°GD$) of any particular trait. Heritability is a population characteristic; it is *not* a term that is applied to the inheritance of a specific trait in any particular individual of the population. In this broad sense, heritability of a particular trait in a specific population sample is defined as

$$h_b^2 = °GD = \frac{V_G}{V_G + V_E}.$$

That is, the degree of genetic determination of a trait is the ratio of genetic variation to genetic plus environmental variation (Fuller and Thompson 1978).

One way to assess the degree of genetic determination is to use inbred strains and some genetic crosses in a hypothetical study of wheel-running activity (Table 5–2). Two strains of mice *(Mus musculus),* A and B, can be tested, along with a strain of mice from crosses of A and B, called the F_1. Each strain of mice — A, B, and F_1 — is homogeneous; hence any variation in wheel-running activity must be attributed to V_E. If we make a cross of F_1 mice, producing individuals (F_2) that are no longer homogeneous, the variance in wheel-running activity then includes both genetic and environmental variation: $V_G + V_E = V_T$. We can estimate the degree of genetic determination as V_G/V_T. In this example, $°GD$ works out to 0.39; that is, 39 percent of the variation in wheel-running activity of these mice can be attibuted to genetic influences.

In other instances we may be interested in predicting the efficiency of selection (Fuller and Thompson 1978). In such cases what we measure is called the realized heritability, or h_r^2. The major method of estimating h_r^2 utilizes artificial selection. We can consider, for example, selection for larger litter size in hamsters (Figure 5–4). The base stock of hamsters has a mean litter size of 8.4 pups. The **selection differential,** S, is defined as the difference between the mean litter size of the parents that are selected for breeding the next generation and the mean litter size of the base stock. Here the selected male and female parents have a mean litter size of 13.6 pups, so $S = 13.6 - 8.4 = 5.2$.

The **response to selection,** R, is defined as the difference between the mean litter size of the base stock and that of the offspring generation produced by mating selected males and females. If the offspring generation has a mean litter size of 10.1 pups, then $R = 10.1 - 8.4 = 1.7$. The ratio of the response to selection to the selec-

TABLE 5-2   Calculation of degree of genetic determination (°GD) for wheel-running activity in mice that are a cross of two inbred strains
Tests lasted twenty-four hours.

Mouse Strain	Mean Number of Wheel Revolutions	Variance
A	1,107	112
B	5,680	418
F_1 (A × B)	5,235	325
F_2 (F_1 × F_1)	4,745	465

$$V_E = \frac{112 + 418 + 325}{3} = 285$$

$V_G = 465 - 285 = 180$
$°GD = 180/465 = 0.39$

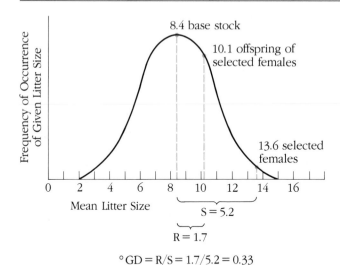

FIGURE 5-4   Measurement of realized heritability
In this example h_r^2 was calculated using artificial selection and computations of selection differential (S) and response to selection (R) for litter size in hamsters.

tion differential (R/S) is a measure of the h_r^2. In our example $R/S = 1.7/5.2 = 0.33$, so about one-third of the variation in litter size for this particular stock of hamsters can be attributed to genetic factors. Clearly, if a small value for S produces a large response for R, then the heritability estimate will be high; conversely, if a large selection differential produces only a slight shift in R, then the h_r^2 estimate will be low.

Estimates of realized heritability can be compared for various traits; however, such comparisons have inherent problems. As we noted earlier, conclusions in behavior genetics are limited by constraints, including the population and strain(s), the type of test, the environmental conditions. Another, perhaps less problematic, use for estimates of realized heritability is the prediction of response to selection. If we have calculated h_r^2 and the selection differential is known, the response to selection for a particular trait can be predicted.

The field of quantitative behavior genetics has received extensive treatment in recent decades. Falconer (1960), Roberts (1967), and Fuller and Thompson (1978) provide detailed discussions of measurement of heritability and related topics.

CROSS-FOSTERING

The technique of cross-fostering helps us differentiate species-specific behaviors and environmentally influenced behaviors. If we transfer neonatal animals from the parent female to another female of the same species or strain, or to a female of a different species or strain (if she will successfully rear the young), we can then compare animals with similar genetic composition but with different rearing environments to assess the relative importance of the effects of the genotype and of the maternal-care environment on certain behavioral traits. If genetically similar animals, reared under different conditions by either a biological parent or by a foster parent, exhibit similar specific behaviors, then we can conclude that genetic control of those behaviors is fairly rigid. However, if the behaviors differ significantly, we can conclude that these behaviors are strongly influenced by environment.

TRANSFERS WITHIN A SPECIES.   Broadhurst (1965) conducted a study of whether the defecation rates of rats subjected to novel or stressful environments were affected by postnatal maternal influences. He used selective breeding to obtain two distinct lines of rats: one group exhibited high rates of defecation, a sign of "emotionality," and the other exhibited low rates (Figure 5–5). To test whether the greater "emotionality" of the mothers in one group led to behavioral differences in their progeny, he bred rats of both groups and, at the time of birth, cross-fostered some pups of each subline to mothers of the opposite subline. Young pups that had been cross-fostered were later tested. The rates of defecation of cross-fostered animals were found to be the same as the rates of defecation of their genetic parents. We can, therefore, conclude that the different defecation rates of the two sublines can be attributed to genetic factors rather than to postnatal environmental factors.

TRANSFERS BETWEEN SPECIES.   In a different type of cross-fostering study, Quadagno and Banks (1970) tested the effects of early environment on certain behaviors of two rodent species — an inbred strain of the house mouse *(Mus musculus)* and the pygmy mouse *(Baiomys taylori).* Pups of both species were reared under one of four treatment conditions: (1) *Mus* pups reared by *Mus* dams (control) or (2) by *Baiomys* dams, and (3) *Baiomys* pups reared by *Mus* dams or (4) by *Baiomys* dams (control). Each litter consisted of three pups. In normal litters all three pups and the dam were

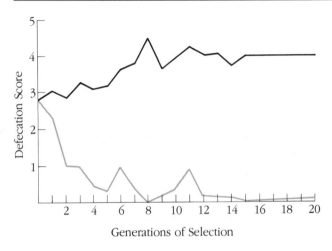

FIGURE 5–5   **Selection for high and low rates of defecation in rats exposed to a novel, stressful situation**   When pups born to mothers of each line are cross-fostered to mothers of the other line and are later tested for defecation rates, the observed rates correspond to those of their genetic parents.
Source: P. L. Broadhurst, "The Choice of Animal for Behaviour Studies," *Laboratory Animals Centre Collected Papers* 12 (1963):65–80. Reprinted with permission of the publisher and the author.

from the same species; in cross-fostered litters one pup was transferred from the opposite species and the other two sibs were from the same species as the mother. Thus, the environmental stimulation included not only maternal influences but also possible sibling influences. Note that any possible confounding effects from having litters of different sizes were eliminated.

For the first twenty-one days of life, the mice were reared in the litters as described, and then each mouse was housed individually until testing began at sixty-five days of age. Each mouse was given a series of social and behavioral tests. The tests produced several significant results, only a few of which are mentioned here. First, general-activity scores for cross-fostered *Baiomys* were higher than for control *Baiomys,* and general-activity scores of cross-fostered *Mus* were lower than for control *Mus.* Some environmental factor, possibly handling effects by the dams or activity influences of sibs from the foster species, must account for this difference. Second, in paired encounters involving different combinations of conspecific-reared and cross-fostered mice, individuals that had been cross-fostered exhibited more positive affiliative responses to members of the other species. Again, an environmental factor probably accounts for this effect. Third, mating tests of mice from the different treatment groups showed that mice had not lost their ability to mate with members of their own species. Sexual behavior is apparently less susceptible to the effects of early experience with heterospecifics and is under greater genetic influence than are some other behaviors.

MUTATIONS

Mutant strains, in which a behavioral effect can be traced to a particular allele, can be used to study the effects of a single gene on morphology and behavior. As an example we can consider the mating behavior of fruit flies *(Drosophila)* that have a single mutant allele. Mutant-type adults have yellow bodies; wild-type adults have gray bodies. Flies with the yellow mutant gene mated with wild-type females have reduced reproductive success, compared with wild-type male–wild-type female matings (Bastock and Manning 1955; Bastock 1956). We can trace this decrease in fertility to distinct differences in the mating patterns of the mutant flies. When male fruit flies court females, they engage in wing-vibration displays (courtship behavior). In the mutant flies the pattern of wing vibration is altered; in mutant males the bouts of wing vibration are shorter and occur less frequently. The effect of these changes is reduced stimulation for the courted female. Further tests have confirmed that the difference in mating success is due to the difference in mating behavior and not to other factors, such as variations in body color or scent.

Individual genes or groups of genes that act as a unit do not necessarily exert just a single effect, but may have multiple effects; this phenomenon is called **pleiotropism.** In the case of the fruit flies, the yellow mutant allele affects at least one morphological characteristic — body color, and one behavioral trait — courtship display.

MODES OF INHERITANCE

Behavior geneticists also analyze the mode of inheritance of a behavioral trait. When two animals that show distinct differences for a behavioral trait are mated, how is the behavior of the offspring affected? We can consider the wheel-running activity of mice, diagramed in Figure 5–6. Suppose that individuals of one strain of mice run an average of 100 m/hr, and mice of a second strain run only 20 m/hr. If the progeny produced by mating males with females of the opposite strain show virtually the same wheel-running speed as either parent strain—that is, either 20 or 100 m/hr—then there is a *dominant mode* of inheritance. A behavioral response by the progeny that falls somewhere in between that of both parents (e.g., 60 m/hr) indicates an *intermediate mode* of inheritance. In some instances the wheel-running speed of the progeny may be greater than that exhibited by either parent strain (e.g., 120 m/hr) and is an example of **overdominance,** or **hybrid vigor.**

   Schröder and Sund (1984) investigated the inheritance and learning of a water escape behavior in mice. The apparatus consisted of a small tank of water with a wire mesh platform just above the water level at one end of the tank. Mice were given a series of five trials, one per day for five days, to measure the time it took to escape from the water onto the platform. Two parental strains, C57B1 and Balb/c, were chosen because they differed widely in their ability to learn this task (Figure 5–7). When mice of the two strains were cross-bred to produce the F_1 generation and these progeny were tested, they exhibited significantly faster learning of the water escape response than either parental strain. This is an example of overdominance.

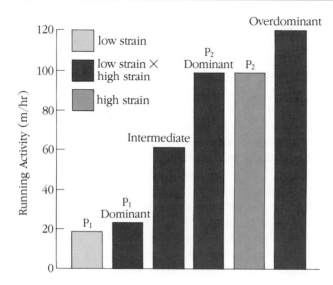

FIGURE 5–6   Modes of inheritance for wheel-running activity in mice When two strains of mice, low (P_1) and high (P_2), which exhibit significant differences in wheel-running behavior, are mated, three modes of inheritance can be exhibited: dominance—progeny exhibit behavior like one parent or the other; intermediate—progeny exhibit slow wheel-running activity in between levels recorded for the two parent strains; and overdominance (hybrid vigor)—progeny exhibit more wheel-running activity than either parental type.

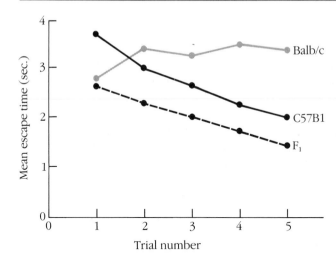

FIGURE 5-7 Learning curves for a water escape response for two strains of mice and their F₁ hybrid cross. The results show a faster learning response for the F₁ stock than for either parental strain.
Source: Data taken from Figure 3 on page 227 of J. H. Schröder and M. Sund. 1984. Inheritance of water-escape performance and water-escape learning in mice. *Behavior Genetics* 14:221–233.

TWIN STUDIES

Identical, or **monozygotic,** human twins develop from the splitting of a single fertilized egg; fraternal, or **dizygotic,** twins develop from two separate fertilized eggs. Because monozygotic twins have the same genotype, they provide scientists with a unique research opportunity. Identical twins are more alike in their characteristics and behavior traits than are fraternal twins. For example, correlations of intelligence test scores between ordinary siblings or fraternal twins are about 0.50; test-score correlations for monozygotic twins are between 0.73 and 0.90. (Erlenmeyer-Kimling and Jarvik 1963). Correlations between scores of unrelated individuals range from 0.00 to 0.30. When analyzing data like these, we must be aware that rearing environments for identical and fraternal twins are probably different. The differences in test scores must be interpreted in light of the possible variations in rearing environments, the composition and size of the sample population, and the type of intelligence test used for gathering the data.

The genotypes of a pair of monozygotic twins are identical—they are like mice from an inbred strain. If a pair of monozygotic twins are reared apart from one another—as sometimes happens for a variety of reasons—it is possible to examine the differential aspects of varied environments on behavior. This will be particularly true if the twins were separated at or near birth, so that the environmental effects will have occurred over nearly all the postnatal life of the individuals.

Cavalli-Sforza and Bodmer (1971) and Ehrman and Parsons (1981), using data from Shields (1962) and Newman et al. (1937) have summarized several studies on monozygotic and dizygotic human twins. For monozygotic twins, reared together or apart, and dizygotic twins, values were tabulated for hereditary and environmental influences on various traits in humans. Across all three types of twins, and within each type, the heritability values for height are higher than are those for weight. The heritability values are, in turn, generally somewhat higher for weight than for measures of IQ or personality. All of these heredity component values were positive; they ranged from about 0.30 to over 0.80. In contrast, estimates of the environmental component for these same traits were more variable, with a mixture of negative and positive values (total range from about -0.60 to $+0.80$). Ehrman and Parsons (1981) note that for both the hereditary and environmental components, variations may be due to the nature of the tests and differences between the sample populations. Another potential problem in interpreting data from twin studies is that environmental conditions are not actually the same for monozygotic and dizygotic twins (Smith 1965). In general, monozygotic twins are treated more similarly than are dizygotic twins — by parents, other relatives, and peers.

The heritability of IQ has been the subject of some controversy in recent decades. Jensen (1967, 1973), Shockley (1969), and others have attempted to present data to support racial differences in IQ and to argue that heredity plays a more dominant role in perpetuating these differences than does environment. Hirsch and his colleagues (Hirsch 1975, 1981; Hirsch, McGuire, and Vetta 1980) have provided sound explanations of some of the fallacies and other problems inherent in the arguments of Jensen, Shockley, and others. A quote from Hirsch's 1981 article provides the best summary of this viewpoint (p. 33):

The key to "establishing the relative roles of heredity and environment" has been believed erroneously to be the heritability estimate. But heritability estimates cannot be made for human intelligence measurements, because the heritability coefficient is undefined in the presence of either correlation or interaction between genotype and environment, both of which occur for human intelligence. When correlation exists, either (1) between genetic and environmental contributions to trait expression, or (2) between environmental contributions to trait expression in both members of a parent-child or sib pair, heritability is not defined. Furthermore, when heritability can be defined, for example in well-controlled plant and animal breeding experiments, it has *no* relevance to measured differences in average values of trait expression between different populations: heritability estimates throw no light upon intergroup comparisons! Also, heritability estimates provide no information about ontogeny and are thus irrelevant to the formulation of public policy on education and social conditions.

GENES TO BEHAVIOR

How are the processes that take place at the level of the gene and DNA translated into physiology and behavior? These processes can affect behavior in developing organisms as part of **epigenesis,** the interaction of the genetic program and the experiences and environment of the organism. Biochemical events involving genes can also affect behavior in adult organisms through the turning on and off of specific genes or gene complexes, leading to the synthesis of particular proteins. A schematic representation of the major components in an organism's system and their feedback relationships is shown in Figure 5–8. The model attempts to show the pathways that connect genes and behavior in both developing and adult organisms.

Some investigators have used **mosaics,** organisms whose tissues are of two or more genetically different kinds, to test aspects of the gene-to-behavior sequence; mosaics permit us to observe both anatomical and behavioral anomalies combined in the same animal. Hotta and Benzer (1972, 1973) exposed fruit flies to chemicals that increased the frequency of mutations and produced various "abnormal" types of flies. They then used genetic techniques in mating the flies to produce mosaics, which had some normal and some mutant tissues — for example, one normal wild-type red eye and one mutant white eye (Benzer 1973) — as well as some affected behaviors — for example, courtship.

Benzer then sought to determine what tissue parts must be mutant for the abnormal behavior to be expressed. He tested many fruit flies for the presence or absence of various mutant parts, analyzed the resulting statistics, and made a map of the early embryonic fly on which he located the **foci** — the groups of cells that differentiate into specific structures and organs. We can use these maps to trace the effects of particular mutant genes on behavior through the structures and physiological processes they affect.

GENETICS AND EVOLUTION

GENE FREQUENCY

We have looked at ways genes affect behavior, but there is another aspect to the gene-behavior relationship. Behavior may significantly affect the frequency and expression of certain genes in a population (Ford 1964; King 1967; Oliverio 1983). Changes in the gene pool, one of the outcomes of natural selection, can, in turn, affect behavior.

For example, variations in **courtship behavior** — the processes of mate selection and synchronization of activities that lead to mating — may alter gene frequencies. Merrell (1949, 1953) used wild-type and mutant strains of fruit flies to test this possibility. Four mutant strains — designated yellow, cut, raspberry, and forked — had sex-linked recessive alleles, expressed only in the male phenotypes and detected by examination of external characteristics. He started populations of flies in bottles with a mutant gene frequency of 0.5. Thus for each generation of flies, a departure from random mating would be indicated by an increase or decrease of the mutant gene in males. After several generations, the number of raspberry, cut, and yellow mutants decreased; the number of forked mutants showed no significant shift. Other tests demonstrated that although fertility or viability did not differ in

FIGURE 5–8   Model illustrating relationship between genes and environment in control of behavior

Conclusions: Relationships between genes and behavior: (1) Genes do not make *traits;* many steps occur between DNA segments and behavioral capacities of an animal. (2) Genotype (sum of genetic information) is a molecular code for regulated construction of proteins. (3) Raw materials obtained from the environment such as protein building blocks are broken down to incorporate the amino acids into species specific proteins. (4) Therefore, the key role of genes is enzyme production which in turn regulates entrance and exit of substances into cells. (5) Enzymes regulate reactions and thus control development.

Source: Adapted from several sources.

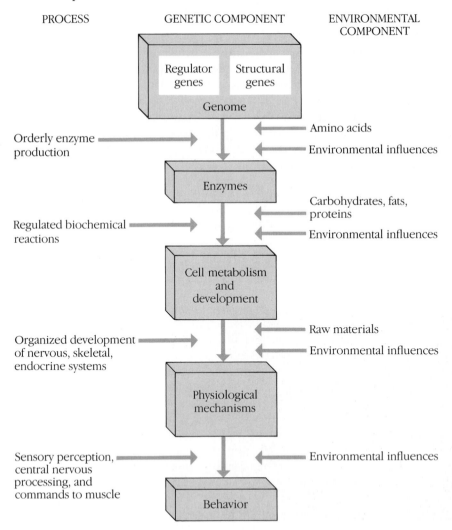

any of the mutant types, mating success differed significantly. The tests thus implicated some aspect of courtship and mating behavior as the critical factor leading to differences in productivity and, hence, to shifts in gene frequencies.

ADAPTATIONS

Other investigations have demonstrated that gene-controlled behaviors may vary between groups of animals of the same species. Some of these differences are adaptive to habitat characteristics and others are adaptive to stages of a population cycle.

HABITAT ADAPTATIONS.   Prairie deermice *(Peromyscus maniculatus bairdi)* of the midwestern United States inhabit only fields and are never caught in forested areas. In contrast, woodland deermice *(P. m. gracilis)* inhabit only forested areas in south-central Canada and the extreme northern and northeastern regions of the United States. Mice of these subspecies exhibit several behaviors and preferences that are predictable from the habitats they occupy (Figure 5–9). Woodland deermice prefer a temperature of 29.1°C; grassland deermice select a temperature of 25.8°C, in agreement with the cooler habitat of the latter subspecies (Olgivie and Stinson 1966). In addition, the two subspecies have different reactions to sand. In a one-day test, grassland mice removed a median of 5.9 lb of sand (from a hopper) in contrast to 0.1 lb for woodland mice (King and Weisman 1964). Again this result seems logical because grassland deermice are terrestrial, living in burrows in the ground; woodland deermice are semi-arboreal, sometimes nesting in trees.

FIGURE 5-9   Deermice *Peromyscus leucopus* **in natural habitat**
Source: Photography by Hal Harrison from Grant Heilman.

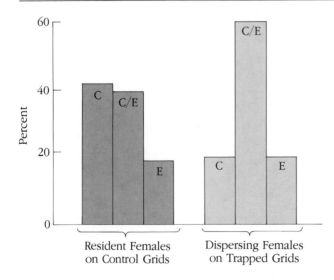

FIGURE 5–10 Transferrin genotypes in female meadow voles Genotypes of dispersing females are compared with those of resident females in the autumn during the increasing phase. C, C/E, and E represent the three transferrin genotypes.
Source: C. J. Krebs et al., "Population Cycles in Small Rodents," *Science* 179 (1973):35–41. Copyright 1973 by the American Association for the Advancement of Science. Reprinted with permission of the publisher and the author. Data from Myers and Krebs (1971).

POPULATION CYCLES. The populations of meadow voles *(Microtus pennsylvanicus)* rise and fall in cycles that last three or four years (Chitty and Chitty 1962; Krebs 1966). During the low phase of the cycle in a particular area, voles are difficult to find. In contrast, at the time of peak populations, there may be over 200 voles per hectare. Trapping data indicate that many animals emigrate during the increase phase of the population cycle; at peak density, however, few animals emigrate, and losses in numbers must be attributed to deaths. In voles, blood proteins, whose production is regulated by specific gene loci, can be measured by electrophoresis. **Electrophoresis** is a technique in which substances are separated from one another on the bases of their electric charges and molecular weights.

One such locus, transferin *(Tf)*, has two possible alleles, designated Tf^c and Tf^e (Myers and Krebs 1971; Krebs et al. 1973). Comparison of the frequencies of heterozygous and homozygous alleles in female voles from dispersing and nondispersing populations reveals a striking difference (Figure 5–10). The Tf^c/Tf^e allele combination is significantly more frequent in the dispersing population. Additional data indicate that 89 percent of the loss of heterozygous females from populations during the increase phase of the cycle is attributable to emigration (female voles dispersing from the local population). We should keep in mind that, in this instance, no direct gene-to-behavior sequence has been clearly delineated; however, changes in gene frequencies have been correlated with phases of the population cycle. The foregoing example also illustrates how rapidly the genotypic makeup of populations can change.

SUMMARY

Behavior genetics is concerned with two categories of questions: first, the role and relative importance of inheritance in development and regulation of behavior, and second, the role of genetics in the evolutionary process. The basic formula used to express the genotypic and environmental contributions to phenotype variation is $V_T = V_E + V_G + V_I$. In other words, the phenotypic variation is the sum of genotypic, environmental, and interactive forces. While the assumption that $V_I = 0$ is often made, we should keep in mind that the gene-environment interaction represented by V_I is a critical parameter in most situations.

We can use a variety of methodological techniques to assess the relative importance of genetic and environmental parameters in regulating behavior. Inbreeding, or brother-sister matings for many generations, is used to develop homogeneous strains of laboratory animals. It permits us to hold genetic factors constant while we manipulate environmental variables, or to hold environment constant and examine inherited differences in behavior between different strains or species. By using the variance values for a trait from two parental strains, and the F_1 and the F_2 crosses, we can obtain an estimate of the heritability (h_b^2) for a trait — this is heritability in the broad sense or $°GD$. We can calculate an estimate of realized heritability (h_r^2), which is useful in predicting the effectiveness of selection, by computing the ratio of response to selection (R) to the selection of differential (S). Cross-fostering, or the transfer of neonatal animals from the parent female to another female of the same species or strain or to a female of a different species or strain, is used to determine which behaviors are species-specific and which are affected by the maternal environment.

Another area of investigation in behavioral genetics is the determination of the mode of inheritance — which can be either a dominant mode, an intermediate mode, or an overdominant mode (also called hybrid vigor). In addition, studies of mutant strains seek to determine the processes that take place at the level of the gene and the results of these processes, which we see translated into morphology, physiology, and behavior.

Finally, behavior geneticists are concerned with the ways behavior affects the frequency and expression of genes and how changes in the gene pool can affect behavior as well as morphology.

Discussion Questions

1. How would you use the techniques of behavior genetics to study fitness as measured by reproductive success?

2. What effect would plasticity (flexibility) of phenotype have on the effectiveness of artificial selection for a trait?

3. Stamm (1954) investigated the genetics of food-hoarding behavior in two strains of rats. The two parent strains were Irish (mean hoarding score = 5.8) and black-hooded (mean hoarding score = 12.5). A cross between the two parental strains produced a new F_1 strain of rats with a mean hoarding score of 13.2, which was not significantly different from that of the black-hooded parental strain. What is the mode of inheritance of this hoarding trait in these rat strains?

4. Two populations of fish of the same species inhabit neighboring lakes that have identical environmental conditions, including fauna and flora. The population in one lake is genetically homozygous; the population in the second lake is heterozygous. Chemical pollution from nearby fields enters both lakes and results in rapidly shifting, less stable conditions and changes in the fauna and flora. What can you predict about the ability of the two populations of fish to survive and maintain their numbers in each pond, given the environmental changes?

5. Recall that the formula for calculating realized heritability is $h_r^2 = R/S$, with R representing the response to selection and S the selection differential. Below are the values needed to calculate R, S, and h_r^2 for an experiment involving measurement of the amount of sand dug by mice from five strains — A, B, C, D, and E — each of which was selected for one generation for a tendency to dig sand. The data given are the mean grams of sand dug from a trough in twenty-four hours; ten mice were tested for each mean shown. After calculating the h_r^2 for each strain, answer the following questions: What can you conclude about the h_r^2 in these mice? What restrictions would you put on these conclusions?

	Strain				
	A	*B*	*C*	*D*	*E*
Original population	62.6	41.5	10.8	17.6	31.5
Selected mice used for breeding	70.2	45.8	14.2	21.6	36.8
Offspring generation	63.5	44.6	11.2	20.8	33.8

Suggested Readings

Craig, J. V. 1981. *Domestic Animal Behavior.* Englewood Cliffs, N.J.: Prentice-Hall.
One of the first books covering the developing field of
applied animal ethology — particularly of interest to those
with interests in veterinary medicine, animal science, and
zoos.

DeFries, J. C., and R. Plomin. 1978. Behavioral genetics. *Ann. Rev. Psychol.* 29:473–516.
Comprehensive and sometimes a bit technical treatment of
behavior-genetic analysis. Only minimal background
needed to make this useful reading.

Ehrman, L., and P. A. Parsons. 1976. *The Genetics of Behavior.* Sunderland, Mass.:
Sinauer.
Current textbook with emphasis on basic concepts. Exten-
sive use of *Drosophila,* house mice, and humans as examples.
Best source for the beginning student who wishes to pursue
this topic in more depth.

Manning, A. 1976. The place of genetics in the study of behaviour. In P. P. G.
Bateson and R. A. Hinde (eds.), *Growing Points in Ethology.* Cambridge: Cambridge
University Press.
Historical context and definition of the role of behavior
genetic analysis in the larger framework of behavior
analysis are the strong points of this book chapter. Covers
the types of research strategies emphasized by those using
genetics as a tool for the study of behavior.

References

Bastock, M. 1956. A gene mutation which changes a behavior pattern. *Evolution* 10:421–439.

Bastock, M., and A. Manning. 1955. The courtship of *Drosophila melanogaster. Behaviour* 8:85–111.

Benzer, S. 1973. Genetic dissection of behavior. *Sci. Amer.* 229:24–37.

Broadhurst, P. L. 1965. The inheritance of behavior. *Science J.* 24:39–43.

Cavalli-Sforza, L. L., and W. F. Bodmer. 1971. *Genetics of Human Populations.* San Francisco: W. H. Freeman.

Chitty, D., and H. Chitty. 1962. Population trends among the voles at Lake Vyrnwy, 1932–60. *Symp. Theriologicum, Brno.* 1960:67–76.

Cooper, R., and J. Zubek. 1958. Effects of enriched and restricted early environments on the learning ability of bright and dull rats. *Can. J. Psychol.* 12:159–164.

Erlenmeyer-Kimling, L., and L. F. Jarvik. 1963. Genetics and intelligence: A review. *Science* 142:1477–1478.

Ehrman, L., and P. A. Parsons. 1981. *Behavior Genetics and Evolution.* New York: McGraw-Hill.

Falconer, D. S. 1960. *Introduction to Quantitative Genetics.* New York: Ronald.

Ford, E. B. 1964. *Ecological Genetics.* New York: Wiley.

Fuller, J. L., and W. R. Thompson. 1978. *Behavior Genetics.* St. Louis: Mosby.

Hirsch, J. 1967. *Behavior-Genetic Analysis.* New York: McGraw-Hill.

———. 1975. Jensenism: The bankruptcy of "science" without scholarship. *Educ. Theory* 25:3–27.

———. 1981. To "unfrock the charlatans." *Sage Race Rel. Abstr.* 6:1–67.

Hirsch, J., and J. C. Boudreau. 1958. Studies in experimental behavior genetics. I. The heritability of phototaxis in a population of *Dro-*

sophila melanogaster. *J. Comp. Physiol. Psychol.* 51:647–651.

Hirsch, J., T. R. McGuire, and A. Vetta. 1980. Concepts of behavior genetics and misapplications to humans. In J. S. Lockard (ed.), *Evolution of Human Social Behavior.* New York: Elsevier.

Hotta, Y., and S. Benzer. 1972. The mapping of behavior in *Drosophila* mosaics. *Nature* 240:527–535.

———. 1973. Courtship in *Drosophila* mosaics: Sex-specific foci for sequential action patterns. *Proc. Nat. Acad. Sci. USA* 73:4154–4158.

Jensen, A. R. 1967. Estimation of the limits of heritability of traits by comparison of monozygotic and dizygotic twins. *Proc. Nat. Acad. Sci.* (USA) 58:149–156.

———. 1973. *Educability and Group Differences.* New York: Harper & Row.

King, J. A. 1967. Behavioral modification of the gene pool. In J. Hirsch (ed.), *Behavior-Genetic Analysis.* New York: McGraw-Hill.

King, J. A., and R. G. Weisman. 1964. Sand-digging contingent upon bar-pressing in deermice (*Peromyscus*). *Anim. Behav.* 12:446–450.

King, J. A., D. Maas, and R. G. Weisman. 1964. Geographic variations in nest size among species of *Peromyscus. Evolution* 18:230–234.

Krebs, C. J. 1966. Demographic changes in fluctuating populations of *Microtus californicus. Ecol. Monogr.* 36:239–273.

Krebs, C. J. et al. 1973. Population cycles in small rodents. *Science* 179:35–41.

Lynch, C. B., and J. P. Hegmann. 1972. Genetic differences influencing behavioral temperature regulation in small mammals. I. Nesting by *Mus musculus. Behav. Gen.* 2:43–54.

Merrell, D. J. 1949. Selective mating in *Drosophila melanogaster. Genetics* 34:370–389.

———. 1953. Selective mating as a cause of gene frequency changes in laboratory populations of *Drosophila melanogaster. Evolution* 7:287–296.

Myers, J. H., and C. J. Krebs. 1971. Genetic, behavioral and reproductive attributes of dispersing field voles *Microtus pennsylvanicus* and *Microtus ochrogaster. Ecol. Monogr.* 41:53–78.

Newman, H. H., F. N. Freeman, and K. J. Holzinger. 1937. *Twins: A Study of Heredity and Environment.* Chicago: University of Chicago Press.

Olgivie, D. W., and R. H. Stinson. 1966. Temperature selection in *Peromyscus* and laboratory mice. *J. Mammal.* 47:655–660.

Oliverio, A. 1983. Genes and behavior: An evolutionary perspective. *Adv. Stud. Behav.* 13:191–217.

Pfaffenberger, C. J., J. L. Fuller, B. E. Ginsberg, and S. W. Bielfelt. 1976. *Guide Dogs for the Blind: Their Selection, Development and Training.* Amsterdam: Elsevier.

Plomin, R., J. C. DeFries, and J. C. Loehlin. 1977. Genotype-environment interaction and correlation in the analysis of human behavior. *Psych. Bull.* 84:309–322.

Quadagno, D. M., and E. M. Banks. 1970. The effect of reciprocal cross-fostering on the behavior of two species of rodents, *Mus musculus* and *Baiomys taylori ater. Anim. Behav.* 18:379–390.

Roberts, R. C. 1967. Some concepts and methods in quantitative genetics. In J. Hirsch (ed.), *Behavior-Genetic Analysis.* New York: McGraw-Hill.

Schröder, J. H., and M. Sund. 1984. Inheritance of water-escape performance and water-escape learning in mice. *Behav. Genet.* 14:221–234.

Shields, J. 1962. *Monozygotic Twins Brought Up Apart and Brought Up Together.* London: Oxford University Press.

Shockley, W. B. 1969. Human quality problems and research taboos. In J. A. Pintus (ed.), *New Concepts and Directions in Education.* Greenwich, Conn.: Educational Records Bureau.

Smith, R. T. 1965. A comparison of socioenvironmental factors in monozygotic and dizygotic twins, testing an assumption. In S. G. Vandenberg (ed.), *Methods and Goals in Human Behavior Genetics.* New York: Academic Press.

Stamm, J. S. 1954. Genetics of hoarding. I. Hoarding differences between homozygous strains of rats. *J. Comp. Psychol.* 47:157–161.

6

THE NERVOUS SYSTEM
AND BEHAVIOR

Every living organism is continuously bombarded by stimuli from the environment. To survive, animals must be equipped with sensory receptors to receive information with some type of nervous system to sort out and interpret stimuli, and with effector or motor systems and the endocrine system to produce behavioral responses in reaction to appropriate stimuli. In this chapter we examine the ways in which nervous systems control and generate behavior. Chapter 7 explores the role of the endocrine system.

We look first at how the basic units of the nervous system — the neurons — function in transmitting impulses. Next we consider the systems of sensation and perception and the filtering mechanisms animals have evolved for sorting out relevant from less important stimuli. We then explore the evolution of different types of nervous systems and their relationship to the behavior capacities of various animal types. Finally, we turn to an examination of research techniques and the kinds of findings they generate regarding the nervous system and behavior.

THE NERVOUS SYSTEM

The nervous system carries messages throughout an organism by means of a basic structural unit, the **neuron,** which has evolved many variations on a similar theme

(Figure 6–1). Typical neurons consist of a **soma** (cell body) with a nucleus, an **axon** that has threadlike **synaptic processes** on one end, and a series of **dendrites** on the other end. The position of the cell body and the numbers and kinds of dendrites vary depending on the type of nerve cell.

COMMUNICATION OF IMPULSES

Communication within the nervous system consists of the passage of an electrical impulse through two separate stages: (1) movement of the impulse along the neuron and (2) transmission of the impulse from one neuron to another. An electrical impulse is conducted along a neuron from the dendrites through the axon by the movement of electrically charged atoms, called ions, across cell membranes. Normally there are more sodium (Na^+) ions outside the neuron and more potassium (K^+) ions inside (Figure 6–2). Nerve cells must constantly expend energy to maintain this ionic balance because ions have a tendency to move across the membranes to establish equal proportions of both ions inside and outside the cell. Thus Na^+ ions are actively transported to the exterior and K^+ ions are actively moved into the cell.

FIGURE 6–1   Three types of neurons
Neurons are the basic structural building blocks of nervous systems. Shown here are a generalized neuron, a receptor or sensory cell, and a brain cell from a vertebrate.

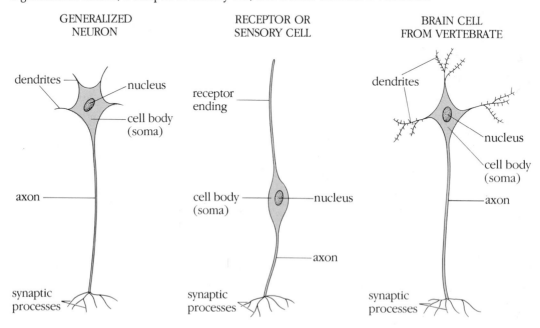

FIGURE 6-2   Nerve cell at rest and during passage of impulse
The resting state is maintained by active transport of Na$^+$ ions out of and K$^+$ ions
into the cell. When a stimulus triggers a wave of permeability changes in the
membrane, Na$^+$ ions flow in and soon thereafter, in order to balance this inflow of
positively charged ions, K$^+$ ions flow out of the cell. After a brief interval the resting
state is restored.

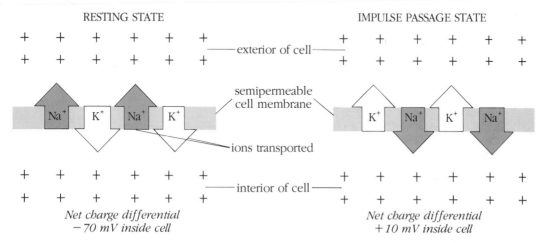

This process results in an unstable ion balance, with a net charge differential of about
-70 millivolts (mV) across the semipermeable cell membrane and a negative charge
inside the cell.

An impulse is initiated when a stimulus reaches dendrites at one end of the
neuron, changes **membrane permeabililty,** and alters the ion balance. Changes in
permeability occur at the point of stimulus contact and in immediately surrounding
regions of the membrane. This permits sodium ions to flow into the cell. If only a
slight change in ion concentrations takes place, the ion balance is soon restored.
However, if the charge differential drops to about -50 mV or less, an **action-potential** occurs, and an impulse is triggered (Figure 6–3). Sodium ions flood into the cell
and produce a net positive charge ($+10$ mV) inside the cell membrane (Figure 6–2).
These changes trigger a series of similar membrane changes in neighboring regions
and, through a chain reaction process, the impulse passes along a neuron.

To restore normal ion balance in an area through which an impulse has just
passed, several events take place. First, the membrane becomes permeable to K$^+$
ions, which flow outward to the surrounding medium. Then the cell's machinery
actively pumps Na$^+$ ions out of the cell and K$^+$ ions into the cell and eventually
restores the -70 mV charge differential. During the time it takes to restore the
normal resting balance — the **refractory period** — another stimulus cannot produce
an impulse.

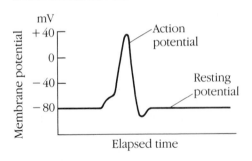

FIGURE 6-3   Action potential
The resting membrane potential is about −75 mV due to the excess positive charges (Na⁺ ions) on the outside. When the threshold is reached, the neuron depolarizes rapidly and reaches a membrane potential of about +35 mV before beginning to repolarize. During repolarization the membrane passes through a refractory period when it hyperpolarizes and has a potential of about −85 mV.

An impulse is transmitted from the axon endings of one cell to the dendrites of the next cell at a cell junction called a **synapse.** The events that occur at a typical synapse are shown diagrammatically in Figure 6–4. An impulse causes synaptic vesicles to migrate to the end of an axon, where they release chemicals, called **neurotransmitters,** into the intracellular space (called the **synaptic cleft** or **gap**). These chemical messengers affect the permeability of dendrite membranes and thus initiate an impulse, which is conducted along the next neuron by the processes just described. In this way an impulse is passed from neuron to neuron.

Neuromodulators are chemicals that are released by neurons that can affect a population of neurons without initiating action potentials and without synaptic contact. Neuromodulators act at synapses to facilitate or inhibit impulse transmission.

PROPERTIES OF NERVE CELLS

The activities of nerve cells are unique. First, each neuron either fires or does not fire; the production of an impulse is an all-or-none phenomenon. There are few gradations in the intensity of nerve impulses; moreover, the propagation process is virtually identical for all neurons. Transmission along neurons in most organisms is unidirectional; impulses cross synaptic junctions in only one direction. These properties of neurons reduce ambiguity, because the same simple message is carried by all types of nerve cells. A. L. Hodgkin (1971) provides a more complete discussion of the conduction of nerve impulses.

If nerve impulses are all similar, how do neurons transmit any information about the nature of the stimulus or its strength? Different types of environmental information are picked up by different sensory receptors and carried along separate neural paths to particular areas within the central nervous system for decoding and interpretation. The separate, but integrated, systems for different types of information — coming, for example, from the eyes or from the ears — prevent confusion about whether the stimulus was originally photic (relating to light) or auditory

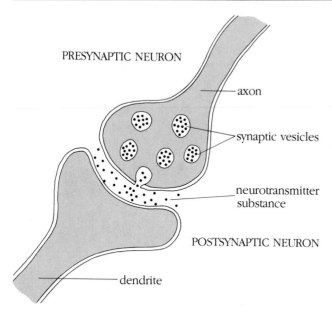

PRESYNAPTIC NEURON

axon

synaptic vesicles

neurotransmitter substance

POSTSYNAPTIC NEURON

dendrite

FIGURE 6-4   Transmission of neural impulse between adjacent nerve cells
Several steps occur in the transmission of a neural impulse between two adjacent nerve cells. First, synaptic vesicles migrate to the end of an axon. Second, vesicle contents, neurotransmitters, are released into the intracellular space, where they migrate to dendrites of the second neuron. Third, the neurotransmitter substances trigger a new action potential in the post-synaptic neuron.

(relating to sound). The strength of stimulation may be communicated in at least two ways: the same neuron may fire repeatedly, with brief refractory periods; or several neurons, all carrying the same type of information, may fire simultaneously. Thus sensory information may be encoded by the frequencies with which individual neurons fire and by the number of neurons firing.

SENSATION AND PERCEPTION

SENSORY EQUIPMENT

The term **sensation** refers to the process of transducing environmental stimulation, or energy—for example, sound, light, heat, mechanical forces—into electrical impulses. The events involved in sensation of external stimuli are largely peripheral; they occur at the body surface. To receive this environmental information, animals have evolved a variety of sense organs, often specialized to respond to the demands of that organism's particular environment. There are also sensory receptors located within muscles and organs of the body which provide information on such things as muscle tension, body position, and internal conditions. We will be concerned here primarily with the sensory equipment that receives external cues, often called **exteroreceptors.** Examples of exteroreceptive sensory systems include the electric organ

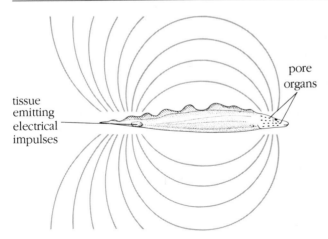

FIGURE 6-5   Impulses of electric fish

Electric fish *(Gymnarchus niloticus)* use weak electrical impulses emitted by a specialized tissue in the tail region to find their way about in murky water. The electrical field around the fish, indicated by lines in this figure, is continually monitored by pore organs located around the head region. An object in the water disturbs the symmetry of the field and thus can be detected by the fish.

and sensory system for certain fish living in murky waters; the echolocation system in bats, involving high frequency pulses and specialized ears to receive returning echoes, which are used for navigation and prey location; and recent evidence suggesting that pigeons may have a specialized system for sensing geomagnetic forces, useful in orientation (Walcott et al. 1979).

The electric fish *(Gymnarchus niloticus)*, inhabits muddy rivers in parts of tropical Africa, where navigation by visual clues would be difficult. Special organs, located near the rear of the fish's body, emit a continuous stream of very weak electrical impulses (Figure 6–5), and porelike structures on the head contain sensory receptors that are stimulated by extremely small changes in electrical impulses (Hopkins 1974; Machin and Lissman 1960). The electrical impulses set up an electrical field around the fish, and the sensory receptors detect moving or stationary objects in the environment as distortions in this electrical field. The electric fish can thus navigate with little difficulty in an environment where visual senses would not

FIGURE 6-6   Chemical structure of bombykol

The female silkworm moth *(Bombyx mori)* emits a chemical substance called bombykol that serves to attract males from a considerable distance for the purpose of mating. This chemical is an effective stimulus even at very low concentrations in the air. (C = carbon, H = hydrogen, O = oxygen)

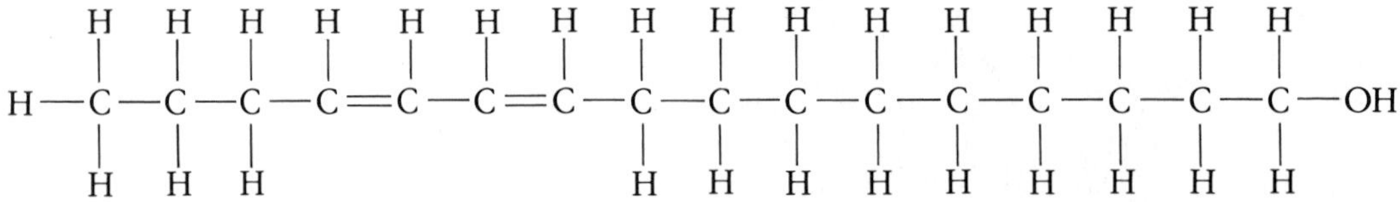

be sufficient. We should note that neither the stimulus nor the machinery for picking up these types of electrical stimuli are part of the human sensory world.

Silkworm moths *(Bombyx mori)*, like humans, are stimulated by chemical sensory cues, but the quality of the stimulus and the extremely fine acuity of the moth's chemical sense organs are clearly not within human sensory capabilities. The female silkworm moth has a special way of communicating her availability as a reproductive partner (Figure 6–6); a specialized gland on her abdomen emits a chemical substance called bombykol (Beroza and Knipling 1972). A male moth (Figure 6–7), flying downwind from the female, can detect this specific odor even when the concentration is as low as several molecules of odor per million molecules of air. Fine hairs on the male's featherlike antennae are stereochemically specialized to detect the odor. Once the male detects the scent, he flies upwind toward the odor source to locate and mate with the female. Males of other moth species are apparently not affected by the chemical attractant; most other animals do not respond to its presence in the air.

PERCEPTION SYSTEMS

Once information has been coded into electrical impulses, it may be conveyed throughout the animal's nervous system. **Perception** is the analysis and interpretation of sensory information. This decoding process is the function of groups of neurons, found in bundles, called **ganglia,** in many lower animals and in larger aggregations of nerve cells in the central nervous systems of higher organisms, including vertebrates. In the simplest type of nervous system, little or no decoding takes place; a stimulus-response system is wired into the basic plan of the animal. The way an animal perceives a particular stimulus is a function of the type of sensory

FIGURE 6–7   Male silkworm moth
The male moth, here the giant silkworm moth *Hyalophora euryalis,* has specialized receptors on its antennae for detecting the chemical attractant bombykol.
Source: Photo by E. S. Ross.

information it receives, the structure of its nervous system, and the past experiences that have been permanently encoded in its central nervous system.

When we investigate the sensory and perceptual worlds of diverse types of animals, we need to make both physiological and behavioral measurements. If we place electrodes in nerve tracts, which conduct impulses from a sensory end organ toward decoding centers, we can record whether neural impulses are initiated when we provide stimulation. For example, when we place electrodes in the cochlear nerve of a turtle and play pure tones near the turtle's ear, we obtain a response record for auditory thresholds (Figure 6–8). The turtle *(Pseudemys scripta)* is maximally sensitive to airborne sound in the range of 200 to 600 cycles per second (Gulick and Zwick 1966).

An experimental assessment of perception must be designed so that the behavioral task an animal performs indicates both the reception and the interpretation of the stimuli by the animal. Octopodes can be trained, by means of food rewards and mild electric shocks, to make choices among cylinders having varying patterns of grooves on the basis of differences in the total amount of grooved area (Figure 6–9). However, they cannot distinguish the patterns of the grooves (Wells and Wells 1957). Octopodes can learn to distinguish cylinders *a* and *d* but cannot discriminate between *a, f,* and *g.* Techniques used to explore the sensory and perceptual capacities of various animals must be designed with some care to take into account the animal's response capabilities and behavior patterns.

An animal species can thus be characterized by its distinctive sensory equipment for receiving input from the environment and by its nervous structure for analyzing and interpreting the information received by the sensory system. In investigating the behavior of any organism, we must understand both its sensory system and its perceptual world.

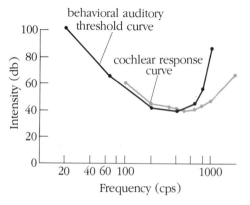

FIGURE 6-8   Auditory sensitivity of the turtle
Each curve is based on data from four turtles.
Source: W. L. Gulick and H. Zwick, "Auditory Sensitivity of the Turtle," *Psychological Record* 16 (1966):47–53. Reprinted with permission.

FIGURE 6-9   Tactile discrimination in octopodes
The proportion of grooved surface in the cylinders decreases from left to right.
Beside each object the number indicates the total percentage of the surface that is
grooved. Octopodes have difficulty distinguishing between objects with the same or
nearly the same total percentage of grooved surface (a, b, and g) regardless of the
pattern of the grooves. They can, however, distinguish cylinders with different
amounts of grooved surface (a and d).
Source: M. J. Wells and J. Wells, "The Function of the Brain of *Octopus* in Tactile
Discrimination," *Journal of Experimental Biology* 34(1957):133. Reprinted with
permission.

Proportion of Grooved to Flat Surface

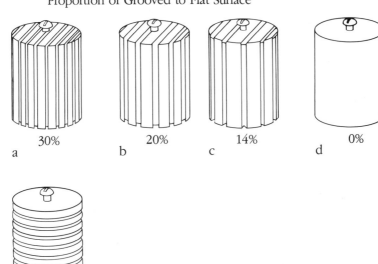

   30%   20%   14%   0%
 a   b   c   d

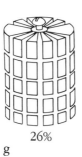

  50%   30%
 e   f

     26%
    g

SENSORY FILTERS

In connection with sensory and perceptual processes, ethologists developed the concept of **sensory filters.** As we noted at the beginning of the chapter, every animal is continuously bombarded by an enormous number of diverse stimuli. Receiving, processing, and responding to all of these stimuli would be impossible. One of the primary functions of the sensory and nervous systems is to select only pertinent stimuli. To accomplish this selectivity, animals have evolved at least two major types of information-filtering systems, one peripheral and the other central.

PERIPHERAL FILTERS.   Peripheral filters function at the level of the sensory receptors. Each animal species can receive information from a limited number of sensory modes. The quantitative range of sensitivity within a particular mode may also limit reception of environmental stimuli. Thus, not all animals can sense ultraviolet light as, for instance, bees can; and few organisms are equipped, as is the electric fish, with the machinery to transduce weak electrical current into neural impulses. The limitations of the peripheral sensory system act as an initial screen, and only a limited number of the stimuli that impinge on an organism result in messages sent to decoding centers in the organism's nervous system.

     Certain species of noctuid moths (underwing moths of the genus *Catocala*) have special hearing receptors — tympanic membranes located on each side of the body in the thoracic region (Figure 6–10a) (Roeder and Treat 1961; Roeder 1970). Two neurons, A_1 and A_2 connect each "ear" to the central nervous system (Figure 6–10b). These moths are capable of detecting and evading capture by predatory bats.

FIGURE 6-10   Tympanic membranes in moths
The tympanic membrane of noctuid moths is located in the thoracic region (a) and is connected to the central nervous system by two neurons, A1 and A2 (b).

POSITION OF
TYMPANIC MEMBRANE

CUTAWAY VIEW OF
TYMPANIC MEMBRANE

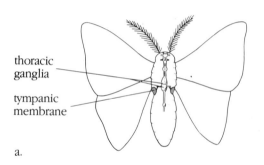

thoracic
ganglia

tympanic
membrane

a.

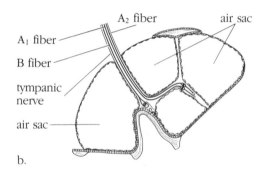

A_1 fiber

B fiber

tympanic
nerve

air sac

A_2 fiber

air sac

b.

Roeder placed small electrodes into each of the neurons and recorded nerve impulses while he directed bat vocalizations and artificially created sounds toward the moth. By varying the frequency and intensity of these sounds and recording the nerve impulses, he was able to draw several conclusions about peripheral sensory filters. First, when sound frequency is varied, there is no change in pattern from either neuron; the moth cannot detect differences in frequency. When the sounds are soft, the A_1 neuron responds; the A_2 does not respond. When the sounds are loud, both neurons respond, but the A_1 fires more often than it does with low-intensity (soft) sounds. Thus some aspects of incoming sounds are filtered out; for sounds that are passed on to the central nervous system, different patterns of firing in the two neurons encode the sound stimulus.

CENTRAL FILTERING PROCESSES.   Central filtering processes within the nervous system sort out incoming information, select relevant or important stimuli for further action, and eventually produce a particular response. Evidence for the existence of these processes has come from several sources.

When an animal is presented with a variety of stimuli, all of which are detected by its sense organs, it responds selectively — that is, its behavior is based on only one or a few of the stimuli presented. For example, when an egg rolls out of its nest, a brooding herring gull *(Larus argentatus)* retrieves it by pulling it back into the nest with its bill. Tinbergen (1953) tested herring gulls to determine which stimulus qualities of the egg — its size, shape, color, or pattern (Figure 6–11) — might trigger the retrieving response.

Tinbergen measured preference by placing pairs of model eggs on the edge of the nest and recording which egg the bird chose to retrieve first. He then conducted the test with eggs that varied in size, shape, color, and color pattern and found that the most critical aspects in the retrieving response were the egg's size and shape. The gull's visual system is quite capable of receiving and transducing incoming stimuli from the eggs into the messages sent to the central nervous system, but the gull's

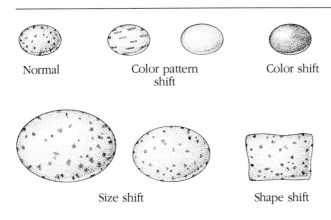

Normal   Color pattern shift   Color shift

Size shift   Shape shift

FIGURE 6–11   Eggs used to test retrieval behavior in herring gulls The eggs shown here, which vary in size, shape, color, or color pattern were used by Tinbergen to test the stimulus qualities that are most important in eliciting egg retrieval behavior in herring gulls. Tinbergen found that size and shape were more critical than other factors.

behavioral response is directed at specific stimuli. Thus Tinbergen concluded that some type of central filtering process must be in force during the decoding of the visual input. This central filtering process affects the gull's preference for a particular stimulus quality of the egg.

Animals exhibit different responses to the same stimulus; the response depends on their internal conditions. Our everyday experience can demonstrate how the central filtering process operates on the auditory mode. In a crowded college dormitory, with a stereo playing the latest popular music at a high volume and many conversations taking place, two students are talking seriously about campus politics. Although they are receiving a large quantity of auditory sensory input, they are able to hear one another and to concentrate on the topic at hand because central filtering processes act on the incoming nerve impulses from the ear and sort out only the relevant stimuli for further processing.

EVOLUTION OF THE NERVOUS SYSTEM

The topic of evolutionary changes in the structures and properties of animal nervous systems and the changes or trends in behavior associated with them is vast and complex. In this section we take a brief look at the subject to get some idea of the ties between the nervous system, evolution, and behavior; generalizations are, of necessity, rather broad.

SIMPLE SYSTEMS

Unicellular organisms do not possess a formal structure that could be called a nervous system, but they do exhibit a general phenomenon, characteristic of most cells, called irritability. **Irritability** is the wavelike passage of a stimulus from one point of a cell to another. Cells respond in a nonspecific fashion to external stimuli. In the process of evolution, irritability became channeled and refined; and specific tissues and structures developed for receiving and transmitting information within the organism. The development of the nervous system was critical for the evolution of multicellular animals and for complex, flexible, and integrated behavior sequences.

The **nerve net** (Figure 6–12) represents one of the earliest specialized constructions of neural tissues for conducting messages between cells and is characteristically found in coelenterates, such as sea anemones (e.g., *Stephanauge*), hydra (e.g., *Obelia*), and jellyfish (e.g., *Aurelia*). However, some organisms with a nerve net possess another form of nervous system as well and many higher animal species retain some form of nerve net—for example, in the gut of vertebrates where peristaltic contraction involves diffuse conduction. The nerve net appears as a random

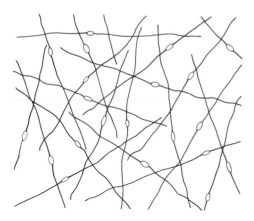

FIGURE 6-12   Nerve net from sea anemone
The many neuronal fibers in the sea anemone appear to be randomly oriented, but even this degree of organization allows for the important property of irritability.

arrangement of nerve fibers. Synapses occur at junctions between the nerve fibers, on neurons, but, unlike synapses in the nervous systems of other organisms, those in coelenterate nerve nets are nonpolarized and conduct impulses in either direction. An impulse that begins at one point passes to most other neurons throughout the net because each neuron is through conducting. Only later, evidently, did synapses evolve as part of a system that conducts impulses in only one direction and along discrete, closed channels.

The nerve net regulates simple behaviors, such as posture changes and feeding responses, and may also be a good mechanism for localized responses; distances between receptors and effectors are short, and responses need not be mediated through a central nervous system. Starfish (class Asteroidea), for example, have small pincers located on their backs that help to keep small organisms from settling there. The actions of these pincers are controlled by local nerve nets.

RADIALLY SYMMETRICAL NERVOUS SYSTEM

Two trends mark the evolution of slightly more complex nervous systems: First, different types of neurons became specialized to serve particular functions; and, second, arrangements of neurons became more ordered, forming nerve tracts and integrative centers. Following the development of nerve nets, the next major change was the development of the radially symmetrical nervous systems characteristic of starfish (Figure 6–13) and sea urchins. Starfish have a circular nerve ring, located in the central portion of the body, with **nerve tracts** (collections of axons) extending into each arm. Within each arm is a network of sensory and motor neurons, called a **plexus.** The radially symmetrical nervous system permits more flexible and complex behaviors than does the nerve net system.

Some of the responses exhibited by starfish are reflexive, involving only

FIGURE 6-13 Starfish nervous system
The starfish is characterized by a radially symmetrical body plan. Its nervous
system, as shown here, contains a central ring and nerve tracts with sensory and
motor fibers that connect the central ring with each arm.
Source: Photo by Runk/Schoenberger from Grant Heilman.

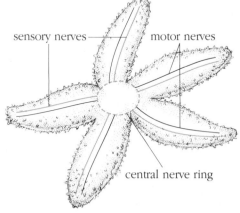

peripheral sensory and motor neurons; other responses involve integrated behavior
mediated through the central ring. Most important, coordinated movements of the
five arms are controlled, in part, through impulses that travel through sensory
neurons and interneurons to motor neurons. These pathways necessarily involve
the central ring. Thus, the distances between sensory receptors and muscle-effectors
is increased in radially symmetric nervous systems. Alternation of excitation and
inhibition, which is the process of stimulation and the blocking of stimulation to
muscle tissue via motor neurons, produces extension and retraction of the arms and
results in locomotion. Starfish are capable of righting themselves if placed upside
down and of wrapping themselves around a bivalve shellfish (e.g., a clam) in order
to open the shell and obtain food; the interneurons and central ring mediate the
coordinated movements necessary to perform these tasks.

Animals with radially symmetrical nervous systems also evolved more specialized types of sensory receptors, including systems for receiving contact, taste,
and general chemical signals. The combination of refined systems for sensing the
environment and coordinated responses means that starfish and related organisms
possess a greater degree of flexibility and diversity of behavioral responses than do
simpler forms, but they are still relatively limited in their total capacity for adaptation
through the nervous system.

BILATERALLY SYMMETRICAL NERVOUS SYSTEM

The bilateral nervous system, characteristic of many types of worms, molluscs, and arthropods (including insects and crustaceans) as well as vertebrates, exhibits clear advances over the radially symmetrical system. A bilaterally symmetrical organism generally possesses a distinct head and tail and has a segmented body (Figure 6–14). We can observe several evolutionary developments in the nervous systems of these animals.

The first development is the refinement and diversification of sensory organs and systems. For instance, sensory receptors are very important in some insects for mechanisms of spatial distribution and mate finding. Male house crickets *(Acheta domesticus)* use sounds produced by stridulation—the rubbing of specialized structures on the wings to produce a chirping sound—to establish and maintain distinct territories (exclusive use of an area) or to attract females (Heiligenberg 1966; Bentley 1969; Bentley and Hoy 1970). Specialized sensory receptors for registering the chirping sounds are located in the cricket's head region.

Most insects have evolved highly complex visual receptors called **compound eyes,** which have no lenses and are composed of many separate light receptor units called **ommatidia** (Figure 6–15). Each ommatidium is stimulated by a point of the object being viewed, and an overall image is comprised of a mosaic of many such points. Octopodes, a group of molluscs, have evolved a visual receptor with a lens, similar in some respects to the eyes of vertebrates. Other examples of specialized sensory receptors are the antennae of some moths, which are used for detecting bat vocalizations, and the taste receptors of blowflies.

The second development in bilaterally symmetrical organisms is the formation of an articulated skeleton—an external (in invertebrates) or internal (in vertebrates) support structure. The evolution of skeletal structures was accompanied by changes in effector or muscle systems. Concurrently, patterns of muscle innervation and control were evolving, making possible coordinated movements of various body parts and more rapid and refined responses by the organism.

The third, and perhaps the most important, evolutionary development seen

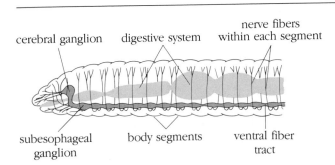

cerebral ganglion   digestive system   nerve fibers within each segment

subesophageal ganglion   body segments   ventral fiber tract

FIGURE 6-14   Earthworm nervous system

Earthworms are bilaterally symmetrical organisms, with bodies divided into many segments. The earthworm nervous system has several important features: the concentration of neurons in ganglia toward the anterior or head region, the ventral fiber tract, and sensory and motor neurons that connect to the ventral nerve cord within each body segment.

FIGURE 6–15 Compound eye and ommatidium of insect
The compound eye of many insects is comprised of many individual photo-receptive cells called ommatidia. Each ommatidium "sees" one point of the visual object and together these points from the ommatidia form a mosaic image.

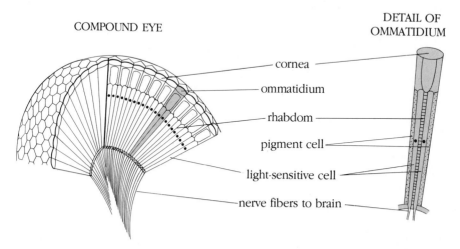

COMPOUND EYE

DETAIL OF OMMATIDIUM

cornea

ommatidium

rhabdom

pigment cell

light-sensitive cell

nerve fibers to brain

in bilaterally symmetrical organisms is the **centralization** of neural processes and control of behavior. Two separate, but related, changes are involved in centralization. In earthworms *(Lumbricus terrestris)* nerve tracts coalesced into a distinct, ventral nerve cord (see Figure 6–14). Within each body segment, sensory neurons enter the nerve cord from receptors, and effector neurons exit to muscle systems. With this system some local control is preserved, but the possibility of coordination of activity among body segments is increased.

In many invertebrates neural tissues are concentrated in ganglia located in anterior body segments. Since an organism that possesses a head and a tail most frequently proceeds anteriorly, a great deal of incoming information originates in front of it. Thus having the necessary neural machinery near the source of stimulation, rather than in the tail region, is advantageous. Moreover, successful monitoring of the external environment requires more types and greater numbers of sensory receptors; therefore more and more neural tissue became concentrated in the anterior segments.

GENERALIZED VERTEBRATE NERVOUS SYSTEM

A sequence of forms, including generalized vertebrate nervous systems, developed from bilaterally symmetrical organisms. The evolutionary changes in vertebrate nervous systems are characterized by (1) further concentration and enlargement of the central nervous system, (2) development of many interconnections among nerve

FIGURE 6-16   Vertebrate nervous systems
Nervous systems of vertebrates are characterized by centralization, encephalization, and the development of many interconnections within the nervous system. Differences among vertebrate species in the size of the brain and cranial capacity are evident here in the drawings of reptile, bird, cat, and human nervous systems.

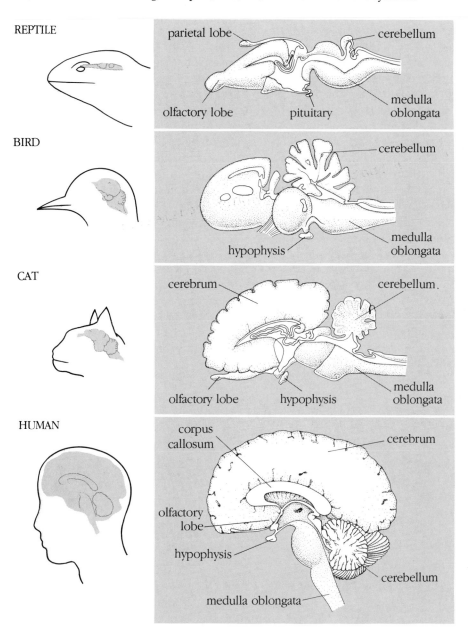

cells, and (3) **encephalization,** which encompasses changes in size and configuration of the anterior regions of the nervous system leading to the evolution of the brain. A number of these changes can be seen in Figure 6–16.

A variety of behavioral phenomena characteristic of vertebrates are intimately tied to these changes in the nervous system. First, the complexity of behavior patterns and the general flexibility of behavioral responses of vertebrates exceeds that of other animal groups. Second, because of changes in the structure of the nervous system and in the morphology of neurons themselves, vertebrates as a group are generally capable of faster responses than are invertebrates though some hard-wired responses in invertebrates (e.g., escape responses of cockroaches) occur as fast as any vertebrate response. A third consequence — a result of the formation of true brains through encephalization and the enlargement of anterior portions of the nervous system — is a greater capacity for the storage of information. This feature of the vertebrate nervous system has enormous ramifications for behavior in terms of retention of past experiences, which may influence future behavior. Finally, a further result of encephalization is the capacity to form associations between past, present, and possible future events. The ability to predict behavioral responses in particular situations and to deal with stimuli in the abstract reaches its fullest development in humans.

Another important feature of the vertebrate nervous system is the evolution of the **limbic system.** This is comprised of the septum, the cingulate gyrus, the hypothalamus, the amygdala, the hippocampus, and the fornix (see Figure 6–17).

FIGURE 6–17   Primate limbic system
A variety of separate structures comprise the limbic system (grey areas) in primates. The monkey brain shown here is representative of higher vertebrates.

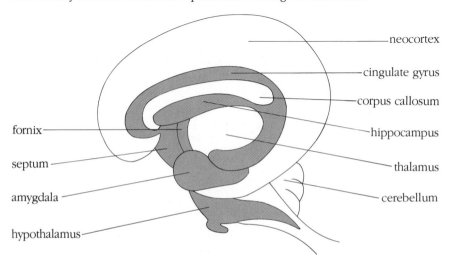

The limbic system is sometimes called the **rhinencephalon** because many of the system's functions depend on olfaction (smell). The limbic system mediates behavioral functions that pertain to need gratification, such as sex, feeding, and actions related to "emotions."

METHODS OF INVESTIGATING THE NERVOUS SYSTEM

Investigators have experimented with different techniques to explain the functioning of the neural mechanisms that control behavior. In each of the following subsections, we look at an experimental technique used to investigate the relationships between the nervous system and behavior and discuss the results of research that has used the technique. For some experimental sequences several different techniques have been employed to explore the relationships between the nervous system and behavior.

RECORDING NEURAL ACTIVITY

One direct method of discovering relationships between neural mechanisms and behavior is to implant electrodes in an organism's nervous system to record neural impulses directly. By recording impulses and simultaneously observing behavior, we can draw correlations between patterns of neural discharges and aspects of particular behavioral sequences.

Lettvin and his colleagues (1959) performed an intriguing series of studies utilizing the electrode-recording technique (see also Maturana et al. 1960). They placed frogs *(Rana pipiens)* in front of a gray hemisphere, which was 35 cm in diameter and positioned in front of the frog's eye. Each frog was prepared so that the investigators could make electrode recordings from ganglion cells associated with the frog's retina. Then they used magnets inside the hemisphere to move objects the frog could see and simultaneously recorded neural impulses passing through the ganglia. Lettvin and his associates were able to identify five major types of ganglion cells:

1. Moving edge detectors, which respond to any moving edge that passes across the frog's visual field;
2. Sustained contrast detectors, which fire repeatedly when a moving object passes into the visual field and stops there;
3. Net-dimming detectors, which respond when the general level of illumination, particularly at the center of the visual field, is decreased;

4. Convexity detectors, which fire when any small object passes across the
   visual field and when any dark body that has a convex leading edge passes
   through the field of view;
5. Darkness detectors, which fire more frequently as the level of illumination
   is decreased.

Each of these types of detectors can presumably be related to the habitat and
behavior of the frogs. Thus the convexity detectors have become known as "bug
detectors" because they are probably involved in detection and capture of prey.

In related studies on toads' recognition of prey, Ewert (1980) demonstrated
that wormlike stimuli, represented by rectangles moving past the toad in the direction of the long axis, produced the greatest number of turning and prey capture
movements by the toad. Attempts to relate these behavioral responses directly to
response patterns of ganglion cells in the retina have proven generally unsuccessful.
However, further tests revealed that the information carried from retinal cells is
conducted to at least five areas in the toad's brain. One of these areas, a subpopulation of cells called T5(2), is of particular interest. These cells exhibit selective firing
activity when wormlike stimuli are presented to the toad. That is, there is a good
correlation between the pattern of neuronal firing in this group of cells and the
behavioral response profile when the length, velocity and other attributes of the
stimulus presented to the toad are varied.

Are the T5(2) cells necessary for prey recognition? Are they alone sufficient
for prey recognition? Given our current techniques, only a partial answer to these
questions is possible. If the T5(2) cells are destroyed on one side of the brain, stimuli
moving on the opposite side of the toad's visual field do not evoke prey-capture
behavior (Ewert 1980). As in other vertebrates, much of the visual information from
each eye is transmitted to the opposite side of the brain's optic area. Data from frogs
indicate that ablations such as destroying the T5(2) cells on one side of the brain do
not result in blindness, nor do they result in any related motor incapacity (Ingle 1973,
1976; Comer and Grobstein 1981). But our ablation techniques are a bit too crude to
target only the T5(2) cells, and other cells in this region may also have been destroyed. Thus we can tentatively conclude only that some cells — possibly the T5(2)
cells — are necessary for prey recognition.

A possible test of the sufficiency of the T5(2) cells to recognize prey uses an
electrode to stimulate cells in this subpopulation with a small electric current. When
this is done in the absence of any visual stimuli, the toads respond as if they had
sighted a worm. Thus, some tectal neurons are at least sufficient to trigger the
orienting response. Again, we must caution that not just T5(2) cells may be involved
— the current may also have stimulated some neighboring cells. Only more refined
techniques and additional experiments will completely answer our questions.

A third series of experiments, conducted primarily with the mudpuppy *(Necturus maculosus)*, reveals more about retina response to moving objects (Dowling
1976; Werblin and Dowling 1969). For each ganglion cell there is a corresponding
region of the retina called a **receptive field.** When tiny spots of light are shown on
the retina, recordings show ganglion cells are either excited and fire, or they are

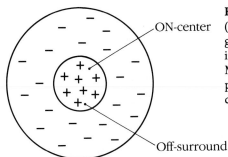

Receptive field
of one cell

(a)

FIGURE 6-18 Receptive fields of mudpuppies
(a) When light is shown on the retina the activity of ganglion cells is excited if the light strikes on-center and inhibited if it shines on the off-surround neurons. (b) Mapping a series of receptive fields results in an overlapping pattern, with fields of the same or different sizes, depending on the organism.

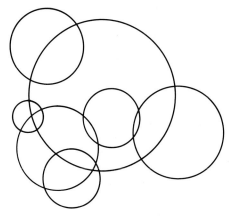

Series of receptive fields

(b)

inhibited. In general, vertebrate retinal ganglion cells have receptive fields comprised of a pair of concentric circles. When light shines on the inner circle, the cell is excited (Figure 6–18a). This is termed the **on-center.** If light is directed at regions of the outer circle as well, the cell's response is less than to an on-center region alone. Thus, the outer portion of the field exerts an inhibitory effect and is termed the **off surround.** All vertebrates thus far examined have, in addition, some retinal ganglion cells with the opposite properties; these are called **off-center on-surround neurons.** When the receptive fields of a series of neighboring ganglion cells are mapped, the resulting pattern is a series of overlapping circles (Figure 6–18b). Moving stimuli like the mock bugs or worms presented to frogs or toads would thus be "seen" by these animals in passing across the retina as a series of excitations and inhibitions.

TRANSECTION

When a blowfly *(Phormia regina)* is hungry, chemoreceptors on its legs are stimulated by potential food sources it encounters. In extends its proboscis, thereby stimulating its oral taste receptors, and it begins to feed. But when does a blowfly stop eating? When is it satiated?

In a fascinating series of experiments, Dethier and his associates (Dethier 1962, 1976; Dethier and Bodenstein 1958; Dethier and Gelperin 1967) used the technique of transection to analyze the feeding behavior of the blowfly (Figure 6–19). They **transected** (cut) certain nerves of the blowfly to observe the effects of transection on a particular behavior pattern — food intake — and thus to identify the neural events that are involved in the determination of satiety.

Dethier and his colleagues tested several unsuccessful hypotheses before they reached their ultimate conclusions. They first thought hunger and satiety might be directly related to blood sugar levels, so they injected hungry flies with sugar solutions, but the flies were not satiated with the sugar solutions. When a blowfly feeds, ingested material travels first to a blind sac called a crop, where it is stored before it goes through the process of digestion. To test whether hunger is related to the fullness of the crop, they performed microsurgery to tie off the crops of a number of flies. This procedure prevents food from accumulating. But the flies remained hungry after this operation and continued to feed. More determined than ever to solve the problem, Dethier and his co-workers hypothesized that hunger and satiety might be directly controlled by whether the hindgut, the major portion of the alimentary canal of the fly, was loaded with food. They tested this idea by delicately giving the flies "enemas," thus filling the hindgut. Again, the flies remained hungry.

Dethier and his colleagues then considered the only remaining portion of the digestive system — the foregut, where the intial step in digestion takes place — as the site of a satiety mechanism. By using very refined, miniaturized surgical techniques (involving a paraffin block to hold the fly, forceps, and a razor blade), they were able

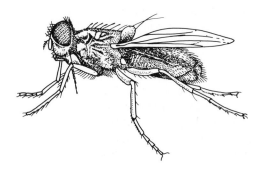

FIGURE 6–19   Blowfly feeding
When the blowfly is hungry, sensory mechanisms on its feet detect the presence of sugars, and the proboscis is extended to obtain nourishment.

to transect a small nerve that connects the foregut and the brain. They hoped to find out whether this nerve carries the message regarding the fullness of the foregut, which, in turn, halts the feeding behavior. When hungry blowflies with this transection were given an opportunity to feed, they too began to eat; but they did not stop eating. From these data and related information, Dethier and his colleagues concluded that as long as there is food in the blowfly's foregut, stretch receptors located there send messages to the brain that inhibit further eating behavior. Severing the nerve makes the fly permanently hungry. Additional research has shown that stretch receptors located in the body wall also send messages to the brain that act to inhibit further feeding when the fly is satiated.

Burghardt and Pruitt (Burghardt 1966, 1969; Burghardt and Pruitt 1975) have used transection to investigate the feeding responses of garter snakes (*Thamnophis sirtalis*). Garter snakes respond with both tongue flicks and attacks to pieces of cotton that contain extracts of prey items (e.g., earthworms or slugs). The question was whether garter snakes detect the prey via the olfactory system or via the vomeronasal organ, a separate set of chemosensitive receptors located in the nasal cavity of some vertebrates. To test it, Halpern and Frumin (1979) performed the following operations on the snakes: (1) transection of the bilateral olfactory nerves, (2) transection of the bilateral vomeronasal nerves, or (3) a sham operation which included all aspects of the transection surgery except actually cutting the nerve. The results showed that while 100 percent of the snakes in all three groups had responded with attacks prior to surgery, only 10 percent of snakes that had the vomeronasal nerve cut responded after surgery. Response levels remained at 100 percent for snakes in the other two test groups. The vomeronasal organ was thus shown to be important for this type of food discrimination in this species of garter snake.

NEURAL STIMULATION

Another method of investigating questions about neural control of behavior is by means of **electrode stimulation** — that is, placing electrodes directly into specific areas of the peripheral nervous system or the brain and passing an electrical current through the electrodes. If the stimulation evokes a recognizable sequence of behavior, which is normally part of that species' repertoire, then the electrode is probably located in or near a nerve pathway or brain area that controls or generates the observed behavior.

Willows (1967), for example, used this technique on the marine mollusc *Tritonia*. Instead of having a brain to act as the major decoding center, these organisms possess ganglia. Electric current passes through electrodes in these ganglia and within individual neurons in the ganglia (Figure 6–20) will evoke a particular behavior. By providing electrical stimulation to specific nerve cells, Willows elicited turning and swimming movements (Figure 6–20). Others (Brazier 1968) have used

FIGURE 6-20   Testing *Tritonia* through stimulation
Use of stimulation through electrodes to evoke distinct behavior patterns is one
method for mapping which nerves and decoding centers affect particular behavior
patterns. An apparatus (a) used to suspend the marine nudibranch *Tritonia* in sea
water permits placement of electrodes into specific ganglia or neurons. Turning and
swimming movements (b) can be produced when specific neurons are stimulated.

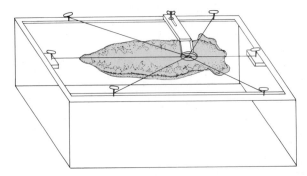

Tritonia suspended in apparatus for electrode stimulation

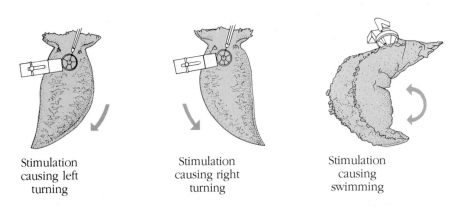

| Stimulation causing left turning | Stimulation causing right turning | Stimulation causing swimming |

this technique to map the neural pathways and parts of the brain responsible for
generating certain motor patterns in a range of species from insects to mammals.

Electrical stimulation of a rat's lateral hypothalamus through implanted electrodes elicits behaviors such as eating, drinking, aggression, and grooming (Panksepp 1971; Valenstein 1969; Valenstein, Cox, and Kakolewski 1970), even when
careful examination confirms that the electrodes are in virtually identical locations.
To further explore this issue, Bachus and Valenstein (1979) placed electrodes in the
hypothalamus of a group of rats *(Rattus norvegicus)*. They then preselected rats that
exhibited drinking behavior when electrical stimulation was provided. Bachus and
Valenstein used electrolytic lesions through the same electrode in order to destroy
tissue around the electrode's tip. As a further test, they gave some rats considerable

prelesion stimulation experience, and gave other rats minimal experiences. After the lesioning, they again gave electrical stimulation and found that the stimulation evoked drinking behavior in spite of the tissue damage. Bachus and Valenstein concluded from this study that individual differences in drinking behavior in response to electrical stimulation in the hypothalamus through implanted electrodes cannot be attributed solely to variation in the specific location of the electrodes in the hypothalamus.

TELEMETRY.   Stimulation by radio signals, called **telemetry,** is a variation on the electrically evoked response technique and has the advantage of permitting test animals to move about freely without restraint, even with the electrodes in place. One of the first investigators to utilize this technique, José Delgado, employed it to investigate aggressive behavior in rhesus monkeys *(Macaca mulatta)* (Delgado 1963, 1966). He determined the dominance hierarchy (rank ordering of animals from most dominant to most subordinate) by observing four rhesus monkeys housed in an enclosed room equipped with perching platforms. He then temporarily removed the top-ranking monkey and implanted electrodes in the areas of the brain known to elicit and inhibit aggression. He attached a radio-receiver pack to the monkey's back. When he sent a radio signal to the pack, it produced five seconds of electrical stimulation to the monkey's brain through a designated electrode. Stimulation of one area of the brain led to inhibition of aggression (see Figure 12–10). Stimulation of a different area of the brain, the postero-ventral nucleus, caused the dominant monkey's aggression to increase. (A lower-ranking monkey learned to press a lever, located at the side of the room, that activated the radiostimulator and inhibited aggression by the dominant male.) This technique has been used effectively in recent years to study the neural control of social aggression in primates (Herndon, Perachio, and McCoy 1979).

SELF-STIMULATION.   A second variation of the electrode stimulation technique permits the subject animal to provide **self-stimulation** by pressing a lever or some other device. Rats will learn to push a lever to receive electrical impulse stimulation to electrodes implanted in specific areas of the brain (Olds 1956). When electrodes are placed in the amygdala or septum, the rats press the lever frequently, which suggests that the stimulation is rewarding or pleasurable. When electrodes are implanted in other locations, the rats press the lever very infrequently; the stimulus effect is probably less pleasurable or even punishing.

Investigators have used the self-stimulation technique to study aspects of the neural bases of behavior—for example, responses to fear and tension by male rhesus monkeys. Maxim (1977) prepared monkeys by implanting electrodes at self-stimulation sites in the hypothalamus. When he presented the monkeys with a snake, a fear stimulus, they pressed the bar for self-stimulation more frequently. When he paired an implanted male with a nonimplanted male, the implanted monkey pressed the bar more frequently in response to strong dominance behaviors by the nonimplanted male. More bar presses were also recorded when either the implanted or nonimplanted male was about to engage in intense play wrestling. Bar pressing was not related to other activities, such as eating, drinking, or playing with objects. Self-stimulation appears, in this instance, to have a fear-reducing effect; the rate of bar pressing was related to the potential threat or amount of dominance behavior exhibited by another monkey.

Research on brain neurochemistry has also uncovered a class of compounds called **endorphins** (Watkins and Mayer 1982). These chemicals appear to be the body's own opiates, which exert a pain-relieving effect. Endorphins have been related to several behavioral and physiological phenomena, including hibernation and water balance, and they may also function as neurotransmitters. Acupuncture treatment apparently induces the release of excess endorphins in the brain which then have an anesthetic-like effect on the patient. The self-stimulation effects may be interpreted as related to the release of endorphins.

LESIONS AND SPLIT-BRAIN TECHNIQUES

Using electrical currents, chemicals or microknives to destroy specific brain areas, or to cause **lesions,** is more drastic than the other techniques discussed thus far. If lesion treatment results in behavioral modification or deficit, we can draw tentative conclusions regarding the capacities of the remaining brain tissue. Lesions in different regions of the brains of rhesus monkeys produce different social contact behaviors (Girgis 1971). If tissue is destroyed in the medial portion of the amygdaloid nucleus, monkeys avoid social contact through flight or withdrawal. Lesions in the lateral regions of the amygdaloid nucleus make the monkeys placid and nonaggressive. Arbib (1972) has reviewed considerable literature on the use of lesions to explain the functioning of neural pathways and of specific regions of the brain in an attempt to produce a model of how brains operate as decoding and integrative centers.

The lesion technique was used early by Lashley (1929) to study the effects of brain damage on behavior, primarily in rats. He noted that cortical lesions produced deficiencies in tasks such as learning visual discriminations. Rats with varying degrees of ablation of the cortex exhibited functional losses, affecting their ability to learn new tasks, or even losing previously learned discriminations. However, the passage of time often led to considerable recovery — animals again could learn and retain the visual discriminations of cube sizes or black and white patterns. These observations led Lashley to his hypotheses of **mass action** and **equipotentiality.** Simply stated, this means that different portions of the cortex have equivalent capacities, and one region can substitute for another in the performance of various sensory, motor, and association tasks.

Beginning after World War II and extending into the 1970s work by Roger Sperry (1961, 1974) challenged the Lashley hypothesis. Studies using rats, cats, and humans have given us a different perspective on how the brain functions. In particular, these investigations involved human patients and related animal models in which some or all of the connections between the two major cortical hemispheres were severed. A variety of functions are separated by such surgery. The left cerebral hemisphere, to which the right half of the visual field projects, is primarily involved in language (speech, writing) and in calculations (Figure 6–21). Functions relating to

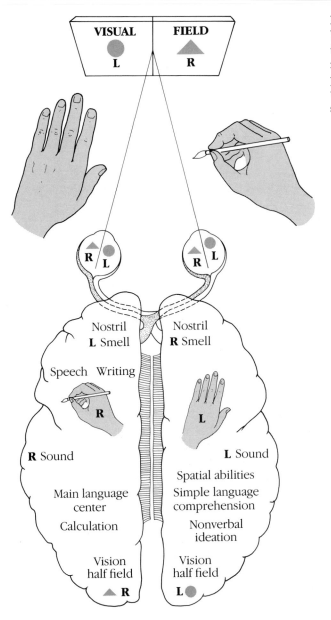

FIGURE 6-21   Split-brain patients
When the connections between the two cerebral hemispheres are severed, neuroanatomical observations and postoperative testing demonstrate that various functions are localized in one side of the brain or the other.

spatial abilities and a variety of nonverbal capacities are localized in the right cerebral hemisphere, to which the left half of the visual field projects. Following surgery which severs the intercerebral connections — those that connect the two brain hemispheres with one another — normal transfer between hemispheres does not occur. Objects touched by one hand cannot be located by the other hand, and subjects cannot recognize an object they have just been shown if that object is then presented to the other half of the visual field. Sperry was a co-recipient of the 1981 Nobel Prize in Physiology of Medicine for his pioneering work in neurobiology.

The foregoing studies and much additional information support the notion that there are indeed areas with specialized functions within the brain cortex. Hebb (1949) proposed that groups of cell assemblies, comprised of up to thousands of cells, may be the sites of localized function where information is stored and retrieved. It is possible there may be some degree of equipotentiality within the cells of such assemblies. In Chapter 10 we examine further the problem of how memory traces are made and stored in the brain.

An important study of lesions or ablations of neural tissue was reported by Anand and Brobeck (1951) in their experiments on the effects of hypothalamic lesions on food intake in rats. When they made a lesion in the ventromedial hypothalamus of the rat, the rat continued to eat virtually nonstop, a condition called **hyperphagia,** and became obese. In contrast, when they made a lesion in the lateral hypothalamus, the rat reduced or ceased its eating activity, a condition called **hypophagia,** or **aphagia.** Thus the medial hypothalamus contains a "feeding" center, and the lateral hypothalamus has a satiety center. Investigations that utilized electrical stimulation confirmed these findings. Stimulation of the ventromedial region of the hypothalamus stops feeding behavior; stimulation of the lateral region of the hypothalamus initiates feeding behavior.

Powley and Keesey (1970) provided additional data and an alternative interpretation of the effects of hypothalamic lesions on feeding behavior in rats. They concluded that a lesion in the lateral hypothalamus lowers the point around which weight balance is maintained, or the set point for weight control. Aphagia after surgery results from the rat's attempt to reduce its body weight to the new set point, after which normal regulation follows. The aphagia is not the result of the surgical recovery period, as previous investigators claimed. Powley and Keesey caused presurgical weight reduction in male rats by starving them for a brief period before lesioning. Rats that underwent prelesion starvation exhibited only brief hyperphagia after surgery and then regulated normally around the new set point. Weight-control mechanisms appear to be active in these rats immediately after surgery; the rats are not simply undergoing weight loss as a result of surgical recovery or destruction of a control center.

As we shall find out in more detail in Chapter 7, hormones trigger the neural activity that underlies maternal behavior in recently parturient females (Terkel and Rosenblatt 1972; Zarrow, Gandelman, and Denenberg 1971), but maintenance of the maternal behavior also has a nonhormonal basis (Numan, Leon, and Moltz 1972; Rosenblatt 1970). Do specific localized areas of the brain play a particular role

TABLE 6-1   Mean values for three measures of maternal behavior in MPOA-lesioned and sham-operated female rats

	MPOA-Lesioned Female Rats		Sham-Operated Female Rats	
	Preoperative	*Postoperative*	*Preoperative*	*Postoperative*
Nursing (sec) (30-min test)	1,551	4	1,578	1,323
Retrieving (percent)	100	0	100	89
Nest building (percent)	96	0	94	76

Source: Data from Numan (1974).

in maintaining maternal behavior? When Numan (1974) made lesions in the medial preoptic area (MPOA), data from observations of maternal behaviors in both MPOA-lesioned and sham-operated rats (Table 6-1) indicated that the maternal behavior of the lesioned rats was drastically impaired. The MPOA and/or the connections to the anterior hypothalamus that were destroyed by the lesion are, therefore, important for normal maternal behavior.

Experiments on rats, dogs, and other animals have shown that MPOA lesions in males result in a cessation of sexual behaviors (Hart 1974; Giantonio, Lund, and Gerall 1970). No investigators have reported a return of sexual activity in male animals with MPOA lesions. However, Twiggs, Popolow, and Gerall (1978) suggest at least one additional consideration critical for correct interpretation of such lesion studies. Male rats that were given MPOA lesions before puberty and that had social interactions with peers did exhibit normal adult male sexual behavior patterns. Thus it appears that enhanced social experience can offset the effects of MPOA lesions. In addition, the age at which such lesions are made could be a critical factor.

FUNCTIONAL NEUROANATOMY

The study of **functional neuroanatomy** — the size, structure, and arrangement of cells within the nervous system and particularly the brain — has resulted in several important findings concerning the brain's role in behavior.

Working primarily with cats and monkeys, Hubel and Wiesel (1961, 1977, 1979) explored the mechanisms by which light stimuli are processed by the brain — specifically, how visual information is encoded in the cortex. They investigated how the receptive fields of retinal ganglion cells are translated into projections in the cortex. Hubel and Wiesel discovered several levels of projection of increasing complexity within the cortex; these appear to correspond to a layered arrangement of the retinal cells. There is also a correspondence between the receptive fields of the retinal ganglion cells and the responses of particular cells in the visual projection areas of the brain. In some way, at some levels in the visual cortex, responses are transformed so that cells respond to oriented line segments rather than to spots of light.

Hubel and Wiesel shared the 1981 Nobel Prize with Sperry for their fine work on functional architecture of the visual system. Significant progress is currently being made toward untangling the relationships between incoming stimuli that influence behavior and the very organized neuroanatomy of the brain.

Another example of the importance of understanding the connections between brain organization and behavior concern the songs of canaries *(Serinus canarius)* (Nottebohm 1975, 1980, 1981; DeVoogd and Nottebohm 1981; Paton and Nottebohm, 1984; Nottebohm and Nottebohm 1976). Canaries produce sounds in the syrinx during expiration as air is forced out of the lungs and air sacs. Male birds sing; females generally do not. When a young canary matures at one year of age, it learns a song repertoire. Each succeeding year, the bird learns a new repertoire. The muscles of the syrinx that control the frequency and amplitude of sound produced are innervated by the hypoglossus nerve. Canary songs are comprised of phrases (Figure 6–22) made up of syllables, each containing one or more elements. Nottebohm has shown that the left hypoglossal nerve and left cerebral hemisphere exert dominant control over the song, though shifts in hemispheric dominance are possible. Thus, much of their ensuing research has involved the dominant side of the brain.

Nottebohm and colleagues have uncovered some fascinating aspects of the canary song control system.

FIGURE 6–22   Fragment of a song from an adult male canary
A canary song has three different phrases. Each syllable in the first and second phrase is composed of two elements, whereas syllables in the third phrase consist of a single element. The vertical axis represents sound frequency (in kHz), and the horizontal axis is in seconds.
Source: From F. Nottebohm and M. Nottebohm (1976). Left hypoglossal dominance in the control of canary and white-crowned sparrow song. *J. Comp. Physiol.* 108:171–192.

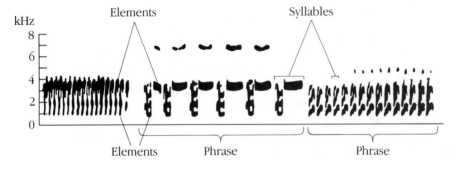

1. Within a cortical nucleus implicated in song control, the neurons increase dendritic growth when the birds are treated with testosterone.
2. The same nucleus is 76 percent larger in male birds in the spring—when they sing—than in the fall, when the birds are not singing.
3. Female birds treated with testosterone show similar anatomical changes and start to sing.
4. Experiments with adult canaries using radioactively labeled thymidine (a marker for DNA synthesis) indicate that new neurons are being formed in one of the several nuclei involved in song control. That these new neurons are parts of functional circuits was confirmed by using electrodes to detect impulses when auditory stimuli were presented.

These findings indicate that neuroanatomical changes occur on a seasonal basis that corresponds to the annual behavioral cycle of the birds. As a new song repertoire is learned each spring, new connections, possibly involving newly generated neurons, are formed in brain areas associated with song control. These data also demonstrate a plasticity in brain structure with respect to the addition of new cells and connections which could be important in facilitating our understanding of other brain-behavior interrelationships.

PSYCHOPHARMACOLOGY AND NEUROTRANSMITTERS

In recent years the identification of various neurotransmitters (e.g., acetylcholine, serotonin) from the brain and the development of synthetic drug compounds chemically related to these neurotransmitters have permitted investigators to explore relationships between the nervous system and behavior. This new area of brain research, called **psychopharmacology,** incorporates chemistry, physiology, and psychology.

Lorden and Oltmans (1978) studied learned taste aversion in rats. Rats that become ill after they consume a particular food or liquid eat less of that food or avoid it altogether on subsequent presentations. Lorden and Oltmans conditioned rats to avoid saccharin water by giving them injections of lithium chloride to make them ill. Rats with brain lesions in a particular nucleus still learned the aversion. However, lesioned rats that were given an injection of a chemical precursor of serotonin (a brain neurotransmitter substance) prior to conditioning failed to learn the taste aversion to saccharin water. The chemical treatment may have acted within the brain to decrease sensitivity to the lithium chloride injections.

The copulatory behavior of male rats involves a series of mounts with intromissions leading to ejaculation. This is followed by a refractory period during which the males are not affected by external stimuli (e.g., presentation of a receptive female). In a series of three experiments McIntosh and Barfield (1984a,b,c) investigated the role of neurotransmitters in affecting the duration of the post-ejaculatory refractory period in mole rats. The first test concerned serotonin. Rats were given a pre-treatment sexual behavior test, followed by injections of a chemical that blocks serotonin synthesis in the brain. When post-treatment sexual behavior was compared to pre-treatment behavior, there was a significant reduction in the length of the refractory period — 458 seconds on average before treatment, and 272 seconds after treatment.

In a similar sequence McIntosh and Barfield tested for the possible importance of dopamine in affecting the length of the refractory period. In this instance, they used a chemical that blocks the receptor sites for dopamine on the post-synaptic membrane. The effect was a significant lengthening of the post-ejaculatory refractory period — from an average of 326 seconds to 546 seconds. A third test examined the role of norepinephrine, another neurotransmitter, on the refractory period. Intraperitoneal administration of a chemical that inhibits brain norepinephrine synthesis resulted in significant lengthening of the refractory period — from an average of 343 seconds to 526 seconds.

From these investigations McIntosh and Barfield derived several conclusions. Normally a serotongergic system inhibits the resumption of mating after ejaculation. Dopamine and norepinephrine pathways are involved in the arousal of copulatory behavior and in the maintenance of the normal post-ejaculatory refractory period.

Other investigators have explored the role(s) of various drugs that affect the normal processes of synaptic transmission within the central nervous system. For example, chlorpromazine, a tranquilizer, acts to block the post-synaptic membrane receptors for the neurotransmitters acetylcholine and epinephrine. Nicotine from tobacco mimics the effects of acetylcholine and facilitates synaptic transmission. Barbituates such as secobarbitol and pentobarbitol, which are sedatives, act by interfering with the synthesis and release of serotonin and norepinephrine.

CANNULATION

Hollow electrodes or fine tubes (cannulas) can be inserted into specific areas of the brain to introduce substances. Davidson (1966) used this **cannulation** technique to test a hypothesis regarding androgen and male rat sexual behavior: Does sexual behavior result from direct action of hormones on specific brain regions? Davidson castrated sexually experienced male rats, then gave the treatment group implants of crystalline testosterone propionate (TP, a synthetic androgen) in various areas of the brain and gave the control group implants of cholesterol in several areas of the brain. Implants of TP in the hypothalamus and median preoptic areas brought about renewed sexual activity; implants of TP in other regions of the brain had lesser effects or no effect on sexual behavior; and implants of cholesterol had no effect at all on sexual behavior. Davidson concluded from these data that sexual behavior is dependent on testosterone activation in the hypothalamic–medial preoptic area of the brain.

TRANSPLANTATION

One final technique which should be mentioned is the **transplantation** of neural tissue from one organism to another, or occasionally from one location to another within the same organism. Transplantation of preoptic tissue from neonatal male rats into the preoptic area of female littermates resulted in several changes in the behavior of the recipients (Arendash and Gorski 1982). The females with transplanted tissue exhibited increased levels of both masculine and feminine sexual behavior as adults. The ongoing development of the nervous system of the female rat is affected by the presence of male brain tissue. The transplantation technique has also been used in recent years by investigators exploring the control of biological clocks. An extended example of the use of the technique involving activity rhythms in cockroaches is in Chapter 8.

POTENTIAL PROBLEMS

Before leaving the topic of neurobiology and behavior we should consider some of the potential problems presented by the techniques just described. A problem common to most of these techniques — how to determine exactly which nerves were cut or used for recording impulses or which cells were stimulated or lesioned — has two partial solutions. First, we can use X-rays to ascertain the locations of electrodes in a neuron or in the brain. However, to ascertain the exact position of an electrode is sometimes difficult. A second, more common, solution is to use histological techniques to map the positions of electrodes or lesions. We can make thin sections of brain or other neural tissue after we complete an experiment; we can then stain the section, examine mounted slides under a microscope, and pinpoint the areas from which recordings were made or the regions that were lesioned. Although the electrode is not left in place for this procedure, electrode paths through neural tissue generally remain visible after the electrodes have been withdrawn.

   Two other problems are associated with lesion studies and, in some instances, with electrode implants for recording or stimulation. It is tricky to perform a lesion on an animal in such a way that only the desired neural tissue is destroyed. Examination of slides made with histological techniques often reveals that either less or more tissue has been destroyed than was intended. During electrode implanting, the passage of the electrode through neural tissue may destroy some cells or otherwise impair their capacity to function normally. The possible side effects of this tissue destruction may not be accounted for in the design of an experiment, in the hypothesis, or in the explanation of results.

Lesioning an area to destroy tissue or unintentionally destroying nerve cells when implanting an electrode can lead to yet another problem. Once neural pathways have been interrupted by tissue damage, degeneration — the death or functional impairment of those cells that are connected to neurons that have already been destroyed or severely damaged — often begins. Degeneration of additional tissue, beyond that accounted for in the initial lesion or nerve transection, may have effects on behavior that are not predicted by the experimental hypothesis. It may be difficult to completely and accurately assess the influence of these side effects on behavior.

Many investigations of nervous systems use animal subjects that are anesthetized or restrained in some manner. It is impossible to be sure that the responses and neural discharge patterns of a partially or completely anesthetized or restrained animal are identical to those of a normal, awake, free-moving animal. Some techniques, like radiostimulation, have provided at least a partial solution to this problem.

Finally, if a particular area of the brain is lesioned because we hypothesize that this area is involved in certain specific behaviors, we must also ask whether lesions in other areas of the brain might produce a similar behavioral result — that is, are the lesion effects general or specific? To test this possibility, we may need to perform control lesions in additional regions of the brain not thought to be associated with the behavior being investigated.

Because of the problems associated with the techniques used to study the neural regulation of behavior, investigators must proceed with caution and must consider alternative explanations, when they draw conclusions from data gathered in an experiment, that take into account the possible side effects of treatments. As animal behaviorists, we must remain constantly aware of the limitations and consequences of the methods and techniques we employ in our research investigations.

SUMMARY

Neurons, each consisting of a cell body, axon, and dendrites, are the basic structural units of nervous systems and conduct electrical impulses that result from changes in membrane permeability and active transport of Na^+ and K^+ ions. Transmission of impulses across synaptic junctions between neurons occurs through a chemical process involving neurotransmitters.

Organisms possess mechanisms for transducing energy from different environmental stimuli into neural impulses — a process defined as sensation. The interpretation (perception) of these sensory neural impulses takes place in aggregations of cells called ganglia or in the central nervous system.

The sensory and perceptual worlds of animals differ from our own in three general ways: Some organisms (e.g., electric fish) have totally different modes of sensation. Some organisms (e.g., silkworm moths) may share the same mode as humans, but the range over which they can transduce stimuli from the environment may differ. Finally, some organisms (e.g., octopodes) may transmit the same sensory information to the central nervous system, but they may differ in their perceptual interpretation of this information. Organisms have sensory filtering mechanisms that help to select only pertinent stimuli to receive, process, and respond to.

Several key developments characterize the evolution of the nervous system from a phenomenon called irritability in unicellular organisms through nerve nets, the radially symmetrical nervous system, the bilaterally symmetrical nervous system, to the generalized vertebrate nervous system. These developments include changes in the numbers and types of sensory receptors; formation of specific neural pathways through aggregations of individual neurons into fiber tracts; and encephalization, the concentration of neural material near the anterior end of the body. These developments result in changes in the flexibility and complexity of behavior.

Investigators can use several techniques to study the functional processes within different portions of the nervous system. They can make direct recordings of neural impulses through implanted electrodes. They can transect nerves and observe the resulting effects on behavior. They can use electrode stimulation, or the sending of an electrical current to an electrode implanted in the nervous system, to trace neural pathways associated with specific behaviors. Two variations of this latter method are telestimulation, in which a radio receiver is attached to the subject's head, and self-stimulation, in which an animal can press a lever or other device to receive stimulation.

Investigators can also use chemicals or electrical currents to destroy specific areas of the brain and thus to assess the role of these areas in affecting behavior patterns. They can inject neurotransmitter substances or drugs which affect the neurotransmitters. Finally, researchers can introduce chemicals via a cannula into specific areas of the brain to assess effects of chemical substances on behavior.

These techniques have permitted us to begin to formulate some theories of how the brain functions in relation to stimulation of the animal and the resulting, observed behaviors. We know that sensory systems like the eye receive and encode visual stimuli in highly specific ways — for example, ganglion cells in the retina are characterized by an on-center off-surround receptive field arrangement. Information is carried to the brain along discrete pathways. The structure and arrangement of cells within the brain is highly ordered, with stimuli from each modality and region of the body projecting to specific cortical regions. Some functions appear to be predominant in one hemisphere of the vertebrate brain, while other functions are represented primarily in the other hemisphere.

Further studies relating behavior and brain structure have produced information on the functional flexibility of the brain of canaries. Seasonal changes in behaviors such as singing appear to be related to structural changes within specific areas of the brain. We are rapidly expanding our knowledge of the ways specific neurotransmitters affect behaviors and of the close relationships between hormones and the neural control of behavior.

Throughout our investigation of the neural control of behavior we must constantly bear in mind that our conclusions should reflect an understanding of the limitations and problems associated with the research techniques employed.

Discussion Questions

1. Peripheral and central sensory filters are the mechanisms by which an organism selects among incoming stimuli. You are attending the evening program of thoroughbred horse racing at a local track and have $200 to spend on bets, food, drink, and so on. Describe the peripheral and central filters that are operative on you in this situation.

2. The sensory and perceptual systems of various organisms may differ in at least three ways: (a) organisms may have different sensory equipment and may be capable of receiving different types of stimuli; (b) organisms may have the same sensory equipment but different thresholds, so that they receive a particular stimulus over different ranges; or (c) organisms may interpret the same stimulus in different ways. Can you think of specific examples of these differences in the various sensory modalities?

3. As we noted in the text, a number of changes—including changes in numbers and types of sensory receptors, the aggregation of neurons into fiber tracts, and encephalization—characterize the evolution of nervous systems. Changes in behavioral capacities are associated with these evolutionary changes. In terms of flexibility and adaptation, what advantages and disadvantages do the following animals have: unicellular organisms, worms, insects, amphibians, mammals.

4. Permanent electrode implants in regions of the brain have been used to elicit particular behavior patterns in animals. Passing an electrical current through an electrode implanted in a particular region of a fish brain produces attack behavior. Below are the results of a hypothetical study of electrical stimulation and attack behavior in fish. Explain why each treatment group is necessary to the overall design. The results are shown as the number of attack responses elicited from ten fish of each treatment type, each of which was given ten electrical impulses in the presence of another conspecific male.

Electrode Location	*No. of Attack Responses (max. = 100)*
No electrodes (normal fish)	77
Electrodes in three brain regions, but no electrical stimulation	73
Electrodes in brain region A stimulated	11
Electrodes in brain region B stimulated	81
Electrodes in brain region C stimulated	96

Should any additional treatment groups be included? Discuss the possible interpretations of these data.

Suggested Readings

Atrens, D., and I. Curthoys. 1982. *The Neurosciences and Behavior.* Sydney: Academic Press Australia.
An introductory text—short, but packed with good information. Somewhat elementary, but well-written and with excellent drawings.

Bentley, D., and M. Konishi. 1978. Neural control of behavior. *Ann. Rev. Neurosci.* 1:35–60.
Current review of the field, with emphasis on insects and birds. Both authors have done extensive research at the interface between studies of the nervous system and investigations of functional aspects of observed behavior patterns.

Camhi, J. 1984. *Neuroethology.* Sunderland, Mass.: Sinauer.
An up-to-date treatment of this subject with a solid emphasis on the behavioral aspects of the relationship between events in animal nervous systems and the observed actions. Excellent for the novice as well as for the advanced student.

Dethier, V. G. 1962. *To Know a Fly.* Englewood Cliffs, N.J.: Prentice-Hall.
Delightful and humorous introduction to theory and method in animal behavior through Dethier's work on feeding in blowflies. "Must" reading for all. Dethier has been a major contributor to modern comparative psychology.

Northcutt, R. G. 1981. Evolution of the telencephalon in nonmammals. *Ann. Rev. Neurosci.* 4:301–350.
Review article covers evolution of the forebrain in fish, reptiles, amphibians, and birds. Draws connections between evolutionary changes in overall brain structure/function and correlative changes in behavior. Heavy on terminology, but packed with information.

References

Anand, B. K., and J. R. Brobeck, 1951. Hypothalamic control of food intake in rats and cats. *Yale J. Biol. Med.* 24:123–140.

Arendash, G. W., and R. A. Gorski. 1982. Enhancement of sexual behavior in female rats by neonatal transplantation of brain tissue from males. *Science* 217:1276–1278.

Arbib, M. A. 1972. *The Metaphorical Brain.* New York: Wiley-Interscience.

Bachus, S. E., and E. S. Valenstein. 1979. Individual behavioral responses to hypothalamic stimulation persist despite destruction of tissue surrounding electrode tip. *Physiol. Behav.* 23:421–426.

Bentley, D. 1969. Intracellular activity in cricket neurosis during the generation of song patterns. *Zeit. Vergleich. Physiol.* 62:267–283.

Bentley, D., and R. R. Hoy. 1970. Postembryonic

development of adult motor patterns in crickets: A neural analysis. *Science* 170:1409–1411.

Beroza, M., and E. F. Knipling. 1972. Gypsy moth control with the sex attractant pheromone. *Science* 177:19–27.

Brazier, M. A. B. 1968. *The Electrical Activity of the Nervous System.* Baltimore: Williams & Wilkins.

Burghardt, G. M. 1966. Stimulus control of the prey attack response in naive garter snakes. *Psychnom. Sci.* 4:37–38.

———. 1969. Comparative prey attack studies in newborn snakes of the genus *Thamnophis*. *Behaviour* 33:77–144.

Burghardt, G. M., and C. H. Pruitt. 1975. Role of the tongue and senses in feeding of naive and experienced garter snakes. *Physiol. Behav.* 14:185–194.

Burtt, E. T. 1974. *The Senses of Animals.* London: Wykeham.

Comer, C., and P. Grobstein. 1981. Tactually elicited prey acquisition behavior in the frog, *Rana pipiens*, and a comparison with visually elicited behavior. *J. Comp. Physiol.* 142:141–150.

Davidson, J. M. 1966. Activation of the male rat's sexual behavior by intracerebral implantation of androgen. *Endocrinology* 79:783–794.

Delgado, J. M. R. 1963. Cerebral heterostimulation in a monkey colony. *Science* 141:161–163.

———. 1966. Aggressive behavior evoked by radio-stimulation in monkey colonies. *Amer. Zool.* 6:669–681.

Dethier, V. 1962. *To Know a Fly.* Englewood Cliffs, N.J.: Prentice-Hall.

Dethier, V. G. 1976. *The Hungry Fly.* Cambridge: Harvard University Press.

Dethier, V. G., and D. Bodenstein. 1958. Hunger in the blowfly. *Zeit. Tierpsychol.* 15:129–140.

Dethier, V. G., and A. Gelperin. 1967. Hyperphagia in the blowfly. *J. Exp. Biol.* 47:191–200.

Dethier, V. G., and E. Stellar. 1961. *Animal Behavior: Its Evolutionary and Neurological Basis.* Englewood Cliffs, N.J.: Prentice-Hall.

DeVoogd, T., and F. Nottebohm. 1981. Gonadal hormones influence dendritic growth in the adult avian brain. *Science* 214:202–204.

Dowling, J. E. 1976. Physiology and morphology of the retina. In R. Llinas and L. Precht (eds.), *Frog Neurobiology.* Berlin: Springer-Verlag.

Ewert, J. P. 1980. *Neuroethology.* Berlin: Springer-Verlag.

Giantonio, G. W., N. L. Lund, and A. A. Gerall. 1970. Effects of diencephalic and rhinencephalic lesions on the male rat's sexual behavior. *J. Comp. Physiol. Psychol.* 73:38–46.

Girgis, M. 1971. The role of the thalamus in the regulation of aggressive behavior. *Int. J. Neurol.* 8:327–351.

Gulick, W. L., and H. Zwick. 1966. Auditory sensitivity of the turtle. *Psychol. Rec.* 16:47–53.

Halpern, M., and N. Frumin. 1979. Roles of the vomeronasal and olfactory systems in prey attack and feeding in adult garter snakes. *Physiol. Behav.* 22:1183–1189.

Hart, B. L. 1974. Medical pre-optic-anterior hypothalamic area and sexual behavior of male dogs. *J. Comp. Physiol. Psychol.* 86:328–349.

Hebb, D. O. 1949. *The Organization of Behavior.* New York: Wiley.

Heiligenberg, W. 1966. The stimulation of territorial singing in house crickets (*Acheta domesticus*). *Zeit. Vergleich. Physiol.* 49:459–464.

Herndon, J. G., A. A. Perachio, and M. McCoy. 1979. Orthogonal relationship between electrically elicited social aggression and self-stimulation from the same brain sites. *Brain Research* 171:374–380.

Hodgkin, A. L. 1971. *Conduction of the Nervous Impulse.* Springfield, Ill.: Charles C Thomas.

Hopkins, C. D. 1974. Electric communication in fish. *Amer. Scientist* 62:426–437.

Hubel, D. H., and T. N. Wiesel. 1962. Receptive fields, binocular interaction and functional architecture in the cat's visual cortex. *J. Physiol.* 160:106–154.

———. 1977. Functional architecture of macaque monkey visual cortex. *Proc. Royal Soc. London* Series B 198:1–59.

———. 1979. Brain mechanisms of vision. *Sci. Amer.* 241(Sept.):150–162.

Ingle, D. 1973. Two visual systems in the frog. *Science* 181:1053–1055.

———. 1976. Spatial vision in anurans. In K. V. Fite (ed.), *The Amphibian Visual System: A Multidisciplinary Approach.* New York: Academic Press.

Lashley, K. S. 1929. *Brain Mechanisms and Intelligence.* Chicago: University of Chicago Press.

Lettvin, J. Y., et al. 1959. What the frog's eye tells the frog's brain. *Proc. Inst. Radio Engineers* 47:1940–1951.

Lorden, J. F., and G. A. Oltmans. 1978. Alteration of the characteristics of learned taste aversion by manipulation of serotonin levels in the rat. *Pharmacol. Biochem. Behav.* 8:13–18.

Machin, K. E., and H. W. Lissman. 1960. The

mode of operation of the electric receptors in *Gymnarchus niloticus. J. Exp. Biol.* 37:801–811.

Maturana, H. R., et al. 1960. Anatomy and physiology of vision in the frog *(Rana pipiens). J. Gen. Physiol. Suppl.* 43:129–175.

Maxim, P. E. 1977. Self-stimulation of a hypothalamic site in response to tension or fear. *Physiol. Behav.* 18:197–201.

McIntosh, T. K., and R. J. Barfield. 1984a. Brain monoaminergic control of male reproductive behavior. I. Serotonin and the post-ejaculatory refractory period. *Behav. Brain Res.* 12:255–265.

———. 1984b. Brain monoaminergic control of male reproductive behavior. II. Dopamine and the post-ejaculatory refractory period. *Behav. Brain Res.* 12:267–273.

———. 1984c. Brain monoaminergic control of male reproductive behavior. III. Norepinephrine and the post-ejaculatory refractory period. *Behav. Brain Res.* 12:275–281.

Nottebohm, F. 1975. Vocal behavior in birds. In J. R. King and D. S. Farner (eds.), *Avian Biology,* vol. 5. New York: Academic Press.

———. 1980. Brain pathways for vocal learning in birds: a review of the first 10 years. In J. M. Sprague and A. N. Epstein (eds.), *Progress in Psychobiology and Physiological Psychology* 9:85–124.

———. 1981. A brain for all seasons: Cyclical anatomical changes in song control nuclei of the canary brain. *Science* 214:1368–1370.

Nottebohm, F., and M. Nottebohm. 1976. Left hypoglossal dominance in the control of canary and white-crowned sparrow song. *J. Comp. Physiol.* 108:171–192.

Numan, M. 1974. Medial preoptic area and maternal behavior in the female rat. *J. Comp. Physiol. Psychol.* 87:746–759.

Numan, M., M. Leon, and H. Moltz. 1972. Interference with prolactin release and maternal behavior of female rats. *Horm. Behav.* 3:29–38.

Olds, J. 1956. A preliminary mapping of electrical reinforcing effects in the rat brain. *J. Comp. Physiol. Psychol.* 49:281–285.

Panksepp, J. 1971. Aggression elicited by electrical stimulation of the hypothalamus in albino rats. *Physiol. Behav.* 6:321–329.

Paton, J. A., and F. Nottebohm. 1984. Neurons generated in the adult brain are recruited into functional circuits. *Science* 225:1046–1048.

Powley, T. L., and R. E. Keesey. 1970. Relationship of body weight to the lateral hypothalamic feeding syndrome. *J. Comp. Physiol. Psychol.* 70:25–36.

Roeder, K. D. 1970. Episodes in insect brains. *Amer. Scientist* 58:378–389.

Roeder, K. D., and A. E. Treat. 1961. The detection and evasion of bats by moths. *Amer. Scientist* 49:135–148.

Rosenblatt, J. S. 1970. Views on the onset and maintenance of maternal behavior in the rat. In L. R. Aronson et al., eds., *Development and Evolution of Behavior.* San Francisco: Freeman.

Sarnat, H. B., and M. G. Netsky. 1974. *Evolution of the Nervous System.* New York: Oxford University Press.

Smith, J. E. 1950. Some observations on the nervous mechanisms underlying the behavior of starfish. *Symp. Soc. Exp. Biol. Med.* 4:196–220.

Sperry, R. W. 1961. Cerebral organization and behavior. *Science* 133:1749–1757.

Sperry, R. W. 1974. Lateral specialisation in the surgically separated hemispheres. In F. O. Schmitt and F. G. Worden (eds.), *The Neurosciences: Third Study Program.* Cambridge: MIT Press, pp. 5–20.

Terkel, J., and J. S. Rosenblatt. 1972. Humoral factors underlying maternal behavior at parturition: Cross transfusion between freely moving rats. *J. Comp. Physiol. Psychol.* 80:365–371.

Tinbergen, N. 1953. *The Herring Gull's World.* London: Collins.

Twiggs, D. G., H. B. Popolow, and A. A. Gerall. 1978. Medial preoptic lesions and male sexual behavior: Age and environmental interactions. *Science* 200:1414–1415.

Valenstein, E. S. 1969. Behavior elicited by hypothalamic stimulation: A preoptency hypothesis. *Brain Behav. Evol.* 2:295–316.

Valenstein, E. S., V. C. Cox, and J. Kakolewski. 1970. A reexamination of the role of the hypothalamus in motivation. *Psychol. Rev.* 77:16–31.

Walcott, C., J. L. Gould, and J. L. Kirschvink. 1979. Pigeons have magnets. *Science* 205:1027–1029.

Watkins, L. R., and D. J. Mayer. 1982. Organization of endogenous opiate and nonopiate pain control systems. *Science* 216:1185–1192.

Wells, M. J., and J. Wells. 1957. The function of the brain of *Octopus* in tactile discrimination. *J. Exp. Biol.* 34:131–142.

Werblin, F. S., and J. E. Dowling. 1969. Organization of the retina of the mudpuppy, *Necturus maculosus.* II. Intracellular recording. *J. Neurophysiol.* 32:339–355.

Willows, A. O. D. 1967. Behavioral acts elicited by stimulation of single, identifiable brain cells. *Science* 157:570–574.

Zarrow, M. X., R. Gandelman, and V. H. Denen-

berg. 1971. Prolactin: Is it an essential hormone for maternal behavior in the mammal? *Horm. Behav.* 2:343–354.

7

HORMONES AND BEHAVIOR

Hormones are chemical substances produced either by specialized ductless glands located in various parts of the body or by neurons, called neurosecretory cells, within the nervous system, in which case they are called **neurosecretions.** Hormones are carried by the circulatory system, and neurosecretions travel along nerve axons or in the blood. Both are messengers to various target organs, and they influence such processes as growth, metabolism, water balance, and reproduction. In this chapter we consider how the endocrine system acts as a behavior-regulating mechanism in both invertebrates and vertebrates through the production of hormones and how behavior can influence the endocrine system.

For example, a male rat placed with a sexually mature female rat will mount the female within a matter of seconds and copulate (Figure 7–1). A castrated male rat takes much longer to initiate mounting (Figure 7–2). The testes are a source of androgens, hormones that affect reproductive behavior. Injections of synthetic androgens can replace the hormone normally produced by the testes and restore mount latencies to near normal levels.

Many insects, like grasshoppers (*Melanoplus*), hatch from eggs and reach adult form through a series of nymphal stages (Figure 7–3). Between the stages, the insect molts—that is, it sheds its exoskeleton and undergoes other external and internal changes. If we remove the prothoracic gland, which secretes **ecdysone**—the hormone that controls molting—in an early nymphal stage, the normal developmental sequence stops. If, however, a prothoracic gland from another grasshopper is transplanted or if ecdysone is supplied artificially, the sequence of nymphal

FIGURE 7-1   Male rat mounting female rat
The male may perform only a mount or a mount with intromission. The sexual
performance and timing of sexual behavior patterns in male rats is influenced by the
presence of testicular hormones.
Source: Photo by Ronald J. Barfield.

stages resumes. The hormone changes are also correlated with certain behavioral
transitions. Both the dispersal of grasshoppers, individually or in large numbers, and
the mating behavior are, in part, under hormonal control.

We take a general look at the endocrine systems of invertebrates and vertebrates in the first part of this chapter. The major portion of the chapter explores the
ways in which hormones influence behavior; by examining various studies we gain
an appreciation of the effects of hormones on behavior and look closely at the
techniques that have been used to explore these relationships. Our primary focus is
on the mechanisms of the hormonal system. In later chapters we will discuss more
thoroughly the ecological and evolutionary significance of endocrine systems.

COMPARISON OF THE ENDOCRINE AND
NERVOUS SYSTEMS

It is important to understand that both the nervous system and the endocrine system
are feedback systems that form key parts of the body's mechanisms for interfacing
with the environment, and both systems are critical for adaptation. In general, the
nervous system provides relatively more rapid and specific responses to external and
internal events; the endocrine system acts more slowly and often produces more
general effects. The nervous system provides a means of fine-tuning responses to
specific external conditions or stimuli. The endocrine system regulates more generalized responses to prevailing external and internal conditions.

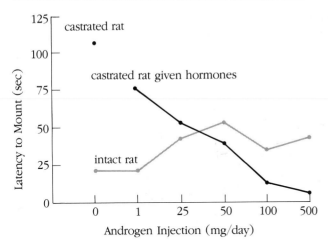

FIGURE 7-2   Sexual behavior in castrated rats

The average latency to the first mount varies in intact rats, castrated rats, and rats given daily injections of doses of synthetic androgen.

Source: F. A. Beach and A. M. Holz-Tucker, "Effects of Different Concentrations of Androgen upon Sexual Behavior in Castrated Rats," *Journal of Comparative Physiology and Psychology* 42 (1949):433–453.

INVERTEBRATE ENDOCRINE SYSTEMS

Different types of invertebrates — for example, echinoderms (e.g., starfish), segmented worms (e.g., leeches), molluscs (e.g., clams), and crustaceans (e.g., lobsters) — possess different endocrine systems. Perhaps the role of hormones in affecting behavior is most clearly understood for insects; one such example is the grasshopper.

Neurosecretions or hormones that affect behavior are secreted in five major locations in grasshoppers: (1) **neurosecretory cells** in the brain; (2) paired **corpora cardiaca** just behind the brain, near the aorta; (3) paired **corpora allata** located

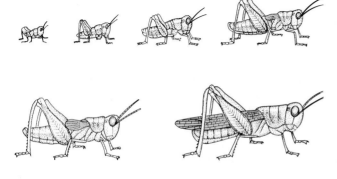

FIGURE 7-3   Grasshopper development

Hoppers, young grasshoppers that hatch from eggs, proceed through a series of five to seven molts, changing in body size and in internal and external characteristics with each molt until they attain the adult stage in several months. Molting and other changes in physiology and behavior are controlled largely by the neuroendocrine secretions produced by the brain and various specialized glands within the insect.

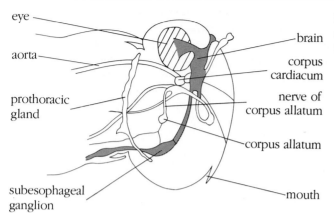

FIGURE 7-4   Brain neurosecretory cells and glands of the head region in the grasshopper

alongside the esophagus; (4) a single, elongated **prothoracic gland** at the rear of the head, near the neck membrane; and (5) male and female **gonads** in posterior body regions (Figure 7–4). With some variations and exceptions, most insect species have these structures.

One important characteristic of invertebrate neuroendocrine structures is that they all either involve neurons directly, as with the neurosecretory cells, or they are closely tied to the nervous system through nerves connecting with the brain and with one another. The glands are also interconnected through the circulatory system. This dual system of interrelationships of the neuroendocrine glands and the nervous system is an important feature of hormonal systems in both vertebrates and invertebrates; these mechanisms, which exert controlling influences on behavior, have apparently evolved in close harmony.

In invertebrates, a neurosecretion or a hormone does not necessarily exert a direct influence on behavior. Instead, the neurosecretion or hormone may have another endocrine gland as its target; these secretions are termed **trophic neurosecretions** or **trophic hormones.** For example, in many insects neurosecretions from the brain exert a trophic effect on the prothoracic glands, which in turn produce and secrete the hormone ecdysone that controls molting, the developmental sequence of insects (Figure 7–3). Other trophic hormones influence dispersal and mating behavior.

VERTEBRATE ENDOCRINE SYSTEMS

During the course of evolution, many changes have taken place in animal endocrine systems. In particular, the endocrine systems of vertebrates have evolved into two major parts. (1) The **pituitary gland** is located near the hypothalamus on the ventral side of the brain and has close connections to several central nervous system structures (Figure 7–5). The pituitary gland and the hypothalamus are closely connected and together form an important bridge between the nervous and endocrine systems. Two types of connections exist: the posterior pituitary is comprised, in part, of neurons that originate in the hypothalamus, and the anterior pituitary is linked to the hypothalamus via the hypothalamic-pituitary portal system. The pituitary produces both trophic hormones, which affect other endocrine glands, and direct-acting hormones. (2) A series of endocrine glands, including, in particular, the **thyroid, pineal gland, adrenals, pancreas,** and **gonads** (Figure 7–6), are located throughout the body. The trophic hormones released by the various parts of the pituitary are peptides, with a basic structure consisting of chains of amino acids. The hormones from the adrenal glands, testes, ovaries, and placenta are steroid hormones, organic compounds with a common basic structure of rings of carbon atoms and side chains, and with varying additional side chains.

FIGURE 7–5   Hypothalamus and pituitary of a generalized vertebrate neuroendocrine system

The products of various portions of the pituitary gland regulate other endocrine glands and secrete hormones that directly influence physiology and behavior. Posterior pituitary cells are parts of nerve cells that extend from the hypothalamus, which store and release neurosecretions. Neurosecretions, released in the hypothalamus, are carried by blood vessels to the anterior pituitary, where they directly influence hormone production and release.

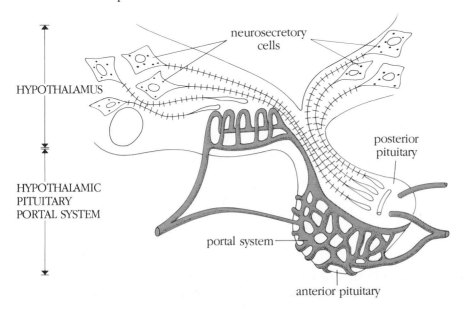

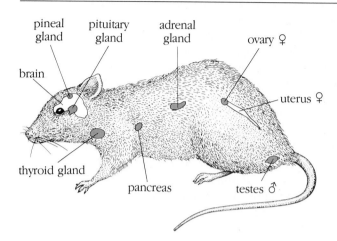

pineal gland   pituitary gland   adrenal gland   ovary ♀

brain

uterus ♀

thyroid gland

pancreas   testes ♂

FIGURE 7-6   Endocrine glands of the rat
Different glands receive either nervous system communication, circulatory system communication, or both from other parts of the body.

ENDOCRINE GLAND SECRETIONS

Several pituitary gland secretions exert controlling effects on behavior and related physiological processes in vertebrates (see Table 7–1). Oxytocin and vasopressin are produced by hypothalamic neurons and are stored by nerve terminals in the posterior pituitary *(pars nervosa)*; and they are then released as neurosecretions into the bloodstream. Oxytocin acts to stimulate uterine contractions that may facilitate sperm movements in the female genital tract after copulation and that also help to expel the fetus during parturition. Oxytocin also stimulates the ejection of milk from the mammary glands. Vasopressin influences the physiology of the kidneys; it alters urine concentration and thus helps to regulate water balance. For example, excretion of highly concentrated urine and retention of body water are physiological and behavioral adaptations of many desert mammals, such as camels, kangaroo rats, and gerbils.

The intermediate pituitary *(pars intermedia)* secretes melanophore-stimulating hormone (MSH), which affects the concentration or dispersion of pigment granules in chromatophores, or color cells, found in many vertebrates, particularly fish, reptiles, and amphibians. In the absence of MSH, pigment granules remain clumped; MSH stimulation leads to dispersion of the granules and a color change. For example, adult, male, three-spined stickleback fish (*Gasterosteus aculeatus*) normally have pale-colored sides. But when two males engage in a territorial boundary display, release of MSH triggers dispersion of pigment granules, and the fish's sides take on a bright blue color. Color changes in vertebrates may serve as communications signals, or they may produce coloration patterns that render an animal inconspicuous against a particular background (camouflage).

Four hormones that indirectly affect behavior are secreted by the anterior pituitary *(pars distalis)*; three of these are trophic hormones that exert their effects on other endocrine glands. In females, follicle-stimulating hormone (FSH) and luteinizing hormone (LH) affect the cycle of egg maturation in the ovaries, sexual receptivity, conception, and pregnancy. In males FSH and LH control sperm production and secretion of male hormones, or androgens. A third trophic secretion, adrenocorticotrophic hormone (ACTH), affects the production and secretion of adrenal cortex steroid hormones. Prolactin, a pituitary hormone present in birds and mammals, is important for maternal behavior and influences milk production in mammals and crop milk accumulation in some birds.

The pineal gland, located within the brain, secretes several hormones in the form of indoleamines, proteins, and polypeptides. The most important of these for the study of behavior is probably melatonin. Melatonin functions in the modulation of reproduction and annual patterns of breeding activity in mammals (Reiter 1980), among other things. We will postpone further discussion of the role of the pineal and melatonin until Chapter 8 where we discuss biological rhythms.

The gonads, stimulated by trophic secretions from the pituitary, produce

TABLE 7-1   Sources of hormones that affect behavior, hormones produced, and regulatory and physiological effects

Source	Hormone	Regulatory and Physiological Effect
Pineal	Melatonin	Annual reproductive cycle
Posterior pituitary	Oxytocin Vasopressin	Milk ejection; parturition; water balance
Anterior pituitary	Luteinizing hormone (LH)	Corpora lutea formation; progesterone secretion; Leydig cells stimulation (androgen secretion)
	Follicle-stimulating hormone (FSH)	Follicle development (♀); ovulation (with LH and estrogen); spermatogenesis
	Prolactin	Milk secretion; bird parental behavior
	Adrenocorticotrophic hormone (ACTH)	Adrenal cortical steroid secretion
Intermediate pituitary	Melanophore-stimulating hormone (MSH)	Color change
Adrenal cortex	Steroids	Water balance; metabolism; electrolyte balance
Adrenal medulla	Adrenaline Noradrenaline	Blood sugar level; stress reaction
Testis	Androgens	Testis development; spermatogenesis, secondary sex characteristics
Ovary and placenta	Estrogens	Uterine growth; mammary gland development
	Progestogens	Gestation

androgens (from testes) and estrogens and progestogens (from ovaries). During pregnancy, progestogens secreted by the placenta play a key role in maintaining gestation. These hormones affect reproduction, maternal behavior, and aggression as well as some secondary sex characteristics that may be important communication signals. Adrenal hormones are important in maintaining the body's water balance, metabolism, and electrolyte balance. Adrenaline and noradrenaline, from the adrenal medulla, play important roles in emergency stress reactions.

For all hormones, the specificity of action is dependent upon the specificity of receptor sites on the target tissues. This is true for both peptide and steroid hormones and when the target tissues are other endocrine glands or other non-endocrine target tissues. The bloodstream may be carrying numerous "messages" in the form of a variety of hormones, but the effects on physiology and behavior occur only when a particular hormone makes contact with the appropriate receptor sites. Thus, the testes secrete androgens into the bloodstream, and the hormones are carried throughout the body. The effects of the androgens may involve such diverse processes as seminal vesicle growth, changes in secondary sexual characteristics, and influences on behavior through target tissues in the brain. In each instance there are appropriate receptors for the androgens located in the appropriate tissues.

ENDOCRINE GLAND INTERACTIONS

FEEDBACK LOOPS.   Many endocrine gland secretions are characterized by feedback loops, one of which is diagrammed in Figure 7–7. Pituitary secretion of FSH and LH is controlled by **releasing factors** that travel via the hypothalamic-pituitary portal system from the hypothalamus. The trophic hormones FSH and LH are then carried in the blood to the testes, where they stimulate the gonadal processes of spermatogenesis in the seminiferous tubules and the production and release of testosterone from the interstitial cells. Testosterone in turn is carried in the bloodstream to other locations, including the sex accessory glands (e.g., the seminal vesicles, where semen production is stimulated) and also back to the hypothalamus. Specialized hypothalamic sensory cells, acting as part of the body's homeostatic machinery, continuously monitor the levels of various chemicals, including testosterone or the metabolic products of testosterone, in the blood. Thus, when we castrate an animal the amount of testosterone present decreases, but, at the same time, the levels of FSH will rise. To which effect should we then attribute any observed changes in behavior—the lowered levels of testosterone or the heightened levels of FSH? This illustrates one of the difficulties in conducting research on hormones affecting behavior.

The levels of circulating testosterone influence the secretion of hypothalamic-releasing factors to the pituitary in a negative feedback relationship. As the testosterone level in the blood increases, the secretion of hypothalamic-releasing factors decreases. Conversely, as the testosterone level in the blood decreases, more secretion from the releasing factors enters the hypothalamic-pituitary portal system, travels to the anterior pituitary, and causes an increase in output of FSH and LH.

FIGURE 7-7   Feedback loops in endocrine physiology of a male mammal or bird
Feedback loops are a key feature of hormone-behavior relationships. Some of the
pathways involve direct effects (e.g., FSH and LH act to stimulate the testes). Other
pathways involve negative feedback effects (e.g., testosterone in the bloodstream
passing through the hypothalamus is monitored by specific receptors). High levels
of testosterone may result in reduced output of the releasing factors, which affect
the anterior pituitary.

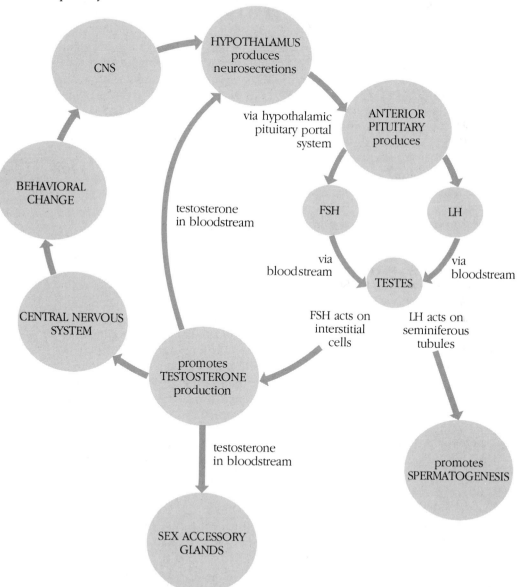

There are also feedback loops involving FSH and LH and the female gonadal hormones, estrogen and progesterone. These effects may be exemplified by tracing the sequence of events in the estrous cycle of a female mammal. At the start of the cycle, releasing factors from the hypothalamus stimulate release of FSH and LH from the pituitary; FSH predominates at this time. FSH and LH stimulate the growth of follicles (or of only one follicle in some species) within the ovary, and the production of estrogen from the follicles. When the levels of estrogen in the blood reach a peak, indicating the follicles are mature, there is a negative feedback effect on the release of FSH from the pituitary. However, there is a positive feedback effect on LH; more is released with the rising levels of estrogen, and LH becomes the predominant pituitary hormone. In mammals which ovulate spontaneously, the follicles rupture and the ova are released at about the time of the transition to LH dominance. The ruptured follicles secrete estrogen and progesterone under the continuing influence of LH and prolactin. However, rising blood levels of progesterone exert a negative feedback influence which diminishes the pituitary release of LH. Estrogen and progesterone are also responsible for changes in behavior. Species vary in their reproductive physiology and endocrinology, but, in general, the estrous phase of the cycle, when the female mammal is receptive to males, is influenced by the two ovarian steroid hormones. The complex physiological and behavioral changes are the product of several endocrine pathways, involving both positive and negative feedback effects.

Similar feedback loops exist for ACTH and adrenal hormones. The feedback principle is important for a complete understanding of hormone interactions. The feedback loops can also be influenced by factors in the environment (e.g., daylength) to alter or set hormone levels.

SYNERGISM AND ANTAGONISM.   Two other endocrine interactions are synergism and antagonism. Female estrogens and progestogens, which together affect sexual behavior in vertebrates, are often in circulation simultaneously. Sexual receptivity in female rats depends on the synergistic action of the two hormones (Beach 1976). In contrast, some hormones, when they occur in circulation together, act antagonistically. Male pigeons initially act aggressively toward a potential mate, yet eventually their aggression is inhibited as a result of the antagonistic interactions of testosterone and progestogens.

EXPERIMENTAL METHODS

Investigators have used several techniques to explore the interrelationships between hormones and behavior. These include:

1. Extirpation, or removal, of a particular endocrine gland to assess the absence of a specific hormone on behavior;

2. Hormone replacement therapy, in which the investigator injects a specific hormone into an animal or transplants a gland from another animal to replace one previously removed by surgery;
3. Blood transfusion to transfer the "hormonal state" of one animal to another in order to observe the behavioral effect;
4. Bioassays to indirectly assess circulating hormone levels by measuring a secondary characteristic, such as a skin gland, that is dependent on a particular hormone;
5. Radioimmunoassay to directly measure circulating levels of a hormone through the use of immunological methods; and
6. Autoradiography to localize the sites at which hormone uptake occurs.

Each of these techniques will be explored in more detail in the sections that follow.

ACTIVATIONAL EFFECTS

Hormonal influences on behavior can be divided roughly into two categories: activational and organizational effects. In activational effects, hormones act as triggering influences on the expression and performance of behavior patterns. Direct activational effects occur where hormone secretion or inhibition of secretion leads to a relatively rapid response. Indirect activational effects require more complex sequences of stimulation and hormone secretion. The organizational effects of hormones are manifested during an organism's development. For example, sex differentiation and patterns of growth for body tissues are, in part, under hormonal control.

SECONDARY SEXUAL CHARACTERISTICS

In addition to their direct activational effects on behavior, hormones can affect secondary sexual characteristics. The characteristic cock's comb becomes greatly decreased in castrated roosters. Male cats, which spray urine probably as a marking behavior, often cease to spray after their testes are removed. In these two examples a change in a secondary sexual characteristic (cock's comb) and a sexually related behavior (urine spraying) affects the behavioral processes of communication.

AGGRESSION AND SEXUAL BEHAVIOR PATTERNS

When male ring doves (*Streptopelia risoria*) are castrated, they exhibit decreased levels of aggression, courtship, and copulation behaviors. When castrated birds are treated with implants of crystalline testosterone propionate in specific sites in the hypothalamus, the normal levels of these behaviors are restored (Barfield 1971; Hutchinson 1969, 1971, 1978). Experiments like this clearly demonstrate the activational effects of testosterone on sexual and aggressive behaviors and suggest that specific brain sites may be stimulated by testosterone, which influences sexual behavior.

Other research has shown that the presence or absence of testosterone influences aggressive behavior in birds and mammals such as ring doves, roosters, mice, rats, and domestic cats (Tollman and King 1956; Barfield et al. 1972; Leshner 1975; Guhl 1961; Bennett 1940). Intact males are more aggressive, have shorter latencies to initiate fighting behavior, and fight more frequently than do castrated subjects of the same species.

Both male and female Syrian golden hamsters (*Mesocricetus auratus*) have paired sebaceous flank glands, with which they mark (by depositing sebum from the gland) objects in their environment (Figure 7–8). The flank glands are androgen dependent; measurement of their size and pigmentation can be combined in an index that serves as a bioassay measurement of relative levels of circulating androgens (Vandenbergh 1971, 1973; Drickamer and Vandenbergh 1973; Drickamer, Vandenbergh, and Colby 1973). When groups of four intact male hamsters that were kept isolated since weaning and were equal in body weight were allowed free social interaction in a large pen, a significant positive correlation ($r = 0.77$) between the outcomes of encounters and the index of gland size and pigmentation emerged. The key feature of this experiment was that the investigators measured the glands before they placed the hamsters together and could predict the outcomes of social encounters on the basis of the bioassay of relative levels of circulating androgens. In a related experiment, investigators gave castrated male hamsters injections of testosterone propionate. When they gave each of groups of four males a different dose and then allowed the four to interact, they could again predict the outcomes of social encounters ($r = 0.81$) on the basis of dose level and the corresponding gland index.

FIGURE 7–8   Hamster flank glands
Hamsters have paired sebaceous flank glands which are used to mark objects in the environment with sebum. The size and pigmentation of the glands are androgen dependent. A normal, intact male (left), and (right) a castrated male.
Source: Photo from John G. Vandenbergh.

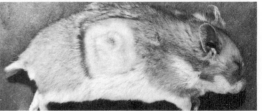

Interestingly, investigators could also predict outcomes of social interactions among female hamsters on the basis of the measurement and pigmentation of the females' flank glands (Drickamer and Vandenbergh 1973).

SEXUAL ATTRACTION

In certain species of cockroaches (e.g., *Periplaneta*) and moths (e.g., *Bombyx*), the females release pheromones, which act as a sex attractant for males. A **pheromone** is a chemical that is produced by an animal and released to the external environment, where it may be received by conspecifics and may affect their physiology or behavior. If a female cockroach's corpus allatum is surgically removed after she has molted to the adult stage, she does not produce pheromones when she becomes reproductively active (Barth 1965). Females treated in this way are incapable of attracting males and do not mate. However, if a corpus allatum from another adult female is transplanted into the test subject, she is again capable of pheromone production. Apparently a hormone that is produced by the corpus allatum acts on peripheral glandular tissues to stimulate production of the sex-attractant pheromone. Interestingly, if a female is gonadectomized as an adult, the processes of pheromone release, mate attraction, and mating are not affected.

Why has this system of mate attraction evolved in some moths, cockroaches, and some other insects? There are several possible answers to this question. First, emission of a species-specific sex-attractant chemical may be a species-isolating mechanism. If several closely related species are reproducing at the same time, it will be advantageous, to avoid gamete waste, if insects can locate potential mates of their own species. Thus if each species has a different sex-attractant pheromone, individuals will readily locate and mate with partners of their own species. Second, some pheromones may serve to synchronize the reproductive activities of the two sexes of a given species. Third, in a related fashion, chemical attractants could serve a sexual excitatory function by bringing both members of the pair to peak reproductive activity simultaneously, thereby ensuring a higher total reproductive success. Finally, of course, the attractant ensures that a male will find a female and mate with her.

ECLOSION

The process whereby the adult form of an insect emerges from the pupa after metamorphosis is called **eclosion** and is another activational effect controlled hormonally. Many moth species eclose at a species-specific time of day. The eclosion hormone, produced by neurosecretory cells in the brain, plays a critical role in this process (Truman 1971; Truman and Riddiford 1970). If the eclosion hormone is injected into pupae that are near the end of metamorphosis, eclosion behavior, such as abdomen movements and wing spreading after emergence, can be activated at any time of day. Moths that have had their brains removed usually emerge; therefore the presence of the eclosion hormone is not an absolute requirement for eclosion to take place. The process, however, is not as coordinated in brainless subjects, and some activities (e.g., wing spreading) are usually absent. Thus although the hormone may not be necessary for eclosion, it does appear to be necessary for proper coordination of the sequence.

LIFE STAGES

Adult male desert locusts *(Schistocera gregoria)* fail to exhibit sexual behavior after the corpora allata have been removed. When corpora allata from other adult males are transplanted into allatecomized males, sexual behavior is restored (Lohrer 1961; Pener 1965). However, similar investigations have revealed that the corpora allata are not needed for sexual behavior in certain grasshoppers (Barth 1968).

Several locust species exhibit both solitary and gregarious phases; young hoppers (refer to Figure 7–3) that are reared in isolation show moderate levels of activity and do not engage in sustained flights; hoppers reared under crowded conditions do make long flights (Johnson 1969). Some evidence supports the tentative conclusion that this difference has a hormonal basis. First, solitary hoppers have larger prothoracic glands than do gregarious forms (Carlisle and Ellis 1959). Second, adult locusts that develop from solitary hoppers retain the prothoracic glands, but the glands are absent in adults that develop from gregarious hoppers. Third, if homogenates of prothoracic glands are introduced into gregarious hoppers, general activity is reduced (Carlisle and Ellis 1959), and if prothoracic glands from solitary adults are transplanted into gregarious adults, sustained flight activity is diminished (Michel 1972).

MOLTING

Studies of hormones and neurosecretions in invertebrates other than insects have revealed common patterns of effects. In some types of worms and molluscs and in crustaceans, investigators have demonstrated the presence of one or more neurosecretory or endocrine glands whose secretions affect sexual differentiation and maturation of gametes and stimulate reproduction (Charniaux-Cotton and Kleinholz 1964; Wells and Wells 1959; Golding 1972). In crustaceans several behavior-related phenomena are under partial endocrine control. Many crustaceans (as well as other arthropods) molt—or shed their exoskeletons—periodically as they grow. Removal of both eyestalks in these animals shortens the interval between molts. If the crustaceans are given extracts from a particular neurosecretory gland, molting is prevented. Clearly the gland produces a molt-inhibiting factor. It is also possible that some cells in the eyestalks are involved in the production and secretion of a substance that shortens the interval between molts (Kleinholz 1970).

COLOR CHANGE

As we saw earlier, a melanophore-stimulating hormone (MSH) affects color changes in fish. We can also see MSH action in short-tailed weasels *(Mustela erminea),* which undergo seasonal changes in pelage, or coat, color during spring and fall molts (Rust 1965; Rust and Meyer 1969). In the spring, MSH secretions increase, and new brown hairs replace the white winter coat. During the fall months MSH secretion is inhibited by the action of another hormone, melatonin, secreted by the pineal gland. The developing hairs at this time of year are not pigmented, and the weasel's coat returns to its white winter color. These seasonal shifts in coat color may be both behaviorally and functionally significant because the result matches the general background color of the weasel's environment. Camouflage coloration enables the weasel to hunt without being conspicuous and to be "hidden" from potential predators.

In crustacea, chromatophores contain a variety of colored pigment cells, and organisms of many species can approximately match the color of the background substrate (Fingerman 1965, 1973).

ORGANIZATIONAL EFFECTS

Studies of rats, guinea pigs, mice, and rhesus monkeys have provided clear evidence that certain hormones exert critical effects on sex differentiation during early development (see reviews by Toran-Allerand 1978; Dörner and Kawakami 1978). Most of these investigations have concentrated on the effects of gonadal hormones on the organizational effects of later adult sexual and aggressive behaviors (Young 1961; Young et al. 1964; Harris and Levine 1965; Barraclough 1967; Barraclough and Gorski 1961; Davidson 1966a; Levine and Mullins 1966; Luttge and Whalen 1970; see also review by Feder 1981).

If a male rat is castrated within the first four or five days after birth, he does not show normal sexual behavior as an adult. If a neonatally castrated male rat is given estrogen and progesterone as an adult, he exhibits female sexual responses, such as the lordosis posture that a receptive female adopts to permit mounting and intromission by a male (Figure 7–9). A male rat that is castrated as an adult and then given estrogen and progesterone does not show female sexual behaviors.

FIGURE 7–9   Lordosis posture in female rat
For mounting and intromission by males, female rodents generally adopt a lordosis posture, characterized by shifting the tail to one side, raising the hindquarters, and lowering the abdomen. Since this behavior is characteristic of receptive females it can be used as a reliable test of feminization.

In another experiment, neonatal male rat pups were injected with estrogen. Histological examinations of the rats revealed some degeneration of the seminiferous tubules, where sperm are produced. The investigators noted that although these males showed mounting behavior when placed with a receptive female, the behavior was irregular — the mounts were often incorrectly oriented, and no ejaculation occurred.

Female rats that were treated within the first four or five days after birth with either artificial androgen (TP) or artificial estrogen (EB) did not show normal estrous cyclicity as adults and did not respond by exhibiting lordotic behavior when injected with estrogen and progesterone. When these neonatally injected females were given injections of TP as adults, they exhibited malelike sexual behaviors. In fact, the neonatal injection of androgen alone will produce malelike sexual behaviors in adults of some inbred strains of mice (Manning and McGill 1974). Females given estrogen injections neonatally sometimes showed irregular estrous cycles; females injected with TP neonatally exhibited permanent vaginal cornification. (Vaginal cornification, the sloughing off of the epithelium lining, is produced by the action of estrogens on the lining of the vaginal tract and is a sign of estrus.) Masculinized females also exhibited propensities and latencies to fight comparable to those of adult males (Edwards 1968).

Similar studies of the organizational effects of hormones on female behavior have been conducted on guinea pigs (Phoenix et al. 1959) and rhesus monkeys (Goy 1970). Female progeny of females that were treated with androgens when they were pregnant have external genitalia that are masculinized (smaller vaginal opening and hypertrophied clitoris), and they exhibit malelike sexual behaviors. These data, in combination with the previously cited studies of rats, point up an important aspect of early injections of hormones. There is a critical time period during which hormone injections must occur for certain behaviors to be affected. In rats and mice this critical period occurs during the initial four or five days after birth; in guinea pigs and rhesus monkeys the critical time period occurs prenatally. Neonatal injections do not have the same effects on guinea pigs and rhesus monkeys that they do on rats and mice. Rhesus monkeys and guinea pigs attain a stage of development at some time prenatally that rats and mice attain only postnatally. The effects of the injected hormones thus occur at about the same relative stage of development in these animals.

In addition to organizational effects on male and female sexual behavior, there are organizational effects that influence nonreproductive behaviors. Meany and colleagues (1982) investigated the influence of glucorticoids on play-fighting behavior in Norway rat pups. Doses of glucorticoids administered neonatally to male rat pups supresses the normally high levels of play-fighting found in males. Similar treatments given to female rat pups did not influence the level of play-fighting behavior. Thus, there appears to be a sex-dependent, organizational effect of glucorticoids on the development of play-fighting in rats.

Some organizational effects of hormones on sexual behaviors have been suggested by some kinds of data for humans (Money 1980; Money and Ehrhardt 1972). Two effects involve masculinization of genetically female fetuses. Mothers who had been given progestogens to help maintain pregnancy often gave birth to females whose appearance was masculinized. (This treatment has since been discontinued.) Among treated nonhuman animals, some newborn females have a hereditary disease that results in high levels of androgen output from the adrenal glands — this disease in a pregnant woman can masculinize a female fetus. In a third instance, genetic males may have target tissues that cannot respond to the androgens secreted from their own testes — this results in a more feminine appearance. It is important to note that in humans social experiences often predominate in shaping behavior — individuals respond to the sex of rearing regardless of the genetic sex.

The female of a pair of heterosexual twins in cattle is called a **freemartin,** and is unable to produce her own offspring in adulthood. In the early part of this century scientists recognized that some type of hormone action was most likely involved in these effects; presumably there was, at some time during early embryonic life, a circulatory pathway link connecting the blood flow of two fetuses of genetically opposite sexes. This process could result in the transfer of some androgens to the female fetus which would alter her program of development, masculinizing the heifer and resulting in a sterile or barren individual (Cole 1916; Lillie 1916). We now know that this hypothesis is substantially correct.

More recently investigators have tested the effects of postnatal treatment with androgen on behavior in heifers (Bouissou 1978; Bouissou and Guadioso 1982). Heifers were given 100 days of treatment with testosterone propionate beginning when they were three months old — prior to puberty. When these heifers were tested three to twelve months after cessation of treatment, they tended to be more dominant than untreated control animals. The treated heifers were not, however, more aggressive than controls, nor were there any differences between treated and untreated heifers with respect to physical appearance or normal sexual cycles. The investigators concluded that there were some persistent effects of the early hormone administration: the treatment resulted in heifers that exhibited less fear (and presumably more boldness) in encounters with other cows. Similar effects of androgen treatment resulting in higher social rank have also been demonstrated in mares (Jussiaux and Trillaud 1979) and hinds of the red deer (Fletcher 1973).

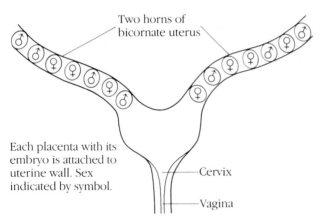

Two horns of
bicornate uterus

Each placenta with its
embryo is attached to
uterine wall. Sex
indicated by symbol.

Cervix

Vagina

FIGURE 7–10   Rat uterus
Both rats and mice have bicornate uteri. The often numerous fetuses of a pregnancy are arranged sequentially in each of the two horns of the uterus as shown here for the rat. Pups of one sex may be positioned between two pups of their own sex, two of the opposite sex or between fetuses of opposite sexes. The location of a female next to a male fetus results in masculinization of the genetic female.

Lastly, the intrauterine position of a female fetus in rats and mice can influence her genital morphology and sexual behavior (Clemens 1974; vom Saal and Bronson 1978). Both rats and mice ovulate multiple eggs at each cycle and bear litters of sizes ranging from 1 to 20 pups. Both species have bicornate uteri; fetuses are arranged sequentially in each arm of the uterus (Figure 7–10). Thus, fetuses of either genetic sex can be positioned between two other fetuses of the same sex, between two of the opposite sex, or between two fetuses of opposite sexes. In both rats and mice there is a brief period late during gestation and around the time of parturition when testicular androgens are produced and released. Clemens (1974) proposed that fetal females could be masculinized by exposure to testosterone from neighboring litter mates. Anogenital distance, it turns out, is a reliable measure of androgenic exposure. For both rodent species anogenital distances for genetic females positioned in utero next to males are larger than for females positioned between two females. In addition, certain behavioral and physiological traits are affected. In rats, females that have been masculinized due to in utero position exhibit more mounting behavior (Clemens, Gladue, and Coniglio 1978), while in mice the estrous cycles are longer and levels of aggressive behavior are increased for females whose position was next to a male (Gandelman, vom Saal, and Reinisch 1977; vom Saal and Bronson 1980a,b).

From all of these studies we can reach some tentative conclusions about the effects of gonadal hormones on the organization of behavior in mammals. Gonadal hormones (either estrogens or androgens) secreted at the appropriate critical time during early development, affect the undifferentiated brain by exposure to the hormone to produce a malelike adult organism, or one exhibiting malelike behavior, regardless of the individual's genetic sex. Early gonadal hormone secretion apparently sensitizes the brain to hormones that circulate in the blood later. In the absence of gonadal hormone secretion, a typical cyclical female pattern develops (see recent review by Arnold and Gorski, 1984). Again, the absence of hormones establishes certain patterns of sensitivity in critical areas of the brain, such as the hypothalamus, so that later hormone circulation produces particular physiological and behavioral responses. Both the type of sensitivity in specific areas of the brain and the sensitivity to certain temporal patterns of levels of different chemicals circulating in the blood are affected (Dörner and Kawakami 1978).

Other investigations have implicated thyroid and adrenal hormones in the organizational processes during development. Thyroidectomized rats exhibit characteristics of cretinism, slower general body growth, delayed sexual maturation, and retarded general development of the nervous system. They are slower in their actions, and they learn more slowly and with greater difficulty.

Studies of adrenal glands have demonstrated that young rats given a few minutes of handling each day during early infancy show less extreme responses to stressful situations as adults (Levine 1967, 1968). When handled rats are placed in a stress situation as adults, they secrete less adrenal steroid hormones and show less fear response to presentations of novel stimuli. Handled pups secrete more adrenal steroid hormones in response to shock or ACTH injections. Early in development these higher levels of early adrenal steroid hormones may affect the brain mechanisms related to stress responses and establish permanent changes that continue into adulthood. These findings exemplify another principle regarding organizational effects on behavior; some of the effects organize behavior that do not require hormones for activation in adulthood, whereas other effects do require hormones later in life for activation of the behavior.

Finally, some studies have shown that two invertebrate hormones, MH (molting hormone, or ecdysone) and JH (juvenile hormone), interact in the control of growth and metamorphosis in insects. When levels of JH in the blood are high and MH is also present, the insect continues to grow and differentiate but will not molt to the adult stage. If, however, MH acts alone, molting is induced and the insect will metamorphose and differentiate into the adult form. There is additional evidence from studies of some insect species that the differentiation and maturation of sexual organs depend, in part, on gonadal hormones (Fraenkel 1975; Gilbert 1974; deWilde 1975).

ENDOCRINE-ENVIRONMENT-BEHAVIOR INTERACTIONS

Some activational effects involve complex interactions among behaviors, hormones, and specific environmental stimuli. We discuss in some detail three examples — reproductive sequence in ring doves, parturition and maternal behavior in rats, and reproduction in lizards — that illustrate this interrelationship.

REPRODUCTIVE SEQUENCE IN RING DOVES

Figure 7–11 illustrates the reproductive sequence in ring doves. A male begins courtship displays shortly after being placed with a female. The failure of castrated males to court females indicates the importance of a continuous supply of androgens

---

FIGURE 7–11   Reproductive behavior cycle of the ring dove
This cycle provides an example of indirect environmental determinants of behavior. The sequence involves (1) courtship and copulation, (2) nest building, (3) egg laying, (4) incubation, and (5) feeding crop milk to the young squabs after they hatch. The cycle then repeats itself.

1. Courtship and copulation

5. Feeding young squabs

2. Nest building

4. Incubation

3. Egg laying

for the initiation of the cycle. Male courtship stimulates pituitary release of FSH in the female dove; FSH, in turn, brings about follicle development in the ovaries. The follicles secrete estrogen, which affects uterine growth and development. Within a day or two, the birds begin nest construction; during this phase of the cycle, they copulate and continually add to their nest. The presence of a nest stimulates the production and secretion of progesterone in females. Progesterone in both sexes promotes incubation behavior after eggs are laid. Egg laying is activated, in part, by secretion of LH by the female's pituitary.

Incubation, maintained by progesterone secretion, lasts for fourteen days; the male and female take turns on the nest. Under the influence of the presence of eggs in the nest and as a result of stimulation from incubation behavior, both the male's and the female's pituitary glands secrete prolactin. Prolactin acts to inhibit FSH and LH secretion, and all sex behavior ceases. Prolactin also stimulates crop development and the production of crop milk — a nutrient-rich fluid secreted in the gullets of both males and females — and may also help to maintain incubation behavior. When young squabs hatch after two weeks, the parents immediately feed them with crop milk. During the next ten to twelve days, both parents continue to feed the young with crop milk; however, feeding behavior wanes toward the end of this period, probably because the secretion of prolactin decreases. As prolactin decreases, the pituitary secretes FSH and LH, the same pair of doves resumes courtship, and the sequence begins again.

At each stage of this sequence, the internal state of each bird interacts with external variables to produce the observed behavior pattern. The variables consist of (1) the hormonal state of both the male and female dove, including feedback loops; (2) the behavior of each member of the pair that stimulates changes in the hormonal levels and behavior of its mate; and (3) environmental cues, such as nests and eggs, that influence hormonal and behavioral changes in both (Figure 7–12).

Daniel Lehrman and his colleagues conducted many ingenious experiments

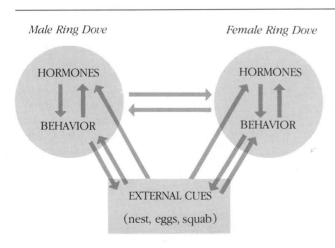

Male Ring Dove   *Female Ring Dove*

HORMONES   HORMONES

BEHAVIOR   BEHAVIOR

EXTERNAL CUES
(nest, eggs, squab)

FIGURE 7–12 Interrelationship between hormones and behavior in ring doves
Hormones and behaviors of each individual, interactions between individuals, and external cues affect the synchrony of reproductive behavior patterns. Bidirectional arrows indicate feedback relationships, and unidirectional arrows indicate direct effects.

to clarify the interactions and changes that together comprise the reproductive sequence in ring doves. A description of a few of their experiments gives us a glimpse of the logic and experimental method they employed (see Lehrman 1955, 1958a, 1958b, 1959, 1961, 1964, 1965; Lehrman, Brody, and Wortis 1961; Erickson and Lehrman 1964; Cheng 1979).

To determine whether the presence of a mate or of nesting material affects incubation behavior of female ring doves, they used three experimental groups: (1) control females housed alone, (2) females housed with a mate only, and (3) females housed with a male and nesting material. They assessed the results of these pairings in terms of the percentages of test females in each group that exhibited incubation behavior when presented with a nest containing eggs (Figure 7–13). Control females never incubated eggs. By days 6, 7, and 8 after pairing, increasing percentages of females in the second group, those housed with only a male, incubated eggs; by day 8, 100 percent of females caged with both a male and nesting material incubated test eggs. They concluded that the presence of both a male and nesting material is necessary for complete incubation behavior in a female. Similar research with male ring doves shows that the presence of both a female and nesting material is necessary for a male to show complete incubation behavior.

A combination of several techniques has been used by Silver and her colleagues to explore the stimuli that affect males and possible corresponding hormonal changes during the early portion of the ring dove reproductive cycle (Silver 1978;

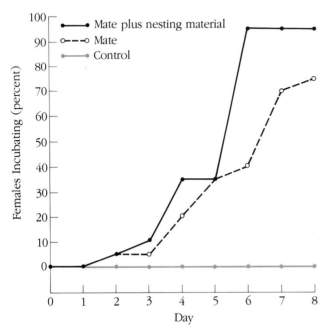

FIGURE 7-13 Incubation behavior in ring doves

Development of incubation behavior in female ring doves is effected by association with a mate or with a mate plus nesting material. Each point on this graph is derived from tests of twenty different birds.
Source: Data from D. S. Lehrman, P. N. Brody, and R. P. Wortis, "The Presence of the Mate and of Nesting Material as Stimuli for the Development of Incubation Behavior and for Gonadotrophin Secretion in the Ring Dove *(Streptopelia risoria)*," *Endocrinology* 68 (1961):507–516. Reprinted with permission.

FIGURE 7–14 Ring doves courting
Female ring doves separated from their male partners by a glass partition can still induce both behavioral and hormonal changes in the males. Males in this situation begin to court the female and exhibit increased levels of plasma testosterone.
Source: Photo from Rae Silver.

O'Connell et al. 1981a,b). Male doves were presented with varying stimuli, and their behavior was observed and radioimmunoassay measurements were made of plasma hormone levels. Males presented with intact, courting females had higher levels of testosterone than males paired with ovariectomized females. Males exposed to females caged behind glass partitions had testosterone levels comparable to those for males given free access to the females (Figure 7–14). Thus, it appears that the gonadal condition of the female can influence the male's response. When surgically deafened males are exposed to intact females their testosterone levels were lower than those for normal males given the same exposure. Some social stimuli appear to be important early in the reproductive sequence, resulting in high male testosterone levels and in stronger behavioral responses to females.

By about the eighth day after pairing, when incubation begins, the male's testosterone levels have declined to a pre-courtship, baseline level. Courting males were exposed to either an incubating female or another courting female. Those in the first group exhibited reduced testosterone levels within a few days, whereas testosterone levels remained high in the second group. Also, the effects of nests and nest material on testosterone levels was tested. Males whose nests were destroyed each day, or who were given no nest material at all, had higher testosterone levels than males given nest material and left undisturbed. Male doves in the last group had testosterone levels that dropped to the baseline level by the eighth day.

The external stimuli and behavior of the female doves appear to be influencing male behavior and male hormone levels at several early stages in the sequence. In the male's behavior, transitions appear to be strongly mediated by context stimuli; whereas for females, as we noted previously, hormonal changes play a dominant role in behavioral shifts from one stage to the next.

In another study male ring doves were allowed to court and mate with females and to participate in nest construction. After the eggs had been laid and incubation had begun, each male was placed behind a partition so that he could see his mate sitting on the eggs but could not himself incubate the eggs. These males, stimulated only by the visual cues from an incubating mate, underwent normal crop development and fed the young squabs normally after hatching. Males that were not permitted to watch their mates incubate eggs failed to develop crops or to feed the young. Although participation in nest building is necessary for crop development, direct involvement in incubation is not.

Why has such a finely tuned system of complex interactions among behavior, hormones, and external cues evolved in ring doves? The almost lockstep system that characterizes the reproductive cycle of ring doves, in which each stage in the sequence is predicated on certain interactions and cues from preceding stages, ensures that as the cycle proceeds both members of the pair are in synchrony and both are in the proper "frame" to perform the required behaviors as needed. Since the system involves the dual participation of two birds and necessitates that their activities be coordinated throughout, the reproductive success of the pair is guaranteed only if both perform certain acts in synchrony. For example, females that laid eggs before a nest was completed would contribute litle to future generations. If the male developed a crop or began to produce crop milk just as the female laid the eggs, long before any squabs had hatched, and the male ceased to produce crop milk during the two weeks the young are normally fed, only the female would be capable of supplying the squabs with nourishment. Her food supply might be insufficient, and the reproductive energy of both members of the pair would have been wasted.

PARTURITION AND MATERNAL BEHAVIOR IN RATS

Maternal behavior in the laboratory rat is similar in some respects to the cyclical reproductive behaviors of ring doves —namely, the internal state of the female rat is a response, in part, to external stimulation from the presence of a nest and pups. Three major events of the maternal cycle will illustrate the interaction between internal state and external stimulation: (1) parturition and the events surrounding the birth of a new litter; (2) nest building, which occurs both before and after parturition; and (3) the period of lactation (Lehrman 1961; Zarrow 1961; Rosenblatt and Lehrman 1963; Richards 1967; Lott and Rosenblatt 1969; Rosenblatt 1970; Lubin et al. 1972; Rosenblatt, Siegel, and Mayer 1979).

Progesterone, sometimes called the pregnancy hormone, helps maintain proper internal conditions prior to parturition. Internal changes in hormone secretions and shifting external cues occur in the days just prior to birth and at the time of birth. Previously low levels of estrogen begin to increase, and progesterone levels may decrease; the overall effect is a shift in the estrogen/progesterone ratio. These hormonal changes help to trigger parturition and, after birth, the retrieving and nursing behaviors that characterize a mother's treatment of her newborn pups. Estrogen acts synergistically with oxytocin secreted by the pituitary to promote the secretion of milk from the mammary glands. Prolactin from the pituitary also acts to promote milk production (Masson 1948; Lott 1962; Grota and Eik-Nes 1967).

To demonstrate that hormonal changes that occur around the time of birth are at least partially responsible for producing changes in maternal behavior, Terkel and Rosenblatt (1968, 1972) used a system of chronically implanted heart catheters in rats. They mounted catheters on the rats' necks and connected them to a swivel pump that permitted the investigators to shunt (transfuse) blood from one rat to another. When blood was shunted from females that were newly parturient to virgin females that were given young pups, the virgin females exhibited maternal behaviors, such as retrieving pups and crouching over the pups to nurse, after fourteen to fifteen hours of exposure to the pups.

Some of the events that occur during parturition—for example, the birth process itself and the consumption of birth fluids or the placenta—do not have to take place for the female to exhibit maternal behaviors. If a female is prevented from experiencing the events associated with normal birth through the removal of the fetuses by Caesarian section and she is shortly thereafter presented with newborn pups, she still exhibits characteristic maternal behaviors (Moltz, Robbins, and Parks 1966).

Nest-building behavior begins to increase in intensity four to six days prior to parturition. The construction of the first nest, called a prepartum nest, is apparently a behavior that is under partial hormonal regulation, although investigators have obtained conflicting results on this point. Several experiments have demonstrated that changes in the estrogen/progesterone ratio are an important trigger for increases in nest building. After pups are present, a female builds a litter nest with higher sides (Figure 7–15); the trigger for this behavior seems to be the pups' presence.

Finally, investigations have shown that maternal behavior during lactation and up to the time of weaning, at three to four weeks of age, depends on close behavioral synchrony between mother and pups. Most of the interactive behaviors between mother and pups operate through nonhormonal channels. For example,

FIGURE 7-15 Female rat with pups in litter nest
Prior to parturition, the female rat constructs a prepartum nest, a flat mat of material. After giving birth, she builds a litter nest with higher sides, partially due to stimulation provided by the presence of her pups.
Source: Photo by Ronald J. Barfield.

virgin females and males have been induced to show retrieving responses and the nursing crouch used by a maternal female to permit suckling behavior by her pups. The test subjects exhibited these maternal behaviors when they were presented with stimulus pups each day for up to several weeks; they did not exhibit maternal behaviors at the first presentation of pups. In addition, a female that has been presented with a replacement litter up to the midpoint of her lactation period will often continue to lactate (Nicoll and Meites 1959). As development of a normal litter proceeds, pup behaviors interfere with maternal behavior — that is, make it difficult for the female to perform the maternal behaviors. Maternal rats retrieve pups frequently up to the second week of lactation, and then this behavior declines. At first pups are nursed only in the nest and with the mother's assistance, but later nursing can be initiated by pups and can take place away from the nest, wherever pups and mother encounter each other. After three to four weeks of nursing, FSH increases again in the female, and the estrous cycles resume (Bruce 1961; Rosenblatt 1967; Terkel and Rosenblatt 1971).

REPRODUCTION IN LIZARDS

Using primarily the green anole lizard *(Anolis carolinensis),* Crews (1975, 1977, 1979, 1980; Crews and Greenberg 1981) and his colleagues conducted field and laboratory investigations of the behavioral endocrinology of reproduction and the significance of various endocrine-behavior events as adaptations to the lizards' environments. Louisiana populations of the green anole, which investigators have studied most extensively, exhibit a four-part annual cycle:

1. From late September to late January, the lizards are quiescent and live under the bark of trees and under rocks.
2. In February the males emerge from dormancy and establish breeding territories.
3. In March the females become active; mating begins in late April, and by May they are laying one egg every ten to fourteen days, a pattern that they continue for several months.
4. In August both sexes enter a refractory period of about one month; during this time environmental and social cues that bring about gonadal recrudescence in the spring are no longer effective.

Combined laboratory and field techniques have led to significant conclusions about reproduction in the green anole lizard. The seasonal rise in temperature (Licht 1973) and the male's courtship behavior (Crews, Rosenblatt, and Lehrman 1974) affect ovarian development and egg laying in the female. The extension of the male's dewlap during displays is the key both to mate selection by the female and to the

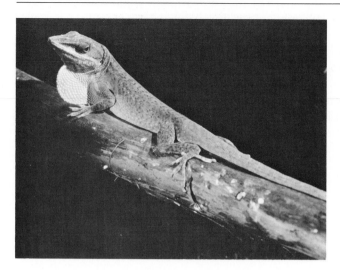

FIGURE 7-16   Dewlap display of male *anolis* lizard
The male *Anolis* lizard extends the dewlap toward the female as part of his courtship display. The dewlap display patterns are important for species-specific mate selection. This behavior pattern also appears to be critical for stimulating ovarian activity in the female.
Source: Photo from David Crews.

promotion of ovarian activity (Figure 7–16). Females are receptive to advances by males only when conception is most likely to occur. This receptivity is regulated, in part, by estrogen secreted by the developing follicle. Mating activity inhibits further receptivity by the female; the inhibition begins within twenty-four hours after mating and lasts for the duration of that cycle.

Several points regarding the adaptive significance of this reproductive pattern emerge from the data. The mating inhibition in females is adaptive because copulation in this species of lizard is prolonged and usually takes place in the open — for example, on tree limbs — where the mating pair is vulnerable to predation. Field data support the contention that more lizards are captured when they are mating than when they are engaged in other activities. The restriction of receptivity and mating to the time when fertilization is most likely to occur is thus evolutionarily adaptive (Valenstein and Crews 1977). Cessation of reproduction during the refractory period that precedes the winter dormancy period is adaptive in at least two ways. First, it ensures that young do not hatch at a time of diminished food resources and poorer environmental conditions. Second, the female is able to build up fat reserves for the winter rather than devote energy intake toward reproduction.

Sexual behavior in both male and female lizards depends on gonadal hormones, a pattern that we have noted in many other vertebrates. Investigators found that when they gave female lizards an injection of progesterone, followed by an injection of estrogen twenty-four hours later, a synergistic effect that induced female receptivity resulted. We noted a similar synergistic effect for these same two hormones on receptivity in rats (see Beach 1976, for review). Data obtained by autoradiography on the uptake of sex steroids by specific areas of the brain augment these parallel findings in lizards and rats. Experimenters inject the animal with radioactively labeled estrogen; several hours later they sacrifice the animal; section the brain, using a freezing microtome; and place the sections on special glass plates that have been treated with a photographic emulsion. They leave the plates in the dark for periods ranging from days to months depending upon the exact procedure used; and then they develop the plates to ascertain the specific sites where uptake occurred. The results for rats and lizards are strikingly similar — in both species, concentrations occur in the septum and preoptic regions (McEwen et al. 1979; Morrell, Kelley, and Pfaff 1975). These and related findings of comparative investigations point toward an exciting avenue for future research on common patterns of endocrinological events and interactions between hormones and behaviors in different classes of vertebrates.

HORMONE-BRAIN RELATIONSHIPS

In the previous chapter and in this one we have already mentioned some of the ongoing research on the mechanisms by which the endocrine and nervous systems interact with one another. A few additional comments will provide an appropriate finale to our examination of these major control systems for behavior.

      For many of the hormones that influence behavior, we now know there are regions of the brain with cells that possess specific receptor sites for those hormones. The experimental procedures involve a combination of autoradiography — to precisely pinpoint the receptor locations, and the use of electrodes — to measure electrophysiological activity from the same sites (Morrell and Pfaff 1981; Pfaff 1981). Thus, for example, when radioactively labelled sex steroids are given to rats (Pfaff and Keiner 1973), frogs (Kelley, Morrell, and Pfaff 1975), or chaffinches (Zigmond, Nottebohm, and Pfaff 1973), the label becomes concentrated in similar brain areas in all three classes. The preoptic area and portions of the limbic system and hypothalamus are among the sites which commonly have the highest amounts of the labelled steroid. When electrode recordings are made from cells in these locations, some important relationships between hormones, behavior, and neural activity emerge. Firing rates for neurons in the highly labelled portions of the limbic system and hypothalamus vary during the course of the rat estrous cycle — generally, they are highest shortly before ovulation (Kawakami, Terasawa, and Ibuki 1970; Terasawa and Sawyer 1969). When ovariectomized female rats are given injections of estrogen, firing rates in these same sites are elevated (Cross and Dyer 1972). Thus, there appears to be some clear connection between the hormone and neural activity.

      The subject of neural substrates and hormone sensitivity in mammals has been reviewed by Komisaruk (1978). Clearly, we are now approaching the time when we will understand the mechanisms relating hormones, the nervous system, and behavior at the level of the individual cells and the biochemistry within and among those cells.

SUMMARY

The endocrine, or hormonal, system has evolved in concert with the nervous system, and the two systems are closely interrelated in their regulation of behavior. Hormones are chemical substances produced by specialized ductless glands or by neurosecretory cells. They influence the processes of growth, metabolism, water balance, and reproduction. Most of the endocrine structures in invertebrates, as exemplified here by the insects, involve neurons directly or they are closely tied to the nervous system. Trophic hormones in both invertebrates and vertebrates affect the production and secretion of other hormones. Feedback loops, particularly in vertebrates, control the release of hormones and neurosecretions and monitor hormonal levels. Synergism and antagonism between hormones are important determinants of the effects of those hormones on behavior.

Three major regions of the vertebrate pituitary gland produce hormones that affect behavior. The posterior pituitary secretes oxytocin, which affects uterine muscles and milk ejection, and vasopressin, which plays a key role in regulating water balance. The central portion of the pituitary secretes MSH, a hormone that affects pigments, and thus skin color or hair color. The anterior pituitary secretes three trophic hormones—FSH and LH, which affect the gonads and reproduction, and ACTH, which influences production and release of hormones from the adrenal cortex. The anterior pituitary also produces prolactin, a hormone that affects maternal behavior, milk production, and related processes in birds and mammals. The gonads and adrenals are other endocrine glands that produce and secrete hormones affecting behavior.

Some hormones exert activational, or triggering, effects on behavior. Examples of activational effects include coat color changes in the weasel, the release of pheromones and the process of eclosion in insects, molting in crustaceans, sexual and courtship behaviors in rats and ring doves. The effects of other hormones are organizational because these effects are manifested during development. In particular, sexual differentiation is an organizational effect in some mammals because it is under some degree of hormonal control. Some hormones can exert both activational and organizational effects.

Extensive research has revealed that there are complex environment-hormone-behavior interactions, as exemplified here by the reproductive sequences in ring doves, maternal behavior in rats, and the seasonal reproductive pattern of lizards. The hormonal effects within individual animals, the effects of behavior on hormones, and the interactions between the hormonal/physiological state of the animal and features of its environment produce observed behavior patterns.

The research techniques investigators use to uncover these hormone-behavior relationships include the removal of endocrine glands, hormone replacement therapy, blood transfusions, autoradiography, and assays for hormone levels. Using several of these techniques, investigators have explored the mechanisms by which the endocrine and nervous systems are interrelated in controlling behavior.

Discussion Questions

1. Both male and female golden hamsters possess sebaceous flank glands that produce sebum, a substance that is used to mark objects in their environment. Devise three or four experimental questions (hypotheses) you would want to ask in studying the hormonal regulation and control of this scent-marking behavior. For each question identify the treatment groups you would use for testing and justify the need for each.

2. The following summary table is from the review of early organizational effects of hormones on behavior by Adkins-Regan (1981). A number of reptiles have been tested using various techniques to manipulate the early hormone environment. What general conclusions can you make about organizatonal effects on sex structures in these reptiles? How do these results compare with those discussed in the chapter on mammals?

Summary of experiments on sex differentiation in Reptilia

Species	Age at Treatment	Treatment	Gonads	Gonaducts	Other Sex Structures
			Effect on[a]		
Several	Embryos	Gonadectomy		M →F	
Several	Embryos	Estrogens	m → f		
Lacerta vivipara	Embryos	Gonadectomy		M → F	F → M
	Embryos	Parabiosis			M → F
	Embryos	Androgens			f → m
	Embryos	Estrogens	m → f	M → F	M → F
Anguis fragilis	Embryos	Testosterone		F → M	
Anolis carolinensis	Juveniles	Estrogen			M → F
Sceloporus sp.	Juveniles	Estrogen			M → F
Alligator mississippiensis	Juveniles	Testosterone		m → f	f → m
Crocodylus niloticus	Juveniles	Testosterone	f → m	m → f	
Thamnophis sirtalis	Embryos	Testosterone		m → f	
Malaclemmys centrata	Juveniles	Testosterone		m → f	f → m
	Juveniles	Estradiol	m → f	m → f	
Chrysemys marginata	Embryos	Testosterone	f → m		
Emys orbicularis	Embryos	Estradiol or testosterone	m → f		
Emys leprosa	Recently hatched	Androgens		m → f	f → m
	Recently hatched	Estrogen	m → f	m → f	
Testudo graeca	Embryos	Estradiol	m → f	m → f	

Source: From E. Adkins-Regan (1981), "Early organizatonal effects of hormones," Pp 159–228 in N. Adler (ed.), *Neuroendocrinology of Reproduction.* New York: Plenum Press, p. 188.

[a] F → M indicates extensive or complete masculinization of the character; M → F indicates extensive or complete feminization. f → m indicates slight or partial masculinization; m → f indicates slight or partial feminization.

3. The reproductive sequences of ring doves and lizards and the maternal behavior of rats illustrate the feedback loops involved in the interactions of internal hor-

monal conditions, behavior, and environmental cues. What other behavior sequences can you describe that have similar complex feedback interactions?

4. The testes of twenty-five males of a bird species were weighed in each of four seasons. To organize the data, the birds were divided into groups of five on the basis of their social dominance interactions; the five top-ranking birds were placed in category I, the next five in category II, and so forth. The table below shows average testes weights in milligrams for the birds in each category and in each of the four seasons. What general conclusions can you draw from these data? What conditions of interpretation would you place on these conclusions?

Weights of birds' testes (in milligrams)

	Male Dominance Category				
	I	*II*	*III*	*IV*	*V*
Spring	7.6	7.0	6.8	6.4	5.8
Summer	8.0	7.7	7.2	6.9	6.2
Fall	4.1	3.9	3.8	3.9	3.8
Winter	2.6	2.5	2.6	2.4	2.5

Suggested Readings

Adkins-Regan, E. 1981. Early organizational effects of hormones: An evolutionary perspective. In N. Adler (ed.), *Neuroendocrinology of Reproduction.* New York: Plenum.
The most comprehensive summary to date of the organizational effects of hormones on behavior. Covers the entire taxonomic spectrum from coelenterates to mammals. Excellent references.

Beyer, C., ed. 1979. *Endocrine Control of Sexual Behavior.* New York: Raven Press.
A volume of review articles by various authorities. Covers topics ranging from behavior-hormones-environment interactions to biochemical endocrinology. Deals exclusively with vertebrates and primarily with mammals. Excellent reference source; has details of methods used in analyzing interactions between hormones and behavior.

Crews, D. 1980. Interrelationships among ecological, behavioral, and neuroendocrinological processes in the reproductive cycle of *Anolis carolinensis* and other reptiles. *Adv. Stud. Behav.* 11:1–74.
Crews's investigations on lizards parallel those of Lehrman and Hinde on avian systems. Strong ecological and evolutionary flavor. Provides integrated approach of the animal in its natural habitat and laboratory studies at the chemical level.

Hutchinson, J. B. 1976. Hypothalamic mechanisms of sexual behavior, with special reference to birds. *Adv. Stud. Behav.* 6:159–200.
Comments on research conducted at the interface between the hormonal and nervous systems. Combines a review approach with presentation of new research findings.

Nemeroff, C. B., and A. J. Dunn. 1984. *Peptides, Hormones and Behavior.* Lancaster, England: MTP Press.
A volume of contributed papers covering a range of topics pertaining to hormones and behavior. Chapter topics range from biochemistry of hormone production to animal models for hormones affecting behavior.

Truman, J. W., and L. M. Riddiford. 1974. Hormonal mechanisms underlying insect behavior. *Adv. Insect Physiol.* 10:297–352.
Good summary of relationship between hormones and behavior for invertebrate organisms. Good bibliography. Solid starting point for techniques and methods of studying insect behavior from neural and hormonal perspectives.

References

Adkins-Regan, E. 1981. Early organizational effects of hormones. In N. Adler (ed.), *Neuroendocrinology of Reproduction.* New York: Plenum.

Arnold, A. P. and R. A. Gorski. 1984. Gonadal steroid induction of structural sex differences in the central nervous system. *Annual Review of Neuroscience* 7:413–442.

Barfield, R. J. 1971. Activation of sexual and aggressive behavior activated by androgen implanted into the male ring dove brain. *Endocrinology* 89:1470–1476.

Barfield, R. J., D. E. Busch, and K. Wallen. 1972. Gonadal influence on agonistic behavior in the male domestic rat. *Horm. Behav.* 3:247–259.

Barraclough, C. A. 1967. Modifications in reproductive function after exposure to hormones during the prenatal and postnatal period. In L. Martini and W. F. Ganing, eds., *Neuroendocrinology,* vol. 2. New York: Academic Press.

Barraclough, C. A., and R. A. Gorski. 1961. Evidence that the hypothalamus is responsible for androgen-induced sterility in the female rat. *Endocrinology* 68:68–79.

Barth, R. H. 1965. Insect mating behavior: Endocrine control of a chemical communication system. *Science* 149:882–883.

———. 1968. The comparative physiology of reproductive processes in cockroaches. Part I. Mating behavior and its endocrine control. *Adv. Reprod. Physiol.* 3:167–201.

Beach, F. A. 1976. Sexual attractivity, proceptivity and receptivity in female mammals. *Horm. Behav.* 7:105–138.

Beach, F. A., and A. M. Holz-Tucker. 1949. Effects of different concentrations of androgen upon sexual behavior in castrated male rats. *J. Comp. Physiol. Psychol.* 42:433–453.

Bennett, M. A. 1940. The social hierarchy in ring doves. II. The effect of treatment with testosterone propionate. *Ecology* 21:148–165.

Bouissou, M. F. 1978. Effect of injections of testosterone propionate on dominance relationships in a group of cows. *Horm. Behav.* 11:388–400.

Bouissou, M. F. and V. Gaudioso. 1982. Effect of early androgen treatment on subsequent social behavior in heifers. *Horm. Behav.* 16:132–146.

Bruce, H. M. 1961. Observations on the suckling stimulus and lactation in the rat. *J. Reprod. Fertil.* 2:17–34.

Buchsbaum, R. 1938. *Animals without Backbones.* Chicago: University of Chicago Press.

Carlisle, D. B., and P. E. Ellis. 1959. La persistance des glandes ventrales céphaliques chez les criquets solitaires. *Comptes Rendu* 249:1059–1060.

Charniaux-Cotton, H., and L. H. Kleinholz. 1964. Hormones in invertebrates other than insects. In G. Pincus, K. V. Thimann, and E. B. Astwood (eds.), *The Hormones*, vol. 4. New York: Academic Press.

Cheng, M. F. 1979. Progress and prospects in ring dove research: A personal view. *Adv. Stud. Behav.* 9:97–130.

Clemens, L. G. 1974. Neurohormonal control of male sexual behavior. In W. Montagna and S. Sadler (eds.), *Reproductive Behavior.* New York: Plenum.

Clemens, L. G., B. A. Gladue, and L. P. Coniglio. 1978. Prenatal endogenous androgenic influences on masculine sexual behavior and genital morphology in male and female rats. *Horm. Behav.* 10:40–53.

Cole, L. J. 1916. Twinning in cattle with special reference to the free-martin. *Science* 43:177.

Crews, D. 1975. Psychobiology of reptilian reproduction. *Science* 189:1059–1065.

———. 1977. The annotated *Anole:* Studies on the control of lizard reproduction. *Amer. Scientist* 65:428–434.

———. 1979. Neuroendocrinology of lizard reproduction. *Biol. Reprod.* 20:51–73.

———. 1980. Interrelationships among ecological, behavioral and neuroendocrine processes in the reproductive cycle of *Anolis carolinensis* and other reptiles. *Adv. Stud. Behav.* 11:1–75.

Crews, D., J. S. Rosenblatt, and D. S. Lehrman. 1974. Effects of unseasonal environmental regime, group presence, group composition and males' physiological state on ovarian recrudescence in the lizard *Anolis carolinensis. Endocrinology* 95:102–106.

Crews, D., and N. Greenberg. 1981. Function and causation of social signals in lizards. *Amer. Zool.* 21:273–294.

Cross, B. A., and R. G. Dyer. 1972. Ovarian modulation of unit activity in the anterior hypothalamus of the cyclic rat. *J. Physiol.* (Lond.) 222:25P.

Davidson, J. M. 1966a. Characteristics of sex behavior in male rats following castration. *Anim. Behav.* 14:266–272.

———. 1966b. Activation of the male rat's sexual behavior by intracerebral implantation of androgen. *Endocrinology* 79:783–794.

deWilde, J. 1975. An endocrine view of metamorphosis, polymorphism and diapause in insects. *Amer. Zool.* (supplement 1):13–28.

Dörner, G., and M. Kawakami, eds. 1978. *Hormones and Brain Development.* New York: Elsevier North-Holland.

Drickamer, L. C., and J. G. Vandenbergh. 1973. Predictors of dominance in the female golden hamster *(Mesocricetus auratus). Anim. Behav.* 21:564–570.

Drickamer, L. C., J. C. Vandenbergh, and D. R. Colby. 1973. Predictors of dominance in the male golden hamster *(Mesocricetus auratus). Anim. Behav.* 21:557–563.

Edwards, D. A. 1968. Mice: Fighting by neonatally androgenized females. *Science* 161:1027–1028.

Erickson, C. J., and D. S. Lehrman. 1964. Effect of castration of male ring doves on ovarian activity of females. *J. Comp. Physiol. Psychol.* 58:164–166.

Feder, H. H. 1981. Experimental analysis of hormone actions on the hypothalamus, anterior pituitary and ovary. In N. Adler (ed.), *Neuroendocrinology of Reproduction.* New York: Plenum.

Fingerman, M. 1965. Chromatophores. *Physiol. Rev.* 45:296–339.

———. 1973. Behavior of chromatophores of the fiddler crab *Uca pugilator* and the dwarf crayfish *Cambarellus shufeldti* in response to synthetic *Pandalus* red pigment concentrating hormone. *Gen. Comp. Endocr.* 20:589–592.

Fletcher, T. J. 1978. The induction of male sexual behavior in red deer *(Cervus elaphus)* by the administration of testosterone to hinds and estradiol 17B to stags. *Horm. Behav.* 11:74–88.

Fraenkel, G. 1975. Interactions between ecdysone, bursicon and other endocrines during puparium formation and adult emergence in flies. *Amer. Zool.* 15 (supplement 1):29–41.

Gandelman, R., F. S. vom Saal, and J. M. Reinisch. 1977. Contiguity to male fetuses affects morphology and behavior of female mice. *Nature* 266:722–724.

Gilbert, L. I. 1974. Endocrine action during insect growth. *Rec. Prog. Horm. Res.* 30:347–384.

Golding, D. W. 1972. Studies in the comparative neuroendocrinology of polychaete reproduction. *Gen. Comp. Endocr.* (supplement 3):580–590.

Goy, R. W. 1970. Experimental control of psychosexuality. *Phil. Trans. Roy. Soc. Lond.*, series B. 259:149–162.

Grota, L. J., and K. B. Eik-Nes. 1967. Plasma progesterone concentrations during pregnancy and lactation in the rat. *J. Reprod. Fertil.* 13:83–91.

Guhl, A. M. 1961. Gonadal hormones and social behavior. In W. C. Young (ed.), *Sex and Internal Secretions.* Baltimore: Williams & Wilkins.

Harris, G., and S. Levine. 1965. Sexual differentiation of the brain and its experimental control. *J. Physiol.* 181:379–400.

Hutchinson, J. B. 1969. Changes in hypothalamic responsiveness to testosterone in male Barbary dove *(Streptopelia risoria). Nature* 222:176.

———. 1971. Effects of hypothalamic implants of gonadal steroids on courtship behavior in Barbary doves *(Streptopelia risoria). J. Endocr.* 50:97–113.

———. 1978. Hypothalamic regulation of male sexual responsiveness to androgen. In J. B. Hutchinson (ed.), *Biological Determinants of Sexual Behavior.* New York: Wiley.

Johnson, C. G. 1969. *Migration and Dispersal of Insects by Flight.* London: Methuen.

Jussiaux, M., and C. Trillaud. 1979. Comparaison entres des techniques de vasectomie et d'androgenisation pour la détection des chaleurs. *Cereopa, Journee d'Etude,* March 1979.

Kawakami, M., E. Terasawa, and T. Ibuki. 1970. Changes in multiple unit activity of the brain during the estrous cycle. *Neuroendocrinology* 6:30–48.

Kelley, D. B., J. I. Morrell, and D. W. Pfaff. 1975. Autoradiographic localization of hormone-concentrating cells in the brain of an amphibian, *Xenopus laevis.* I. Testosterone. *J. Comp. Neurol.* 164:47–62.

Kleinholz, L. H. 1970. A progress report on the separation and purification of crustacean neurosecretory pigmentary-effector hormones. *Gen. Comp. Endocr.* 14:578–588.

Komisaruk, B. I. 1978. The nature of the neural substrate of female sexual behaviour in mammals and its hormonal sensitivity: Review and speculations. In J. B. Hutchison (ed.), *Biological Determinants of Sexual Behaviour.* Chichester, England: Wiley.

Lehrman, D. S. 1955. The physiological basis of parental feeding behavior in ring doves *(Streptopelia risoria). Behaviour* 7:241–286.

———. 1958a. Effect of female sex hormones on incubation behavior in the ring dove *(Streptopelia risoria). J. Comp. Physiol. Psychol.* 51:142–145.

———. 1958b. Induction of broodiness by participation in courtship and nest-building in the ring dove *(Streptopelia risoria). J. Comp. Physiol. Psychol.* 51:32–36.

———. 1959. Hormonal responses to external stimuli in birds. *Ibis* 101:478–496.

———. 1961. Hormonal regulation of parental behavior in birds and infrahuman mammals. In W. C. Young (ed.), *Sex and Internal Secretions.* Baltimore: Williams & Wilkins.

———. 1964. The reproductive behavior of ring doves. *Sci. Amer.* 211:48–54.

———. 1965. Interaction between internal and external environments in the regulation of the reproductive cycle of the ring dove. In F. A. Beach (ed.), *Sex and Behavior.* New York: Wiley.

Lehrman, D. S., P. N. Brody, and R. P. Wortis. 1961. The presence of the mate and of nesting material as stimuli for the development of incubation behavior and for gonadotrophin secretion in the ring dove *(Streptopelia risoria). Endocrinology* 68:507–516.

Leshner, A. I. 1975. A model of hormones and agonistic behavior. *Physiol. Behav.* 15:225–235.

Levine, S. 1967. Maternal and environmental influences on the adrenocortical responses to stress in weanling rats. *Science* 156:258–260.

———. 1968. Hormones and conditioning. *Nebr. Symp. Motiv.* 16:85–101.

Levine, S., and R. F. Mullins. 1966. Hormonal influences on brain organizaton in infant rats. *Science* 152:1585–1592.

Licht, P. 1973. Influence of temperature and photoperiod on the annual ovarian cycle in the lizard *Anolis carolinensis. Copeia* 1973:465–472.

Lillie, F. R. 1916. The theory of the free-martin. *Science* 43:611–613.

Lohrer, W. 1961. The chemical acceleration of the maturation process and its hormonal control in the male of the desert locust. *Proc. Roy. Soc. Lond.,* series B. 153:380–397.

Lohrer, W., and F. Huber. 1966. Nervous and endocrine control of sexual behavior in a grasshopper *(Comphocerus rufus* L., Acridinae). *Soc. Exp. Biol. Symp.* 20:381–400.

Lott, D. F. 1962. The role of progesterone in the maternal behavior of rodents. *J. Comp. Physiol. Psychol.* 55:610–613.

Lott, D. F., and J. S. Rosenblatt. 1969. Development of maternal responsiveness during pregnancy in the rat. In B. M. Foss (ed.), *Determinants of Infant Behavior,* vol. 4. London: Methuen.

Lubin, M., et al. 1972. Hormones and maternal behavior in the male rat. *Horm. Behav.* 3:369–374.

Luttge, W. G., and R. E. Whalen. 1970. Dihydrotestosterone, androstenodione, testosterone: Comparative effectiveness in masculinizing and defeminizing reproductive systems in male and female rats. *Horm. Behav.* 1:265–281.

Manning, A., and T. E. McGill. 1974. Early androgen and sexual behavior in female house mice. *Horm. Behav.* 5:19–31.

Masson, G. M. 1948. Effects of estradiol and progesterone on lactation. *Anat. Rec.* 102:513–521.

McEwen, B. S., et al. 1979. The brain as a target for steroid hormone action. *Ann. Rev. Neurosci.* 2:65–112.

Meany, M. J., J. Stewart, and W. W. Beatty. 1982. The influence of glucocorticoids during the neonatal period on the development of play-fighting in Norway rat pups. *Hormones and Behavior* 16:475–491.

Michel, R. 1972. Étude expérimentale de l'influence des glandes prothoraciques sur l'activité de vol du Criquet Pèlerin *Schistocerca gregaria*. *Gen. Comp. Endocr.* 19:96–101.

Moltz, H., D. Robbins, and M. Parks. 1966. Caesarian delivery and the maternal behavior of primiparous and multiparous rats. *J. Comp. Physiol. Psychol.* 61:455–460.

Money, J. 1977. Human hermaphroditism. In F. A. Beach (ed.), *Human Sexuality in Four Perspectives.* Baltimore: Johns Hopkins University Press.

Money, J., and A. A. Ehrhardt. 1972. *Man and Woman, Boy and Girl: Differentiation and Dimmorphism of Gender Identity from Conception to Maturity.* Baltimore: Johns Hopkins University Press.

Morrell, J. I., D. B. Kelley, and D. W. Pfaff. 1975. Sex steroid binding in the brains of vertebrates: Studies with light microscopic autoradiography. In K. M. Knigge et al. (eds.), *The Ventricular System in Neuroendocrine Mechanisms.* Basel: Karger.

Morrell, J. I., and D. W. Pfaff. 1981. Autoradiographic technique for steroid hormone localization. In N. Adler (ed.), *Neuroendocrinology of Reproduction.* New York: Plenum.

Nicoll, C. S., and J. Meites. 1959. Prolongation of lactation in the rat by litter replacement. *Proc. Soc. Exp. Biol. Med.* 101:81–82.

O'Connell, M. E., C. Reboulleau, H. H. Feder, and P. Silver. 1981a. Social interactions and androgen levels in birds. I. Female characteristics associated with increased plasma androgen levels in the male ring dove *(Streptopelia risoria)*. *Gen. Comp. Encodrinol.* 44:454–463.

O'Connell, M. E., R. Silver, H. H. Feder, and C. Reboulleau. 1981b. Social interactions and androgen levels in birds. II. Social factors associated with a decline in plasma androgen levels in male ring doves *(Streptopelia risoria)*. *Gen. Comp. Endocrinol.* 44:464–469.

Pener, M. P. 1965. On the influence of corpora allata on maturation and sexual behavior of *Schistocerca gregaria. J. Zool. Lond.* 147:119–136.

Pfaff, D. W. 1981. Electrophysiological effects of steroid hormones in brain tissue. In N. Adler (ed.), *Neuroendocrinology of Reproduction.* New York: Plenum.

Pfaff, D. W., and M. Keiner. 1973. Atlas of estradiol-concentrating cells in the central nervous system of the female rat. *J. Comp. Neurol.* 151:121–130.

Phoenix, C. H., et al. 1959. Organizing action of prenatally administered testosterone propionate on the tissues mediating mating behavior in the female guinea pig. *Endocrinology* 65:369–382.

Reiter, R. J. 1980. The pineal gland: A regulator of regulators. *Prog. in Psychobiol. and Physiol. Psych.* 9:323–356.

Richards, M. P. 1967. Maternal behavior in rodents and lagomorphs. In A. McLaren (ed.), *Advances in Reproductive Physiology*, vol. 2. New York: Academic Press.

Rosenblatt, J. S. 1967. Nonhormonal basis for maternal behavior in the rat. *Science* 156:1512–1514.

———. 1970. Views on the onset and maintenance of maternal behavior in the rat. In L. R. Aronson et al. (eds.), *Development and Evolution of Behavior.* San Francisco: Freeman.

Rosenblatt, J. S., and D. S. Lehrman. 1963. Maternal behavior of the laboratory rat. In H. L. Rheingold (ed.), *Maternal Behavior in Mammals.* New York: Wiley.

Rosenblatt, J. S., H. I. Siegel, and A. D. Mayer. 1979. Progress in the study of maternal behavior in the rat: Hormonal, nonhormonal, sensory, and developmental aspects. *Adv. Stud. Behav.* 10:226–311.

Rust, C. C. 1965. Hormonal control of pelage cycles in the short-tailed weasel *(Mustela erminea bangsi). Gen. Comp. Endocr.* 5:222–231.

Rust, C. C., and R. K. Meyer. 1969. Coat color, molt and testis size in male short-tailed weasels treated with melatonin. *Science* 165:921–922.

Silver, R. 1978. The parental behavior of ring doves. *Amer. Sci.* 66:209–215.

Terasawa, E., and C. H. Sawyer. 1969. Changes in electrical activity in the rat hypothalamus related to electrochemical stimulation of adenohypophyseal function. *Endocrinology* 85:143–151.

Terkel, J., and J. S. Rosenblatt. 1968. Maternal behavior induced by maternal blood plasma injection into virgin rats. *J. Comp. Physiol. Psychol.* 65:479–482.

––––––. 1971. Aspects of nonhormonal maternal behavior in the rat. *Horm. Behav.* 2:161–171.

––––––. 1972. Humoral factors underlying maternal behavior at parturition. *J. Comp. Physiol. Psychol.* 80:365–371.

Tollman, J., and J. A. King. 1956. The effects of testosterone propionate on aggression in male and female C57BL/10 mice. *Anim. Behav.* 4:147–149.

Toran-Allerand, C. D. 1978. Gonadal hormones and brain development: Cellular aspects of sexual differentiation. *Amer. Zool.* 18:553–565.

Truman, J. W. 1971. Physiology of insect ecdysis. I. The eclosion behavior of saturniid moths and its hormonal release. *J. Exp. Biol.* 54:805–814.

Truman, J. W., and L. M. Riddiford. 1970. Neuroendocrine control of ecdysis in silkmoths. *Science* 167:1624–1626.

Turner, C. D., and J. T. Bagnara. 1976. *General Endocrinology.* Philadelphia: Saunders.

Uvarov, B. 1966. *Grasshoppers and Locusts.* New York: Cambridge University Press.

Valenstein, P., and D. Crews. 1977. Mating-induced termination of behavioral estrus in the female lizard *Anolis carolinensis. Horm. Behav.* 9:362–370.

Vandenbergh, J. G. 1971. The effects of gonadal hormones on the aggressive behavior of adult golden hamsters. *Anim. Behav.* 19:589–594.

––––––. 1973. Effects of gonadal hormones on the flank gland of the golden hamster. *Horm. Res.* 4:28–33.

vom Saal, F. S., and F. H. Bronson. 1978. In utero proximity of female mouse fetuses to males: Effect on reproductive performance during later life. *Biol. Reprod.* 19:842–853.

––––––. 1980a. Variation in length of the estrous cycle in mice due to former intrauterine proximity to male fetuses. *Biol. Reprod.* 22:777–780.

––––––. 1980b. Sexual characteristics of adult female mice are correlated with their blood testosterone levels during prenatal development. *Science* 208:597–599.

Wells, M. J., and J. Wells. 1959. Hormonal control of sexual maturity in *Octopus. J. Exp. Biol.* 36:1–33.

Young, W. C. 1961. The hormones and mating behavior. In W. C. Young (ed.), *Sex and Internal Secretions.* Baltimore: Williams & Wilkins.

Young, W. C., R. W. Goy, and C. H. Phoenix. 1964. Hormones and sexual behavior. *Science* 143:212–218.

Zarrow, M. X. 1961. Gestation. In W. C. Young (ed.), *Sex and Internal Secretions.* Baltimore: Williams & Wilkins.

Zigmond, R. E., F. Nottebohm, and D. W. Pfaff. 1973. Androgen-concentrating cells in the midbrain of a songbird. *Science* 179:1005–1006.

8

BIOLOGICAL TIMEKEEPING

Many moths and butterflies emerge from their coccoons at dawn when the still, somewhat moist air provides the best conditions for slowly drying their unfolding wings. Bats emerge from their caves or roosts at dusk each evening to forage for insects or fruits, and return before dawn to sleep during the daylight hours. These activities and a host of other animal behavior patterns occur on a regular daily basis. Many songbirds of the north temperate deciduous forest fly southward during the autumn, overwinter in tropical or subtropical zones, and begin their return flight northward as spring approaches. Woodchucks emerge from their winter dens in spring, remain active for four to six months, and return to hibernation by late autumn. During their months above ground they mate, produce offspring, rear the young, and prepare for another winter. These and other animal activities occur on an annual cycle. Still other animal behavior patterns are observed with frequencies which approximate cyclic features in the environment, e.g. tides, lunar phases. **Biological rhythms** are those animal activities and behaviors which can be directly related to distinct environmental frequencies. Biological rhythms are the external manifestations of and are regulated by biological clocks. **Biological clocks** are internal timing mechanisms that involve both a *self-sustaining physiological pacemaker* and an *environmental cyclic synchronizer (Zeitgeber)*.

In this chapter we first examine the characteristics of biological rhythms. Next is a section on locating the pacemakers and understanding their physiology: the mechanistic or "how" questions. We can also ask questions about the functional significance of biological rhythms within ecological contexts: the "why" questions.

We would expect natural selection to favor individuals whose peak activity periods — feeding, for example — coincide with peak activity of their prey. The final section of the chapter deals with the significance of biological clocks.

BIOLOGICAL RHYTHMS

Each biological rhythm is comprised of repeating units called **cycles.** The length of time required to complete an entire cycle is the rhythm's *period;* twenty-four hours is the period in the example shown in Figure 8–1. The magnitude of the change in activity rate during a cycle — the difference between peaks and troughs — is the **amplitude.** Any specified, recognizable part or portion of a cycle is called a **phase;** the designated portion of the cycle shown in Figure 8–1 would be an active feeding phase.

Biological rhythms may be characterized by several specific properties. First, whereas temperature changes alter the rate of most chemical reactions and cellular processes, biological rhythms are **temperature-compensated.** Generally the rate of a chemical reaction doubles for each 10°C increase in temperature. However, biological rhythms are relatively insensitive to change in temperature (Sweeney and Hastings 1960; Rawson 1960). This is significant, for if biological rhythms were

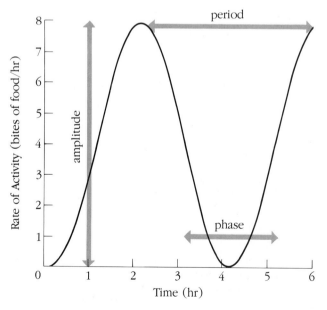

FIGURE 8-1   Biological rhythm The characteristics of a biological rhythm include the period, phase, and amplitude. Each of these characteristics can vary for the same animal over time, between animals of the same species, or between animals of different species.

speeded up or slowed down by ambient temperature changes, they would not keep accurate time.

Second, biological clocks are generally unaffected by metabolic poisons or inhibitors that block biochemical pathways within cells. We might expect that application of a metabolic poison, such as sodium cyanide, would alter the period of a biological rhythm, yet this does not happen.

A third property is that the periods of biological rhythms occur with approximately the same frequency of one or more environmental features. (The Latin word *circa*—about or approximately—usually is part of the name of this kind of rhythm, as we will see.) Fourth, as we explore further later in the chapter, biological rhythms are self-sustaining, maintaining approximately their normal cyclicity even in the absence of environmental cues. Fifth, and last, biological rhythms can be entrained to environmental cues. The self-sustaining pacemaker mechanism(s) may be set and adjusted according to input from the external environment.

TYPES OF RHYTHMS

EPICYCLES. Different organisms exhibit a variety of biological activities with varying frequencies and periods (Table 8–1). Some of these patterns are of relatively short duration and are generally termed **epicycles.** Lugworms *(Arenicola marina),* living in burrows on sand flats in the intertidal zones, feed every six to eight minutes. Some small mammals like meadow voles *(Microtus pennsylvanicus),* which are active pri-

TABLE 8–1   Summary of biological rhythms, with periods ranging from a few minutes to several years

Type of Cycle	*Organism*	*Behavior*
Epicycles (variable)	Lugworm Meadow vole	Feeding (every 6–8 min) Feeding/resting (every 15–120 min during daylight)
Tidal (12.4 hours)	Oyster Fiddler crab	Opening of shell valves Locomotion/feeding
Lunar (28 days)	Midge (marine insect) Grunion (marine fish)	Mating/egg laying Egg laying
Circadian (24 hours)	Deermouse Fruit fly	Drinking/general activity Emergence of adults from pupa
Circannual (12 months)	Woodchuck Chickadee Robin	Hibernation Reproduction Migration/reproduction
Intermittent (variable–days up to several years)	Desert insect Lion Shiner (river fish)	Reproduction (triggered by rain) Feeding (triggered by hunger) Reproduction (triggered by flooding)

marily during the daylight hours, show bursts of activity followed by periods of rest in short cycles that vary from twelve to twenty minutes up to about two hours. These epicycles do not generally fit the definition provided for biological rhythms, and we will not consider them further here except to note that they are cyclic and, as the two examples indicate, they are potentially important activity rhythms in the lives of some organisms.

TIDAL RHYTHMS.   A primary environmental feature of seacoasts is the ebb and flow of the tides. Tidal rhythms affect activity periods in many organisms that inhabit these zones. Tides are the result of gravitational forces associated with the moon and exhibit a 12.4 hour period from one low-tide phase to the next. Many oyster species (family Ostreidae) open their shell valves in accordance with tidal rhythm (Figure 8–2). They are open to feed when they are covered with water, but remain closed when the tide is low to prevent dessication (Brown 1970).

LUNAR RHYTHMS.   Based on the approximately twenty-eight-day cycle of the moon, lunar rhythms are clearly related to the tidal rhythm. Some marine insects, like the midge *(Clunio marinus),* coordinate their eclosion, mating, and egg-laying activities with the lunar cycle. They lay their eggs at very low tide, thereby ensuring that the larvae will hatch in the proper marine environment (Neumann 1966). Grunion *(Leuresthes* spp.), a marine fish, spawn during the spring tides on the sand beaches of California and use the moon as a timing cue for these activities (Walker 1949).

CIRCADIAN RHYTHMS.   Many organisms exhibit biological rhythms, governed by self-sustaining internal pacemakers, of about twenty-four hours duration. These we call **circadian rhythms.** Within the daily cycle, some animals exhibit peak activity during the daylight hours **(diurnal);** some are active primarily at night **(nocturnal);** and still others exhibit peak activity around dusk and/or dawn **(crepuscular).** Activity periods may shift seasonally. Many bird species that are year-round residents of northern temperate zones are primarily crepuscular throughout late spring and summer, but they shift to a more diurnal pattern during the colder winter months, thus avoiding the very cold temperatures of early winter mornings. Circadian activity periods may also show age-dependent shifts. Young woodchucks restrict most of their activity to early evening hours, whereas adult woodchucks are more diurnal in their pattern. Similarly, young dragonflies fly during the few hours just after dawn, but adults of the species fly mostly in the middle of the day (Corbet 1957, 1960).

CIRCANNUAL RHYTHMS.   As the term implies, **circannual rhythms** are behavioral and physiological patterns, governed by self-sustaining internal pacemakers, that occur with a period of about one year. Some mammals enter a condition of deep sleep and reduced metabolic activity, or **hibernation,** during the winter months. By doing so they avoid the harsh conditions of winter. Many bird species escape the rigors of winter in northern and temperate climates by migrating to southern latitudes. The annual life cycle of many insects that live where there are seasonal

FIGURE 8–2 Oysters at low tide
Oysters exhibit a rhythm that is regulated by the ebb and flow of the tides. They remain with their shells open and feed on microorganisms in the water when the tide is in, and they close their shells to prevent desiccation when the tide is out. When oysters are translocated inland, where there are no tides, they continue to open and close their shells as if there were tides.
Source: Photo by B. J. Miller, Fairfax, Virginia/BPS.

climatic shifts incorporate a **diapause phase** — a period of dormancy — during the more rigorous portions of the climatic cycle. For example, silkworm moths (*Bombyx* spp.) and mosquitos (*Aedes* spp.) lay eggs that are dormant during the winter (Kogure 1933; Vinogradova 1965); nymphs of the dragonfly *(Tetragneura cynosura)* over-winter as larvae forms and complete development the following spring (Lutz and Jenner 1964); both the parasitic wasp *(Nasonia vitripennis)* and its host, the flesh fly *(Sacrophaga argyrostoma)* enter diapause as larvae (Saunders 1978); still other insects go through diapause in the pupal stage or as adults.

ENDOGENOUS PACEMAKER

One of the critical characteristics of biological rhythms is the existence of an internal self-sustaining pacemaker. What is the evidence for the existence of such **endogenous** clock mechanisms? As we examine the following sequence of information in support of the existence of such internal chronometers bear in mind that what we are measuring are generally the overt, observable manifestations of the clock — the periodic changes in physiology and behavior — not the mechanism itself.

FREE-RUNNING RHYTHMS. One indirect method of establishing the existence of some sort of endogenous timer for daily rhythms involves placing an animal in constant environmental conditions, for example, constant darkness. When this is done, many organisms exhibit activity rhythms, called **free-running rhythms,** with a period different from that of any known cyclic environmental variable. The pattern of activity of a flying squirrel *(Glaucomys volans)* housed in constant darkness for

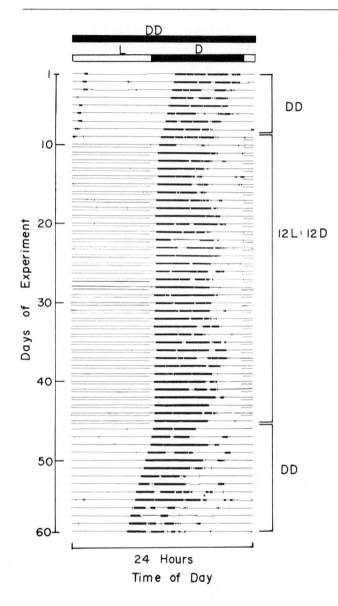

FIGURE 8–3   Free-running periods and entrainment
When a flying squirrel is placed in constant darkness, it adopts a free-running rhythm with a period different from any known cyclic environmental variable. When the squirrel later is placed in an environment with a varying cue, for example, a light-dark cycle, the squirrel's activity onset becomes locked onto the lights-off signal.
Source: Photo from Patricia DeCoursey.

several weeks is shown in Figure 8–3 (DeCoursey 1960, 1961). Since the animal's activity has a clearly rhythmic pattern in the absence of any obvious cyclic cue, there is evidence for some endogenously based system of timekeeping.

For most species studied, the free-running periods follow what has become known as **Aschoff's Rule** (Aschoff 1960, 1979). When animals are kept in constant darkness their activity rhythm continues with a period of nearly twenty-four hours, but it **drifts** slightly, becoming somewhat shorter (as in the flying squirrel) or somewhat longer each day. Aschoff's Rule states that the direction of this drift and the rate of drift away from the twenty-four hour period are a function of the light intensity and of whether the animal is normally diurnal or nocturnal. For nocturnal animals like the flying squirrel, housing under constant dark conditions results in a free-running rhythm with a period of shorter than twenty-four hours: the activity begins slightly earlier each day. Conversely, for a diurnal animal housed in the dark, the free-running period is slightly longer each day: the activity begins slightly later each day. There are some exceptions to this rule, but in general the data show a variety of animal species conform to the pattern.

ISOLATION. Some experiments have clearly ruled out "learning" or other similar influences as a mechanism for biological rhythms. Birds and some reptiles that hatch from eggs can be kept under constant conditions in an incubator from a time prior to hatching until after hatching. If the newly hatched organisms exhibit circadian rhythms, a major component of biological rhythms would appear to be inherited and endogenous. Hoffman's (1959) studies of lizards have supported the endogenous control hypothesis. Hoffman maintained lizard eggs under one of three conditions, eighteen-hour days consisting of nine hours of light and nine hours of dark; twenty-four-hour days, with twelve hours of light and twelve hours of dark; and thirty-six-hour days with eighteen hours of light and eighteen hours of dark. Animals from all three groups, hatched and maintained under constant conditions, exhibited free-running activity periods of 23.4 to 23.9 hours. We can therefore conclude that a component of the biological clock mechanism in these lizards appears to be inherited, remains unaffected by various rearing regimes, and is thus endogenous.

TRANSLOCATION. Additional support for the hypothesis of endogenous control of biological rhythms has come from **translocation** experiments. Honeybees (*Apis* spp.) are known to visit particular feeding sites at the same time each day. This behavior is functionally significant because many flowers that provide bees with food are open only at specific times of day. A German scientist, Renner (1957, 1959, 1960), working in a specially designed room with constant conditions, trained bees to leave a hive to forage at a specific time each day. He trained the bees in Paris and then flew them, during the night, to New York, where he placed them in a similar enclosed room with constant conditions. On their first day in New York, the bees began to forage at a time identical to that which would have been expected if they had remained in Paris—that is, twenty-four hours after their last foraging.

  In related studies, bees that inhabited outdoor hives were translocated from Long Island to California. Investigators found that, although the bees initially foraged at the same time as at the original site, they gradually adjusted to local time by foraging later and later each day (Figure 8–4).

  A key aspect of current biological clock models illustrated by these experiments is that while the clocks are endogenous, external conditions can cause the

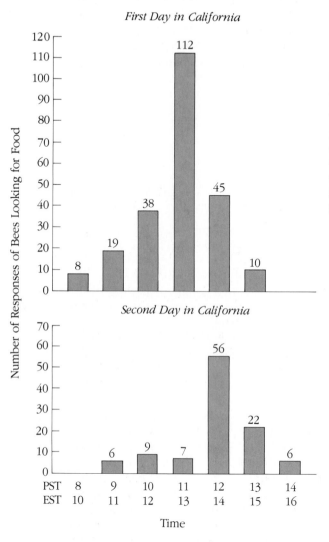

First Day in California

Second Day in California

FIGURE 8-4   **Visiting frequency of bees**

Bees were trained to forage at a particular time period (1 P.M. to 2:30 P.M. EST) in New York and were then flown to California at night. During the first two days at the new site, the bees gradually shifted the time of their feeding. These data support the notion of an endogenous biological rhythm that shifts in response to local conditions.

Source: Based on data from Renner (1960).

biological rhythms to be reset. Similar relocation studies of circadian skin color changes in fiddler crabs and shell valve-opening time in oysters reveal that these animals, too, react initially as if they were still in the original location, but gradually their clocks become reset to local time. The jet lag experienced by humans making long-distance airline flights is another example of this phenomenon.

VARIATION IN PERIOD.   A final source of evidence for the endogenicity of biological rhythms comes from the variation that exists in the natural activity periods of most organisms. For example, measurement of the activity periods of two frogs *(Rana pipiens)* under constant conditions reveals circadian rhythms of twenty-three hours, ten minutes and twenty-four hours, thirty-three minutes. These rhythms do not match the period of any known environmental variable. The animals must have internal chronometers that, due to individual differences, lead to deviations from a twenty-four-hour rhythm.

ZEITGEBERS

Many organisms exhibit circadian and circannual periodicities that appear to be closely adjusted to patterns of daylength, temperature fluctuation, or other environmental patterns. The mechanism whereby the period of a rhythm occurs repetitively and coincides approximately with the presence of some external stimulus is called **entrainment.** As we noted in the introduction, cues that provide information to animals about periodicity of environmental variables are called *Zeitgebers* ("time givers"). *Zeitgebers* are the entraining agents defined as those cyclic environmental cues that can entrain free-running, endogenous pacemakers. *Zeitgebers* can influence rhythms by affecting both the phase and the frequency.

Returning to the example in Figure 8–3, when the flying squirrel is provided with a daily cycle of light and dark, its activity rhythm generally conforms to the light cycle — the onset of activity occurs at about the same time each evening. The squirrel is thus said to be entrained to the daily light-dark cycle. For most endothermic vertebrates thus far tested, the light-dark cycle is the critical cue for entrainment. Further, in most terrestrial organisms, the daily light-dark cycle is the *Zeitgeber*. This is not surprising, since the daily light cycle is usually the most consistent environmental cue in the terrestrial habitat. For animals in the intertidal zone, the ebb and flow of the tides may be a prime *Zeitgeber*. Ectotherms such as lizards and insects that are unable fully to control their own body temperatures may use temperature or light cues as *Zeitgebers*. When entrained, they exhibit increased activity under warmer conditions and decreased activity during the cooler portions of the daily cycle.

Other environmental cues that may serve to entrain some biological rhythms in some organisms are summarized in Moore-Ede, Sulzman, and Fuller (1982). Cycles of food availability, but not water availability, can entrain activity rhythms in several species of small mammals (Richter 1922; Moore 1980). In other mammals, including humans, social cues may be involved in the entrainment of biological rhythms. Two groups of four volunteer subjects each were isolated in separate, but identical chambers (Vernikos-Danellis and Winget 1979). Each individual maintained an activity rhythm that was synchronous with the others in that room. However, the average free-running periods for the two rooms differed; 24.4 hours in one room and 24.1 hours in the other. To avoid any confounding effect from self-selected light-dark cycles, both groups of subjects were maintained in constant illumination. To further test the possible role of social cues in this synchronization of biological clocks within, but not between, the rooms, one subject was moved from one group to the other. That subject soon went through a phase shift and became synchronized with the activity rhythm of the new group. The exact nature of the social cues involved in the entrainment of circadian rhythm in humans is not yet known.

Social cues may also act in concert with other environmental cues to entrain biological rhythms. When individuals who have just flown across six time zones are required to remain in their hotel rooms, they entrain more slowly to the new local time than do individuals who are permitted to leave their rooms and who thus obtain more social-environmental cues from their new surroundings.

If we bring an animal such as a chipmunk *(Tamias striatus)* into the laboratory and with artifical lighting reverse the light and dark portions of the chipmunk's natural cycle, the animal will shift its activity phase to the light portion of the cycle within a few days. The ability to manipulate the activity phase of animals' daily rhythms has been used by zoological parks to display some nocturnally active animals. By having special quarters that are kept darkened when the zoo is open to visitors and illuminated when the zoo is closed, the public is able to see these animals as active creatures, instead of looking at them only as they sleep. As investigators we can use this type of manipulation to conduct behavioral observations under darkened conditions with nocturnal animals during our own diurnal activity phase.

The entrainment of behavioral and physiological rhythms to the cycles of various environmental cues and the phase shifting which takes place when a key environmental parameter is altered are evidence for the *Zeitgeber* component of the system of biological timekeeping.

MODEL

Using information thus far obtained on biological rhythms we can attempt to depict what is known in the form of a model (Figure 8–5). The model system accounts for the two major elements known to be important for biological timekeeping, an endogenous self-sustaining pacemaker and a system for entrainment to environmental *Zeitgebers*. The model utilizes light and a circadian rhythm as an example; however, the same model may be used with other types of rhythms, involving entrainment to tidal, annual, or other cues.

LOCATION AND PHYSIOLOGY OF THE PACEMAKER

Great progress has been made during the past several decades on locating the pacemakers for biological rhythms and studying their physiology. Much of this work has been summarized by Brady (1982); Menaker, Takahashi and Eskin (1978); Takahashi and Zatz (1979); Aschoff (1981); and DeCoursey (1983). The work has been conducted using cockroaches, sea hares, birds, and rodents.

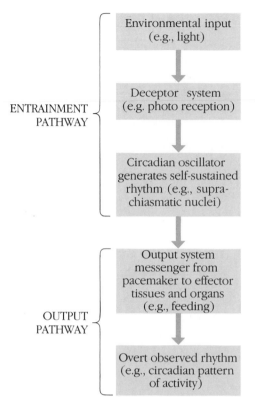

ENTRAINMENT
PATHWAY

Environmental input
(e.g., light)

Deceptor system
(e.g. photo reception)

Circadian oscillator
generates self-sustained
rhythm (e.g., supra-
chiasmatic nuclei)

OUTPUT
PATHWAY

Output system
messenger from
pacemaker to effector
tissues and organs
(e.g., feeding)

Overt observed rhythm
(e.g., circadian pattern
of activity)

FIGURE 8-5 Model of a mammalian pacemaker system portraying the various component parts and how they may function together, resulting in the overt behavioral and physiological rhythms which can be observed

COCKROACHES. One research objective of studies of clock mechanisms has been to determine if they are primarily neural or hormonal. Harker's work (1960, 1964) suggested that hormonal and neurosecretory products play key roles in the cockroach's biological clock mechanism (Figure 8–6). Harker made one cockroach *(Leucophaea moderae)* arhythmic by keeping it in a constant condition—namely, that of continuous light. She then interconnected the blood system of this cockroach with that of a cockroach she had kept in a normal light/dark cycle and that showed regular circadian periodicity. The arhythmic recipient cockroach adopted the regular circadian periodicty of the donor. However, other studies (Brady 1967a, 1967b; Roberts 1966) have failed to replicate Harker's work and instead have shown that cockroaches kept in constant conditions can maintain regular circadian periodicities; the recipient cockroach in Harker's studies may simply have been reverting to the free-running rhythm.

Decapitation of a cockroach results in arhythmic activity; the surgery apparently removes the source of the key oscillator but not the ability to exhibit rhythmic locomotor patterns. Experimenters have used these headless organisms in implantation and transection experiments to test various organs and tissues as candidates for

FIGURE 8-6   Parts of the neural and hormonal systems of the cockroach
This schematic diagram illustrates a compilation of studies of biological clock phenomena. The connected spheres represent the ganglia of the central nervous system.
Endocrine tissues that can be removed without affecting biological rhythms are
shown as dotted boxes. Arrows designate organs that have been transplanted into
arhythmic, headless recipients from rhythmic donors. *0* indicates a host with
implant that shows no detectable rhythm. Cuts in nerve trunks are shown as heavy
broken lines. Cuts that are made at B, E, and F or by splitting the protocerebral
lobes (part of the interior of the brain) bilaterally appear to stop the rhythm, but
cuts at A, D, C, or by splitting the pars interocerebralis (part of the interior of the
brain) do not stop the rhythm.
Source: J. Brady, "How Are Insect Circadian Rhythms Controlled?" *Nature 223*
(1969):781–784. Copyright © 1969 Macmillan Journals Limited. Reprinted with
permission.

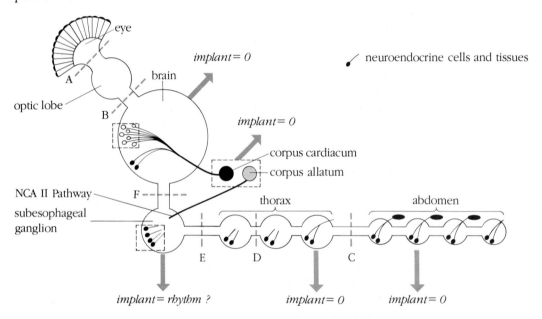

endocrine or neural clocks. Brains, corpora allata, and corpora cardiaca transplanted
from hosts exhibiting rhythmic activity to headless cockroaches have all failed to
produce regular rhythmic activity in the recipients. The relatively unimportant role
of neuroendocrine secretions of the brain in cockroach clock mechanisms has been
ascertained by surgical transection operations shown at locations B, E, and F in
Figure 8–6.

Brady (1969), in further transection studies, has demonstrated that the optic
lobes play a key role in cockroach circadian rhythms. Surgery at point A does not
affect periodicity, but transection at point B, which destroys the influence of the

optic lobes, eliminates rhythmic activity. The pacemaker thus appears to involve the optic lobes and possibly the lateral brain neurosecretory cells.

Information about periodicity is hypothetically communicated to effector organs in the body through three possible pathways: (1) hormones in the blood, (2) hormones in the nerves, or (3) electrical impulses in the nerves. Cutting the NCA II pathway (Figure 8–6) and removing the corpora cardiaca without affecting rhythmicity would indicate that the first alternative is unlikely, since lateral brain neurosecretory cells send axons only to the corpora cardiaca, where their secretions must either be released or travel down the NCA II pathway to the sub-esophageal ganglion. Since surgery on the NCA II pathway has no effect on rhythmicity, the only remaining pathways involve the circumesophageal connections (Figure 8–6); however, no neurosecretory activities have been demonstrated for these connections, and thus we may also rule out the second alternative.

These steps bring us to a third experiment: disconnecting the brain from the remainder of the nerve cord. If rhythmicity is lost when we separate the brain and nerve cord, the third alternative would appear to be correct. Severing the nerve cord posterior to the thorax (point C) does not affect the rhythm. However, cutting the nerve cord between the thoracic ganglia (point D) reveals a graded effect: the more anteriorly the cut is made, the higher the proportion of cockroaches that exhibit disrupted activity rhythms. Thus a pacemaker located in the optic lobes appears to communicate with effector tissues, such as muscles and organs, through electrical impulses traveling to the thoracic region of the ventral nerve cord.

Further studies of the optic lobes of the cockroach have led to the tentative conclusion that the critical area for the pacemaker lies primarily in cell bodies in the inner region of the lobes (Roberts 1974; Sokolove 1975). Another series of experiments (Page, Caldarola, and Pittendrigh 1977) revealed that there are separate pacemakers in each optic lobe, which interact with one another. In addition, the presence of either compound eye alone can serve to entrain the pacemakers in both optic lobes.

In a recent study Page (1982) has extended our understanding of cockroach circadian pacemakers one step further. The optic lobes of a cockroach were surgically removed and were replaced by the optic lobes from a second cockroach with a different circadian rhythm. After a lapse of some four to eight weeks the recipient cockroach again exhibited a circadian activity pattern — and the free-running period of the restored rhythm matched that of the donor cockroach. These findings support the hypothesis that the optic lobes contain a circadian pacemaker, and that after a lapse of some weeks new neural connections, between the optic lobes and the brain regenerated in the recipient cockroach. This resulted in the reestablishment of a recognizable pattern of circadian activity.

SEA HARES. Experiments on the sea hare *(Aplysia)* have provided additional evidence about the relationships between neural structures, the optic system, and biological rhythms (Jacklet 1969, 1973; Jacklet and Geronimo 1971). Sea hares' isolated eyes, kept in total darkness in sea water, exhibit a circadian rhythm of optic

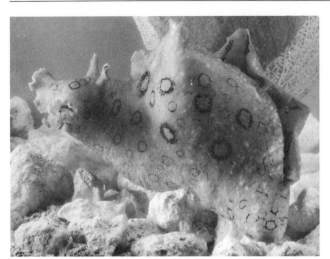

FIGURE 8-7   Sea hare
The sea hare's eyes are embedded in the integument of its head. When we remove the eyes and place them in sea water, they continue to exhibit a circadian pattern of optic nerve impulses.
Source: Photo by Runk/Schoenberger from Grant Heilman.

nerve impulses (Figure 8–7). Eyes taken from sea hares maintained in conditions of twelve hours of light and twelve hours of dark prior to the eye excision show a peak firing frequency at "dawn" on the day after removal. Eyes from sea hares kept in constant light prior to surgery exhibit regular rhythms of optic nerve impulses. Thus some type of oscillator, possibly related to neurosecretory processes, must be present in the eye. The phase of the eye rhythm is readjustable at each dawn, and the oscillator in the eye possibly controls other rhythms elsewhere in the sea hare. A clock mechanism that adjusts to day-to-day changes in daylength through a pacemaker in the eye could be adaptive for an organism that lives in a variable environment.

The circadian firing frequency in the sea hare's optic nerve apparently results from a coupled, interacting population of 950 cells in the retina. Two findings support this conclusion. First, cells can be removed from the population, down to a critical level of about 20 percent of the original total, without altering the basic period and amplitude of the rhythm (Jacklet 1973). Second, below that level, removal of additional cells results in a shortening of the rhythm's period and a dampening of the amplitude. When the number of cells has been reduced to 20 or fewer, the rhythm becomes ultradian—beyond a normal daily pattern. Apparently, a population of noncircadian oscillators located in the retina produce a circadian rhythm in the optic nerve when they are coupled to each other. Circadian rhythms in effector systems in the body may be regulated, in part, by this optic system oscillator. The cockroach appears to have two coupled pacemakers in each optic lobe; sea hares have pacemakers in their eyes (Hudson and Lickey 1977).

Further work on the *Aplysia* system has resulted in an increased understanding of some of the biochemical events involved in the circadian oscillator in this species (Eskin and Takahashi 1983). Using isolated eye preparations from sea hares, these investigators determined that the amount of cyclic adenosine 3', 5'-monophosphate (AMP) formed in homogenates of the eye was affected by the dose of serotonin, a brain neurotransmitter, present, but not by other neurotransmitters. The action of the serotonin is mediated through the activation of adenylate cyclase, an enzyme critical in the AMP synthesis pathway. The investigators also demonstrated that forskolin, a specific activator of adenylate cyclase activity, produced both advance and delay phase shifts in the timing of the circadian rhythm from the isolated *Aplysia* eye. From these findings it is possible to conclude that cyclic AMP is important in mediating phase shifts in the pacemaker located in the sea hare eye.

VERTEBRATES. Investigations of biological clock mechanisms in rats *(Rattus norvegicus)* have led to the finding that a direct neural connection exists between the retina and the hypothalamus, a pathway that terminates in the suprachiasmatic nuclei (SCN), a specific region of the hypothalamus (Hendrickson, Wagoner, and Cowan 1972; Moore and Lenn 1972). When investigators made lesions to cut off connections to these nuclei, the rats lost circadian rhythms for drinking behavior and wheel-running activity (Moore and Eichler 1972; Stephan and Zucker 1972). Additional studies on rats have shown that other circadian rhythms are stopped or radically altered by lesions of the SCN (see Rusak and Zucker 1979; Moore 1979).

These studies on the SCN lead some investigators to hypothesize that the SCN are the pacemaker(s). There is now ample evidence to indicate this is not the case. For example, bilateral SCN lesions in monkeys, which result in a circadian arhythmicity of the activity-rest cycle for the animal, do not eliminate the ongoing circadian rhythm for body temperature (Fuller et al. 1981). Thus, there must be at least one other pacemaker located somewhere else in the animal. Several bits of evidence suggest that another pacemaker may be located in the ventromedial nucleus of the hypothalamus (VMH). Evidence suggests some form of mutual coupling between the various pacemakers; there are neural pathways connecting the SCN and VMH. Also, it is possible to entrain the second pacemaker by food; food entrainment persists after removal of the SCN (Stephan, Swann and Sisk, 1979a, 1979b; Boulos, Rosenwasser and Terman 1980). Further, Krieger (1980) has shown that circadian rhythms retained after destruction of the SCN are, in fact, lost when the VMH is destroyed.

Investigations of rhythms in birds and mammals have implicated the pineal gland as a probable receptor of light stimuli that entrain and affect circadian patterns (Gaston 1971; Menaker and Zimmerman 1976). Recent reviews have supported the hypothesis that both neural processes (Block and Page 1978) and hormonal-neuroendocrine products (Zucker, Rusak, and King 1976) are involved in the mechanisms underlying biological rhythms in vertebrates.

The pineal gland in mammals lies within the brain, near the midline, but in some earlier forms of terrestrial vertebrates this structure was positioned at the top surface of the brain and served as a third or median eye. (We have all seen the mythical movie monsters with the third eye on the forehead!) In today's reptiles, birds, and amphibians the pineal gland is located just under the skull—indeed, it is still sensitive to light in many of these organisms. It is not surprising then that the pineal plays a key role in regulating certain rhythms based on photoperiod (the light cycle).

The most important of the variety of endocrine products from the pineal may be melatonin, an indoleamine (Reiter 1980). Daily subcutaneous injections of melatonin given to pinealectomized male hamsters induce regression of the gonads, mimicing the effect of shortened photoperiods in these animals (Tamarkin, Brown, and Goldman 1975; Reiter 1974a). A series of investigations using hamsters, rats, mice, and other mammals have repeatedly demonstrated the antigonadotrophic effects of melatonin. In these mammals the pineal normally receives photoperiod information via neural circuitry from the eyes. Reiter (1974b) has proposed that one key function for the pineal may be to control, in part, the annual rhythm of reproduction. Thus, the seasonal changes in photoperiod may be translated into physiological effects via the pineal and its endocrine products.

Additional evidence has also been accumulated on the control mechanisms of circannual rhythms, primarily in vertebrates. The data collected so far largely concern correlated responses, probably several steps removed from the actual clock mechanism. In ground squirrels *(Spermophilus tridecemlineatus),* selected mixtures of dialysates (materials that pass through the membrane in dialysis) and their residues obtained from the blood of other ground squirrels or from woodchucks *(Marmota monax)* accelerate or impede the induction of hibernation (Dawe and Spurrier 1972). The brown fat found in some animals, which was thought for many years to contain chemicals responsible for inducing hibernation (Johansson 1959), has been shown to contain, instead, a substance that produces arousal (Smith and Hock 1963). When turtles *(Testudo hermanni)* are housed outdoors, the levels of two compounds from their pineal gland—serotonin and melatonin—exhibit both circannual and circadian rhythms (Vivien-Roels, Arendt, and Bradtke 1979). The turtles synthesize serotonin during the day and melatonin at night; this pattern of synthesis disappears entirely during hibernation. During the breeding season concentrations of both chemicals and the amplitude of circadian fluctuations increase. Additional investigations are needed to determine the relationship between the concentrations of these chemicals and the observed circannual and circadian rhythms.

SIGNIFICANCE OF BIOLOGICAL TIMEKEEPING

ECOLOGICAL ADAPTATIONS

Cloudsley-Thompson (1960), Pittendrigh (1960), Enright (1970), and Daan and Aschoff (1982) have summarized the ecological and evolutionary significance of biological rhythms. Physical factors of the environment—such as light, temperature, and humidity—can be critical for some organisms, particularly those that have an integument that loses water to the atmosphere, those that cannot fully regulate their own body temperature, and those that live in temperate, arctic, or desert environments, where physical factors undergo dramatic seasonal or daily variation.

Many insect species live in environments in which the photoperiod, or day-length, differs from the periods of physical factors in the environment, such as temperature, moisture, and chemical substances, that may have either positive or deleterious effects. Photoperiod exerts no direct beneficial or harmful influence but may serve as an adaptive timing cue, or a predictor of environmental conditions that are critical for survival (Beck 1968).

Field observations and laboratory experiments have shown that woodlice exhibit a circadian rhythm that is regulated by photoperiod — that is, they are active at night and quiescent during the day (Figure 8–8). At night, temperatures are lower and humidity is higher than during the day. Cloudsley-Thompson (1952, 1960) found that during the day woodlice respond negatively to light and positively to moisture; thus they tend to remain in dark, damp portions of the test chamber during the daytime. At night woodlice are photonegative, but their positive reactions to humidity are not as pronounced. Under conditions of extremely low relative humidity, the woodlice become weakly photopositive.

Together these behavioral reactions can be seen as adaptive traits related directly to water conservation and the nocturnal habits of woodlice. Since response to moisture is weaker at night, woodlice can move into and through dry places where they would never be found during the day. The woodlouse's stronger photonegative response at night ensures that when daylight comes, the insect will go into hiding. Finally, weakly photopositive responses when humidity is very low means that the woodlouse can move out of its resting place to seek a moister environment, even during daylight hours, when prevailing conditions dictate such a move to ensure survival — for example, when remaining in a dry location could lead to the woodlouse's death through desiccation.

DIURNALITY

For animals that have an impervious integument — birds, mammals, and organisms that live in environments where conditions remain relatively constant throughout

---

FIGURE 8-8 Woodlouse
Woodlice are small nocturnal crustaceans that rest during the day in dark, damp places to avoid desiccation but move about freely at night. Their circadian rhythm is correlated with the day/night cycle, but the periodicity is in part a function of temperature and moisture conditions.

the day, like the ocean—a number of biotic factors, including reproduction, feeding habits, interspecific competition, and predation, help to determine the timing of daily activities. Whether an animal is nocturnal, diurnal, or crepuscular appears to depend on its responses to various biotic factors. Determination of the ecological significance of circadian rhythms in birds and mammals is exceedingly difficult and complex; few concrete answers have, as yet, emerged.

For example, let us consider the evolution of primates from primitive insectivores, an evolutionary sequence that is represented today by some species of primates in Africa and Asia (Simons 1972)—lorises, tarsiers, and some lemurs (Figure 8–9). Primates usually exhibit a diurnal activity phase in their circadian rhythms; the shift from nocturnal to diurnal activity during primate evolution can be traced, in part, to the effects of certain biotic factors.

Rodents probably evolved concurrently with early primates from insectivorelike predecessors and rapidly became dominant in the terrestrial habitat. The

---

FIGURE 8–9   Loris and hamadryas baboon
Today's primates have evolved from insectivorelike predecessors. The loris (left) is a descendant of one of the earliest forms to evolve; note the large eyes for night vision. The hamadryas baboon (right) has returned to a largely terrestrial existence, possibly aided by its large size and social defense mechanisms against predation.
Source: Photos copyright © by Zoological Society of San Diego.

resulting strong competition between rodents and early primates for food, nesting sites, and living space may have brought about two adaptive changes that characterize primate evolution: a shift by primates to arboreal habitats and a phase shift in the timing of their activity peak to the daylight hours. In addition, if the first change involved a move to the trees — and the evidence supports this contention — then there may have been additional selection pressure for diurnal activity because arboreal locomotion is safer in daylight.

An alternative explanation of the evolution of diurnality in primates is that the insectivorelike stem group, from which the primates evolved, was originally comprised of small, diurnal, arboreal animals. Predation might have led to natural selection for a nocturnal activity phase, since adequate cover during the daylight hours exists only in the high forest canopy. Later, as the true primates evolved, the food habits of some species shifted to a diet containing more fruits and leaves, and body size increased. With increased body size, the primates were better able to defend themselves or escape from predators, and some species adopted a terrestrial lifestyle. Having at least two equally plausible explanations (there may be others) for the diurnal activity of primates illustrates the problem of the use of post-hoc reasoning to account for the evolution of observed present-day behavior.

HIBERNATION

Circannual rhythms in invertebrates and vertebrates have clear ecological significance. Some mammals, such as ground squirrels, chipmunks, and jumping mice, exhibit an annual pattern of hibernation (Kayser 1965; Pengelley and Asmundson 1974). From mid-fall until mid-spring, these mammals enter a physiological state in which their normal endothermic body temperature (ca. 37°C) falls to within a few degrees of the environmental level. Hibernators begin preparation for the winter season well in advance of the arrival of colder conditions. Some hibernating mammals, such as woodchucks, alter their diet by late summer and add a layer of body fat, which provides extra insulation and a food supply for the winter. Other animals, such as chipmunks and hamsters, which inhabit a shallow burrow system during the spring and summer, extend their system deeper or dig special burrow systems with a hibernation chamber. These winter sleeping sites are always located deep beneath the ground surface, below the frost line, so there is no danger of being frozen during the coldest periods of winter. The physiology of hibernation and some closely related behavioral phenomena have recently been reviewed by Lyman et al. (1982).

The most obvious adaptive value of a circannual rhythm of hibernation is that hibernators avoid many problems of survival; they do not have to find and possibly compete for limited food supplies, and they do not have to maintain a warm and suitable nest site during adverse winter conditions. We should note, however, that other mammals in the same regions do not hibernate but have adopted alternative strategies for survival. Some mice, for example, produce great numbers of progeny during the summer months; a portion of these will survive the rigors of winter to breed the next spring. Other mice live in communal nests during the winter and gain warmth from each other. Some mammals, like raccoons, do not actually hibernate, but during the coldest periods of winter, they may remain in their dens for several days in a state of quiescence or torpor.

Perhaps the most important aspect of the circannual rhythms in mammals that do hibernate is that they are able to anticipate the coming of winter. During late summer and fall their behavior changes; they shift their dietary habits and prepare hibernation burrow systems. If these animals waited until the food supply has disappeared or the ground was frozen or covered with snow, it would be too late.

Hibernating mammals must mate soon after emergence from the winter den so that their progeny can develop before the onset of the next winter season. When the animals emerge, both males and females are in nearly prime reproductive condition; some of their physiological systems have "awakened" before the end of hibernation. These internal changes are apparently triggered by an internal clock — either an endogenous mechanism or a clock that is set exogenously when the animal goes into hibernation in the fall.

MIGRATION

The annual rhythm of migratory behavior exhibited by birds that breed in temperate and polar regions but winter in more southern latitudes offers interesting parallels to hibernation. By flying south for the winter, these birds escape severe winter conditions and food shortages; the lack of sufficient physiological adaptation generally precludes their remaining in northern climates for the entire year. When they return to the north in the spring, these birds, like hibernating mammals, must reproduce soon after their arrival at summer breeding locations to ensure that their progeny are fully developed by the time of their return flight south in the fall. Possession of a circannual clock mechanism permits these birds to anticipate the coming of fall in time to prepare physiologically for and to initiate migration. At the birds' wintering grounds to the south, where fewer dramatic seasonal changes in the environment occur, the same clock mechanism "tells" the birds when it is spring so that they can start their northbound flight. Studies have shown that the bird species are entrained by light and possibly in some cases by temperature *Zeitgebers*. As we shall see in Chapter 14, biological clocks also play a role in the orientation and navigation of many birds.

Some marine organisms exhibit circannual rhythms. Longhurst (1976) studied vertical distribution and daily vertical migration in plankton and found that daily vertical movements are true circadian rhythms, controlled by an internal clock mechanism, but that the nature and timing of the vertical migration of plankton varies with species, sex, age, and even location. Light is the principal, but not the only, cue that can affect these rhythms. Three ecological and evolutionary explanations of vertical migration have been proposed (McLaren 1963; Longhurst 1976). First, migrating upward at night—a common, but not universal, feature of the vertical movement of plankton—may be a means of avoiding predation. Marine fish that are potential predators are active in the upper, better-lighted zone of the ocean during the day and move to a lower zone at night. Thus, the plankton are in the upper zone at night when the fish that would most likely consume them are not. However, many of these species are at a depth during the day where enough light still penetrates to permit some predators to be active. Moreover, some of the plankton that rise to the upper level or to the surface at night are bioluminescent and thus announce their presence. Second, by migrating vertically between two levels of water, plankton may use the current that is generated as a means of dispersion. Third, the plankton may derive bioenergetic or nutritional benefits from these daily vertical movements.

SUMMARY

Animals exhibit patterns of activity and inactivity that recur at regular intervals. The timing of these behaviors is apparently controlled by biological pacemakers, regulated by circadian, lunar, or circannual rhythms. Biological rhythms involve two elements: an endogenous self-sustaining pacemaker and external cues which can entrain the pacemaker. Evidence for the existence of an internal chronometer comes from several sources. When an animal is placed in an environment with constant conditions, it will exhibit an activity rhythm with a period different from that of any known environmental variable; these are called free-running rhythms. Animals hatched and reared in isolation and under constant conditions exhibit circadian (daily) free-running rhythms. Animals that are translocated from one time zone to another will, initially, exhibit an activity phase matching that of their original environment. After some period of time in the new location they will show a phase shift and will adjust the time of their activity to the new local time. Last, the activity rhythms for several animals of a species differ from one another; there are individual differences in the chronometers.

The activity rhythms of animals can be entrained to a variety of environmental cues called *Zeitgebers*. The most prominent and consistent environmental variable is the light-dark cycle. It is therefore not surprising that many daily and annual rhythms are entrained to the light-dark cycle in a wide variety of animals. Other rhythms may be entrained to tidal or lunar cues, which are prominent in some environments. The availability of food and social cues have also been demonstrated as probable *Zeitgebers* for rhythms in some organisms.

Research on cockroaches, sea hares, birds, and rodents has begun to provide evidence on the location of the pacemakers and to discover how they function. Most biological clocks are relatively unaffected by either temperature changes or metabolic inhibitors. Evidence to date implicates neural, neuroendocrine, and hormonal events in the pacemaker systems of different organisms. For both cockroaches and sea hares the experimental results indicate that there are pacemakers in the optic lobes and eyes. In mammals the evidence suggests the existence of at least two pacemaker systems, in the suprachiasmatic nuclei and in the ventromedial hypothalamus. The pineal gland, and in particular the melatonin produced there, have been implicated in the regulation of annual cycles of reproduction in some vertebrates.

The ecological and evolutionary significance of biological timekeeping can best be assessed with respect to physical variables such as daylight, temperature, and moisture and to related biotic factors, like competition for limited resources, feeding habits, and predation. For example, the evolution of diurnality among primates may have resulted from their competition with nocturnal rodents. An alternative theory is that as primates evolved larger body sizes, they may have been better able to cope with potential predators, and thus could adopt a diurnal activity phase that enhanced feeding opportunities. Hibernation by some mammals and migration by some birds in temperate and polar climates are evolutionary adaptations to avoid the rigors of the winters in those regions.

Discussion Questions

1. Inventive research techniques, such as translocation, the use of magnets to assess the role of geomagnetic forces, and microsurgical techniques to explore clock mechanisms, have been employed in the study of circadian biological rhythms. If you were about to embark on the study of the clock mechanisms of circadian rhythms and the factors that influence these rhythms, what new techniques might you employ?

2. Let us suppose that a new species of small mammal, the yellow-striped scamp, has just been discovered in the woodlands of northern Manitoba. Scamps are both arboreal and terrestrial; they are about 10 cm long, including the 4 cm tail; and they eat both seeds and insects. What types of laboratory and field observations and experiments would you conduct to describe the biological rhythms of this species? In your answer consider behavior patterns and underlying physiological mechanisms and the adaptive significance of the observed rhythms.

3. Certain behavior patterns (e.g., feeding) are regulated by circadian clocks, while other behavior patterns (e.g., hibernation) may be regulated by circannual clocks. What other behavior patterns that may be regulated by circadian or circannual rhythms can you name? What is the adaptive significance of each of the behaviors you have named in regard to their being regulated, at least in part, by a biological clock mechanism?

4. The chart below represents the singing behavior of a wood thrush over a four-teen-day period. For the first seven days the bird was housed in constant dark-ness. For the second seven days the bird was subjected to a daily light cycle with the lights on at 0600 hours and off at 1800 hours. What is the period of the activity rhythm in constant darkness and in the light-dark cycle? What changes has the light-dark cycle brought about in the frequency and phase of the activity period?

Actogram for singing of a thrush; note the animal has two active phases.
Shaded area indicates darkness; black bar denotes singing.

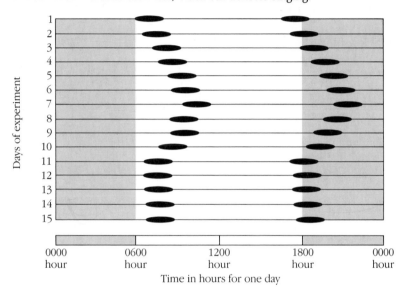

Time in hours for one day

Suggested Readings

Aschoff, J., ed. 1981. *Handbook of Behavioral Neurobiology.* Vol. 4: *Biological Rhythms.* New York: Plenum.
A thorough review summarizing our current knowledge of the various biological clock phenomena.

Block, G. D., and T. L. Page. 1978. Circadian pacemakers in the nervous system. *Ann. Rev. Neurosci.* 1:19–34.
Review of knowledge on pacemaker locations and related nervous system physiology. Some historical perspective and good bibliography.

Brady, J. 1982. *Biological Timekeeping.* Cambridge: Cambridge University Press.
A basic textbook on biological rhythms. Includes solid elementary explanations of clock processes and good references to current material.

DeCoursey, P. J. 1983. Biological timing. *Biology of the Crustacea.* New York: Academic Press. 7:107–162.

An excellent review article in two parts. The first portion gives an up-to-date review of the various theories and explanations for biological clock phenomea. The second portion deals with the application of these ideas in crustaceans.

Menaker, M., J. S. Takahashi, and A. Eskin. 1978. The physiology of circadian pacemakers. *Ann. Rev. Physiol.* 40:501–526.

Article concentrating on research involved with locating biological clocks and defining clock mechanisms in invertebrates and vertebrates. Short but useful summary of this work, with extensive references.

Saunders, D. S. 1974. Circadian rhythms and photoperiodism in insects. In M. Rockstein (ed.), *Physiology of Insects,* vol. 2. New York: Academic Press.

Book chapter by knowledgeable investigator. Comprehensive summary of what is known on the topic. Some discussion of ecological relevance of biological clock phenomena in insects.

References

Aschoff, J. 1960. Exogenous and endogenous components in circadian rhythms. *Cold Spr. Harb. Symp. Quant. Biol.* 25:11–28.

———. 1963. Comparative physiology: Diurnal rhythms. *Ann. Rev. Physiol.* 25:581–600.

———. 1979. Circadian rhythms: Influences of internal and external factors on the period measured in constant conditions. *Zeit. Tierpsychol.* 49:225–249.

———. 1981. *Handbook of Behavioral Neurobiology.* Vol. 4: *Biological Rhythms.* New York: Plenum.

Beck, S. D. 1968. *Insect Photoperiodism.* New York: Academic Press.

Block, G. D., and T. L. Page. 1978. Circadian pacemakers in the nervous system. *Ann. Rev. Neurosci.* 1:19–34.

Boulos, Z., A. M. Rosenwasser, and M. Terman. 1980. Feeding schedules and the circadian organization of behavior in the rat. *Behav. Brain Res.* 1:39–65.

Brady, J. 1967a. Control of the circadian rhythm of activity in the cockroach. I. The role of the corpora cardiaca, brain and stress. *J. Exp. Biol.* 47:153–163.

———. 1967b. Control of the circadian rhythm of activity in the cockroach. II. The role of the subesophageal ganglion and ventral nerve cord. *J. Exp. Biol.* 47:165–178.

———. 1969. How are insect circadian rhythms controlled? *Nature* 223:781–784.

———. 1979. *Biological Clocks.* Baltimore: University Press.

Brown, F. A. 1970. Hypothesis of environmental timing of the clock. In F. A. Brown, J. W. Hastings, and J. D. Palmer, eds., *The Biological Clock: Two Views.* New York: Academic Press.

Cloudsley-Thompson, J. L. 1952. Studies in diurnal rhythms. II. Changes in the physiological responses of the woodlouse *Oniscus aspellus* (L) to environmental stimuli. *J. Exp. Biol.* 29:295–303.

———. 1960. Adaptive functions of circadian rhythms. *Cold. Spr. Harb. Symp. Quant. Biol.* 24:361–367.

Corbet, P. S. 1957. The life history of the emperor

dragonfly, *Anax imperator* Leach. *J. Anim. Ecol.* 26:1–69.

———. 1960. Patterns of circadian rhythms in insects. *Cold Spr. Harb. Symp. Quant. Biol.* 25:357–360.

Daan, S., and J. Aschoff. 1982. Circadian contributions to survival. In J. Aschoff, S. Daan, and G. A. Groos (eds.), *Vertebrate Circadian Systems.* Berlin: Springer-Verlag.

Dawe, A. R., and W. A. Spurrier. 1972. The blood-borne "trigger" for natural mammalian hibernation in the 13-lined ground squirrel and the woodchuck. *Cryobiology* 9:163–172.

DeCoursey, P. J. 1960. Phase control of activity in a rodent. *Cold. Spring Harb. Symp. Quant. Biol.* 25:49–56.

———. 1961. Effect of light on the circadian activity rhythm of the flying squirrel, *Glaucomys volans. Zeit. Physiol.* 44:331–354.

———. 1983. Biological timekeeping. *Biology of Crustacea* 7:107–162.

Enright, J. T. 1970. Ecological aspects of endogenous rhythmicity. *Ann. Rev. Ecol. Syst.* 1:221–238.

Eskin, A., and J. S. Takahashi. 1983. Adenylate cyclase activation shifts the phase of a circadian pacemaker. *Science* 220:82–84.

Fuller, C. A., R. Lydic, F. M. Sulzman, H. E. Albers, B. Tepper, and M. C. Moore-Ede. 1981. Circadian rhythm of body temperature persists after suprachiasmatic lesions in the squirrel monkey. *Am. J. Physiol.* 241:R385–391.

Gaston, S. 1971. The influence of the pineal organ on the circadian activity rhythm in birds. In M. Menaker (ed), *Biochronometry.* Washington, D.C.: National Academy of Sciences.

Harker, J. E. 1960. Endocrine and nervous factors in insect circadian rhythms. *Cold. Spr. Harb. Symp. Quant. Biol.* 25:279–287.

———. 1964. The physiology of diurnal rhythms. *Camb. Monogr. Exp. Biol.* 13:1–114.

Hendrickson, A. E., N. Wagoner, and W. M. Cowan. 1972. Autoradiographic and electron microscopic study of retino-hypothalamic connections. *Zeit. Zellforsch.* 125:1–26.

Hoffman, K. 1959. Die Aktivitätsperiodik von im 18- und 36-Stunden-tag erbrüteten Eidechsen. *Zeit. Vergleich. Physiol.* 42:422–432.

Hudson, D. J., and M. E. Lickey. 1977. Weak negative coupling between the circadian pacemakers of the eyes of *Aplysia. Neurosci. Abst.* 3:179.

Jacklet, J. W. 1969. Circadian rhythm of optic nerve impulses recorded in darkness from isolated eye of *Aplysia Science* 164:562–563.

———. 1973. Neuronal population interactions in a circadian rhythm in *Aplysia.* In J. Salanki (ed.), *Neurobiology of Invertebrates.* Budapest: Akademiai Kiado.

Jacklet J. W., and J. Geronimo. 1971. Circadian rhythm: Populations of interacting neurons. *Science* 174:299–302.

Johansson, B. 1959. Brown fat: A review. *Metabolism* 8:221–240.

Kayser, C. 1965. Hibernation. In W. Mayer and R. W. van Gelder (eds.), *Physiological Mammalogy,* vol. 3. New York: Academic Press.

Kogure, M. 1933. The influence of light and temperature on certain characters of the silkworm, *Bombyx mori. J. Dept. Agr. Kyushu Univ.* 4:1–93.

Krieger, D. T. 1980. Ventromedial hypothalamic lesions abolish food-shifted circadian adrenal and temperature rhythmicity. *Endocrinology* 106:649–654.

Longhurst, A. R. 1976. Vertical migration. In D. H. Cushing and J. J. Walsh (eds.), *Ecology of the Seas.* Philadelphia: Saunders.

Lutz, P. E., and C. E. Jenner. 1964. Life-history and photoperiodic responses of nymphs of *Tetragoneura cynosura* (Say). *Biol. Bull.* 127:304–316.

Lyman, C. P., J. S. Willis, A. Malan, and L. C. H. Wang. 1982. *Hibernation and Torpor in Mammals and Birds.* New York: Academic Press.

McLaren, I. A. 1963. Effects of temperature on growth of zooplankton and the adaptive value of vertical migration. *J. Fish. Res. Bd. Can.* 20:685–727.

Menaker, M., J. S. Takahashi, and A. Eskin. 1978. The physiology of circadian pacemakers. *Ann. Rev. Physiol.* 40:501–526.

Menaker, M., and N. Zimmerman. 1976. Role of the pineal in the circadian system of birds. *Amer. Zool.* 16:45–55.

Moore, R. Y. 1979. The anatomy of the central neural mechanisms regulating endocrine rhythms. In D. Krieger (ed.), *Endocrine Rhythms.* New York: Raven Press.

———. 1980. Suprachiasmatic nucleus, secondary synchronizing stimuli and the central neural control of circadian rhythms. *Brain Res.* 183:13–28.

Moore, R. Y., and V. B. Eichler. 1972. Loss of a circadian adrenal corticosterone rhythm following suprachiasmatic lesions in the rat. *Brain Res.* 42:201–206.

Moore, R. Y., and N. J. Lenn. 1972. A retinohy-
pothalamic projection in the rat. *J. Comp.
Neurol.* 146:1–14.

Moore-Ede, M. C., F. M. Sulzman, and C. A.
Fuller. 1982. *The Clocks That Time Us.* Cambridge: Harvard University Press.

Neumann, D. 1966. Die lunare und tägliche
Schlüpfperiodik der Mücke *Clunio*-Steuerung
und Abstimmung auf die Gezeitenperiodik.
Zeit. Vergleich. Physiol. 53:1–61.

Page, T. L. 1982. Transplantation of the cockroach circadian pacemaker. *Science* 216:73–75.

Page, T. L., P. C. Caldarola, and C. S. Pittendrigh.
1977. Mutual entrainment of bilaterally distributed circadian pacemakers. *Proc. Nat. Acad. Sci.
USA* 74:1277–1281.

Pengelley, E. T., and S. J. Asmundson. 1974. Cir-
cannual rhythmicity in hibernating animals. In
E. T. Pengelley (ed.), *Circannual Clocks.* New
York: Academic Press.

Pittendrigh, C. S. 1954. On temperature independence in the clock system controlling emergence time in *Drosophila. Proc. Nat. Acad. Sci.
USA* 40:1018–1029.

———. 1960. Circadian rhythms and the circadian organization of living systems. *Cold Spr.
Harb. Symp. Quant. Biol.* 25:159–184.

Rawson, K. S. 1960. Effects of tissue temperature
on mammalian activity rhythms. *Cold Spr.
Harb. Symp. Quant. Biol.* 25:105–114.

Reiter, R. J. 1974a. Pineal regulation of the hypothalamic-pituitary axis: Gonadotrophins. In E.
Knobil and W. H. Sawyer (eds.), *Handbook of
Physiology, Endocrinology,* vol. 4, part 2. Washington, D.C.: American Physiological Society.

———. 1974b. Circannual reproductive rhythms
in mammals related to photoperiod and pineal
function. A review. *Chronobiology* 1:365–395.

———. 1980. The pineal gland: A regulator of
regulators. *Progress in Psychobiology and Physiological Psychology* 9:323–356.

Renner, M. 1957. Neue Versuche über den Zeit-
sinn der Honigbiene. *Zeit. Vergleich. Physiol.*
40:85–118.

———. 1959. Über ein weiteres Versetzungsex-
periment zur Analyse des Zeitsinnes und der
Sonnenorientierung der Honigbiene. *Zeit.
Vergleich. Physiol.* 42:449–483.

———. 1960. The contribution of the honey bees
to the study of time-sense and astronomical orientation. *Cold Spr. Harb. Symp. Quant. Biol.*
25:361–367.

Richter, C. P. 1922. A behavioristic study of the
activity of the rat. *Comp. Psychol. Monogr.*
1:1–55.

Roberts, S. K. 1966. Circadian activity rhythms in
cockroaches. III. The role of endocrine and
neural factors. *J. Cell Physiol.* 67:473–486.

———. 1974. Circadian rhythms in cockroaches.
Effects of the optic lobe lesions. *J. Comp. Physiol.* 88:21–30.

Rusak, B., and I. Zucker. 1979. Neural regulation
of circadian rhythms. *Physiol. Rev.* 59:449–526.

Saunders, D. S. 1978. Internal and external coin-
cidence and the apparent diversity of photoperiodic clocks in the insects. *J. Comp. Physiol.
"A"* 127:197–207.

Simons, E. L. 1972. *Primate Evolution.* New York:
Macmillan.

Smith, R. E., and R. J. Hock. 1963. Brown fat:
Thermogenic effector of arousal in hibernators.
Science 149:199–200.

Sokolove, P. G. 1975. Localization of the cockroach optic lobe circadian pacemaker with mi-
crolesions. *Brain Res.* 87:13–21.

Stephan, F. K., J. M. Swann, and C. L. Sisk. 1979a.
Anticipation of 24-hour feeding schedules in
rats with lesions of the suprachiasmatic nucleus. *Behav. Neural Biol.* 25:346–363.

———. 1979b. Entrainment of circadian
rhythms by feeding schedules in rats with suprachiasmatic lesions. *Behav. Neural Biol.*
25:545–554.

Stephan, F. K., and I Zucker. 1972. Circadian
rhythms in drinking behavior and locomotor
activity of rats are eliminated by hypothalamic
lesions. *Proc. Nat. Acad. Sci. USA* 69:1583–1586.

Sweeney, B., and J. W. Hastings. 1960. Effects of
temperature upon diurnal rhythms. *Cold Spr.
Harb. Symp. Quant. Biol.* 25:87–104.

Takahashi, J. S., and M. Zatz. 1982. Regulation of
circadian rhythmicity. *Science* 217:1104–1111.

Tamarkin, L., S. Brown, and B. Goldman. 1975.
Neuroendocrine regulation of seasonal repro-
ductive cycles in the hamster. *Abstr. 5th Ann.
Mtg. Soc. Neurosci.,* p. 458.

Vernikos-Danellis, J., and C. D. Winget. 1979.
The importance of light, postural and social
cues in the regulation of the plasma cortical
rhythms in man. In A. Reinberg and F. Halbert
(eds.), *Chronopharmacology.* New York: Perga-
mon.

Vinogradova, Y. B. 1965. An experimental study
of the factors regulating induction of imaginal
diapause in the mosquito *Aedes togoi* Theob.
Entomol. Rev. 44:309–315.

Vivien-Roels, B., J. Arendt, and J. Bradtke. 1979. Circadian and circannual fluctuations of pineal indoleamines (serotonin and melatonin) in *Testudo hermanni* Gmelin (Reptilia, Chelonia). I. Under natural conditions of photoperiod and temperature. *Gen. Comp. Endocr.* 37:197–210.

Walker, B. W. 1949. Periodicity of spawining of the grunion, *Leuresthes tenuis,* Ph.D. Thesis, University of California, Los Angeles.

Zucker, I., B. Rusak, and R. G. King. 1976. Neural bases for circadian rhythms in rodent behavior. *Adv. Psychobiol.* 3:36–74.

PART THREE

BEHAVIORS OF INDIVIDUALS
IN GROUPS

9   Development of Behavior

10   Learning and Motivation

11   Sexual Behavior and Reproduction

12   Aggression

13   Communication

14   Migration, Orientation, and Navigation

DEVELOPMENT OF BEHAVIOR

In this third major section of the book, we examine the behaviors of individuals and groups. The section begins with two chapters (9 and 10) on development and learning, processes which strongly influence the performance of behavior. In the remaining chapters in this section (Chapters 11, 12, and 13) we explore three specific types of behavior: reproduction, aggression, and communication. We progress from an emphasis on processes that control and direct animal activities to an emphasis on functional and evolutionary aspects of behavior.

In the present chapter on the development of behavior we begin by presenting some basic methodological considerations affecting much of the research in this area. We then examine, in chronological sequence, the key events and factors affecting the development of behavior—starting with the embryology of behavior and followed by prenatal events, juvenile events, play behavior, and some comments on continued development into adult life. The chapter concludes with an examination of the nature-nurture issue and the modern epigenetic theory of behavior development.

METHODS IN BEHAVIOR DEVELOPMENT

Many studies of the development of behavior are designed to investigate the mechanisms that underlie behavior patterns observed in adult animals. Such a study may be relatively direct—for example, determining what factors affect nest-site selection by a bird species—or it may be a complex investigation—for example, establishing the ontogenetic determinants of feeding strategies in omnivorous rodents. Some studies apply conditions of either deprivation or enrichment; others alter the quality of specific types of stimulation provided during ontogeny rather than vary the overall quantity of stimulation.

For many investigations of behavior development we use direct observational techniques. (Some of the basic principles discussed in Chapter 3 are pertinent here, and a brief review of that material may be beneficial.) For some studies, particularly those conducted under field conditions, we may employ only observation, with little or no manipulation. These types of investigations provide useful information on the development of traits within the natural ecological and social context. Other studies of behavior development utilize a variety of manipulations to test the effects of various treatments. Examples of these approaches are presented when we examine the chronology of development. First, a bit of elaboration is needed on some aspects of experimental design and testing procedures used by investigators studying the development of animal behavior.

ASPECTS OF EXPERIMENTAL DESIGN

King (1957) set forth seven parameters relevant to the study of early experience. His scheme, or portions of it, can be rather broadly used in designing many of the experiments in behavior development—particularly those that attempt to assess the effects of early experience on behavior.

Parameters of the early experience treatment:

1. Age of the animal at the time the early experience treatment is given.
2. Type or quality of the early experience treatment.
3. Duration or quantity of the early experience treatment.

Parameters of the tests administered to determine the effects of early experience treatment:

4. Age of the animal at the time of testing.
5. Type of test used to assess the effects of early experience treatment.
6. Testing for the persistence of the early experience treatment.

The parameter of the genetics of the species being investigated:

7. Testing different strains or species to determine the effects of early experience on different animals.

TESTING PROCEDURES

We can test the long-term influence or persistence of early experience treatments by two types of experimental design, as outlined in King (1969). One method provides an animal with early experience treatment and tests for the effects of that experience at various ages; we use the same animal for each test. This type is called **longitudinal design** and requires the use of appropriate repeated-measures statistical tests to analyze the results. Alternatively, we can provide an animal with an early experience treatment and test it only once at a later age, because in a second test of the same animal we would not be able to tell whether the observed effect was due to the treatment or to the experience of the first test. We therefore avoid re-using the test subject, but possibly we lose some information about the persistence of early experience treatment on that subject. This second type of design is called **cross-sectional.**

    In some studies of early experience, investigators subject animals to stimulus deprivation or enrichment by manipulating either the quality or quantity of one of three types of stimulation or some combination: (1) sensory, (2) motor, or (3) social. In every case investigators compare the test performances of animals reared under deprived or enriched conditions with the test performances of control subjects reared under "normal" conditions. "Normal rearing" in this context usually refers to the common laboratory environment—a cage that restricts the amount and type of activity of the animal, contrived social contacts, a prepared diet, and constant ambient conditions. These conditions are extremely simplified when compared with a natural environment. Thus we should keep in mind that the control conditions are often stimulus-poor and not at all comparable with the subject's natural surroundings.

    In the chronological sequence which follows we provide some general remarks and several examples of investigations of behavior development at each stage or period.

EMBRYOLOGY OF BEHAVIOR

Some important processes integral to behavior development take place prior to hatching or birth. Investigators have documented a variety of prenatal and postnatal influences on both sensory and motor development and more general effects on subsequent behaviors that are produced by internal and external stimulation of the embryo. Developmental biologists are accumulating considerable evidence concerning chemical and physical processes involved in the epigenesis of the tissues and organs of the body, including the nervous system. **Epigenesis** can be broadly defined as the interaction of genetic inheritance and environment during the course of development of an organism. Integration of information from studies of embryology and studies of the development of behavior is leading both to a new synthesis of what takes place during development and to some new frontiers in research.

NERVOUS SYSTEM DEVELOPMENT

During development continual changes take place in the nervous system, including cell proliferation, migration of neurons, and differentiation of cell types. (See Jacobson 1978 for a full review of this topic.) We are concerned here specifically with the manner in which these embryological events may be related to the ontogeny of discrete, observable, behavioral events. According to Wolff (1981), the neural developmental process can be divided into two major categories of events; (1) the production and distribution of the neurons, and (2) the establishment of appropriate synapses between neurons. Not only is the proliferation of neurons important, but, also, the selective death of neurons is a critical developmental process. Cell death is, in part, a consequence of another process characteristic of nervous system development—overproduction of neurons. This overproduction leads to "competition" for appropriate connections and ultimately to death or survival of neurons based on interactions between the neurons and their targets.

Oppenheim and his colleagues have explored the development of motorneurons which innervate the limbs in embryonic chicks (Oppenheim, Chu-Wang, and Maderut 1978). When chick embryos are immobilized with neuromuscular blocking agents for varying periods between days 4.5 and 9 of incubation, the chicks have increased numbers of motorneurons in the motor columns of the spinal cord when compared to untreated chicks (Pittman and Oppenheim 1979). Apparently the number of motorneurons undergoing cell death during this period is related to muscular activity by the embryo. Treatment started after day 12 of incubation did not influence the rate of cell death. Also, if the immobilization treatment was stopped on about day 10, the cell death rate in treated chicks was simply delayed; they had the same total number of cells as control chicks by days 16 through 18. Thus, it appears that some functional interactions occurring at the developing neuromuscular junctions are central to the process of cell death or survival.

In another study, the developmental effects of removal of the limb bud on one side were compared with the same processes in the intact limb bud which served as a control. The nerve cell death rate was greatly enhanced on the side where the limb bud was ablated. However, the characteristics of developing motorneurons, growing outward from the central nervous system, were no different than normal motorneurons; they exhibited the same cell morphology and axon growth, and their synaptic enzyme systems were normal. Thus it appears that the motorneurons of the motor column of the spinal cord of the chick have the capacity to initiate differentiation. Neurons deprived of their target, the limb bud, are no different from those on the intact side.

In another study of chick development, Bekoff (1978) reports on the use of electrodes to record muscle movements from embryos. A video camera focused on the chick through a window made in the shell was used to record movements. Simultaneously, recordings called electromyograms (EMGs) were made from designated muscles (Figure 9–1). Combining the techniques is an excellent way to study the development of coordinated actions. Patterned output from the EMGs can be

FIGURE 9-1 Ontogeny of coordinated behavior
Schematic diagram of the experimental setup for simultaneously recording behavioral and EMG data from a spontaneously behaving chick embryo. A hole is made in
the shell over the embryo and the egg is placed in a heated, humidified chamber.
Extracellular suction electrodes are placed on individual, identified muscles and the
recorded EMGs are amplified 1000 times, viewed on the oscilloscope screen and
also stored on magnetic tape for later filming and analysis. A stimulator is used to
produce a time mark on one channel of the EMG record at a frequency of 2.5 Hz.
This same signal drives a counter which is in the field of view of the videotape
camera. The videotape records of behavior can then be synchronized on a frame-
by-frame basis with the simultaneously recorded EMG records.
Source: A. Bekoff. 1978. A neuroethological approach to the study of the ontogeny
of coordinated behavior. Chapter 2 in G. M. Burghardt and M. Bekoff (eds.), *The
Development of Behavior.* New York: Garland STPM Press, p. 29.

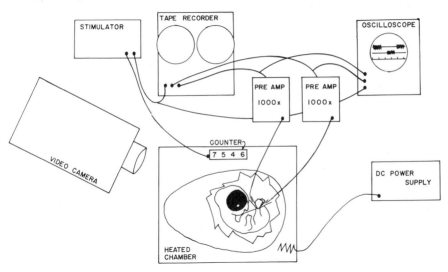

related to behavioral observations of wings, feet, and head. From these data we also
recognize that the neural circuitry that underlies specific coordinated behaviors is
often developed and operative some time before a recognizable pattern emerges
(Bekoff 1976; Bekoff, Stein and Hamburger 1975). Bekoff hypothesizes that complex behaviors in adult animals may be viewed as arising from the coordination of
simpler actions during the course of development.

SENSORY AND MOTOR STIMULATION

Investigators have shown that sensory and motor responses can be evoked and
measured prenatally. We can test whether the sensory and neural systems of an

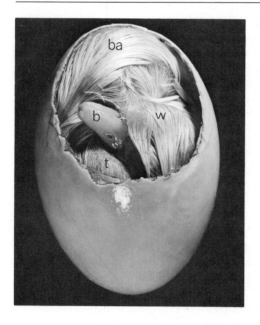

FIGURE 9-2   Bird embryo after removal of portions of eggshell
By using a delicate technique, we can observe development in avian embryos without disturbing the normal ontogenetic processes. (ba = back, b = bill, w = wing, t = thigh)
Source: Photo by R. W. Oppenheim.

embryo at various developmental stages are mature enough to respond to appropriate stimuli; for example, we can measure an embryo's reflex responses to tactile stimulation. Hamburger (1963, 1973) and others studying bird species have produced evidence that both general and specific embryonic movements can be induced by the stimulation of an embryo with a fine brush or blunt probe. The embryos of some bird species begin to respond to stimulation as early as the first week of development. Mammalian embryos, which are much more difficult to study because they are isolated in the uterus, respond with reflex-type movements to stimulation of different body zones (see summary in Barron 1941). Touching the snout of a sixteen-day-old rat fetus produces head movements; touching the side of the face of a human fetus at about eight weeks of gestation produces general movements of the body and limbs.

In studies of the auditory mode, Gottlieb (1968) has shown that, several days before hatching, Peking duck embryos *(Anas platyrhynchos)* respond selectively to sounds of the species' maternal call. Playing recordings of other species' maternal calls produces a quickening of the heart rate, which is a general activation response, but only the species-specific maternal call induces a heightened frequency of bill-clapping responses.

A slightly different, but related, set of experiments has revealed that vocalizations made by prehatching birds may play an important role in synchronizing the hatching time of the birds in a clutch (Vince 1964, 1966, 1969, 1973). Vince first demonstrated that those bobwhite quail *(Colinus virginianus)* embryos that were more advanced in their development accelerated the development of less-advanced embryos in a clutch. Early experiments suggested—and later work confirmed—that a "clicking" vocalization made by the embryos is important for the processes of stimulation and synchronization of hatching among bobwhite quail. Vince has also provided evidence that in some instances slower-developing embryos may retard those that are more advanced. Thus sensory-motor systems that are operational prenatally in certain animal species may have important functional significance.

The motor development of prenatal organisms has been investigated descriptively and experimentally. In a study of duck and chick embryos, Oppenheim (1970, 1972) recorded general activity levels and sequences of body movements from the embryos' first weeks of incubation until hatching. His method (see Kuo 1967; Gottlieb 1968) was to remove a portion of the eggshell to expose the developing embryo (Figure 9–2), which remained viable and continued to develop normally. Figure 9–3 illustrates activity and inactivity of embryos at different stages of development. Oppenheim also analyzed the individual movements of the head and various parts of the body and related the sequences of motor development to pipping of the shell and hatching of the duckling or chick. His work demonstrated the functional relationships between many prehatching motor movements and successful hatching.

Prenatal development of the motor system is not limited to general body, limb, or head movements. Duck and chick embryos are capable of vocalizing several days before hatching (Gottlieb and Vandenbergh 1968). These investigators were able to elicit three types of calls—distress, contentment, and brooding—by prodding the embryonic bird's oral region. All of these calls resemble vocalizations made by birds shortly after hatching and are utilized at that time for communication with the mother.

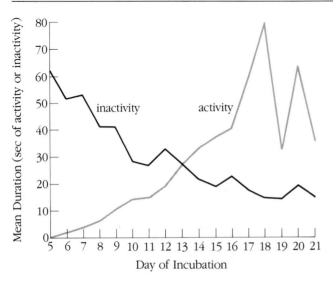

FIGURE 9-3 Measurement of activity in duck embryos
The mean duration in seconds of activity (dotted line) and inactivity (solid line) varies for each day of incubation in duck embryos.
Source: R. W. Oppenheim, "Some Aspects of Embryonic Behavior in the Duck (*Anas platyrhynchos*)," *Animal Behaviour* 18 (1970):335–352. Reprinted with permission from the publisher and the author.

Other investigators have reported on generalized prenatal stimulations that affect later behaviors. Subjecting pregnant female rats *(Rattus norvegicus)* to stress can affect the activity of the young at postnatal ages of 30 to 140 days (Thompson 1957; Ader and Conklin 1963). Offspring of females that were stressed during pregnancy were less active than progeny from unstressed mothers.

Studies of human mothers and neonates (Ferreira 1965; Sontag and Wallace 1935) have revealed differences between infants born to mothers who have undergone stress during pregnancy and infants born to mothers who did not experience the stress. Neonates of stressed mothers were more active, cried more, and gained weight more slowly during the weeks immediately after birth.

PRENATAL MATERNAL EXPERIENCE AFFECTS OFFSPRING BEHAVIOR

Denenberg and Whimbey (1963) conducted an experiment in which young female rat pups were handled by the experimenters for the first twenty days after birth. The pups were removed from the dam by hand, placed on clean wood shavings in a small can for three minutes and then returned to the mother by hand. At maturity these handled females and another group of nonhandled females were bred with colony males. The litters from these dams were all weaned at twenty-one days of age and housed with like-sexed littermates until they were fifty days old. Starting at that age, the young rats of both sexes were tested for activity rates and defecation scores in a 45-inch-square open-field arena. An open-field arena consists of a circular or square pen with lines marked off on the floor. Activity is measured by the number of lines crossed and/or the number of different areas of the pen entered by the test subject in a standard time period. Defecation scores are the number of fecal boluses dropped in the same time period. Open-field tests are designed to measure emotionality — the more "emotional" animals cross fewer lines, enter fewer areas of the pen, and defecate more often (for a review of this method see Archer 1973).

When these measures were compared for rats born to mothers that had been handled in infancy and those born to nonhandled mothers, the former exhibited significantly lower activity levels and more defecation. Thus, treatments administered to female rats prior to the time they were bred had a strong influence on the behavior of their pups. These effects may be mediated through both the prenatal mother-fetus relationship and the interactions between mothers and pups after birth. There could also be some effects attributable to interactions between littermates.

In a related investigation Denenberg and Rosenberg (1967) tested whether the effects just noted could be extended for a second generation to the grandpups of the original handled females. The same general procedures were employed as in the previous experiment, except that the grandpups were tested at twenty-one days of age in the open field. Female grandpups of handled females were significantly less active than female pups descended from nonhandled mothers, but male grandpups were only slightly affected. Thus, the effects of handling female rats in infancy can have effects on behavior two generations later. It should be noted that in this second experiment the mother rats were also provided with a modified housing experience involving objects present on the sawdust-covered floor of the cage for the period from 21 to 50 days of age; this alternative caging may have enhanced the carryover effects to a second generation. The observed effects, particularly on second-generation female rats, could occur through a variety of mechanisms, including physiological influences on the fetus, alterations of the milk provided during lactation, or extra-chromosomal inheritance.

EARLY POSTNATAL EVENTS

The young of some species are born relatively helpless, often with little or no fur or feathers, and they are generally incapable of locomotion or ingesting solid foods; these we call **altricial young.** The young of other species are born in a more advanced stage, capable of locomotion and other behavior patterns, and are often able to consume at least some solid foods; these we call **precocial.** A variety of events at or near the time of birth have important consequences for behavioral processes both immediately and later in life. Among these are social attachments and feeding behaviors. We now examine each of these in more detail using examples of precocial and altricial young.

IMPRINTING

Imprinting may be defined as a process by which an animal learns to make a particular response only to one type of animal or object.

FILIAL IMPRINTING. **Filial imprinting** is the process by which animals develop a social attachment for a particular object. The phenomenon of imprinting has been observed for at least centuries and probably longer. One of the earliest scientific descriptions of imprinting is that of Spalding (1873) and the first thorough investigations of the process were those of Lorenz (1935). Originally this effect was thought to be both instantaneous and irreversible, but we now know that, while the general process occurs in many animals, it is not necessarily instantaneous and it can be reversed or otherwise altered (Hinde 1970; Salzen and Meyer 1967; Salzen and Sluckin 1959).

   In studies of precocial birds the basic experimental method involves exposing the birds to an imprinting stimulus soon after hatching and then testing them at some later time (usually several days later) using the same stimulus and a dissimilar stimulus. Measures of following and approaching behavior are used to assess the success of imprinting. However, to test the possibility that birds may have initially been predisposed to prefer the imprinting stimulus, a second test group is necessary, one in which birds are initially exposed to the second stimulus, then tested. If the majority of the responses are again directed at the imprinting stimulus, then the process is judged to have been successful and both objects are useable imprinting stimuli. For imprinting stimuli, investigators have used live or stuffed birds of the same species, flashing lights, colored spheres or cubes, and stationary colored lights. They have also used the same cues to test the effectiveness of imprinting. A common apparatus used for both imprinting and testing birds is a circular arena with a rotating boom from which the stimulus object can be suspended (Figure 9–4).

FIGURE 9–4   Circular arena apparatus used for filial imprinting and testing the following response of young precocial birds
Note the stuffed bird on the boom, the young duckling, and the lines on the arena floor that can be used to measure the movement of the test bird as well as the proximity of the duckling to the moving object. Here a mallard duckling is following a mallard maternal call in preference to following a moving visual replica of a mallard hen.
Source: Photo from the laboratory of Gilbert Gottlieb.

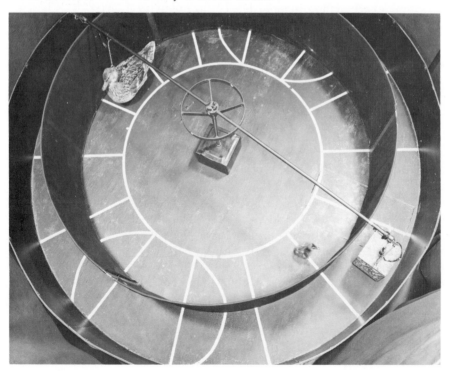

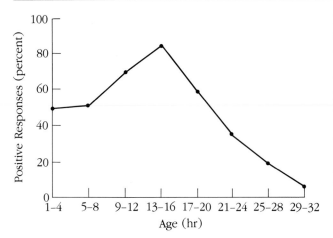

FIGURE 9-5 Critical period for imprinting in mallard ducklings
At each four-hour age interval the percentage of ducklings that show successful imprinting is plotted. The peak percentage occurs at thirteen to sixteen hours after hatching; which we call the sensitive period.
Source: E. H. Hess, "Two Conditions Limiting Control Age for Imprinting," *Journal of Comparative and Physiological Psychology* 52 (1959):516. Copyright 1959 by the American Psychological Association. Adapted by permission of the author and the publisher.

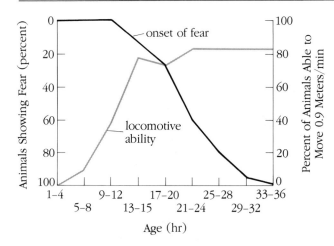

FIGURE 9-6 Locomotion/fear dichotomy
As a young bird develops a fear of novel objects (left axis), it also improves in locomotive ability (right axis). In the middle region of the graph, where the fear response is still low but locomotive ability is relatively high, imprinting is most likely to occur and to have the highest degree of success.
Source: E. H. Hess, "Imprinting," *Science* 130 (1959):133–141. Copyright 1959 by the American Association for the Advancement of Science.

For clarity we will define the period during which the animal can develop an attachment as the **sensitive period,** and the portion of the period during which the performance and reinforcement of the attachment response is greatest is called the **critical period.** In mallard ducks *(Anas platyrhynchos)* (Hess 1959) the sensitive period occurs from about five to twenty-four hours after hatching; the critical period, when imprinting is most successful, encompasses the interval from thirteen to sixteen hours after hatching (Figure 9–5).

One characteristic of imprinting, most often referred to as the **locomotion-fear dichotomy,** may be closely related to the timing of the critical period for imprinting (Figure 9–6). Immediately after hatching, a young duck or chick shows little fear of strange objects, as measured by avoidance and vocalizations, but it is also not yet very mobile. The bird gradually develops locomotive abilities, but at the same time its fear of strange objects increases. Thus, as can be seen in the graph, a period occurs, from roughly thirteen to twenty hours after hatching, when locomotion is high and the fear response is still relatively low; this is the time period for maximally successful imprinting in ducklings.

A final aspect of imprinting concerns the importance of different sensory modalities. In his original investigations of imprinting, Lorenz (1935) found that ducklings responded to both auditory and visual stimuli, but the strongest responses were to the call of the mother. The work of Gottlieb (1971) and his colleagues with precocial birds has revealed that either visual stimuli or auditory stimuli used alone is effective in producing successful imprinting in some species. However, when both auditory and visual cues are provided, the degree of successful imprinting exceeds that achieved with either type of cue used alone. Stimuli of several sensory modes presented simultaneously may more nearly represent natural conditions than do some single-mode experimental regimes.

In goats the adult female parent imprints on her offspring, based primarily upon olfactory and possibly also visual cues (Klopfer, Adams, and Klopfer 1964). The sensitive period for formation of the attachment between does and kids may last for several hours after birth, but the critical time period for formation of the bond is during the initial hour after parturition. During the first postpartum hour, a foster kid can be substituted for the neonate and the doe accepts the alien kid as her own (Klopfer and Klopfer 1968).

Filial imprinting may be significant for several reasons. The imprinting or following response is important in some species, particularly in ducks, which may build nests located at some distance from water, the eventual home of the birds. Within several weeks of hatching the mother leads the ducklings to the water. Imprinting with visual and/or auditory species-specific cues may enable young birds to follow the mother successfully through dense vegetation or other obstacles between the nest and the water. Gottlieb (1963, 1971) has shown the importance of auditory communication in wood ducks *(Aix sponsa).* When the young are ready to leave the nest, located several feet off the ground in a tree hole or other suitable site in a pond or swamp, the mother descends to the water and calls to the ducklings. In response to the call, the young jump up to the nest opening and then drop down into the water to follow the mother.

Recently Miller and Gottlieb (1978) have shown that incubating female mallard ducks emit species-specific vocalizations. These vocalizations are similar to those the mother produces when she calls her young from the nest. In this instance the imprinting process actually begins prior to hatching.

Both visual and auditory imprinting may play an important role in predator avoidance. Quickly following the mother's lead to safety in the seconds between the time a predator is sighted and an attack may mean the difference between life and death. Also, the general recognition of young conspecifics that may occur as the result of imprinting may be significant for the socialization process in young birds and, in some measure, for the general cooperation of conspecifics in a social organization. Associating with conspecifics may be important for survival, in processes like locating food, locating shelter, and migrating.

SEXUAL IMPRINTING.   A different type of imprinting, in which individuals learn selectively to direct their sexual behavior at some stimulus objects (animate or inanimate) but not at others, is termed **sexual imprinting.** Sexual imprinting may serve as a species-identifying and species-isolating mechanism. The sexual preferences of birds have been shown, by appropriate exposure to an imprinting stimulus and later tested, to be imprinted on the stimulus to which they were exposed. Sexual imprinting generally involves longer periods of exposure to the stimulus than filial imprinting. Both precocial birds—turkeys (Schein and Hale 1959) and ducks (Schultz 1965)—and altricial species—zebra finches *(Taeniopygia guttata)* (Immelmann 1965, 1972) and doves *(Columba* spp.) (Craig 1914)—have exhibited mate preferences based on early rearing experiences. Most of these studies have involved cross-fostering of young birds to another species or rearing them with models and post-pubertal testing for mate selection preference.

It is interesting to note that sexual imprinting in mallard ducks and zebra finches differs between males and females of each species. Both species are sexually dimorphic: males have morphological (e.g., color) and behavioral characteristics that trigger sexual responses, and females lack these traits. When male mallards are cross-fostered with other ducks, and zebra finches are cross-fostered with other finches, the males court females of the foster species and not their own; mate selection by female mallards and zebra finches is not affected by the rearing experience. In a recent review of this subject Bateson (1978) has presented a model for testing sexual imprinting that is comprised of four parameters: the age of the animal, the length of exposure to the foster species, the actual stimulus value of the foster species as a potential imprinting stimulus, and the exposure to other species before and after exposure to the foster species.

For most birds in nature, contacts with members of other species during the first few days or even weeks after hatching are limited, thereby ensuring that young birds will imprint on members of their own species and that when they begin to engage in reproductive activities as mature adults, they will court only conspecifics. Thus imprinting as a species-isolating mechanism helps to guarantee that investment of reproductive energy and gametes are not wasted on nonreproductive mating activities.

DEVELOPMENT OF FEEDING AND FOOD PREFERENCES

EFFECTS OF LACTATING MOTHER'S DIET AND CONSPECIFIC ODORS ON OFFSPRING FOOD PREFERENCES.   Is it possible that the food ingested by a lactating female rat influences the later food preferences of her pups? To test this question Galef and

Henderson (1972) experimented with thirty-two pups born to four female rats. After the birth of the pups, the investigators removed the lactating mothers from their home cages and placed them in separate compartments during three one-hour feeding periods each day. They gave two of the females food prepared by Purina and two food made by Turtox. When untested rats are given a choice test they normally prefer the Turtox diet; the foods differ in taste, texture, and color. Galef and Henderson measured the food preferences of the pups for seven days, beginning at seventeen days of age. Figure 9–7 shows that the young rats reared by lactating mothers fed the Purina food ate proportionately more Purina chow than Turtox; and pups reared by lactating mothers fed Turtox preferred Turtox chow.

A rat pup may acquire information about the food its mother is consuming in three possible ways. Young rats may consume feces dropped by the mother; they may ingest or smell particles of food adhering to the mother's fur or oral region; or the flavor or odor of the mother's food may be transmitted in her milk. Further experiments conducted by Galef and Henderson have confirmed that information about diet is transmitted through the mother's milk. The results of these experiments have some interesting implications for our understanding of the development of food habits in young rodents as they leave the natal home site and fend for themselves. Prior to this work the assumption had been that young rats had to learn to locate solid foods without any assistance from adult rats (Barnett 1956). However, the study outlined here and other work by Galef and his colleagues suggest that young rodents may in fact obtain some of this information via their mothers.

At the age of weaning the feeding site selection of young rats is influenced by interactions with adults, and in particular by olfactory cues associated with conspe-

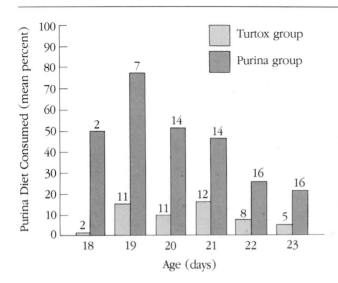

FIGURE 9-7　Feeding preferences in rat pups
The mean amount of Purina chow eaten as a percentage of total food intake by pups reared by mothers fed either a Purina or a Turtox diet varied. The numbers above each bar indicate the sample size for that day. *Source:* Data from B. G. Galef and P. W. Henderson, "Mother's Milk: A Determinant of the Feeding Preferences of Weaning Rat Pups," *Journal of Comparative and Physiological Psychology* 78 (1972):213–219.

cifics (Galef 1977, 1981; Galef and Clark 1971; Galef and Heiber 1976). When presented with a choice, rat pups select a feeding site associated with either conspecifics or their excreta in preference to a clean site. Interestingly, if pups are reared without contact with conspecifics, they select a feeding site without respect to whether it is clean or has conspecific cues present. When pups are reared away from conspecifics, but then are given five days exposure to conspecifics prior to testing, that exposure is sufficient for pups to show a strong preference for feeding sites associated with conspecific stimuli.

TURTLE FOOD PREFERENCES. Burghardt and Hess (1966; Burghardt 1967) investigated the feeding preferences of snapping turtles *(Chelydra serpentia)*. They fed separate groups of newly hatched turtles either horsemeat, fish, or worms for a twelve-day period. When they tested the turtles on day 13, with a choice of three diets, the turtles demonstrated significant preferences for the food on which they had been reared (Table 9–1). The investigators then provided each group of turtles with a different diet for twelve additional days and again tested their diet preference among the three food types. The turtles again exhibited clear preference for the food they had eaten during the twelve-day period immediately after hatching. We can draw two important conclusions from these results: (1) the initial feeding experience may be critical in affecting later dietary preferences in snapping turtles, and (2) effects of the initial feeding experience may be irreversible. Further testing would be needed to confirm and extend these conclusions to the retention into adulthood of the effects of the initial feeding experience, but some type of food imprinting may be occurring — that is, the turtle's food preference may have become fixed even though the turtle will still consume other foods.

FEEDING BEHAVIOR OF GULL CHICKS. The feeding behavior of laughing gull *(Larus atricilla)* has been studied in both field and laboratory settings (Tinbergen and Perdeck, 1950; Hailman 1967, 1969). Tinbergen and Perdeck made observations in the field of the effects of holding cardboard models in front of young gull chicks in their nests. They found that hungry chicks use their bills in a pecking and stroking

TABLE 9-1 Food preferences of snapping turtles after one meal of one of two foods

		First Meal		Second Meal		Choice Test
Group	N	Food	Total Pieces Eaten	Food	Total Pieces Eaten	Number Preferring First-Fed Food
1	12	Horsemeat	26	Worm meat	19	12
2	13	Worm meat	28	Horsemeat	34	8
Totals	25		54		53	20

Source: Data from Burghardt (1967).

FIGURE 9-8   Normal feeding behavior of laughing gull chick
Chicks engage in two separate types of pecking. Here the chick aims an accurately coordinated peck at the beak of a parent, which prompts the parent to regurgitate food. In other instances the chick may peck at food (e.g., fish) the parent has regurgitated.

pattern directed at the bill of a parent, a behavior that induces the adult bird to regurgitate food for the chick (Figure 9–8).

Hailman examined how this food-begging behavior in gull chicks developed. His results indicate that the behavior resulted from an interaction of genes and environment. The genome of the young gull provides the information necessary for the correct maturation of the bird's sensory and motor systems. As it matures the bird learns to peck at the parent's red bill, which contrasts with its black head. By using cardboard models of a gull's head to test pecking behavior, Hailman found that several days were required to attain a record of 75 to 90 percent hits (Figure 9–9). Thus, full development of the begging behavior requires the genetically programmed development of sensory-neural systems to receive and interpret the environmental stimuli, motor capacities for responding, and the experience associated with learning to peck for the food.

JUVENILE EVENTS: BIRDS AND MAMMALS

The period which lasts roughly from fledging (birds) or weaning (mammals) until the animal is fully independent and on the way to maturity can be loosely termed the **juvenile stage.** Development is a continuous process, usually marked by various identifiable events, such as birth or fledging. As investigators we often divide the continuum into stages or periods for convenience. Because animals of different species—and indeed individuals of the same species—often proceed through development at varying rates, the timing and length of these periods will vary. Several pivotal events usually occur during the juvenile period as we define it here, including dispersal from the natal site, puberty, learning appropriate communications signals, various experiences which influence later behavior patterns, and, for birds, the first migration toward the equator. Play behavior, which for some species constitutes an important activity during the juvenile period, will be discussed in more detail later in the chapter.

DEPRIVATION EXPERIMENTS

One method of assessing the possible significance of a particular experience or stimulation during development involves depriving the organism of one or more of these experiences or stimuli and measuring the effects on subsequent behavior. Some of these treatments may actually commence prior to the juvenile period, whereas others start during this stage. The effects may be noticeable and measureable during this stage or later in the life of the organism. Deprivation may be social, sensory, or motor, or some combination of these.

SOCIAL DEPRIVATION IN DOGS.   Fuller (1967) investigated the effects of experiential deprivation on the social behavior of dogs *(Canis familiaris)*. He and his colleagues placed beagle and terrier puppies in isolation for varying lengths of time. They included periodic contact with a human handler or another dog at prescribed intervals in some of the isolation regimes. They then gave the dogs an open-field test and several learning tasks, and observed the subjects for periods when the dogs could make contact with a towel, a ball, or another puppy. The results were similar to the findings for monkeys reported by Harlow and his coworkers (see below): the more severe the deprivation, the more pronounced the dogs' behavioral deficits and abnormalities. The dogs reared under the most restricted regimes sometimes failed even to leave the starting area of an open-field test, were generally less active than dogs from other treatments, and lost most frequently in competition with another puppy.

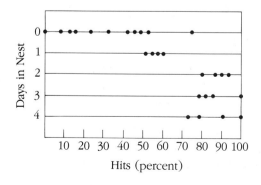

FIGURE 9-9   Accuracy of chick pecking
Gull chicks' improving accuracy in pecking abilities is illustrated here. By the time the chicks have been in the nest two to four days, they exhibit 75 to 90 percent accuracy in directing their pecks at the bill portion of a painted card representation of the parent's head.
Source: Data from J. P. Hailman, "How an Instinct Is Learned," *Scientific American* 221 (1969):100. Reprinted with permission.

Prior to Fuller's work two explanations of the observed behavioral differences between deprivation-reared and normal animals had been proposed. Some investigators ascribed the differences to the deterioration of previously organized neural patterns; this theory is sometimes referred to as the **disuse hypothesis.** Others postulated that the behavioral deficits resulted from a lack of critical information input necessary for proper development of neural patterns and normal behavioral responses; this concept is often called the **loss-of-information hypothesis.**

Fuller proposed a third explanation, which he termed the **stress-of-emergence hypothesis.** Animals that have been in isolation and that are suddenly thrust into a test situation loaded with many novel stimuli may suffer from a stimulus overload—that is, they are faced with a multitude of competing emotional responses. To test this hypothesis Fuller suggested three alternative treatments: (1) providing the isolate-reared dogs several brief, pretest exposures to the testing arena to habituate them to the situation; (2) giving the dogs a tranquilizing drug, like chlorpromazine, before testing to reduce the effects of stimulus bombardment; or (3) giving the dogs several brief periods of handling by the experimenter before testing. These methods have been tried (Fuller 1967) and, with varying degrees of success, have produced some reductions in the stress of emergence. Fuller's hypothesis appears to be valid in some experimental deprivation studies, but we must consider the two earlier hypotheses as primary explanations for the behavioral deficits that result from isolation treatments.

SOCIAL DEPRIVATION IN MONKEYS. From the 1950s onward Harlow and his colleagues have conducted a series of studies on the effects of social deprivation on the behavior of young, adolescent, and adult rhesus monkeys *(Macaca mulatta)* (Harlow 1962; Harlow and Harlow 1962a, 1962b, 1969; Harlow and Zimmerman 1959; Arling and Harlow 1967; Mason 1960, 1961a, 1961b; Mitchell 1970). The investigators usually reared young rhesus monkeys under various conditions for periods of time during the first two years of life. The conditions included:

1. Rearing in total isolation in chambers that remove the infant from all social contacts and most external stimulation.
2. Isolate rearing, but with a cloth or wire surrogate mother (Figure 9–10).
3. Peer group rearing with other monkeys of similar age.
4. Rearing with the mother only.
5. Rearing with the mother, but with varying periods of separation at specified age intervals.
6. Rearing in small social groups in the laboratory (often used as a control condition).

Both the age of the monkey at the time the treatment is initiated and the duration of the rearing experience are critical parameters of social deprivation. This list of rearing treatments is not intended to be exhaustive; researchers have employed other rearing conditions as well as gradations or combinations of these treatments.

FIGURE 9–10   Young rhesus monkey with surrogate mother
When given a choice, most infant rhesus monkeys preferred a cloth surrogate mother to a wire surrogate mother, even when the nursing bottles were attached to the wire mother.
Source: Photo by Harlow, Primate Laboratory, University of Wisconsin.

As we might expect, the severity of the observed effects due to deprivation rearing varies with the degree of social isolation and the duration of the treatment. The first five treatment conditions are listed in approximately the order of degree of deprivation. Behavioral deficits (e.g., failure to perform complete behavior patterns, performing abnormal behavior patterns, lack of responsiveness to conspecifics) are produced in varying intensities, depending upon the severity of the isolation. These effects include withdrawal, fewer social interactions, deficiency in understanding communication, and inability to perform normal sexual behavior. Among the better-known effects of social deprivation are rocking and swaying, self-clasping and other self-directed actions, huddling behavior exhibited by monkeys reared in peer groups, and poor maternal behavior and abusive behavior toward their infants by surrogate-reared females (Figure 9–11)—hence the name "motherless mother."

Studies performed by Harlow and his associates have taught us a great deal about the nature and development of bonds between mothers and their infants and between maternal members of conspecific groupings of primates. We know from this research that several qualities of the rhesus mother, such as contact (even that provided by cloth surrogates), warmth, and some type of movement, are important for normal infant development and development of affection. Young rhesus monkeys spent more time clinging to cloth surrogate mothers than wire mothers, even when the source of milk was associated with the wire surrogates. These investigations have given scientists information that should help provide better rearing conditions for primates in zoos and laboratories.

FIGURE 9-11   Socially deprived rhesus monkey mother
Rearing young rhesus monkeys under varying conditions of social deprivation produces several types of deviant and abnormal behaviors. When mature, the socially deprived female may be a very poor mother, at least with her first infant.
Source: Photo by Harlow, Primate Laboratory, University of Wisconsin.

We should note two additional points in connection with these studies of social deprivation. First, some of the behaviors that Harlow and others have recorded as "abnormal" occur occasionally in rhesus monkeys reared under "normal" conditions (Erwin, Mitchell, and Maple 1973); self-directed aggression has been observed on a number of occasions in nonisolate-reared subjects. Another study reported that in free-ranging rhesus monkeys — or those in their natural habitat, but with some spatial restrictions — only 50 percent of the young born to primiparous females survive to the age of twelve months (Drickamer 1974b). Harlow and his co-workers have also noted that "motherless mothers" exhibit much better maternal behavior with their second infant. Part of the deficiency in maternal care recorded in "motherless mothers" may result from their need to learn how to be good mothers and not strictly to the emotional abnormalities associated with isolate rearing.

Second, and quite significantly, Harlow and his associates (Harlow and Suomi 1971; Suomi 1973; Suomi, Harlow, and Novak 1974) have succeeded in socially rehabilitating isolate-reared monkeys. They accomplished this by exposing six-month-old social isolates to three-month-old normal monkeys, called "therapy monkeys," for two hours per day, three days per week, for one month. The effects produced by some types of early social isolation are not, as was once thought, irreversible, though even rehabilitated monkeys continue to exhibit some behavior defects.

BIRD SONG

In the past several decades an increasing amount of attention has been given by investigators of animal behavior to patterns of bird song — their control, development, and evolutionary significance. We have already explored the control of bird song in Chapter 6, and the functional significance will be explored in Chapter 13. What about the development of song in birds? There appear to be a wide diversity of developmental strategies leading to the diversity of songs and calls with which we are all familiar (Marler and Mundinger 1971; Kroodsma 1978, 1981). From comprehensive analyses of a number of bird species there appear to be at least two major strategies for song development; (1) imitation of the songs of others, particularly of adult conspecifics; and (2) invention or improvisation. Underlying both strategies are the issues of what type of templates exist, which provide some genetic basis for and help to shape or guide the song learning process, and the possible existence of a sensitive phase for development of the song repertoire. Several examples should help explain song development.

The marsh wren *(Cistothorus palustris)* has been used extensively in studies of song learning. Male marsh wrens in nature sing over one hundred types of songs; neighboring males generally sing identical song types; and males often interact by countersinging with one another using the same song type. To test development of song learning males were reared in special housing conditions where the songs they heard could be completely controlled. Males were played specially prepared tutor tapes of songs; each tape contained nine different songs. Males were exposed to one set of tapes for the period 15 to 65 days of age, a second tape for age 65 to 115 days, and a third tape the following spring. The males learned, by imitation, the nine songs on the tape to which they were exposed prior to 65 days of age, but did not learn the songs on the subsequent two tapes. The males never sang any invented songs. Further investigation revealed a more refined estimate for the peak sensitive period, which falls between about days 35 and 55 of age (Kroodsma 1978). In addition, males learned song types played for either three days or nine days. In these additional tests some males improvised some songs, presumably from elements of the songs on the tutor tapes.

The learning situation used for these marsh wrens involved only tutor tapes played over loudspeakers. In another test young wrens were first exposed to a number of song types via tutor tapes and then were given a period of social interaction with adult males with varied song repertoires. The period of social interaction occured in the fall of the first year for some birds and not until the following spring for others. Data on song repertoires for these birds indicate that song learning can occur early from the tapes, or at either of the periods of exposure to adult males. Hence, there is some flexibility in terms of song learning. These findings also make it clear that we should be careful in using only rather artificial stimuli like loudspeakers (see Kroodsma and Pickert 1984). Social interaction has also been shown to be a critical factor in language acquisition in human children (Jerison 1973; Freedle and Lewis 1977).

A longitudinal study of song development has been carried out for male swamp sparrows *(Melospiza georgiana)* that were hatched in the wild and brought into the laboratory. Beginning at 16 to 26 days of age the birds were given song training twice per day for forty days with songs typical for the species (Marler and Peters 1977; Peters, Searcy, and Marler 1980). Recordings were made of the songs of all birds once each week, commencing shortly after training, when birds were just over three months old, and continued until they were over one year of age. When these recordings were analyzed, a seven-stage sequence of song development was discernible (Figure 9–12). The syllables used in training are shown at the top of the figure. Young birds began by singing what Marler and Peters call subsong (stage VII) at an average age of 272 days. They progressed through sub-plastic song (stages V and VI) and began to sing the plastic song (stages II to IV) at an average of 299 days of age. Crystallized song (stage I) began, on average, at 334 days of age. During the course of this developmental sequence, the duration of the song decreased. Syllabic structures began to emerge during sub-plastic song. Analysis of the syllables revealed that about 30 percent were imitations of the training songs and 70 percent were inventions or improvisations. By the time of emergence of the crystallized song the number of syllables sung was only 23 percent of the potential repertoire; both imitated and improvised syllables were included in the crystallized song of most birds (Marler and Peters 1981, 1982). During development it was not unusual to record songs characteristic of more than one stage of the sequence on the same day. However, once the singing of crystallized song began, the birds rarely reverted to an earlier stage. Marler and Peters (1982) hypothesize that the pattern of song development noted for the swamp sparrow may be characteristic of many song bird species.

PUBERTY IN FEMALE HOUSE MICE

Puberty is a critical event in the lives of most organisms — for many, sexual maturation marks the onset of reproductive behavior and the production of progeny. Puberty is also important in terms of the population biology of many species, as we note further in Chapter 17. For house mice *(Mus musculus)* the deme structure generally involves one to several adult males, three to seven females, and their pups and juveniles. Some juvenile females may disperse from the natal site, whereas others may remain within the deme, but virtually all juvenile males disperse (Bailey 1966; Delong 1967; Crowcroft and Rowe 1957; Crowcroft 1973). Within this social context the timing of puberty in female mice has been shown to be significantly affected by various social and environmental factors. Puberty is measured by the occurrence of first vaginal estrus; the heightened levels of estrogen result in cornification of the cells that line the vagina. This phenomenon can be detected by microscopic examination of a vaginal smear.

The presence of a mature male mouse or daily exposure to urine from mature males accelerates the onset of puberty in young female mice (Table 9–2), as does daily exposure to urine from lactating females and urine from singly-caged females in estrus. Daily exposure to urine from group-caged females, regardless of age, results in delays in the onset of puberty. Mice exposed daily to clean bedding or to urine from singly caged diestrus females reach puberty at ages that are intermediate between the acceleration and delay effects produced by exposure to urine from the other sources (Table 9–2).

These effects have been demonstrated in both laboratory and wild stocks of

FIGURE 9–12 Song development in a swamp sparrow

Samples of each of the seven stages of developing song in a male swamp sparrow. The stage is indicated at the left and major stages above. Imitations of the six training syllables portrayed at the top of the figure are identified by number. Training syllable 6 is a song sparrow syllable; the remainder are from swamp sparrows. Syllables were presented in one-part songs, except for syllables 5 and 6, which appeared together in the same "hybrid" song. In nature, swamp sparrows typically have a repertoire of three to four song types, and, although there is some syllable-sharing in local populations, most individual repertoires are unique. The experimental male shown here had a final repertoire of two song types, shown in stages I and II. This crystallized song is virtually identical to the song of the local wild male that served as the original source of the training syllable.

Source: P. Marler and S. Peters. 1982. Structural changes in song ontogeny in the swamp sparrow *Melospiza georgiana. Auk* 99:446–458, p. 448.

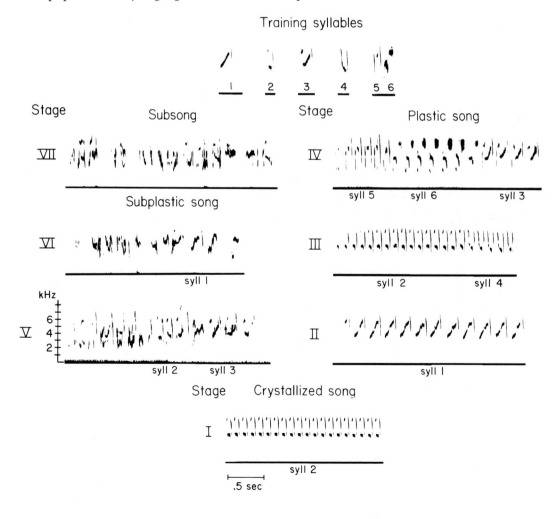

TABLE 9-2 Mean ages and ranges for sexual maturation of female house mice *(Mus musculus)* **given various treatments with urinary chemosignals**
Each young test female was treated starting at 21 days of age, and all test females were housed under the same conditions of photoperiod, food and water availability, and individual caging.

Treatment	*Mean Age at Puberty (days)*	*Age Range (days)*
Control females—exposed to water treatment daily	35	29–42
Females caged with an adult male	27	24–31
Females exposed daily to urine from adult males	31	27–36
Females exposed daily to urine from grouped females	41	36–45
Females exposed daily to urine from estrous females	30	26–36
Females exposed daily to urine from di-estrous females	36	30–41
Females exposed daily to urine from lactating females	30	27–35
Females exposed daily to urine from pregnant females	31	26–36

house mice (Vandenbergh 1967, 1969; Drickamer 1974a, 1979, 1982a, 1983; Vandenbergh, Drickamer, and Colby 1972; Colby and Vandenbergh 1974; Drickamer and Hoover 1978; Drickamer and Murphy 1978), and several of the effects have been replicated in stocks of wild *Mus musculus* maintained in the cloverleaf islands of superhighways (Massey and Vandenbergh 1980, 1981).

When mice are exposed simultaneously to urine from two or three sources the outcome depends on which sources were used. For example, if treatment involves any exposure to urine from grouped females, puberty is delayed in young test females—regardless of what other sources are used. If all donor sources involve urine which accelerate puberty, then the combination treatments result in earlier maturation, but not any earlier than using one of the sources alone (Drickamer 1982b). Diet and daylength have also been shown to affect the timing of puberty in female house mice (Vandenbergh, Drickamer, and Colby 1972; Drickamer 1975).

Hence, a variety of contextual cues from the developing mouse's environment can influence internal physiological events. These effects, in turn, have important consequences for the onset of reproductive behavior in the mice. The chemosignal effects will also influence the length of the interval between a mouse's birth and when it starts to reproduce—called **generation time.** Generation time is a key element in determining the rate of growth or decline in the numbers of young mice entering the population over a given length of time.

JUVENILE EVENTS: INSECTS AND FISH

Much of the research on the ontogeny of behavior in animals has concentrated on birds and mammals. There are, however, a number of excellent reports on behavior development in other animal species.

INSECTS

DROSOPHILA. Fruit flies (genus *Drosophila*) have been used in a wide variety of investigations of the development of behavior; studies have been done both on larvae and on adults utilizing field and laboratory conditions. The primary activity of larvae is feeding. As the larva moves across the food surface it probes with its mouthparts, ingesting food with each cycle of extension and retraction of its body. The larva of *D. melanogaster* go through three molts or **instars** and then pupate before emerging as adult flies. The rate of feeding reaches a peak early in the third instar and then declines during that instar (Burnet and Connolly 1974). Before pupation the larva may migrate to a pupation site away from the food source, or it may remain at the food source during pupation (de Souza, da Cunha, and dos Santos 1970); the pattern varies for different species and sometimes even within a species. Some species may burrow into the soil to pupate, possibly to decrease the risk of predation.

The adults of many *Drosophila* species, particularly females, have been studied more extensively than larvae — adults perform a considerably greater number of behavior patterns. Two major behaviors related to reproduction undergo developmental changes in females, sexual receptivity and oviposition. Manning (1966, 1967) studied sexual receptivity in *D. melanogaster.* As the corpora allata, which release juvenile hormone (JH), and ovaries grow larger, the female becomes receptive. This generally occurs about 48 hours after the **eclosion** of the adult fly from the pupa. Immature females reject courting males that attempt copulation. One hypothesis relating the physiological and behavioral events postulates that JH may affect the brain directly or indirectly, lowering the threshold for sexual receptivity (Schneiderman 1972). If actively functioning corpora allata are implanted in females at the pupal stage, the emergent flies exhibit earlier sexual receptivity and have larger ovaries than nonimplanted controls.

After mating, females become unreceptive again and soon the oviposition behaviors increase. The turning off of receptivity appears due to at least three factors: the presence of sperm in the females' receptacles, the act of copulation itself, and a secretion from the paragonial gland of males (Burnet et al. 1973; Manning 1967). The increase in rates of oviposition is probably due to the increased size of the ovaries, which provides information to the nervous system via stretch receptors, and to the presence of male paragonial gland secretion (Grossfield and Sakri 1972; Merle 1969).

Ringo (1978) discusses the development and maturation processes in females and males of a group of *Drosophila* species inhabiting the Hawaiian Islands. Development takes a longer time in these species, lasting for days or weeks. For males of *D. grimshawi* a series of eight behavior patterns were observed and recorded at four ages during a one-month period following eclosion (Figure 9–13). In general, the diversity of behaviors observed increased with age, and the relative frequency of each behavior increased with age. Males of this species form leks, in which groups of males display communally and mate with females attracted to the lek. For days 15 and 22 there were pronounced increases in sexual and agonistic behaviors — courting, jousting, and abdomen drag (see Spieth 1966 and Ringo 1976 for detailed

FIGURE 9-13   Frequencies ($\overline{X} \pm$ SE) of eight behaviors in *Drosophila grimshawi*
males at ages of 1, 8, 15, 22, and 29 days, post-eclosion
Source: J. M. Ringo. 1978. The development of behavior in *Drosophila*. In G. M.
Burghardt and M. Bekoff (eds.), *Development of Behavior*. New York: Garland STPM
Press, p. 70.

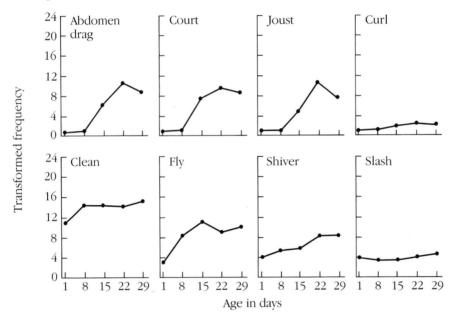

descriptions of the behaviors). The increases in behavior correspond to the time
when the males are most likely to be competing for copulations.

BEES.   Many bee species live in large colonies in which there are **castes**—
physiologically, behaviorally, and often morphologically different forms occurring
together (Michener 1974; Wilson 1971). For example, in the honey bee (*Apis mellif-*
era) and other related species of social bees, there are at least three castes: queens,
drones, and workers. Queens mate, lay eggs, are larger, often do not forage or defend
the colony, and eat proteinaceous food. In contrast, workers generally do not mate or
lay eggs, are smaller, and actively engage in foraging, defense of the colony and
nest-building. The eggs of the queen(s) hatch into larvae, and after these pupate they
emerge as adults.

  One key question concerning development is: How and when does the determination of caste occur? Several factors appear to be important in this process (Wille
and Orozco 1970). Which factors are important varies between major groups of bees,
but in general they include: size of the brood cell, amount of food mass provided to

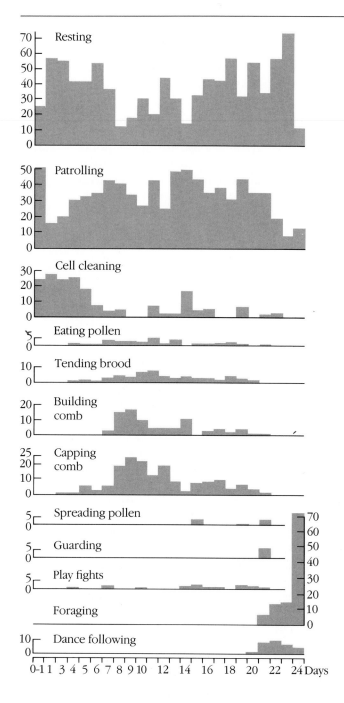

FIGURE 9–14   Time occupied by various activities during the first 24 days of life of a single marked worker honeybee living in a colony The columns of figures at the left and right represent percentages of its time during which the bee engaged in each activity, out of the two to ten hours of observation daily.
Source: Page 305 from C. R. Ribbands. 1953. *Behaviour and Social Life of Honeybees.* London: Bee Research Assoc. Reprinted in D. Michener. 1974. *Social Behavior of the Bees.* Cambridge: Belknap Press of Harvard University Press, p. 127.

the developing larva, quality of the food supplied to the larva, and possibly some chemical cues transmitted with the food. For most species studied, these effects begin immediately upon hatching or very early in larval life (Ribbands 1953; Jung-Hoffman 1966; Darchen and Delage 1970; Weaver 1966). Conditions both within the hive (e.g., loss of a queen) and outside the hive (e.g., changing seasons) can influence the numbers of workers and queens produced.

Most worker bees of these social species undergo changes in behavior as they develop which correspond to changes in their functional roles within the colony or hive (Free 1965). The activity changes for a worker honey bee are shown in the histograms in Figure 9–14. Resting and patrolling occur throughout the twenty-four days surveyed. Many other activities occur in a sequential pattern, beginning with high levels of cell clearing in the early days. This is followed by activities related to the comb and tending the brood. For the last portion of their lifespan the workers become foragers; the average worker bee living in a temperate zone climate survives to the age of six weeks. For many of the activities there is some overlap — bees shift among behaviors of all types on a given day, up to the start of foraging activity.

Several physiological changes are correlated with the shifting patterns of functional roles. Young bees have enlarged hypopharyngeal glands. These produce a major component of the bee milk, part of the diet fed to larvae. Levels of secretion for invertase, an enzyme involved in the conversion of nectar to honey, are highest during the middle portion of the lifespan (Simpson, Riedel, and Wilding 1968). Wax glands are small in the youngest workers, but this is followed by a gradual increase to maximum size by about days 16 to 18, and then the wax glands decline rapidly (Snodgrass 1956). These changes correspond with the higher levels of comb-related activities during this age range. Many other traits follow similar patterns of correlation between developmental events throughout the life of the bee and its changing functional activities in the hive.

Most social bees actively guard and defend their colonies. How do they develop recognition of nestmates and discriminate them from other conspecifics? Work by Breed (1983) indicates that bees will use environmental odor sources, or, if the environmental cues are controlled, they will use genetically based cues to discriminate nestmates from non-nestmates (see also review by Hölldobler and Michener 1980). Investigators have shown that the cues necessary for making these discriminations are acquired prior to emergence as adults.

FISH

While considerably fewer studies of the ontogeny of fish behavior have been conducted than for some other vertebrate groups, there has been some recent progress (Noakes 1978). Perhaps the most extensive investigations were those of Ward and his colleagues (Wyman and Ward 1973; Cole and Ward 1970; Quartermus and Ward 1969). After observations of the orange chromide *(Etroplus maculatus)*, a theoretical model for the ontogeny of numerous behavior patterns was proposed (Figure 9–15).

FIGURE 9-15 A model representing the ontogeny of behavior in *Etroplus maculatus*

The arrows marked "+" indicate facilitation; the solid arrows marked "−" indicate inhibition. The dotted arrows, for diagrammatic simplicity, represent environmental feedback to the organism. The environmental factors implied by the dotted arrows are those indicated feeding into the model from the right. The dotted arrows marked "+" are proposed routes of facilitation between behavioral units.
Source: Wyman, R. L., and J. A. Ward. 1973. The development of behavior in the cichlid fish *Etroplus maculatus* (Bloch). *Zeit. Tierpsychol.* 33:461–491, p. 482.

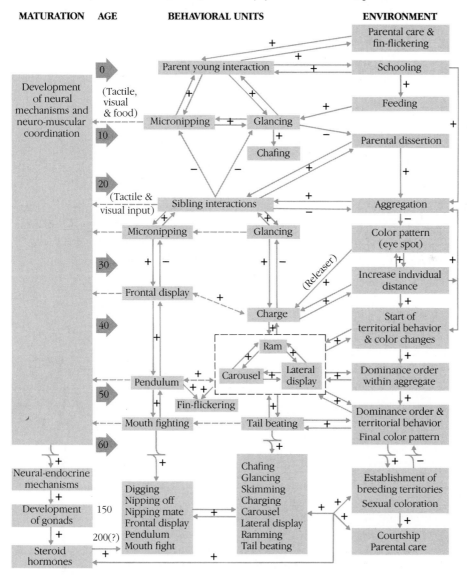

All subsequent behavior patterns of this species are developed from two initial movements: glancing and micronipping. Data for a number of other cichlid species and for some salmonid species seem to fit this model, though extensive additional testing is needed.

PLAY BEHAVIOR

Play behavior has been defined and characterized in many different ways. Fagen defines play as "an inexact term used to denote certain locomotor, manipulative and social behaviors characteristic of young (and some adult) mammals and birds under certain conditions in certain environments" (Fagen 1981, p. 21). We can characterize play in various animal groups according to the actions of the animals and the contexts in which the play behavior is observed. Most investigators recognize at least three types of play, with some overlap. The first type is social play as exemplified by wrestling, chasing, and tumbling activities of the young of many species (Bekoff 1978) (Figure 9–16). A second type provides exercise for developing muscles, locomotor patterns, and other movements. The third type is often labelled "diversive exploration." Generally, this type of play involves sensory inspection of an object followed by extensive repeated manipulation of the object.

    Play behavior of one or more of these types has been recorded in mammals, ranging from rodents and bats to bears, cats, elephants, and whales. Play has also been recorded from a large number of avian species, including raptors, passerines, aquatic or oceanic birds, and parrots. To date only a few anecdotal bits of evidence exist regarding play behavior in other vertebrate groups or among invertebrates. However, additional observations must be made before reaching any firm conclusions limiting play to birds and mammals.

FIGURE 9–16   Canid play behavior
Source: Photo by Heather Parr-Fentress.

FUNCTIONS. Investigators have ascribed a wide range of functions to play behavior (Loizos 1967; Fagen 1981; Bekoff 1974b; Bekoff and Byers 1981). In the most general sense, play functions as practice for adult activities. Animals perform many actions in the course of play which contain elements of behaviors seen in later adult life. Perhaps aggressive behavior is the best example. As young cats or primates engage in various forms of social play, their mock attacks, chases, and mild, noninjurious bites are all partly practice for "real life" use of these same patterns a few months or years later. A second function often ascribed to play behavior is to aid in the process of maturation — growth and development. As young foals or lambs cavort about, alone or in small groups, they use their muscles and develop coordinated movements. A third function of play may be to gain information about the environment; this would be particularly true of diversive play. By exploring and manipulating objects found in their environment, young animals accumulate information that may prove useful later in life. Finally, play can function in the establishment of social relationships with peers and adults. Some of these may be purely affiliations, whereas others could be related to the establishment of dominance-subordinance relationships. For many primate species and some canids, the aggressive play observed in young juveniles gradually becomes more intense and adult-like. The patterns of dominance established in play encounters as juveniles can be retained in later adult life.

CANIDS

One animal group for which play behavior has been studied extensively is the canids (Bekoff 1974a, 1974b, 1974c; Scott and Fuller 1965; Zimen 1972; Moehlman 1979). Bekoff (1974a) studied social play and play soliciting in coyotes *(Canis latrans)*, wolves *(C. lupus)*, and beagles *(C. familiaris)* (Figure 9–16). His observations indicated some species differences and some age-specific trends for several aspects of play behavior, for example, play soliciting and agonistic behavior (Figure 9–17). Bekoff notes that the beagles displayed an early onset for play soliciting and exhibited very little agonistic behavior during play — no fighting occurred at all, only mild threats. Wolves showed moderate levels of play soliciting, with an increase at the last age interval recorded in the sample. Like the beagles, wolves displayed low levels of agonistic behavior, consisting primarily of threats. In contrast, coyotes exhibited higher levels of agonistic behavior throughout the observation period, and a correspondingly lower rate of play-soliciting actions. Coyotes generally establish dominance relationships through fights at an early age (Fox and Clark 1971); this may account for the differences between these animals and the other two species. When all play behaviors were summarized, the beagles were seven times more playful than the coyotes and three times more playful than the wolves. It is interesting to speculate on the possible relationship between these differences in observed play behavior and the differences in social structures of the species tested. Wolves are generally social, group-living animals, though some individuals may lead a solitary existence

FIGURE 9–17   Play-soliciting and agonistic behavior in canids
The median frequency of occurrence (percent) of action patterns observed during
both play soliciting and agonistic interactions in relation to the total number of
actions performed during the stated time periods; t = animals housed together in
pairs at the beginning of this time period; vertical bars = range.
Source: M. Bekoff. 1974. Social play and play-soliciting by infant canids. *Amer. Zool.*
14:323–340, p. 327.

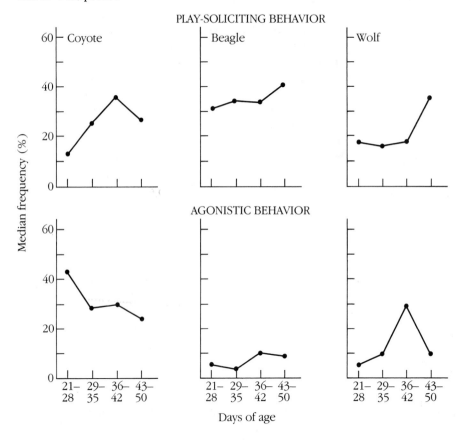

for some periods (Mech 1970). Beagles are at least somewhat social animals, though
we rarely observe such domesticated canids in semi-natural or natural situations
where their feral social structure can be fully recorded. Coyotes, on the other hand,
are generally much more solitary in their social organization, with each animal
roaming a large home range.

In a recent review Price (1984) has suggested neotony as an explanation for
the observation that some domesticated breeds of animals are more playful than
their wild ancestors. Man has selected for the retention of juvenile characteristics in

adulthood. Thus, the beagles in the foregoing example may be more playful than the wolves because of the effects of man's selection.

KEAS

Keas *(Nestor notabilis)* are large parrots that inhabit parts of New Zealand. A variety of play behaviors have been observed and reported for this species (Derscheid 1947; Jackson 1963; Keller 1975, 1976). The most extensive observations were those of Keller on captive keas in zoos. The young keas perform a variety of acrobatic maneuvers, including somersaults, sliding on snow-covered slopes, and hanging by their bills from tree branches or hanging upside down by their feet. Social play in groups is also quite common and involves wrestling and activities which can be described anthropomorphically as resembling "hide-and-go-seek" and "king of the hill." The solicitation to play in a social situation may involve such tactics as adopting a posture normally used in defense with the head down and one foot raised or the stiff-legged walking posture common in some mammals and a few other birds. Last, keas engage in a great deal of object play. They manipulate objects of all sorts using both feet and bills, toss items into the air and fly at them, and they play with snow when it is present.

HUMAN CHILDREN

Numerous studies have been conducted by both developmental psychologists and ethologists on play behavior in young *Homo sapiens* (reviewed by Bruner, Jolly, and Sylva 1976; Fagen 1981). Hutt and Bhanvani (1972) conducted a follow-up study of earlier work by Hutt (1966, 1967a, 1967b, 1970a, 1970b) in order to attempt to make some predictions about differences in play behavior based on longitudinal sampling. A young child confronted with a new toy will first investigate and inspect it (specific exploration) and then play with the toy (diversive exploration). Three to five-year-old nursery school children can be classified into three mutually exclusive categories with regard to their responses to a new toy: non-explorers, who may visually inspect the toy, but do not handle or play with it; explorers, who thoroughly investigate the toy, but fail to do more than that with it; and, inventive explorers, who investigate the toy and then play with it in a variety of innovative ways.

Do these individual differences in exploratory behavior provide any predictive insights with respect to traits in the children at a later age? To test this question Hutt and Bhanvani (1972) obtained data for about fifty children from the original sample of one hundred used to generate the three categories above. The children were seven to ten years of age at the time of the second sampling procedures. Each child was given a series of tests to measure creativity and a personality questionnaire. Each child was rated by parents and teachers on behavior, adjustment, and development. The results provided support for several suggestive conclusions — really more like hypotheses for further testing:

1. Lack of exploring was related to later lack of curiosity and adventure in young boys, and to difficulties in personality and social adjustment in young girls.
2. Children who had been more creative and imaginative in their early play behavior were more likely to be divergent in creativity at the later ages; this was particularly true for boys.

Hutt and Bhanvani note that some of these effects may be attributable to early childhood differences between the sexes; boys are more exploratory in their play activity for a longer period of their life, and girls are socially and linguistically more advanced than boys at nursery school age. We might also note that further longitudinal follow-up testing to assess longer term effects of these early differences would be both appropriate and necessary to strengthen or extend these conclusions.

DEVELOPMENT INTO ADULT LIFE

The development of behavior, as we noted earlier, is a continuum. Many key events take place early in an organism's life. However, this does not mean that developmental processes cease when the organism reaches some particular age. Rather, for some traits, depending upon the species and other conditions, developmental changes continue to occur throughout most or all of the life of the organism. We have already discussed in some detail the worker honeybees that undergo a chronological sequence of changes in physiology and behavior lasting throughout their life. In the preceding section we noted that both young and adults of many species engage in play behavior. Some plasticity for particular traits may be important to the ecology and ultimately to an organism's reproductive success, as the following example illustrates.

White-footed mice *(Peromyscus leucopus noveboracensis)* occupy primarily woodland habitat, but they are generally ubiquitous over a wide range of habitats available within their geographical range in the midwestern and eastern United States; they live in fields, croplands, marshes, and often in human dwellings in addition to being abundant in a variety of types of forests. The prairie deermouse *(P. maniculatus bairdi)* lives in grasslands — usually open fields or croplands — and is almost never captured in wooded areas, brushland, marshes, dwellings, and so on.

Adult mice of these two species were caught and brought into the laboratory where they were presented with a cafeteria arrangement of seeds and grains from the various habitats where the mice had been caught (Drickamer 1970). Mice from both species ate large amounts of corn, but for the other items in their respective diets, some significant differences existed. The *P. leucopus* ate more elm and maple seeds than *P. maniculatus,* whereas the latter species ate more bush clover seeds and wheat. Could there be a relationship between the occupancy of particular habitats and food preferences or feeding strategies utilized by mice of these two species?

In a further test Drickamer (1972) explored the effects of dietary experience on subsequent food choices for both young (21 to 90 days of age) and adult (over 90 days of age) mice of these two species. To circumvent problems of inadequate nutrition that occurred when the mice were given only diets of seeds, laboratory mouse chow was used, but in three different flavors, generated by placing an odor source (a drop of an essential oil such as anise or pine on a small piece of cotton) beneath the food in each dish. Mice were provided with a training experience either as young or as adults, and they were then tested immediately after the two-week training exposure or one month later. The mice were tested in two separate ways to determine whether the training experience influenced their choice of food-odor combination. The results of these tests are summarized in Table 9–3. Preferences for particular combinations were influenced by prior training experience for young *P. maniculatus* and both young and adult *P. leucopus.* Adult *P. maniculatus* were not affected by the training experience. Further tests were used to assess the patterns of visitation to various food sources by young and adult mice of both species. Adult *P. leucopus* changed feeding sites more frequently than young *P. leucopus* or than either young or adult *P. maniculatus.* If the feeding site arrangements are shifted about, the young and adult *P. maniculatus* demonstrate a strong position preference, whereas *P. leucopus* of both ages will either follow the foods to new locations, or change to another food.

How may these observed differences in feeding habits be related to the

TABLE 9-3
Results from tests on two species of *Peromyscus* to determine the effects of training with various food-odor combinations on the selection of diet from a series of food-odor combinations presented cafeteria style or on the chewing of balsa wood pegs to obtain a preferred food-odor combination. '+' indicates that the testing resulted in significant preferences for the food-odor combination with which the mice were trained. '−' indicates no significant effects of the training experience on food-odor combination preference.

	Appetitive	*Consummatory*
Peromyscus leucopus		
Young mice — Tested immediately	+	+
— Tested one month after training	+	−
Adult mice — Tested immediately	+	+
— Tested one month after training	+	−
Peromyscus maniculatus		
Young mice — Tested immediately	+	+
— Tested one month after training	+	+
Adult mice — Tested immediately	−	−
— Tested one month after training	−	−

ecology of these species? Clearly, the adult *P. leucopus* are more flexible in terms of shifting their feeding preferences and in the more diversified strategy which they apparently use when compared to the adult *P. maniculatus*. This suggests that *P. leucopus* may successfully occupy and utilize a wide variety of habitats because of more flexible feeding habits as adults. In contrast, *P. maniculatus,* became more rigid in their food habits and feeding strategy with age; this correlates with their occupancy of a much more limited range of habitat types. The young of both species exhibit some flexibility with regard to feeding, but by the time the animals are about 90 days of age there has been a change for the *P. maniculatus.* Thus we can see that developmental processes can continue into adulthood and that the results of such processes may have important consequences for the life history patterns of various species.

NATURE/NURTURE/EPIGENESIS

For a number of years a controversy existed within animal behavior regarding the forces which shape and determine behavior. Some investigators operated with the hypothesis that observed behaviors were largely under genetic control (nature), whereas others worked with the hypothesis that observed behavior was primarily a function of developmental experiences and environmental influences (nurture). While we often associate the notion of instinct with the ethologists of Europe and the hypothesis of environmental influences with the zoologists and comparative psychologists working in North America, this distinction is too stereotyped. In fact, while the basic concepts for these opposing viewpoints have received more emphasis from one group or the other, research in this century has been conducted using both viewpoints on both sides of the Atlantic Ocean. Both viewpoints have portions of their historical roots in both Europe and North America (see Chapter 2).

Today most animal behaviorists subscribe to the concept of **epigenesis,** the integrated process of behavior development involving the interaction of the genome and experience. We now examine the nature and nurture viewpoints and conclude with a discussion of epigenesis.

GENETIC INFLUENCES

Ethologists, the most important proponents of the theory of genetically preprogrammed (innate) behavior, developed a special terminology for describing this process:

- **Sign stimulus:** an external signal that elicits specific responses from conspecifics.

- **Innate releasing mechanism (IRM):** a neural process, triggered by the sign stimulus, that preprograms an animal for receiving the sign stimulus and mediates a specific behavioral response.
- **Fixed action pattern (FAP):** an innate behavior pattern that is stereotyped, spontaneous, and independent of immediate control, genetically encoded, and independent of individual learning (Tinbergen 1951).

According to the theory, we analyze behavior as a sequence of events: sign stimulus / innate releasing mechanism (IRM) / fixed action pattern (FAP). As an example of the use of this terminology applied to behavior, consider the adult male three-spined stickleback fish *(Gasterosteus aculeatus).* These fish establish territories and engage in aggressive displays and fights at territorial boundaries (Tinbergen 1948, 1951). One way to discover which characteristics (sign stimuli) elicit aggressive responses is to present model fish (Figure 9–18) to a male stickleback and see which model he attacks. The males display various standard, easily recognizable threat and aggressive attack postures (FAPs) in response to sign stimuli (Figure 9–19). Crude models that have a red belly are attacked more often than are normal stickleback models that lack the red underside.

The aggressive posture exhibited by male sticklebacks appears rigidly stereo-

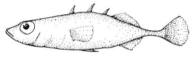
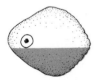

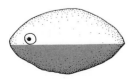

FIGURE 9-18 Fish models used to test aggression in male sticklebacks
The first model of fish resembles the normal three-spined stickleback except that it lacks the red belly characteristics of males of this species; the other four crude models all have red bellies. When these models are presented to a male he attacks the last four models more than the first.

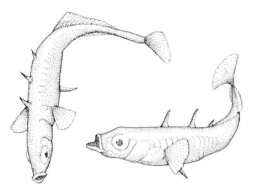

FIGURE 9-19   Aggressive threat displays of male sticklebacks
These postures of the adult male three-spined stickleback fish are stereotyped, are performed in a similar manner by all males of this species, and are called fixed action patterns (FAPs).

typed. This similarity of display patterns by all male sticklebacks may have some strong evolutionary advantages. If each male performed the threat behavior differently, then the male being threatened might not be certain of the meaning of the posture. Confusion concerning an actual threat could lead to misinterpretation and, in turn, to an attack involving physical damage to one or both males.

The hypothesis of genetic control of behavior and the experimental methods used to test various aspects of this hypothesis have been criticized on several grounds (Lehrman 1953, 1970; Moltz 1965). The term *innate* can have two distinctly different meanings to many animal behaviorists. First, it may refer to variations in a trait among individuals in a population. For example, human eye colors are blue, brown, green, and mixtures of these colors, with the color genetically based. In this instance, we might say that differences in a trait are inherited or innate, but that external influences during ontogeny may still affect the development of that trait. Second, some ethologists have used the term *innate* to refer to the notion of fixed development of a specific behavior pattern — that is, the organism exhibits behaviors that are preprogrammed in the genes, as is the case in the fixed action patterns of sticklebacks. Unfortunately these different meanings create a great deal of confusion, and some animal behaviorists use the term indiscriminately. Lehrman and other psychologists, zoologists, and ethologists prefer to use **innate** to refer only to differences between individuals or populations.

Genes code for proteins and not directly for behavior patterns, morphological structures, or physiology. Biochemists and geneticists have discovered that the sequences of molecules in DNA, the chemical that carries genetic information, are codes for the production of specific protein molecules (see Chapter 4) which are involved in the structures and processes within the cells and tissues of the body. To the extent that the genome provides the basic framework for ontogeny, genes have a critical role in determining the eventual behavior patterns exhibited by an organism.

If we define **experience** to include the effects of all interactions between an organism and its environment that influence behavior, then we must consider when a developing animal is capable of receiving internal or external stimuli that affect the organization of brain and the body structures involved in behavior. Lorenz has argued that the neural structures, encoded in the genes, appear first and control both the reception of external and internal stimulation and the behavioral responses of an animal. In support, reports on neurobiological development (see review by Jacobson 1978) have indicated that experience probably does not affect basic neural structure. But recall also the work of Oppenheim, Chu-wang, and Maderut (1978) from earlier in this chapter and the studies of Nottebohm and his colleagues reported in Chapter 6 on canary song, in which evidence is provided that neural structures can be altered by ongoing experiences. Also, Moltz (1965) and Lehrman (1970) have presented arguments that the developmental processes involved in the establishment of the neural circuitry for receiving stimuli and producing responses are a function of both the genome and the experiences of an organism, beginning at conception.

One common method for attempting to demonstrate that a behavior pattern is performed without any prior learning experience is the **isolation experiment.** For example, male stickleback fish reared away from all conspecifics will perform the species-typical zigzag courtship dance correctly the first time they are introduced to a gravid female or model of a gravid female (Cullen 1960). Isolation experiments may involve a range of treatments — from housing in the absence of conspecifics to prevention of motor movements. In considering an isolation experiment it is difficult to be certain exactly from what stimuli an animal is isolated. Rearing away from all conspecifics may not constitute social deprivation only; certain forms of sensory and motor stimulation necessary for the animal's normal development may require social contacts. Moreover, rearing under conditions of restricted motor or sensory stimulation may also deprive an animal of normal social contacts. We should thus be quite cautious in interpreting results of isolation experiments.

Instinct theorists have postulated that genetic encoding accounts for similarities in FAPs among the members of a species. Moltz (1965) has argued in opposition that the stereotyping of a particular behavior pattern may result from consistencies in the environments of the organisms. Thus, for example, the rhythmical activity patterns controlled by biological clocks (see Chapter 8) are, at least in part, functions of external cues from the environment. Thorpe and Jones (1937) experimented with insect species that always lay their eggs on other specific insect hosts. If the eggs of one generation are reared on a different insect host, the next generation of adults may show a preference for laying their eggs on the new host. Other behavior patterns, similarly influenced or channeled by the environment, may seem to be stereotyped (e.g., see King 1968).

Schleidt (1974), who carefully examined aspects of the stereotyping of FAPs, concluded that more quantitative studies of stereotyping are needed. He utilized data from the measurement of gobbling calls of male turkeys to illustrate the variability that exists in an FAP (Figure 9–20). The fixity of an FAP is a relative judgment that depends on the population of animals being sampled, the method of assessing the variation, and the limits of our own perception.

In response to these criticisms of the concept of FAPs, Barlow (1968) developed the concept of **modal action patterns (MAPs).** Barlow defined the MAP as a spatiotemporal behavior pattern that is common to members of a species; different

FIGURE 9–20   Inter-gobble intervals of turkeys
The inter-gobble intervals (IGIs) for 35 samples of 100 gobbles each are plotted as
the cumulative distribution in logarithmic probability coordinates. Data were
obtained from 9 different male turkeys. Data such as these that show variation in a
particular parameter of an FAP are useful in assessing the fixity of such fixed action
patterns.
Source: W. Schleidt, "How 'Fixed' Is the Fixed Action Pattern?" *Zeitschrift für
Tierpsychologie* 36 (1974):194. Reprinted with permission.

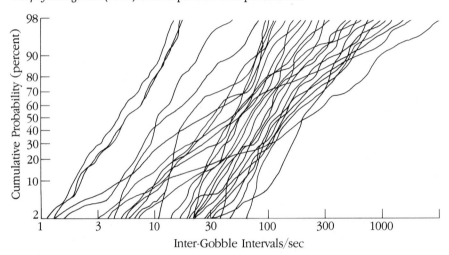

individuals tend to perform the pattern in a recognizably similar or modal fashion.
This sounds very much like the species-typical behavior pattern concept proposed
by Beach (1960). The MAP concept takes into account the possibility of variation
around the modal pattern, the necessity of possessing some flexibility in behavior for
individual adaptation, and the possibility that environmental input or the sign
stimulus can vary the action pattern.

ENVIRONMENTAL INFLUENCES

Proponents of the hypothesis that the environment and experience are of critical
importance in behavior development have supported their viewpoint with evidence
from several types of investigations. First, the emphasis of some investigators on
learning behavior, particularly in the first half of this century, led to the conclusion
that behavior in many animals was largely acquired through experience and practice. Among these investigators E. L. Thorndike, R. M. Yerkes, and J. B. Watson
emphasized the study of animal intelligence and learning processes. Watson (1930)
developed the approach called **behaviorism;** this theoretical and experimental approach operates with the premise that much of behavior results from previous

experience. Later, B. F. Skinner (1938, 1953) espoused and supported via experimentation the idea that most animal actions can be analyzed functionally in terms of combinations of stimulus and response. These S-R connections, according to Skinner, are learned by the organism. Since learning by its very definition implies an important role for experience and practice (see Chapter 10), it is likely that we have historically ascribed a heavy environmental emphasis to animal psychologists investigating learning behavior, when many of them did not actually hold such an extreme position (see Dewsbury 1984). Watson, for example, was very interested in the development of instincts in young animals and the improvement of such behavior with practice.

Second, many comparative psychologists have tested the effects of early experience on later behavior (we have examined some of these studies in this chapter). The sensory, motor, and social aspects of experience have all been manipulated to ascertain the resultant effects on behavior. Some of these investigations involve enrichment, and others utilize deprivation, including isolation. The results often, but not always, demonstrate a key role for experience and environment in the development process. This is not surprising when we realize that it was exactly that which the investigators were manipulating. As with investigations of learning behavior, the very nature of what is being tested provides the impression that those conducting such studies are in fact heavily biased toward an environmental point of view. Many studies of experiential effects also involve genetics, in the form of different species or strains.

Among the criticisms of the studies of learning and experience as environmental influences on the development of behavior is that they have been too often conducted in only laboratory settings and that much of the work has utilized the laboratory rat as a test subject. Beach (1950) presented a numerical analysis of research published in the *Journal of Comparative and Physiological Psychology* during the period from 1930 to 1948. He found that between 60 and 70 percent of the reported studies had used laboratory rats as subjects, and less than 10 percent of the published studies had used invertebrates or nonmammalian vertebrates as test subjects. Hodos and Campbell (1969) have also decried the lack of a truly comparative psychology.

Isolation and deprivation experiments may be criticized because we cannot be certain from what we are isolating the subject animal and we therefore have difficulty interpreting the results.

EPIGENESIS

The modern theory of behavior development is called **epigenesis,** the integrated process of behavior development involving both the genome and environmental influences. According to the epigenetic approach, the expression of the genetic material, which leads to the synthesis of tissues, organs, and thus to behavior patterns, is dependent on the environmental context. Thus in different environmental conditions the same genes may be expressed differently. For example, individuals of a species of fruit fly *(Drosophila)* reared at different temperatures, develop wings capable of normal flight, irregular weak flight, or no flight at all (Harnly 1941). We have already noted in this chapter several ways in which the environmental regime can influence the behavior and sexual selection of various animals.

The general structure of the organism, which develops through the processes of maturation, is dictated in large measure by its genome, but the development of various structures and behaviors is influenced by experience. Genes set the limits, and through interaction with the environment the final product or phenotype is determined. What an animal inherits — that is, what is dictated by the genome — consists of a range of possible expressions of each measureable physical, physiological, and behavioral trait; genes set the limits on the phenotypic expression of traits. For some morphological, physiological, or behavioral traits the prescribed limits of expression may be quite flexible, as dictated by the genetic makeup; whereas for other traits the range of potential expression may be quite narrow.

The theory of epigenesis involves elements of both genetic and environmental viewpoints. Today scientists interested in the development of behavior are exploring questions pertaining to the mechanisms which underlie the appearance of particular behaviors. This often involves working closely with embryologists investigating gene expression and with physiologists and anatomists investigating relationships between structure and function in embryos and neonates. Others interested in the development of behavior are concerned with questions pertaining to life history strategies, the evolution of rates of development and the relationships between patterns of development and the ecological habitats of various species.

SUMMARY

Seven parameters are important in the design of experiments to investigate the effects of early experience on later behavior. These include the age of the animal at the time of early experience treatment, the type and duration of treatment, the age of the animal at the time of testing, and the type of test used, testing for the persistence of effects, and the use of different strains or species to assess relative differences in the effects of experience treatments. Both longitudinal and cross-sectional designs can be used to test the persistence of treatment effects.

The embryology of behavior encompasses investigations involving events that are integral to the process of behavior development and occur prior to birth of hatching. Both sensory and motor responses can be evoked prenatally. Recent studies have linked aspects of development within the nervous system and the development of motor actions by the organism. For some mammals, experiences of the mother (e.g., stress) during pregnancy can affect the later behavior of the progeny.

A variety of events that occur just after birth have important consequences for social behavior processes, both immediately and later in life. Filial imprinting involves the development of a social attachment for a particular object. Imprinting, studied primarily in birds, has been investigated with regard to critical and sensitive periods, a variety of stimulus objects, the locomotion-fear dichotomy and the importance of both the auditory and visual sensory modes. Filial imprinting is significant with respect to the young following the parent(s), predator avoidance, and recognition of conspecifics.

Sexual imprinting involves learning to direct sexual behavior at particular stimulus objects and may serve as a species-identifying mechanism, important in species isolation.

The diet consumed by a female rat may affect the food preferences of her offspring. Also, young rats may be influenced in their selection of feeding sites during weaning by social cues from interactions with conspecifics. Young snapping turtles are strongly influenced in their selection of diet by their first feeding experiences after hatching. Young gulls learn to peck at the bills of their parents to obtain food through a combination of genetic predisposition and practice over time.

A variety of events occur during the juvenile period of development, lasting from fledging or weaning until full maturity and independence are achieved. Investigators have studied the effects of deprivation (social, sensory, motor) and enrichment during the juvenile period on subsequent behavior. Deficits in behaviors due to deprivation may be attributable to loss of information, disuse, or stress of emergence. The severity of the behavioral deficits occasioned by deprivation treatments in dogs and monkeys are related to the degree of deprivation.

Birds learn to sing based on both a genetic template that varies with the species and learning from conspecifics which may take the form of imitation or improvisation. Longitudinal studies of song development indicate that the birds progress through a series of stages in song acquisition; subsong, subplastic song, plastic song, and, finally, crystallized song.

Puberty in house mice is influenced by a variety of social and environmental cues. Some social cues (e.g., urinary chemosignals) accelerate sexual development, whereas others retard the process. Both diet and daylength also influence puberty. The timing of puberty is important both for the onset of reproductive behavior and for the population biology of the mice.

Behavior development has also been explored in a variety of insects and fish. For fruit flies there are clear relationships between internal physiological events and changes in sexual receptivity after eclosion and later, the tendency to oviposit. After honey bees emerge from pupae they progress through a series of functional roles in the colony—again, there are clear developmental correlations between the roles and physiological and morphological changes in the bees. One model for the development of behavior in fish demonstrates how the complex and varied behavior patterns of the young adult derive from a few simple patterns in the newly hatched animal.

Play behavior is an important component of the developmental sequence in many mammals and birds. Studies on canids, keas, and children illustrate the varied types of play: social, exercise/maturation, and exploration. The functions of play behavior include practice for adult activities, aiding the processes of growth and development, learning about the environment, and establishing social relationships.

Development does not end when the young animal becomes independent of its parent(s), but rather continues into adult life. Possessing some capacity to remain flexible in certain behavioral traits may have important consequences for the organism. For example, differences in habitat occupancy of two species of deermouse may be partly a function of the differential dietary flexibility of the adults of these two species.

A critical problem in animal behavior has been the nature/nurture issue — that is, the relative importance of genetic inheritance and of experience for the expression of behavior patterns. Instinct theorists have proposed that fixed actions patterns (FAPs) are, like morphological traits, inherited. FAPs result from preprogrammed neural circuitry that predisposes an organism to make stereotyped responses when it receives specific environmental stimuli. FAPs can be characterized as stereotyped, spontaneous, and independent of immediate external control; they are also genetically encoded and independent of individual learning. The theory that behaviors are inherited has been criticized on the following grounds: (1) genes code for proteins, not for behaviors; (2) isolation experiments are difficult to interpret; and (3) FAPs are not as fixed as was once assumed. The recently developed concept of modal action patterns (MAPs), or species-typical behaviors, accounts for the close similarities between the behavior patterns of members of a species and also for individual variation.

Proponents of the viewpoint that experience and environmental input are central to the processes of behavior development have supported their viewpoint by manipulating experiential conditions and recording the effects on behavior development and by studying learning behavior. That this viewpoint has historically been labelled as emphasizing environmental influences may be largely because the very processes being investigated are by definition environmental. Many investigators exploring environmental influences on behavior development have also manipulated genetics, for example, using different species.

Animal behaviorists today study behavior development in terms of epigenesis — the interaction of the genome and environmental stimuli present during all phases of ontogeny. Genetic inheritance programs the basic characteristics and overall range of flexibility for trait expression in the organism, and environmental influences determine the final nature of behaviors exhibited by the animal. The mechanisms of these interactions are the focus of much current research.

Discussion Questions

1. Careful definition of terms is important in the study of animal behavior. Write out your own definitions of the eight terms listed below. Locate the definitions of these terms in this book and in other books on behavior. How do the definitions differ?

(a) epigenesis   (b) experience   (c) play behavior   (d) critical period
(e) ontogeny   (f) filial imprinting   (g) stress-of-emergence   (h) species-
isolating mechanism.

2. Suppose we are interested in comparing the development of behavior in two
   species of snakes. Specifically we are interested in their habitat preferences. What
   methods would you employ in such an investigation? Assume you will conduct
   your studies in both field and laboratory settings. What advantages and disadvantages can you cite for each of the methods or approaches you have suggested?

3. Many animals we use for studies of behavior are from domesticated laboratory
   stocks or from wild animals that have been kept for varying periods in the
   laboratory. What differences in developmental processes might you expect to
   find between these lab-reared animals and their counterparts in the wild?

4. Pratt and Sackett (1967) raised three groups of young rhesus monkeys with
   different degrees of contact with their peers. They allowed one group no contact;
   the second, only visual and auditory contact; and the third, full normal contact
   with peers. Next they allowed animals of the three groups to interact socially;
   then they gave each individual a three-choice preference test where the alternative choices were conspecifics—one from each of the three different treatment
   conditions. The data in the table represent the preferences of the test subjects.
   What conclusions can you draw from these data?

	Rearing Condition of Stimulus Animal		
Rearing Condition of Test Animal	*No Contact*	*Visual and Auditory Contact*	*Normal Contact*
No contact	156	35	29
Visual and auditory contact	104	214	103
Normal contact	94	114	260

Source: Data from Pratt and Sackett (1967).

The numbers in the table are the mean number of seconds spent by a test monkey
with each of the three possible partners in a choice test.

Suggested Readings

Bateson, P. P. G. 1978. How does behavior develop? In P. P. G. Bateson and P. H.
Klopfer (eds.), *Perspectives in Ethology* 3:55–66.
Theoretical, thought-provoking presentation. Bateson is best
known for his work on imprinting, but this treatment is
broader and more general.

Burghardt, G. M., and M. Bekoff. 1978. *Development of Behavior.* New York: Garland STPM Press.
A book based on a symposium of the Animal Behavior Society. The wide range of topics covered and the short, informative articles provide an excellent "state of the art" summary of this subfield of animal behavior.

Fagen, R. 1981. *Animal Play Behavior.* New York: Oxford University Press.
A thorough synthesis of all aspects of play behavior. The book is quite well written, and presents both detailed examples and comprehensive reference listings. The illustrations are an added bonus.

Hess, E. H. 1973. *Imprinting.* New York: Van Nostrand Reinhold.
Both theory and experimentation of imprinting, primarily on birds, are treated by one of the leading investigators of this phenomenon. Bits and pieces of ecological and evolutionary significance of imprinting processes.

Immelmann, K., G. W. Barlow, L. Petrinovich, and M. Main. 1981. *Behavioral Development.* New York: Cambridge University Press.
This volume resulted from a series of interdisciplinary seminars held in West Germany. Each article is a new contribution; some are presentations of new data, while others are theoretical or review papers. Together they represent one of the most comprehensive collections on the topic of behavior development. There are ample references with each of the twenty-eight separate contributions.

References

Ader, R., and P. M. Conklin. 1963. Handling of pregnant rats: Effects on emotionality of their offspring. *Science* 142:411–412.

Archer, J. 1973. Tests for emotionality in rats and mice: A review. *Anim. Behav.* 21:205–235.

Arling, G. L., and H. F. Harlow. 1967. Effects of social deprivation on maternal behavior of rhesus monkeys. *J. Comp. Physiol. Psychol.* 64:371–377.

Bailey, E. D. 1966. Social interaction as a population regulating mechanism in mice. *Can. J. Zool.* 44:1007–1012.

Barlow, G. W. 1968. Ethological units of behavior. In D. Ingle (ed.), *The Central Nervous System and Fish Behavior.* Chicago: University of Chicago Press.

Barnett, S. A. 1956. Behavior components in the feeding of wild and laboratory rats. *Behaviour* 9:24–43.

Barron, D. H. 1941. The functional development of some mammalian neuromuscular mechanisms. *Biol. Rev.* 16:1–33.

Bateson, P. P. G. 1978. Early experience and sexual preferences. In J. B. Hutchinson (ed.), *Biological Determinants of Sexual Behavior.* New York: Wiley.

Beach, F. A. 1950. The snark was a boojum. *Amer. Psychol.* 5:115–124.

———. 1960. Experimental investigations of species-specific behavior. *Amer. Psychol.* 15:1–18.

Bekoff, A. 1976. Ontogeny of leg motor output in the chick embryo: A neural analysis. *Brain Res.* 106:271–291.

―――. 1978. A neuroethological approach to the study of the ontogeny of coordinated behavior. In G. Burghardt and M. Bekoff (eds.), *Development of Behavior.* New York: Garland STPM Press.

Bekoff, A., P. S. G. Stein, and V. Hamburger. 1975. Coordinated motor output in the hindlimb of the 7-day chick embryo. *Proc. Nat. Acad. Sci. U.S.A.* 72:1245–1248.

Bekoff, M. 1974a. Social play in coyotes, wolves and dogs. *Bioscience* 24:225–230.

―――. 1974b. Social play in mammals. *Am. Zool.* 14:265–436.

―――. 1974c. Social play and play-soliciting by infant canids. *Am. Zool.* 14:323–340.

―――. 1978. Social play: Structure, function and the evolution of a cooperative social behavior. In G. Burghardt and M. Bekoff (eds.), *Development of Behavior.* New York: Garland STPM Press.

Bekoff, M., and J. A. Byers. 1981. A critical reanalysis of the ontogeny and phylogeny of mammalian social and locomotor play: An ethological hornet's nest. In K. Immelmann et al. (eds.), *Behavioral Development.* New York: Cambridge University Press.

Breed, M. D. 1983. Nestmate recognition in honey bees. *Anim. Behav.* 31:86–91.

Bruner, J. S., A. Jolly, and K. Sylva, eds. 1976. *Play: Its Role in Development and Evolution.* New York: Basic Books.

Burnet, B., K. Connolly, M. Kearney, and R. Cook. 1973. Effects of male paragonial gland secretion on sexual receptivity and courtship behaviour of female *Drosophila melanogaster. J. Insect. Physiol.* 19:2421–2431.

Burnet, B., and K. Connolly. 1974. Activity and sexual behavior in *Drosophila melanogaster.* In J. H. F. van Abeelen (ed.), *Genetics of Behaviour.* Amsterdam: North-Holland.

Burghardt, G. M. 1967. The primacy of the first feeding experience in the snapping turtle. *Psychonomic Science* 7:383–384.

Burghardt, G. M., and E. H. Hess. 1966. Food imprinting in the snapping turtle, *Chelydra serpentia. Science* 151:108–109.

Colby, D. R., and J. G. Vandenbergh. 1974. Regulatory effects of urinary pheromones on puberty in the mouse. *Biol. Reprod.* 11:268–279.

Cole, J. E., and J. A. Ward. 1970. An analysis of parental recognition by the young of the cichlid fish, *Etroplus maculatus* (Bloch). *Z. Tierpsychol.* 27:156–276.

Craig, W. 1914. Male doves reared in isolation. *J. Anim. Behav.* 4:121–133.

Crowcroft, P. 1973. *Mice All Over.* Brookfield, Ill.: Chicago Zoological Park.

Crowcroft, P., and F. Rowe. 1957. Social organisation and territorial behaviour in the wild house mouse (*Mus musculus* L.). *Proc. Zool. Soc. Lond.* 140:517–531.

Cullen, E. 1960. Experiment on the effect of social isolation on reproductive behavior in the three-spined stickleback. *Anim. Behav.* 8:235.

Darchen, R., and B. Delage. 1970. Facteur déterminant les castes chez les Trigones. *Compt. Ren. Acad. Sci.* (Paris) 270:1372–1373.

DeLong, K. T. 1967. Population ecology of feral house mice. *Ecology* 48:611–634.

de Souza, H. M. L., A. B. da Cunha, and E. P. dos Santos. 1970. Adaptive polymorphism of behavior evolved in laboratory populations of *Drosophila willistoni. Am. Nat.* 104:175–189.

Denenberg, V. E., and K. M. Rosenberg. 1967. Nongenetic transmission of information. *Nature* 216:549–550.

Denenberg, V. H., and A. E. Whimbey. 1963. Behavior of adult rats is modified by the experiences their mothers had as infants. *Science* 142:1192–1193.

Derscheid, J. M. 1947. Strange parrots. *Avic. Mag.* 53:44–49.

Dewsbury, D. A. 1984. *Comparative Psychology in the Twentieth Century.* Stroudsburg, Penn.: Hutchinson Ross.

Drickamer, L. C. 1970. Seed preferences in wild caught *Peromyscus maniculatus bairdi* and *P. leucopus noveboracensis. J. Mammal.* 51:191–194.

―――. 1972. Experience and selection behavior as factors in the food habits of *Peromyscus:* Use of olfaction. *Behaviour* 41:269–287.

―――. 1974a. A ten-year summary of population and reproduction data for free-ranging *Macaca mulatta* at La Parguera, Puerto Rico. *Folia primat.* 21:61–80.

―――. 1974b. Sexual maturation of female mice: Social inhibition. *Develop. Psychobiol.* 7:257–265.

―――. 1975. Daylength and sexual maturation of female house mice. *Develop. Psychobiol.* 8:561–570.

―――. 1979. Acceleration and delay of first vaginal estrus in wild *Mus musculus. J. Mammal.* 60:215–216.

―――. 1982a. Delay and acceleration of puberty

in female mice by urinary chemosignals from other females. *Develop. Psychobiol.* 15:433–442.

———. 1982b. Acceleration and delay of sexual maturation in female house mice by urinary cues: Dose levels and mixing urine from different sources. *Anim. Behav.* 30:456–460.

———. 1983. Male acceleration of puberty in female mice. *J. Comp. Psychol.* 97:191–200.

Drickamer, L. C., and J. E. Hoover. 1979. Effects of urine from pregnant and lactating female house mice on sexual maturation of juvenile females. *Develop. Psychobiol.* 12:545–551.

Drickamer, L. C., and R. X. Murphy. 1978. Female mouse maturation: Effects of excreted and bladder urine from juvenile and adult males. *Develop. Psychobiol.* 11:63–72.

Erwin, J., G. Mitchell, and T. Maple. 1973. Abnormal behavior in non-isolate-reared rhesus monkeys. *Psychol. Reports* 33:515–523.

Fabricius, E. 1964. Crucial periods in the development of the following response in young nidifugous birds. *Zeit. Tierpsychol.* 21:326–337.

Fagen, R. 1981. *Play Behavior.* New York: Oxford University Press.

Ferreira, A. J. 1965. Emotional factors in the prenatal environment. *J. Nerv. Ment. Disorders* 141:108–118.

Fox, M. W., and A. L. Clark. 1971. The development and temporal sequencing of agonistic behavior in the coyote (*Canis latrans*). *Z. Tierpsychol.* 28:262–278.

Free, J. B. 1965. The allocation of duties among worker honeybees. *Symp. Zool. Soc. Lond.* 14:39–59.

Freedle, R., and M. Lewis. 1977. Prelinguistic conversations. In M. Lewis and L. A. Rosenblum (eds.), *Interaction, Conversation, and the Development of Language.* New York: Wiley.

Fuller, J. L. 1967. Experimental deprivation and later behavior. *Science* 158:1645–1652.

Galef, B. G. 1971. Social factors in the poison avoidance and feeding behavior of wild and domesticated rat pups. *J. Comp. Physiol. Psychol.* 75:341–357.

———. 1977. Mechanisms for the social transmission of food preferences from adult to weanling rats. In L. M. Barker, M. Best, and M. Domjan (eds.), *Learning and Mechanisms in Food Selection.* Waco: Baylor University Press.

———. 1981. Development of olfactory control of feeding-site selection in rat pups. *J. Comp. Physiol. Psychol.* 95:615–622.

Galef, B. G., and M. M. Clark. 1971. Social factors in the poison avoidance and feeding behavior

of wild and domesticated rat pups. *J. Comp. Physiol. Psych.,* 75:341–357.

Galef, B. G., and L. Heiber. 1976. The role of residual olfactory cues in the determination of feeding site selection and exploration patterns of domestic rats. *J. Comp. Physiol. Psychol.* 90:727–739.

Galef, B. G., and P. W. Henderson. 1972. Mother's milk: A determinant of the feeding preferences of weaning rat pups. *J. Comp. Physiol. Psychol.* 78:213–219.

Gottlieb, G. 1963. A naturalistic study of imprinting in wood ducklings (*Aix sponsa*). *J. Comp. Physiol. Psychol.* 56:86–91.

———. 1968. Prenatal behavior of birds. *Quart. Rev. Biol.* 43:148–174.

———. 1971. *Development of Species Identification in Birds.* Chicago: University of Chicago Press.

Gottlieb, G., and J. G. Vandenbergh. 1968. Ontogeny of vocalization in duck and chick embryos. *J. Exp. Zool.* 168:307–326.

Grossfield, J. and B. Sakri. 1972. Divergence in the neural control of oviposition in *Drosophila. J. Insect Physiol.* 18:237–241.

Hailman, J. P. 1967. The ontogeny of an instinct: The pecking response in chicks of the laughing gull (*Larus atricilla* L.) and related species. *Behavior,* Suppl. No. 15:1–159.

———. 1969. How an instinct is learned. *Sci. Amer.* 221:98–106.

Hamburger, V. 1963. Some aspects of the embryology of behavior. *Quart. Rev. Biol.* 38:342–365.

———. 1973. Anatomical and physiological basis of embryonic motility in birds and mammals. In G. Gottlieb (ed.), *Studies on the Development of Behavior and the Nervous System.* New York: Academic Press.

Harlow, H. F. 1962. Development of affection in primates. In E. L. Bliss (ed.) *Roots of Behavior.* New York: Harper & Row.

Harlow, H. F., and M. K. Harlow. 1962a. Social deprivation in monkeys. *Sci. Amer.* 207:136–146.

———. 1962b. The effect of rearing conditions on behavior. *Bull. Menninger Clinic* 26:213–224.

———. 1969. Age-mate or peer affectional system. *Adv. Study Behav.* 2:333–383.

Harlow, H. F., and S. J. Suomi. 1971. Social recovery by isolate-reared monkeys. *Proc. Nat. Acad. Sci. USA* 68:1534–1538.

Harlow, H. F., and R. R. Zimmerman. 1959. Affectional responses in the infant monkey. *Science* 130:421–432.

Harnly, M. H. 1941. Flight capacity in relation to phenotypic and genotypic variations in the wings of *Drosophila melanogaster. J. Exp. Zool.* 88:263–273.

Hess, E. 1959. Imprinting. *Science* 30:133–141.

Hinde, R. A. 1970. *Animal Behaviour,* 2nd ed. New York: McGraw-Hill.

Hodos, W., and C. B. G. Campbell. 1969. *Scala naturae:* Why there is no comparative psychology. *Psychol. Rev.* 76:337–350.

Hölldobler, B., and C. D. Michener. 1980. Mechanisms of identification and discrimination in social hymenoptera. In H. Markl (ed.), *Evolution of Social Behavior: Hypotheses and Empirical Tests.* Deerfield Beach, Fla.: Verlag Chemie.

Hutt, C. 1966. Exploration and play in children. *Symp. Zool. Soc. Lond.* 18:23–44.

———. 1967a. Temporal effects on response decrement and stimulus satiation in exploration. *Brit. J. Psychol.* 58:365–373.

———. 1967b. Effects of stimulus novelty on manipulatory exploration in an infant. *J. Child Psychol. Psychiat.* 8:241–247.

———. 1970a. Specific and diversive exploration. *Adv. Child Dev. Behav.* 5:119–180.

———. 1970b. Curiosity in young children. *Science J.* 6:68–71.

Hutt, C., and R. Bhanvani. 1972. Predictions from play. *Nature* 237:171–172.

Immelmann, K. 1965. Objektfixierung geschlechtlicher Triebhandlung bei Prachtfinken. *Naturwissenschaften* 52:169–170.

———. 1972. Sexual and other long-term aspects of imprinting in birds and other species. *Adv. Stud. Behav.* 4:147–174.

Jackson, J. R. 1963. Nesting of keas. *Notornis* 10:319–326.

Jacobson, M. 1974. A plentitude of neurons. In G. Gottlieb (ed.), *Aspects of Neurogenesis,* vol. 2. New York: Academic Press.

———. 1978. *Developmental Neurobiology,* 2nd ed. New York: Plenum.

Jerison, H. J. 1973. *Evolution of the Brain and Intelligence.* New York: Academic Press.

Jung-Hoffman, I. 1966. Die Determination von Königin und Arbeiterin der Honigbee. *Z. Bienenforschung* 8:296–322.

Keller, R. 1975. Das Spielverhalten der Keas (*Nestor notabilis* Gould) des Zürcher Zoos. *Z. Tierpsychol.* 38:393–408.

———. 1976. Beitrag zur Biologie und Ethologie der Keas (*Nestor notabilis*) des Zürcher Zoos. *Zool. Beitr.* 22:111–156.

King, J. A. 1957. Parameters relevant to determining the effect of early experience upon the adult behavior of animals. *Psychol. Bull.* 55:46–48.

———. 1968. Species specificity and early experience. In G. Newton and S. Levine (eds.), *Early Experience and Behavior.* Springfield, Ill.: Thomas.

———. 1969. A comparison of longitudinal and cross-sectional groups in the development of behavior of deer mice. *An. N.Y. Acad. Sci.* 159:696–709.

Klopfer, P. H., and M. S. Klopfer. 1968. Maternal "imprinting" in goats: Fostering of alien young. *Zeit. Tierpsychol.* 25:862–866.

Klopfer, P. H., D. K. Adams, and M. S. Klopfer. 1964. Maternal imprinting in goats. *Proc. Nat. Acad. Sci. USA* 52:911–914.

Kroodsma, D. E. 1978. Aspects of learning in the ontogeny of bird song: Where, from whom, when, how many, which and how accurately? In G. Burghardt and M. Bekoff (eds.), *Development of Behavior.* New York: Garland STPM Press.

———. 1981. Ontogeny of bird song. In K. Immelmann et al. (eds.), *Behavioral Development.* New York: Cambridge University Press.

Kroodsma, D. E., and R. Pickert. 1984. Repertoire size, auditory templates and selective vocal learning in songbirds. *Anim. Behav.* 32:395–399.

Kuo, Z. Y. 1967. *Dynamics of Behavior Development.* New York: Random House.

Lehrman, D. S. 1953. A critique of Konrad Lorenz's theory of instinctive behavior. *Quart. Rev. Biol.* 28:337–369.

———. 1970. Semantic and conceptual issues in the nature-nurture problem. In L. R. Aronson et al. (eds.), *Development and Evolution of Behavior.* San Francisco: Freeman.

Loizos, C. 1967. Play behavior in high primates: A review. In D. Morris (ed.), *Primate Ethology.* Chicago: Aldine.

Lorenz, K. 1935. Der Kumpan in der Umwelt des Vogels. *J. Ornith.* 83:137–213, 289–413.

Manning, A. 1966. Corpus allatum and sexual receptivity in female *Drosophila melanogaster. Nature* 211:1321–1322.

———. 1967. The control of sexual receptivity in female *Drosophila. Anim. Behav.* 15:239–250.

Marler, P., and P. Mundinger. 1971. Vocal learning in birds. In H. Moltz (ed.), *Ontogeny of Vertebrate Behavior.* New York: Academic Press.

Marler, P., and S. Peters. 1977. Selective vocal learning in a sparrow. *Science* 198:519–521.

———. 1981. Sparrows learn adult song and more from memory. *Science* 213:780–782.

———. 1982. Structural changes in song ontogeny in the swamp sparrow *Melospiza georgiana*. *Auk* 99:446–458.

Mason, W. A. 1960. The effects of social restriction on the behavior of rhesus monkeys. I. Free social behavior. *J. Comp. Physiol. Psychol.* 53:582–589.

———. 1961a. The effects of social restriction on the behavior of rhesus monkeys. II. Tests of gregariousness. *J. Comp. Physiol. Psychol.* 54:287–290.

———. 1961b. The effects of social restriction on the behavior of rhesus monkeys. III. Dominance tests. *J. Comp. Physiol. Psychol.* 54:694–699.

Massey, A., and J. G. Vandenbergh. 1980. Puberty delay by a urinary cue from female house mice in feral populations. *Science* 209:821–822.

———. 1981. Puberty acceleration by a urinary cue from male mice in feral populations. *Biol. Reprod.* 24:523–527.

Mech, L. D. 1970. *The Wolf*. Garden City, N.Y.: Doubleday.

Merle, J. 1969. Fonctionnement ovarien et réceptivité sexuelle de *Drosophila melanogaster* après implantation de fragments de l'appareil génitale mâle. *J. Insect Physiol.* 14:1159–1168.

Michener, C. D. 1974. *Social Behavior of the Bees*. Cambridge: Harvard University Press.

Miller, D. B., and G. Gottlieb. 1978. Maternal vocalizations of mallard ducks *(Anas platyrhynchos)*. *Anim. Behav.* 26:1178–1194.

Mitchell, G. 1970. Abnormal behavior in primates. In L. Rosenblum (ed.), *Primate Behavior: Developments in Field and Laboratory Research*, vol. 1. New York: Academic Press.

Moehlman, P. D. 1979. Jackal helpers and pup survival. *Nature* 277:382–383.

Moltz, H. 1965. Contemporary instinct theory and the fixed action pattern. *Psychol. Rev.* 72:27–47.

Noakes, D. L. G. 1978. Ontogeny of behavior in fishes: A survey and suggestions. In G. Burghardt and M. Bekoff (eds.), *Development of Behavior*. New York: Garland STPM Press.

Oppenheim, R. W. 1970. Some aspects of embryonic behavior in the duck *(Anas platyrhynchos)*. *Anim. Behav.* 18:335–352.

———. 1972. Pre-hatching and hatching behavior in birds: A comparative study of altricial and precocial species. *Anim. Behav.* 20:644–655.

Oppenheim, R. W., W. Chu-wang, and J. L. Maderut. 1978. Cell death of motorneurons in the chick embryo and spinal cord. III. *J. Comp. Neurol.* 177:87–112.

Peters, S., W. A. Searcy, and P. Marler. 1980. Species song discrimination in choice experiments with territorial male swamp and song sparrows. *Anim. Behav.* 28:393–404.

Pittman, R., and R. W. Oppenheim. 1979. Cell death of motorneurons in the chick embryo spinal cord. IV. *J. Comp. Neurol.* 187:425–446.

Pratt, C. L., and G. P. Sackett. 1967. Selection of social partners as a function of peer contact during rearing. *Science* 155:1133–1135.

Price, E. O. 1984. Behavioral aspects of animal domestication *Quarterly Review of Biology* 59:1–32.

Quartermus, C., and J. A. Ward. 1969. Development and significance of two motor patterns used in contacting parents by young orange chromides *(Etroplus maculatus)*. *Anim. Behav.* 17:624–635.

Ribbands, C. R. 1953. *Behaviour and Social Life of Honey Bees*. London: Bee Research Assoc.

Ringo, J. M. 1976. A communal display in Hawaiian *Drosophila* (Diptera: Drosophilidae). *An. Entomol. Soc. Am.* 69:209–214.

———. 1978. The development of behavior in Drosophila. In G. Burghardt and M. Bekoff (eds.), *Development of Behavior*. New York: Garland STPM Press.

Salzen, E. A., and C. C. Meyer. 1967. Imprinting: Reversal of a preference established during the critical period. *Nature* 215:785–786.

Salzen, E. A., and W. Sluckin. 1959. The incidence of the following response and the duration of responsiveness in domestic fowl. *Anim. Behav.* 7:172–179.

Schein, M. W., and E. B. Hale. 1959. The effect of early social experience on male sexual behavior of androgen injected turkeys. *Anim. Behav.* 7:189–200.

Schleidt, W. 1974. How fixed is the fixed action pattern? *Zeit. Tierpsychol.* 36:184–211.

Schneiderman, H. A. 1972. Insect hormones and insect control. In J. J. Menn and M. Beroza (eds.), *Insect Juvenile Hormones*. New York: Academic Press.

Schultz, F. 1965. *Sexuelle Prägung bei Anatiden*. *Zeit. Tierpsychol.* 22:50–103.

Scott, J. P., and J. L. Fuller. 1965. *Genetics and the Social Behavior of the Dog*. Chicago: University of Chicago Press.

Skinner, B. F. 1938. *Behavior of Organisms: An Experimental Analysis*. New York: Appleton-Century-Crofts.

———. 1953. *Science and Human Behavior*. New York: Macmillan.

Simpson, J., I. B. M. Riedel, and N. Wilding. 1968. Invertase in the hypopharyngeal glands of the honeybee. *J. Apicult. Res.* 7:29–36.

Snodgrass, R. E. 1956. *Anatomy of the Honeybee.* Ithaca, N.Y.: Cornell University Press.

Sontag, L. W., and R. F. Wallace. 1935. The movement response of the human fetus to sound stimuli. *Child. Dev.* 6:253–258.

Spalding, D. A., 1873. Instinct with original observations on young animals. *Macmillan's* 27:282–293. Reprinted in *Brit. J. Anim. Behav.* 2:2–11.

Spieth, H. T. 1966. Courtship behavior of endemic Hawaiian *Drosophila. Univ. Texas Publ.* 6615:245–313.

Suomi, S. J. 1973. Surrogate rehabilitation of monkeys reared in total social isolation. *J. Child. Psychol. Psychiat.* 14:71–77.

Suomi, S. J., H. F. Harlow, and M. A. Novak. 1974. Reversal of social deficits produced by isolation rearing in monkeys. *J. Hum. Evol.* 3:527–534.

Thompson, W. R. 1957. Influence of prenatal maternal anxiety on emotionality in young rats. *Science* 125:698–699.

Thorpe, W. H., and F. G. W. Jones. 1937. Olfactory conditioning in a parasitic insect and its relation to the problem of host selection. *Proc. Roy. Soc. London,* Series B 124:56–81.

Tinbergen, N. 1948. Social releasers and the experimental method required for their study. *Wilson Bull.* 60:6–52.

———. 1951. *The Study of Instinct.* Oxford: Oxford University Press.

Tinbergen, N., and A. C. Perdeck. 1950. On the stimulus situation releasing the begging response in the newly hatched herring gull chick (*Larus a argentatus* Ponstopp). *Behaviour* 3:1–38.

Vandenbergh, J. G. 1967. Effect of the presence of a male on the sexual maturation of female mice. *Endocrinology* 81:345–348.

———. 1969. Male odor accelerates female sexual maturation in mice. *Endocrinology* 84:658–660.

Vandenbergh, J. G., L. C. Drickamer, and D. R. Colby. 1972. Social and dietary factors in the sexual maturation of female mice. *J. Reprod. Fertil.* 28:397–405.

Vince, M. A. 1964. Social facilitation of hatching in bobwhite quail. *Anim. Behav.* 12:531–534.

———. 1966. Artificial acceleration of hatching in quail embryos. *Anim. Behav.* 14:389–394.

———. 1969. Embryonic communication, respiration and the synchronization of hatching. In R. A. Hinde (ed.), *Bird Vocalization.* Cambridge: Cambridge University Press.

———. 1973. Effects of external stimulation on the onset of lung ventilation and the time of hatching in the fowl, duck and goose. *Brit. J. Poult. Sci.* 14:389–401.

Watson, J. B. 1930. *Behaviorism.* New York: Norton.

Weaver, N. 1966. Physiology of caste determination. Ann. Rev. Entomol. 11:79–102.

Wille, A., and E. Orozco. 1970. The life cycle and behavior of the social bee *Lasioglossum (Dialictus) umbripenne. Rev. Biol. Trop.* (Costa Rica): 17:199–245.

Wilson, E. O. 1971. *Insect Societies.* Cambridge: Harvard University Press.

Wolff, J. R. 1981. Some morphogenetic aspects of the development of the central nervous system. In K. Immelmann et al. (eds.), *Behavioral Development.* New York: Cambridge University Press.

Wyman, R. L., and J. A. Ward. 1973. The development of behavior in the cichlid fish *Etroplus maculatus* Bloch. *Z. Tierpsychol.* 33:461–491.

Zimen, E. 1972. *Vergleichende Verhaltungsbeobachtungen an Wölfen und Königspudeln.* Munich: Piper.

10

LEARNING AND MOTIVATION

Learning is best defined as relatively permanent modification of behavior that occurs through practice or experience. **Motivation** refers to internal processes that arouse and direct behavior. We cannot measure motivation directly but must make inferences from observed behavior. Motivation is manifested as **drives,** which correspond to various specific **needs** — the basic requirements for an animal to maintain relatively constant internal body conditions **(homeostasis)** necessary for life. Both learning and motivation are critical processes affecting the behavior of animals of all ages.

In this chapter we explore learning and motivation in terms of the underlying processes and their functional significance. First, we examine several types of learning and their characteristics. We then survey learning in a variety of organisms. We next explore the concept of preparedness, followed by a brief review of current knowledge about memory. Finally, we discuss the concept of motivation and the use of this concept to explain internal processes for which we do not yet have thorough explanations.

TYPES OF LEARNING

Our knowledge regarding learning in animals has accumulated from the work of both psychologists and ethologists. A thorough historical review is beyond the scope of this text (see Thorpe 1956; Hilgard and Bower 1975). Two major types of association learning are currently recognized by animal behaviorists: classical conditioning and operant conditioning. There are many similarities and some differences in the key characteristics of these two major types of learning. We also recognize habituation as a distinct type of learning.

HABITUATION

The relatively persistent waning of a response that results from repeated stimulus presentations not followed by any form of reinforcement is termed **habituation.** Habituation is specific to the particular stimuli involved and can be distinguished from fatigue and from sensory adaptation by its relative persistence. **Fatigue** involves a loss of efficiency in the performance of a motor act when that act is repeated in rapid succession. **Sensory adaptation** generally occurs at the level of the peripheral sensory receptors and consists of a slowing down or cessation of nerve impulses transmitted to the central nervous system. While the effects of both fatigue and sensory adaptation last for relatively short time periods, habituation is a persistent central nervous system process that involves changes at the level of the brain or spinal cord.

For example, if, while a group of students is listening to a lecture, a classroom radiator suddenly clanks loudly, everyone will be momentarily startled. If, after a short spell, the radiator again makes its offending noise, the degree of response will be considerably decreased. After several additional clanks and bangs, almost no one will be startled by each new episode of noise and the class will go on as usual. Students and instructor have become habituated to the intermittent noise.

Habituation can be a functionally important aspect of an animal's behavior in its natural surroundings. Young ducklings scurry for cover when any shadow passes overhead, an adaptive response to avoid predators. Gradually the ducks learn, partly through habituation, which types of shadows signal potential danger and which are harmless. Ground squirrels, fiddler crabs, and marine worms all live in burrows. When danger threatens, these animals dash for or withdraw into their burrows for protection. They habituate to specific nonharmful stimuli in their environment but retain escape reactions to threatening or unusual stimuli.

CLASSICAL AND OPERANT CONDITIONING

Classical conditioning starts with a stimulus (the **unconditioned stimulus** or **UCS**) that elicits a specific response (the **unconditioned response** or **UCR**). At approximately the same time as the UCS, a second, neutral stimulus is presented that does not customarily elicit the UCR. When the neutral stimulus and UCS are paired for a number of trials, the response will eventually be elicited by the neutral stimulus

Before conditioning

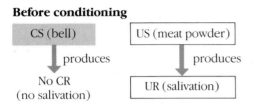

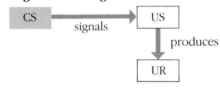

FIGURE 10-1   The sequence of events and relationships among stimuli and responses for classical conditioning

During conditioning

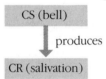

After conditioning

CS (bell)

produces

CR (salivation)

(Figure 10–1). At this point the neutral stimulus (in Figure 10–1, the bell) is termed the **conditioned stimulus (CS)** and the response is the **conditioned response (CR).** Classical conditioning works most effectively if the CS precedes the UCS by a brief time interval for each trial; in fact, little or no conditioned learning may take place if the CS follows the UCS.

The best known example of classical conditioning is that of Pavlov and the salivary responses of dogs (Figure 10–2). In the course of his investigations of digestion and related physiological processes Pavlov described the paradigm for demonstrating classical conditioning. The UCS is meat powder, which, when presented to the dog, elicits salivation (UCR). Presentation of a tone or bell (potential UCS) alone does not elicit the salivary reflex. Now the bell is sounded briefly a short time (one to three seconds) before the meat powder is presented on a number of trials. Soon, presentation of the bell alone is sufficient to elicit the salivation (now a CR). Thus classical conditioning results from the association of two stimuli in the environment. In general, it is also true that for this type of learning the sequence of events occurs regardless of what the test subject does.

Operant conditioning, or as it is sometimes called, **instrumental learning,** involves a wide variety of procedures. In each instance the animal learns to associate its behavior with the consequences of that behavior — that is, the sequence of events

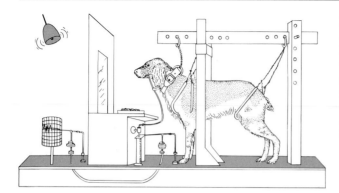

FIGURE 10-2 Pavlov's testing apparatus
Ivan Pavlov discovered classical conditioning through his work on the salivary reflex in dogs. The dog in the restraining apparatus is ready to be tested using the classical conditioning paradigm.

FIGURE 10-3 Skinner box
The interior of the box contains a lever, a light, a food bin, and a grid floor. Additional apparatus for automation and monitoring the rat's behavior is housed behind the back panel and on the left side of the cage. *Source:* Photo by Will Rapport from the office of B. F. Skinner.

is dependent upon the behavior of the animal. Usually some type of reward or punishment is involved. The task performed by the animal may be relatively simple, as in a rat trained to press a lever in a Skinner box (Figure 10–3). Pressing in this example is rewarded with food or water. Another example is when the animal has learned to run a maze to a goal box to receive the reinforcement (Figure 10–4). In nature a weasel may learn to associate the odor of mice with locating and catching a meal.

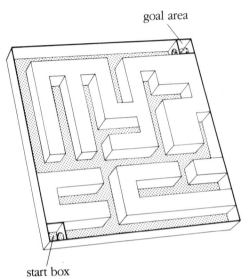

goal area

start box

FIGURE 10-4   Rat maze
A rat may learn a maze by trial-and-error processes, initially making many wrong turns and entering numerous culs-de-sac. Gradually the rat makes fewer errors, until, after a number of trials, it completes the maze without any errors. Also, a rat may perform with fewer errors when it has been given prior exposure to the maze with no rewards present, a process called latent learning.

CLASSICAL AND OPERANT CONDITIONING COMPARED

Many who study learning today postulate that the processes underlying classical and operant conditioning are similar, while others argue that they are different processes. We now examine some apparent similarities and then the apparent differences.

ACQUISITION.   The act of acquiring a response is termed **acquisition.** Several different response measures, some common to both types of learning, are used. A graphic presentation of the response measures is a **learning curve** (Figure 10–5). A sequence of trials given over time often serves as the scale for the horizontal axis. The vertical axis is usually some measure of performance or strength of CR. Among the common measures used are the percentage of conditioned or correct responses, the magnitude of the response, the time required to complete a response, the error rate, or the latency (time interval) between some signal (e.g., the CS) and the response. For several of these (Figure 10–5a) the curve increases over the trials, whereas for other measures the curve decreases (Figure 10–5b). We must emphasize that these are merely measures of performance. It would be incorrect to draw conclusions from these measures regarding the underlying mechanism of learning.

SCHEDULES OF REINFORCEMENT.   **Reinforcement** in classical conditioning consists of presenting the UCS after the CS. In operant conditioning the reinforcer is presented when the subject gives the appropriate response. Reinforcers in operant conditioning may be positive — such as, food, water, or an opportunity to manipulate some object — or they may be negative — such as electric shock or other punishment. The most basic schedule involves providing continuous reinforcement for each trial or response. A variety of other types of schedules provide partial or intermittent reinforcement. **Fixed ratio** schedules require the test subject to make a fixed number of responses to receive reinforcement. **Fixed interval** schedules provide reinforcement for the first response after the passage of a prescribed time interval following the last reinforced response. Similarly, **variable ratio** and **variable interval** schedules program reinforcement to occur after a variable number of responses or for the first response after a variable time interval, respectively. Most of these schedules have been used with the operant conditioning paradigm, but it is possible to use them with classical conditioning as well.

EXTINCTION. **Extinction** refers to the decrease of response rate or magnitude with lack of reinforcement. Eliminating the US for a period of time in classical conditioning or eliminating the reward (punishment) in operant conditioning extinguishes the response. The number of trials or responses which must occur without reinforcement before extinction occurs is a function of several variables. In general, the longer the conditioning procedure has been (the more well learned the response), the harder it is to extinguish that response. Responses which have been reinforced by partial reinforcement during acquisition are the most difficult to extinguish. Animals that have been trained on partial reinforcement schedules "resist" extinction for several reasons: they learned to persist in responding when faced with some nonreinforced responses during training, and there is less difference between partial reinforcement and no reinforcement than between continuous reinforcement and none. Human gambling is an excellent example of a behavior affected by partial reinforcement schedules and which is resistant to extinction.

SPONTANEOUS RECOVERY. If a conditioned response has been extinguished and is then followed by a rest interval of several minutes or up to a day or more, depending

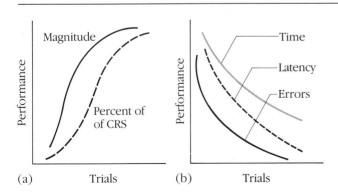

FIGURE 10-5 Characteristic forms of learning curves
(a) The percentage of CRs and response magnitude increase with practice. (b) Latency measures and other time measures decrease as do errors.

upon the species and test situation the animal may exhibit **spontaneous recovery** upon reintroduction to the test situation. Thus, for example, in Pavlov's tests a dog may be producing ten drops of saliva for the bell alone after a number of acquisition trials, but this decreases to only three drops when the UCS is eliminated. After resting for one half-hour, the bell is sounded without the meat powder presentation and the dog produces seven or eight drops of saliva — this is spontaneous recovery. This process can be repeated a variable number of times with decreasing responses. Similarly, when a rat trained to press a bar for food reinforcement is no longer given the reward, the response will be extinguished. After a rest interval of one hour the rat will exhibit spontaneous recovery, pressing the bar many times without receiving a reward, until the response is again extinguished.

GENERALIZATION.   If an animal has been conditioned to respond to a certain stimulus, we find that the response will usually also occur to stimuli similar to that used in the original acquisition trials. In classical conditioning we can vary the pitch of the bell tone; dogs will respond by salivation to tones similar to that of the original bell, but not to tones with quite different pitches. For operant conditioning in humans, consider the learned association between certain types of music accompanying various actions and scenes during a radio or television program. We learn to associate certain cadences and musical motifs with particular effects like danger or sadness or mystery.

DIFFERENCES.   Several key differences exist between classical and operant conditioning. The basic paradigms used to demonstrate the two types of learning differ. In classical conditioning the situation is thought by many to involve a stimulus-stimulus pairing. That is, the CS and UCS are paired. In contrast, operant conditioning involves pairing the stimulus and response. A second difference has already been noted; in classical conditioning the subject does not control the sequence of events, whereas for operant conditioning the sequence is contingent upon the responses of the test animal. Another way to state this is that in classical conditioning the responses are elicited, but in operant conditioning the responses are emitted by the subject. A third difference involves the **shaping** process characteristic of operant conditioning. The experimenter can reinforce particular responses or behavioral actions of the subject and not others, enhancing the frequency of some actions and extinguishing other actions. For example, as a rat is trained to press a lever for a food reinforcer, the experimenter may at first control the process by successively rewarding the rat for being in close proximity to the lever, for exploring the lever, and then for touching the lever, until the rat begins actually to press the bar to obtain the food.

OTHER ASPECTS OF LEARNING

In addition to habituation and the two major types of association conditioning, several other similar processes should be noted and briefly explained.

AVOIDANCE LEARNING. **Avoidance learning** is sometimes called **aversive conditioning** and is really a type of operant conditioning. Many predatory animals learn to avoid distasteful prey by associating sensory cues from these prey with negative aftereffects, such as an upset stomach. Birds learn to avoid consuming monarch butterflies *(Danaus plexippus)* because these insects contain poisons, or emetics, that make the birds ill (Brower 1958; Brower and Brower 1964). Another species of butterfly, the viceroy *(Limenitis archippus),* has evolved coloration patterns like those of the monarch as a predator defense mechanism.

Two separate teams of investigators, working in Washington and California, have reported on the use of an emetic, lithium chloride, as a means of controlling coyote predation on sheep (Gustavson et al. 1976; Ellins, Catalano, and Schechinger 1977). Both teams of investigators laced sheep carcasses with the emetic and placed them where coyotes would probably eat them. It was hoped that the coyotes would become ill soon after consuming the baited sheep, would associate (through aversive conditioning) the illness with the sensory cues (sight, odor, and so on) from the bait, and avoid sheep as prey in the future. Both research teams reported success in reducing predation. However, a more recent report (Sterner and Shumake 1978) has shed some doubt on these conclusions. The report by Ellins and his colleagues provides no information on baseline rates of sheep losses before lithium chloride baiting, the number of baits taken by species other than coyotes, or whether coyote control methods were in use at the time of the study. The work of Gustavson and colleagues has been criticized because other coyote control procedures — trapping, aerial gunning, and the use of poison baits — were being used in the same area at the time of the study.

INSIGHT LEARNING. In **insight learning** the animal makes new associations between previously learned tasks in order to solve a new problem. This, too, is really a form of operant conditioning. The classic example of insight learning is the work of Köhler (1925) on chimpanzees (see also Birch 1945). Chimps learn to connect a series of small poles into one longer pole in order to obtain bananas suspended above the cage floor. Similarly, they learn to stack boxes and climb the stack of boxes to reach the bananas. When the chimps are given a new problem, with the bananas suspended higher above the cage floor, they learn to put the poles together, stack the boxes, and climb the stack of boxes, pole in hand, to knock down the bananas (Figure 10–6).

LATENT LEARNING. Associations made with neither immediate reinforcement or reward nor particular behavior evident at the time of learning have sometimes been labelled **latent learning.** The processes involved are not readily elucidated, but the phenomenon appears to be real, as the following examples should illustrate.

In nature animals probably learn a great deal about their surroundings during the course of their daily activities. This information is of no apparent, immediate functional value but may become important for survival later. Metzgar (1967) has shown how this process might work in nature for the deermouse *(Peromyscus leucopus).* One group of test mice was given exposure to and experience in an enclosed

FIGURE 10-6 Insight learning in chimpanzees
The chimp has previously learned to connect the poles or to stack the boxes in order to obtain a reward of bananas. When the bananas are out of reach even if the chimp uses either the poles or the boxes, the chimp must use insight learning to deduce that a combination both of stacking the boxes and then climbing up them with the pole will provide access to the bananas.

room containing "natural" features such as logs and trees. A second group was held in laboratory cages without experience. Then mice from each group were placed in the enclosed habitat room with an owl present as a predator. Metzgar found that only two of twenty residents with prior experience in the room were caught by the owl, whereas eleven of twenty mice with no prior experience in the habitat were captured. Latent learning apparently provided the residents with more knowledge of the habitat, better enabling them to avoid being prey for the owl.

The predatory digger wasp *(Philanthus triangulum)* which inhabits burrows in sandy soil provides a second example of latent learning. The way these insects relocate their nest burrows after flying away at some distance to capture prey was the subject of studies by Tinbergen and Kruyt (1938; see also Tinbergen 1958). Tinbergen provided landmarks around the wasp's nest holes; upon emergence, the wasp reconnoitered the area and then flew off. If Tinbergen removed or rearranged some landmarks, the returning wasps became disoriented to varying degrees. Additional studies support the conclusion that the entire configuration of landmarks is used by the returning wasp as a guide to the location of its burrow.

OBSERVATIONAL LEARNING.   The tendency to perform an appropriate action or response as the result of having observed another animal's performance in the same situation may involve either classical conditioning or operant conditioning and has been called **observational learning.** For example, Klopfer (1957) demonstrated that ducks *(Anas platyrhynchos)* can learn a discrimination task by observation. He conditioned a group of ducks to feed from one of two dishes by placing the incorrect dish on a wired shock grid. During the conditioning he restrained observer ducks nearby, where they could see the subject ducks learning the discrimination but could not participate. He then released the observer ducks and placed them in the test situation with both food dishes present. The observer ducks avoided the incorrect dish with its wired shock grid.

Imitation occurs when an animal immediately copies the actions of another while they are both in each other's immediate presence (Thorpe 1963). The act performed is one that would not normally be expected in the species' behavioral repertoire. The way a group of Japanese macaques *(Macaca fuscata)* of Koshima Island learned some of their food habits is an excellent example of imitation. Imo, an inventive young female monkey in the group, introduced two new techniques that were copied by other group members through imitation. One new technique was taking sweet potatoes, which the monkeys were fed, to the water to wash them before eating (Figure 10–7). Washing removed the gritty sand adhered to the skins of the potatoes. Also, the salt water may have added some flavor.

Imo also took handfuls of wheat, another food given to the provisioned monkeys, to the water and allowed a little at a time to fall into the water; the sand sank and the wheat floated on the surface of the water. Imo then gathered up and ate

FIGURE 10-7   Japanese macaques washing sweet potatoes
A young female macaque initiated the practice of washing potatoes in water before consuming them—a process that was learned by other macaques through imitation.
Source: Photo by M. Kawai courtesy of Kyoto University Primate Research Institute.

the wheat (Itani 1958; Kawai 1965). Once established, behavior traits such as these may be passed on as a tradition (see Galef 1976; see also Chapter 15 of this text).

IMPRINTING. We have already discussed imprinting as a process of attachment formation and following behavior in Chapter 9. This is sometimes classified as a separate, special type of learning.

LEARNING SETS. One general phenomenon important for an understanding of learning was first elucidated by Harlow and his associates (Harlow 1949, 1951; see also Warren and Barron 1956; and Shell and Riopelle 1957). **Learning sets** are defined as the acquisition of a learning strategy by the animal. Given a series of problems, an animal will transfer some of what it has learned about solving the first problem to the solution of subsequent problems in a series. The learning curves of monkeys given blocks of test problems (Figure 10–8) demonstrate this effect. As the series proceeds, the animal scores an increasingly high percentage of correct responses by the second or third trial on each new problem. In effect, the formation of learning sets occurs as the animal "learns how to learn." Learning sets may also restrict the behavior of an animal. Negative consequences may result if an animal has formed a learning set regarding a particular problem, as it will not be able to shift strategies as readily as an animal without the learning set.

PHYLOGENY OF LEARNING

We cannot retrace the steps in the evolution of learning, but a brief survey of the learning capacities and capabilities of organisms from various animal phyla should provide some insight into the possible generality of any laws of learning and a

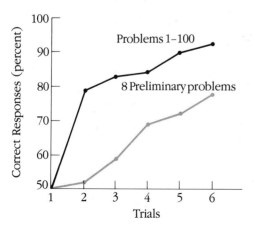

FIGURE 10-8   Discrimination learning curves
Each dot on the graph represents the average percent of correct responses on a particular trial for the 8 problems or 100 problems given to a rhesus monkey. Note the improvement that occurs with successive trials and particularly the large improvement by the second trial in the 100-problem set compared to the small improvement in the set of 8 preliminary problems. *Source:* Data from H. Harlow, "The Formation of Learning Sets," *Psychological Reviews* 56 (1949):51–56.

sampling of the differences in learning ability across the various phyla. Four cautions should be noted. First, there has been a disproportionate examination of learning across the various phyla. Far more studies have been conducted on learning processes in vertebrates than in invertebrates, and even within the vertebrates much of the attention has focused on mammals. Second, there are valid ongoing disagreements among scientists regarding what constitute proper criteria for demonstrating various types of learning and concerning the evaluation and interpretation of research results. Third, exploring the physiological and behavioral aspects of learning in many animals will require more refined techniques and objective, bias-free methods. Fourth, we measure performance and not a genetically determined ability or mechanism. Lots of factors (notably, test design) can influence performance. Additional problems regarding comparative learning studies will be discussed in two subsequent sections of this chapter. With these difficulties in mind we can now survey the known information regarding learning in a number of animal phyla.

PROTOZOA AND COELENTERATA

Solid experimental data exist to demonstrate that protozoans can exhibit habituation responses, but there is no fully accepted evidence for any type of association learning (Corning and von Burg 1973). For example, Patterson (1973) demonstrated that for *Vorticella convallaria* the contraction response exhibited to both mechanical and electrical stimuli can be habituated with repeated stimulus presentations. Gelber (1952, 1965) claims to have demonstrated that *Paramecium aurelia* can learn to associate presentations of a fine wire in their environment with food (bacteria). After presenting the wire coated with bacteria on a number of trials, the wire alone seems to elicit an approach response by more paramecia than presentation of the wire alone to untrained animals or to populations trained with an uncoated wire. An alternative interpretation of these findings was provided by Jensen (1957, 1965). He hypothesized and tested the notion that the unicellular paramecia are merely responding to a food-rich zone in their environment, created by the presence of the bacteria-coated wire. The debate over these alternative interpretations has not been resolved.

For coelenterates there is also ample evidence for the habituation response, but little evidence for association learning. In the early years of this century investigators generally assumed that coelenterates exhibited only certain involuntary, stereotyped reflexes. We now know that there are endogenous neural rhythms in many coelenterates and that, rather than being totally "passive," many coelenterates are capable of spontaneous active responses in the course of interacting with their environment (Rushforth 1973). One possible demonstration of a form of association learning (Ross 1965) is based on the fact that sea anemones (genus *Stomphia*) respond to chemostimulation from certain starfish (e.g., *Dermasterias imbicata*) by stretching their bodies, detaching from the substrate, and "swimming" away (Figure 10-9). The animal soon lands and eventually reattaches to the substrate. The chemostimulation is paired with gentle pressure applied near the base of the anemone.

FIGURE 10-9 Interaction between starfish and sea anemone
When a starfish *(Dermasterias)* makes contact with the sea anemone *Stomphia cocinea,* the anemone will release from its attachment, "swim" free for a period, and eventually reattach itself to the substrate.

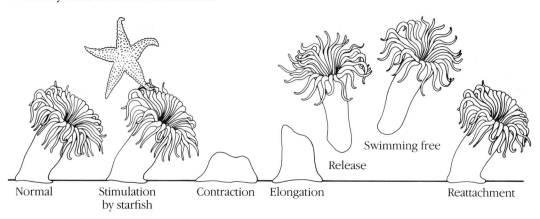

Normal   Stimulation   Contraction   Elongation   Reattachment
          by starfish

Release

Swimming free

With repeated trials the application of the pressure stimulus alone leads to some reduction in the "swimming" response; Ross labelled this as conditioned inhibition. However, the stimulus may have induced some related effect resulting in the anemone movement.

PLATYHELMINTHES AND ANNELIDA

Among the flatworms much research has been conducted with planaria, but little is known about the other groups, some of which (e.g., tapeworms, flukes) are internal parasites. Conclusive evidence exists for habituation and several types of association learning in planarians which have a relatively simple bilateral nervous system. Thompson and McConnell (1955) first reported classical conditioning in these organisms, but it was not until the report by Block and McConnell (1967) that satisfactory control procedures were employed to satisfy most critics. Among the issues involved was whether there was true classical conditioning or whether the observed effects were due, instead, to either sensitization or pseudoconditioning (Dyal and Corning 1973; Corning and Kelly 1973). **Sensitization** is an increase in the strength of a response originally evoked by a CS as the result of pairing with a UCS and a response. **Pseudoconditioning** is the strengthening of a response to a previously neutral stimulus by repeatedly eliciting the response with another stimulus without pairing the presentation of the two stimuli. Elaborate controls, which consist of presenting only the CS, randomly ordering or pairing the CS and UCS, and simultaneously presenting the CS and UCS, have been used in attempts to clarify whether actual classical conditioning of planaria occurred. In general, researchers today agree that planaria are capable of exhibiting some form of conditioned learning.

In the ensuing years, considerable research has been conducted on learning in planaria, resulting in some startling conclusions — and, for a time, a separate journal, *Worm Runner's Digest*, included both serious articles on learning behavior in worms and some humorous articles. Two of the most striking conclusions that arose from these experiments were: (1) Learning, or some component thereof, was passed on to the two regenerated planarians that resulted from severing a pre-trained worm in half. The two regenerated planaria learned a T-maze choice task faster than naive worms or naive regenerates (McConnell, Jacobson, and Kimble 1959); and (2) studies in which naive worms were permitted to cannibalize worms that had previously been trained on a learning task exhibited much faster learning of that task than planaria that consumed naive conspecifics (see review by Corning and Kelly 1973). One potentially serious problem with some of the studies of learning in planarians is that they leave a slime trail as they move along. It is thus possible that using the same apparatus for repeated trials, without cleaning, would confound the results. On the other hand, cleaning the apparatus after each trial sometimes interferes with the behavior of the worms.

Annelids have a segmented nervous system with ganglia in each body segment and some concentration and coalescing of ganglia in the anterior segments. For annelids, habituation has been demonstrated using the backward movements which occur in response to puffs of air (Ratner and Gilpin 1974). Earthworms (*Lumbricus terrestris*) and some other annelids are also capable of both classical and operant conditioning. Most test paradigms used with the worms utilize either aversive conditioning or some form of punishment training such as mild shock. Classical conditioning has been shown by pairing a light stimulus and mild vibrations (Ratner and Miller 1959). Operant conditioning has been demonstrated for annelids in a variety of experiments using a two-choice T-maze apparatus and mild shock for incorrect choices. Like the flatworms, the problem of mucus trails containing chemical cues after repeated trials is a potential problem for investigations using earthworms and other annelids.

MOLLUSCA

The learning capabilities of two major groups of molluscs have been investigated: gastropods (snails and slugs) and cephalopods (squids, octopodes). One great advantage for the scientist working with these organisms is that the nervous systems of many species are readily accessible and contain large neurons and cell bodies. Thus, it is possible to explore some of the neurophysiological processes accompanying learning behavior. Molluscs have nervous systems with a greater degree of centralization and encephalization than flatworms or segmented worms.

Habituation has been studied extensively in snails of the genus *Lymnaea*. Both visual and mechanical stimuli result in a withdrawal movement, in which the snail's shell is drawn downward and forward (Cook 1971). The response habituates with repeated stimulation. Habituation of the gill withdrawal reflex in *Aplysia* occurs with repeated stimulation from a jet of water (Pinsker et al. 1970). Kandel and co-investigators (see Hawkins et al. 1983) have demonstrated differential facilitation of excitatory postsynaptic potentials in the neuronal circuit for this withdrawal reflex. Activity at the synapses between the sensory and motor neurons involved in this reflex arc is facilitated more by tail shock (UCS) if the shock is preceded by spike activity in the sensory neuron than when the spike activity and shock occur in an unpaired pattern, or with the shock alone. Further experiments reveal that this facilitation is presynaptic in origin, involving an influx of Ca^{++} ions (recall Chapter 6) in the affected neurons, compared to unpaired sensory neurons.

Several types of association learning have been reported in various species of both gastropod and cephalopod molluscs. Much of the most interesting work on this group has been conducted with octopodes (see review by Sanders 1973). First, recall the material in Chapter 3 regarding tactile discrimination in the octopus; members of this group can learn to discriminate cylinders based on the amount of grooved surface, but they fail to learn to separate cylinders based on the pattern of grooved surface when the *amount* of grooved area on different cylinders is roughly the same. Through convergent evolution the peripheral aspects of the octopus' visual system are quite similar to the visual system found in vertebrates. However, examination of the results of visual discrimination tests (Figure 10–10) reveals that the interpretations of the stimuli must be quite different than for most vertebrates; octopodes fail to make successful choices when the paired items are mirror images or when there are strong similarities between the arrangements of horizontal and vertical lines of the two stimuli. Last, via conditioned learning, octopodes can be trained to discriminate between various food items (Boycott 1965). If the octopus is presented with a fish and a geometric figure such as a disk, and, upon approaching the fish, the octopus is given a mild shock, the animal learns to avoid the fish. When the disk and fish are presented together, the octopus will retreat. If the fish alone is presented the octopus will attack. Such techniques can be used to train octopodes to approach one type of food (e.g., crabs), while avoiding fish, or vice versa.

Recently, Alkon (1980, 1983) has demonstrated conditioning in the marine snail *(Hermissenda crassicornis)*. The snails are normally positively phototactic during the daylight portion of their daily cycle. Snails were trained in glass tubes filled with sea water and mounted radially on a turntable. Prior to training, the snails were measured for their velocity of movement toward the lighted center portion of the turntable. Then the turntable was rotated, producing centrifugal force for the snails and possibly simulating the water turbulence encountered near the ocean's surface under stormy weather conditions. Thus, the light source and rotation were paired. After training, the velocities were again recorded; trained snails moved toward the center of the turntable with about one-third of the pretraining velocity. Care was taken to perform a number of control procedures to ensure that the observed effect was in fact due to some form of association learning by the snails. By mapping the neural pathways for reception of information from the light source and from the rotational forces, Alkon (1983) was able to produce a wiring diagram for the perti-

FIGURE 10-10 Octopus performance over sixty trials on the discrimination of pairs of shapes differing in orientation only
When an octopus is given discrimination trials with pairs of outline shapes, it can successfully discriminate those that differ in orientation, but not those with the same or nearly the same horizontal and vertical extents.
Source: G. D. Sanders, "The Cephalopode," in W. C. Corning, J. A. Dyal, and A. O. D. Willows, *Invertebrate Learning,* vol. III, 1975, Academic Press, New York.

Problem	Discriminanda	Horizontal extent	Vertical extent	N	Percent correct responses
1	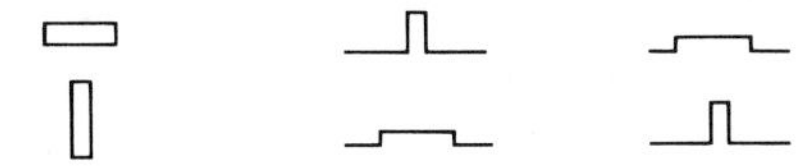			6	81*
2	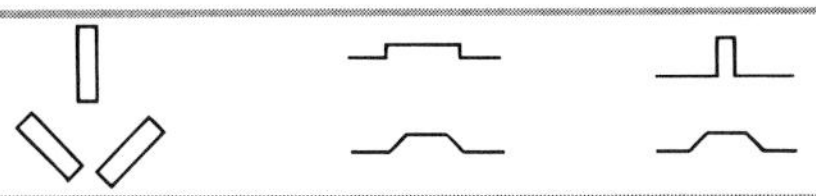			8	71*
3	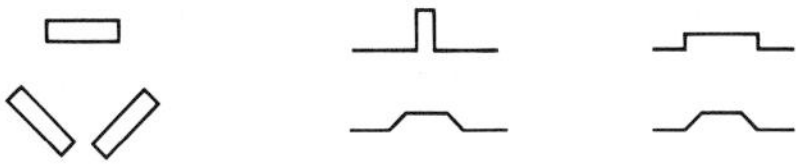			8	65.5*
4	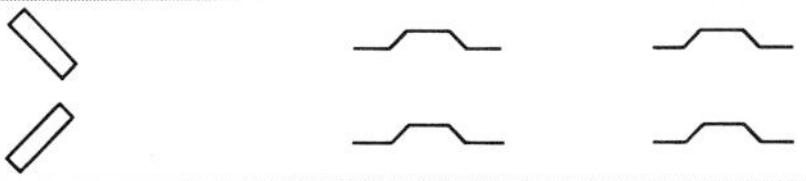			6	50
5	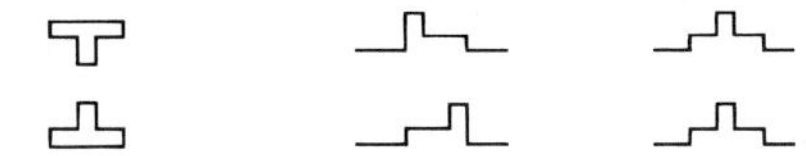			6	59*
6	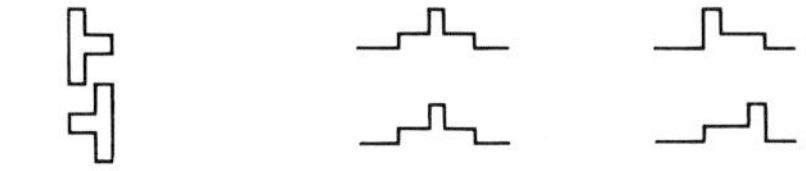			7	56*

*Better than chance level of performance $P < 0.05$.

nent portions of the nervous system of *Hermissenda*. Further, he showed that one of the consequences of training is that particular receptors and cells in the nervous system become either more excited or inhibited, depending upon the conditioning.

ARTHROPODA

The arthropods are a large and diverse group comprising 80 percent of all animal species. Only the vertebrates have been studied more with respect to learning behavior. The arthropods have evolved a wide variety of types of sensory receptors and have a nervous system characterized by an aggregation of ganglia termed a 'brain' located in the anterior segment of the body, with a pair of ventral nerve cords passing to the thorax and abdomen. Three groups of arthropods have been examined more extensively in this regard than the others; the subphylum Chelicerata (e.g., *Limulus,* scorpions, spiders), and the classes Crustacea and Insecta. For each of these groups habituation, classical conditioning, and operant conditioning have been demonstrated in a variety of species. For example, bees can learn to discriminate colors by associating different hues with sugar solutions or with nonsugared water (von Frisch 1967; Wells 1973). Spiders can be trained to associate flies coated with either sugar or quinine with sounds of particular pitches (Walcott 1969). When glass beads are substituted for the flies, the spiders either discard or bite the beads depending upon the pitch of the sound presented with the bead. Various species of crabs and crayfish have been successfully trained in T-mazes (Gilhousen 1927; Datta, Milstein, and Bitterman 1960). Krasne (1973) reports that there may be some transfer of learning from a simple maze to more complex mazes for crabs.

VERTEBRATA

A number of learning studies have been conducted on species from most vertebrate classes. We know considerably more about the complex processes which underlie these phenomena for this phylum (see review by Masterton et al. 1976); this is particularly true for the mammals. Since the rat and rhesus monkey have been used in a large number of these studies, an example for each species will provide some flavor of the research and how it is conducted.

Rats can be trained in an operant conditioning apparatus to press a lever to receive food. They can also be trained in an apparatus with two levers (designated L for left and R for right) to press the levers alternately, LR or RL, for a food reward. Can rats be conditioned to press the levers in a LLRR or RRLL sequence, called double alternation? Investigators (Travis-Niedeffer, Niedeffer, and Davis 1982) tested this question by conditioning rats first on the single lever task, either R or L, followed by the single alternation task, RL or LR. They then rewarded only double-alternation performances. The rats learned this task at a level exceeding chance expectations, and their performance improved over days. It was necessary, however, in some instances, to permit the rats to give extra responses to the first level before pressing the second lever. Thus LLLRR was rewarded the same as LLRR. When the investigators attempted to condition the rats to a sequence of LLRRLLRR or RRLLRRLL (double alternation with a fixed ratio schedule of two repetitions), no rats could successfully perform at better than a chance level. The results are significant because of the new information provided regarding the capacity of the rat to associate a series of responses required to obtain the reward and because previous attempts to condition double-alternation tasks in rats had failed.

One particular apparatus, the **Wisconsin General Test Apparatus (WGTA)** (Figure 10–11) has been used in many of the studies of learning behavior in macaques (see Meyer, Treichler, and Meyer 1965 for the history of this and related techniques). One type of learning problem studied with this apparatus is the **delayed response problem** (Fletcher 1965). The basic procedure starts with the test tray out of reach but in full view of the test animal. Food is placed in one food well, and then two identical objects are placed over the wells. After a prescribed delay the test tray is moved closer to the monkey, which then responds by lifting one object or the other. The four stages in the procedure are the baiting phase, covering phase, delay phase, and response phase. A trial ends when the monkey picks up one of the objects, uncovering either the correct food well, containing a reward, or the empty well. A number of variables can be investigated with this procedure, and other animals, such as cats and raccoons, have been tested with slight modifications of the apparatus. Among the variables that have been manipulated are length of the delay phase, the nature and size of the reward, the nature and the similarity or dissimilarity of the objects used to cover the food wells, and whether the animal is permitted to watch the test tray during the delay phase or, instead, has an opaque screen lowered in front of the tray (Meyer and Harlow 1952). Among the conclusions are these: rhesus monkeys are capable of learning basic discriminations in this procedure with delays of up to 30 seconds or more, more food reward leads to better performance, and imposition of the opaque screen during the delay phase increases error rates by up to 50 percent. One interesting and striking finding in these studies is that the behaviors and performances of individual monkeys differ markedly. In general, monkeys that exhibit hyperactivity in the test situation and those that are more easily distracted during the delay phase exhibit lower levels of performance. Clearly, in studies of learning behavior we must consider the significance of individual differences in performance, regardless of the species being tested, or the task being performed (Warren 1973).

No consistent patterns relating phylogeny and learning among vertebrates

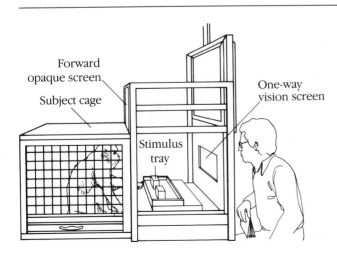

Forward
opaque screen

Subject cage

One-way
vision screen

Stimulus
tray

FIGURE 10-11 Wisconsin General Test Apparatus
An early version of the Wisconsin General Test Apparatus (WGTA) used to test aspects of learning in primates and, with some modifications of the apparatus, cats and raccoons.

have been discovered. For different types of tasks different species may be faster or slower than other species and may perform with a higher or lower error rate. One general finding is the rapid avoidance learning in feeding situations for virtually all vertebrates studied (Domjan 1980). In many instances, one trial is sufficient to establish an association between a food and distasteful or noxious effects that occur anytime from immediately after ingestion up to one-half hour or more after eating. Wild rats that have eaten poisoned bait, but in a sublethal dose, usually become bait shy and avoid that food thereafter, making it difficult to use poisoned food as a rodent control procedure (Barnett 1963). (Recall also our earlier examples of monarch butterflies and coyotes.) While it is highly likely that learned food aversions are important evolutionary adaptations to avoid noxious or harmful foodstuffs, we should be cautious in applying this logic without first carefully examining the nature of each species, its normal life patterns, and possibly the test situations utilized to investigate the food averson learning.

COMPARATIVE LEARNING

Over the past several decades various comparative psychologists have attempted to make comparisons of learning behavior across the various phyla. Most notable of these are the reports by Bitterman (1960, 1965a, 1975) and Dewsbury (1978).

Bitterman's approach has been to look at certain organisms (e.g., fish, rats, turtles, pigeons, and monkeys) and their learning capabilities and then to classify organisms according to whether their learning abilities are like those of his test species. Basically, Bitterman believes different processes occur in different species; different learning phenomena may result from the same process in different organisms; and, conversely, what appear to be similar learning phenomena in different organisms may result from basically different underlying mechanisms. He argues strongly against the principle of **equation**—the attempt to equate situations and procedures for learning in different species. He favors, instead the concept of control by **systematic variation**—examining qualitative differences in learning by investigating a wide variety of species from diverse taxonomic groups. Bitterman's work is significant because it cautions animal behaviorists who would plunge into comparative learning studies without a close look at the capacities of the various animals under study and the variety of methodologies used to test learning phenomena.

Dewsbury (1978) has classified attempts to investigate comparative learning into two categories; quantitative comparisons and qualitative comparisons. Dewsbury first examines the cross-species comparisons for acquisition of learning (see also Brookshire 1970); there appear to be no general patterns across the various phyla for rates of acquisition or for avoidance learning. The development of learning sets for particular conditioned learning responses in different species has been widely used to compare learning in vertebrates. Here again, the data do not support any clear relationship between phylogeny and the exhibition of the learning set phenomenon. Dewsbury summarizes the problems of making such quantitative comparisons under several headings, including: (1) individual differences, which make it difficult to utilize mean performance levels as species-typical (see Rumbaugh 1968); (2) motivational differences, for example, in animals that are food-deprived and then given learning tasks; and (3) a series of species differences that reflect variations in biological constraints on learning. There is little evidence to suggest the possibility of constructing any sort of "scale of intelligence" for vertebrate organisms, let alone comparisons that attempt to include the invertebrate phyla.

Both Dewsbury (1978) and Bitterman (1965b) feel there is some validity in attempting to judge comparative learning by qualitative standards. We have already alluded to Bitterman's scheme for using the general capacities of various vertebrate types. Thus, fish have only rarely been found capable of reversal learning, but both turtles and rats are capable of learning a discrimination task in which the correct solution is changed from one problem to the next. Also, rats given bigger rewards (food) are less resistant to extinction when unrewarded trials are given than are fish; the latter show greater resistance to extinction when they have been trained with larger rewards (food) (Gonzalez and Bitterman 1969; Gonzalez et al. 1972). We should also note that some of the problems with quantitative studies outlined above may also bear on qualitative comparisons.

CONSTRAINTS ON LEARNING

Two general types of constraints are thought to be operative in studies of learning in general and of comparative learning in particular. These are the biological constraints brought to the learning situation by the animal, termed "preparedness" by Seligman (1970), and the methodological constraints imposed by the investigator and test situation (see Hinde and Stevenson-Hinde 1973).

PREPAREDNESS

In recent years the generality of laws of learning has been questioned by Seligman (1970), who introduced the notion of **preparedness** — the genetically based predisposition to learn, which is manifested during development. An animal faced with a particular learning task, whether in nature or in the laboratory, may be prepared, unprepared or contraprepared to perform that task. The constraints and limitations imposed by inheritance affect an animal's relative preparedness to learn (see Hinde and Stevenson-Hinde 1973); in other words, partially due to its genetic make-up, an animal brings certain predispositions and sensory biases to any learning situation. In the evolutionary process, the species was provided with tools and capacities that limit its range of potential responses to specified inputs.

For example, rats (both wild and laboratory strains) are prepared to learn to associate the taste of certain foods with subsequent illness, even when the onset of illness occurs up to several hours after they ingest the food (Barnett 1963; Rozin 1968; Garcia, Hankins, and Rusiniak 1976). This lack of a close timing relationship between the stimulus and response is an interesting and important contradiction of the previously established view that there had to be continuity between the presentation of a stimulus and the positive or negative reinforcement (see also Garcia, Ervin, and Koelling 1966; Garcia and Koelling 1966). Rats learn to avoid foods that make them ill after one or, at most, a few exposures; the evolutionary advantage of rapidly learning to avoid ingesting potentially harmful or poisonous foods should be obvious. Humans also make rapid associations between illness and a food they have ingested. Even though people have actually contracted an illness that is not connected with a food (e.g., the flu) they may still associate something they ate shortly before the onset of the illness with that illness and may maintain that association for some time.

In other instances it appears that even though an animal is capable of learning a particular task, its system is generally unprepared — that is, the animal requires training to complete the learning task. Rats do not naturally press levers for food, but they can be trained (conditioned) to do so. Chimpanzees normally communicate by vocalizations and gestures (van Lawick-Goodall 1968). They can, however, be trained to use some American Sign Language of the Deaf (Gardner and Gardner 1969; Fouts 1973). Chimps learning sign language require a great deal of time and training to become successful at even rudimentary communication with this system (Figure 10–12).

A third possibility is that an animal may be contraprepared to perform a particular task — that is, even with many attempts or a great deal of training, the animal appears to be incapable of performing the task. In an investigation of avoidance or defensive reactions, an animal's responses must be selected from species-specific patterns (Bolles 1970). Thus, pressing a lever to avoid an electric shock is an avoidance task for which the rat is contraprepared, whereas it is not contraprepared to avoid the shock by running away, a more natural tendency. For any particular learning task, the capacities of a specific animal species are somewhere along the continuum formed by preparedness, unpreparedness, and contrapreparedness.

METHODOLOGICAL CONSTRAINTS

Restrictions or constraints operate in many laboratory and field situations in which learning studies are undertaken (Shettlesworth 1972); in fact, the apparatus and the situation being used to test the animal may alter the results. Stimulus cues must be related to the animal's sensory capacities and perceptual world, or *Umwelt*. A decrease in an animal's observed performance may result not from its inability to display a certain action, but rather from its being presented with an inappropriate stimulus — one the animal cannot interpret.

FIGURE 10–12   Chimpanzee signing
Using sign language is not part of the normal behavioral repertoire of chimpanzees. They can, however, be trained to learn a vocabulary of more than 100 words in sign language. This chimp has learned a variety of signs. *Source:* Photo by H. Terrace from Anthro-Photo.

The response tasks an animal must perform should be part of its potential repertoire. In addition, an experimental design must take into account the possibility of sex differences in learning capacities and the probability that animals of different ages may learn at different rates, may possess different amounts or types of prior experience, and may have different sensory/perceptual capabilities.

If we state that a form of learning behavior is present in most animal types, we assume **commonality;** that is, we infer a phyletic, or evolutionary, relationship from the common exhibition of a type of learning. The assumption of the commonality of evolutionary relationships brings with it potential problems and limitations (Hodos and Campbell 1969). Too often we merge data on animals of one lineage with data gathered on animals of a different lineage. To obtain information about the evolutionary history of a behavior pattern, we should use only animals that share a common lineage. Although we may make inferences about a type of learning across divergent lineages, we do so for comparative purposes only and not to imply direct-lineage evolutionary connections among diverse types of animals.

We must always be aware that behaviors do not fossilize. Our discussion of the phylogeny of learning uses living organisms as examples. We should remember that traits considered rudimentary in one lineage may be considered specialized in other lineages.

MEMORY

By definition, learning implies some form of retention of experience. **Memory** refers to the capacity of an organism to form lasting connections based on past experiences. Several theoretical explanations for the processes of memory and information storage and retrieval have been advanced.

THEORIES OF MEMORY

One theory (see Arbib 1972), often called the **dynamic hypothesis,** proposes that experiences and input of sensory information set up persistent electrical activity in the central nervous system. These so-called reverberating circuits, or populations of continuously active neurons, are the suggested basis for information storage in some coded form. When the active neural processes cease, forgetting occurs, and that bit of information is lost.

Other investigators (John 1967; Deutsch 1973) support the hypothesis that learning and sensory input produce permanent changes in biochemical processes or structures within cells, and memory thus involves structural changes. These structural changes are initially brought about by neural activity, but they are retained in permanent storage form after the neural activity produced by an experience has ceased.

STORAGE MECHANISMS

Apparently, two separate memory storage mechanisms exist, or one mechanism has two separate stages: a **short-term memory,** sometimes called the labile phase, and a **long-term memory,** or permanent phase (although forgetting may still occur). Immediately after an event, the experience is stored in short-term memory, possibly through neural activity alone. During this labile phase, various types of interference (e.g. a concussion or sudden blow to the head) may cause loss of the information. After a **memory trace**—the physical manifestation of learning or sensory input within the central nervous system—is transferred to long-term memory by processes that are apparently more chemical and structural in nature, the memory trace becomes relatively permanent. Research on memory processes is still in its early stages; the complexity and exact nature of information storage mechanisms must still be unraveled. Much of the early work on this subject is presented in Lashley (1929) and Hebb (1958). For thorough reviews of the history of theories of memory and with some particular emphasis on human memory see Kety (1982) and Woody (1982).

Localizing and actually characterizing the memory trace or **engram** has proven to be a most intriguing—but difficult—undertaking. As we noted in Chapter 6 when we discussed canary songs, there are apparently correlative changes in dendritic growth and synaptic connections between neurons as the birds learn new song repertoires each spring (Paton and Nottebohm 1984). In addition, ample evidence exists for increased RNA and DNA synthesis accompanying learning and the storage of information. In the final analysis, models for memory must take into account both chemical and structural changes.

MOTIVATION

At the beginning of this chapter we defined motivation in terms of internal processes that are directed by drives, with the attainment of certain goals or needs as rewards. The concepts of motivation and drive are often invoked as **intervening variables** to explain events that occur within an animal and that are not yet fully understood. The external evidence from which motivation and drives are inferred consists of the known stimulus conditions to which an organism has been exposed and of the observed behavioral actions.

Why do we need the concepts of motivation and drives? First, differences in responsiveness to particular stimuli can be measured when the stimulus is presented at different times. If a male European robin *(Turdus migratorius)* is presented with a small cluster of red feathers in midwinter, it will show little interest in the stimulus. But if a few red feathers are presented in the spring, the male robin will exhibit threat displays characteristic of its reaction to another male robin. Second, the strength of stimulus needed to elicit a response may vary over time. A satiated dog may be relatively selective about what items it chooses to eat. The same dog when hungry will accept less palatable food. Third, response frequency or intensity may vary over time. Female rats exhibit differences in general activity levels at different phases of the estrous cycle. Investigators have postulated that motivation and drive can account for these variations in responses. Drives, which are correlated with animal needs such as hunger or thirst, are in effect **homeostatic mechanisms;** satisfying a need helps to maintain the proper internal life-sustaining conditions within the animal. For example, when energy expenditures deplete body resources, an animal needs food. Motivation theory interprets this to mean that the hunger drive is increased, and the animal will begin to search for food. Only when the animal locates and consumes food will it satisfy the specific drive and restore homeostasis.

   The more information scientists gather about the genetic, neural, and hormonal bases of the control of behavior, the less important they are finding the concepts of motivation and drive. They are opening the "black box" that necessitates the postulation of an intervening variable, and they are substituting alternative explanations of the observed behaviors.

LORENZ'S MODEL

One way to understand how motivational concepts have served an important function is to examine a model of motivation. One of the early models is Lorenz's hydraulic system (Lorenz 1950), which includes a reservoir that fills with fluid, a faucet through which the fluid flows into the reservoir, and a valve through which fluid flows out of the reservoir (Figure 10–13). The degree of valve opening is controlled by a pulley connected to a pan on which weights can be added; the more weights in the pan, the wider the valve opens. Since, in Lorenz's conception of the model, the weights are analogous to sign stimuli, the valve opening is controlled both by the building up of "action-specific energy" in the reservoir and by the

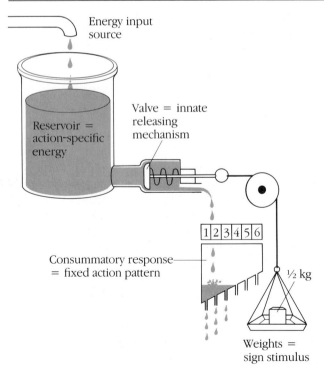

FIGURE 10–13   One early scheme to explain motivation was developed by Lorenz, often called the psycho-hydraulic model
Action-specific energy accumulates in a reservoir until released by the appropriate stimulus, represented by weights on a pan scale, or until the pressure on the valve causes an action pattern to occur spontaneously. The consummatory response or fixed action pattern(s) released vary depending upon how much action-specific energy is released from the valve.
Source: After K. Lorenz. 1950. *Symp. Soc. Exp. Biol.* 4:256.

FIGURE 10-14   Deutsch's model for motivation
This model for motivation, proposed by Deutsch in 1960, involves four separate
components and the environment.
Source: Data from J. Deutsch, *The Structural Basis of Behavior* (Chicago: University of
Chicago Press, 1960), p. 25. Reprinted with permission.

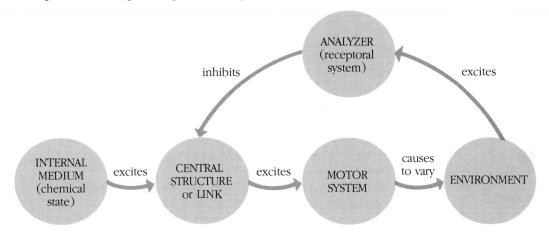

strength of the relevant sign stimuli; the valve is the innate releasing mechanism.
The model can account for certain cyclical changes in behavior (e.g., feeding), but it
fails to account for the sensory feedback that leads to reduction or cessation of an
activity when the goal is achieved (see experiments described below by Janowitz and
Grossman 1949).

DEUTSCH'S MODEL

Another model (Figure 10–14) which was proposed by Deutsch (1960; see also
Toates 1983), postulates a set of components or elements located somewhere in the
central nervous system. Deutsch proposes that mechanisms within the nervous
system monitor the body for specific needs, such as blood levels of various substances, or water, and so on. When conditions in the animal change — that is, when
the homeostatic balance shifts — a central link is activated and the motor system is
stimulated. The animal then exhibits a particular behavior pattern, such as eating or
drinking, as an interaction with the external environment. Behavior results in receptor discharges within the animal that produce an inhibition of the central link and
depress the state that originally stimulated the motor system.

A number of experiments have provided data supporting some correspondence between components of Deutsch's model and specific locations in the brain.
Miller (1957) confirmed that there is a brain site responsible for monitoring water

balance through sensors that detect osmolarity of the blood or other body fluids. By injecting fluids directly into regions of the hypothalamus of goats, he identified specific sites involved in controlling water balance. When he injected small drops of hypertonic saline solution (a hypertonic solution has a higher concentration of the substance than does the surrounding medium) into certain hypothalamic sites, the goats' drinking behavior increased. When he injected an isotonic solution (one that has the same concentration of substance as does the surrounding medium), the drinking behavior was not affected; and when he injected pure water, a hypotonic solution (one that has a lower concentration of the substance than the surrounding medium), the drinking behavior was reduced, even in an otherwise thirsty animal.

Electrical stimulation of particular regions of the hypothalamus of the goat or rat elicits drinking behavior (Anderson and McCann 1955; Greer 1955). In contrast, lesions of the same hypothalamic locus in the dog can greatly reduce water intake (Anderson and McCann 1956). We discussed in Chapter 6 how stimulation or lesions of other regions of the hypothalamus can affect feeding behavior in a manner directly analogous to this pattern.

An experiment on the feeding behavior in dogs by Janowitz and Grossman (1949) illustrates the connection between the analyzer and the central link. They operated on a dog to produce an opening in its esophagus so that any food they introduced into the mouth exited via the animal's throat and did not enter the stomach. The question is, when the animal is hungry and eats, whether the inhibition of feeding is initiated by food in the mouth or in the stomach. According to the Lorenz model, as an animal feeds, specific energy will be discharged and the dog will become satiated. If, however, the system does not discharge through actual feeding behavior, the site of food (sign stimulus) should continue to elicit feeding. When food is passed through the mouth of a test dog, as in the Janowitz and Grossman experiment, further feeding is not inhibited; but when the stomach is filled with food, the animal ceases to feed. Thus the connection between the analyzer and the central link in this system is some type of receptor in the stomach, which transmits messages to the central nervous system to halt further eating. The Lorenz model would not correctly predict the outcome of this experiment, but the Deutsch model does.

The use of a model, like Deutsch's model of motivation, can serve as the basis for exploring the operation of homeostatic mechanisms and stimulates research on the location and verification of the operation of components of the hypothetical scheme. Eventually, as we determine what the actual components are, we will no longer need hypothetical creations like motivation.

SUMMARY

Two important influences on behavior are learning, which is the relatively permanent modification of behavior that occurs through practice or experience, and motivation, which refers to the internal processes that arouse and direct behavior, but which must be inferred because these processes usually cannot be measured directly.

Habituation is the relatively persistent waning of a response that results from repeated stimulus presentations not followed by any form of reinforcement. In classical conditioning a neutral stimulus is paired with an unconditioned stimulus (UCS) to elicit an unconditioned response (UCR). After repeated pairings, the neutral stimulus becomes a conditioned stimulus (CS) and elicits the conditioned response (CR). In operant conditioning, sometimes called instrumental learning, the animal learns to associate a behavior (e.g., performing a particular action or task) with the consequences of that behavior, for example, a food reward (positive reinforcement) or a mild shock (negative reinforcement). Among the many similarities in the processes of classical and operant conditioning and the procedures used to investigate them are acquisition, schedules of reinforcement, extinction, spontaneous recovery, and generalization. They differ in that classical conditioning involves a stimulus-stimulus pairing and the animal does not actually control the sequence of events, whereas for operant conditioning there is a stimulus-response pairing, and the sequence of events is contingent upon the animal's behavior.

Four other types of learning are related to operant conditioning: avoidance learning, insight learning, latent learning, and observational learning. When an animal is presented with a series of similar problems and tested sequentially, the animal will develop a strategy or learning set.

A phylogenetic survey of the capacities of various groups of animals to demonstrate various types of learning reveals some patterns. Protozoans and coelenterates exhibit habituation, but classical and operant conditioning have not yet been demonstrated in these phyla. The flatworms (Platyhelminthes) and segmented worms (Annelida) are capable of habituation responses and also exhibit both classical and operant conditioning in some test situations. Trails of chemical cues left by the organisms may be important for learning in these groups. Molluscs and arthropods are also capable of habituation, as well as classical and operant conditioning. For all of the foregoing groups, very few species have been tested from each phylum, and our knowledge is thus quite limited regarding the overall phylogeny of learning. Vertebrates have been tested more extensively, but even within this phylum there are large gaps in our knowledge. There are no clear relationships between phylogenetic lineage and learning capacities among the vertebrate classes. Vertebrates, along with some invertebrate organisms, do exhibit rapid avoidance learning in feeding situations.

Studies of comparative learning have been both quantitative and qualitative. Quantitative comparisons are impeded by individual differences in learning and motivation and by species differences in the form of biological constraints on learning. An animal presented with a particular learning situation may be prepared, unprepared, or contraprepared. Preparedness is defined in this context as the genetically based predisposition to exhibit certain behaviors. Methodological constraints in the study of learning make it impossible to equate the test conditions for each species. Also, we must be constantly aware of variations in sensory and perceptual worlds among different species.

Memory, the process of information storage, may entail either dynamic reverberating circuitry in populations of neurons, or biochemical or structural changes within cells, or both. Some evidence suggests that long-term, or permanent, memory involves chemical-structural changes, and short-term memory may consist of neural activity.

Motivation and drive are terms used to explain changes in observed behavior due to variations in internal conditions within an animal. As we collect more experimental information about the roles of genetics, neurons, and hormones controlling behavior, we will no longer need these terms nor will we need models explaining motivation, such as those proposed by Lorenz and Deutsch.

Discussion Questions

1. Two important aspects of a young animal's life are its diet and locating a place to live. For each of the organisms listed below indicate what type of learning processes are involved in acquisition of dietary habits and habitat selection: (a) minnow, (b) cardinal, (c) sea anemone, (d) chipmunk, (e) gypsy moth larva. How would you proceed to experimentally test your hypotheses?

2. Drawing on your knowledge of the control of behavior, identify other situations in which we should be able to propose specific experiments to investigate portions of the Deutsch model for motivation. Where possible, provide suggestions for actual experiments.

3. In this chapter we have examined how hypotheses about learning can be tested in applied or natural settings. Think of several ways you might test learning behavior in a natural setting using field methods for each of the following: (a) cockroach, (b) mountain goat, (c) crow, and (d) gorilla.

4. Imagine that we have just located a new species of butterfly. We are interested in obtaining an accurate assessment of the sensory/perceptual world of this organism. Describe the procedures you would use, involving classical and operant conditioning, to conduct the experiments necessary to answer this question.

Suggested Readings

Domjan, M. 1980. Ingestational aversion learning: Unique and general processes. *Adv. Stud. Behav.* 11:276–336.
Comprehensive summary of research on this "hot" topic.
Considers theoretical and practical aspects and has a
methodological perspective. Good reference source for this
topic.

Hinde, R. A., and J. Stevenson-Hinde, eds. 1973. *Constraints on Learning.* New York: Academic Press.
Summary volume on modern learning, largely from ethological and zoological perspectives. Good reading for the scientist or nonscientist. Provides some consideration of ecological and evolutionary issues in the study of learning behavior.

Mackintosh, N. J. 1974. *The Psychology of Animal Learning.* New York: Academic Press.
Traditional treatment of learning theories, but modern and up-to-date in scope. Useful for the specialist and as a general reference.

Plotkin, H. C., and F. J. Odling-Smee. 1979. Learning, change and evolution: An enquiry into the teleonomy of learning. *Adv. Stud. Behav.* 10:1–42.
Relates learning theory and evolution. Written for those who have some background in learning theory and modern evolutionary thought.

Staddon, J. E. R. 1975. Learning as adaptation. In W. K. Estes (ed.), *Handbook of Learning and Cognitive Processes,* vol. 2. Hillsdale, N.J.: Erlbaum.
General discussion of the meaning of learning studies conducted in the laboratory but projected into natural habitats. Some material for thought for those with a primarily psychology background.

References

Alkon, D. L. 1980. Cellular analysis of a gastropod *(Hermissenda crassicornis)* model of associative learning. *Biol. Bull.* 159:505–560.

———. 1983. Learning in a marine snail. *Sci. Amer.* 249(July):70–85.

Anderson, B., and S. M. McCann. 1955. A further study of polydipsia evoked by hypothalamic stimulation in the goat. *Acta Physiol. Scand.* 33:333–346.

———. 1956. The effect of hypothalamic lesions on water intake of the dog. *Acta Physiol. Scand.* 35:312–320.

Arbib, M. A. 1972. *The Metaphorical Brain.* New York: Wiley-Interscience.

Barnett, S. A. 1963. *The Rat: A Study in Behavior.* Chicago: Aldine.

Birch, H. C. 1945. The relation of previous experience to insightful problem-solving. *J. Comp. Physiol. Psychol.* 38:367–383.

Bitterman, M. E. 1960. Toward a comparative psychology of learning. *Amer. Psychol.* 15:704–712.

———. 1965a. The evolution of intelligence. *Sci. Amer.* 212:92–100.

———. 1965b. Phyletic differences in learning. *Amer. Psychol.* 20:396–410.

———. 1975. The comparative analysis of learning. *Science* 188:699–709.

Block, R. A., and J. V. McConnell. 1967. Classically conditioned discrimination in the planarian, *Dugesia dorotocephala. Nature* 215:1465–1466.

———. 1975. The comparative analysis of learning. *Science* 188:699–709.

Bolles, R. 1970. Species-specific defense reactions and avoidance learning. *Psychol. Rev.* 77:32–48.

Boycott, B. B. 1965. Learning in the octopus. *Sci. Amer.* 212(3):42–50.

Brookshire, K. H. 1970. Comparative psychology

of learning. In M. H. Marx (ed.), *Learning: Interactions.* New York: Macmillan.

Brower, J. V. 1958. Experimental studies of mimicry in some North American butterflies. I. The monarch, *Danaus plexippus,* and viceroy, *Limenitis archippus archippus. Evolution* 12:32–47.

Brower, J. V., and L. P. Brower. 1964. Birds, butterflies and plant poisons: A study in ecological chemistry. *Zoologica* 49:137–159.

Cook, A. 1971. Habituation in a freshwater snail *(Limnaea stagnalis). Anim. Behav.* 17:679–682.

Corning, W. C., and S. Kelly. 1973. Platyhelminthes: The turbellarians. In W. C. Corning, J. A. Dyal, and A. O. D. Willows (eds.), *Invertebrate Learning,* vol. 1. New York: Plenum.

Corning, W. C. and von Burg, R. 1973. Protozoa. In W. C. Corning, J. A. Dyal, and A. O. D. Willows (eds.), *Invertebrate Learning,* vol. 1. New York: Plenum.

Datta, L. G., S. Milstein, and M. E. Bitterman. 1960. Habitat reversal in the crab. *J. Comp. Physiol. Psychol.* 53:275–278.

Deutsch, J. A. 1960. *Structural Basis of Behavior.* Chicago: University of Chicago Press.

———. 1973. *Physiological Basis of Memory.* New York: Academic Press.

———. 1973. *Physiological Basis of Memory.* New York: Academic Press.

———. 1978. *Comparative Animal Behavior.* New York: McGraw-Hill.

Domjan, M. 1980. Ingestional aversion learning: Unique and general processes. *Adv. Stud. Behav.* 11:276–337.

Dyal, J. A., and W. C. Corning. 1973. Invertebrates learning and behavior taxonomies. In W. C. Corning, J. A. Dyal, and A. O. D. Willows (eds.), *Invertebrate Learning,* vol. 1. New York: Plenum.

Ellins, S. R., S. M. Catalano, and S. A. Schechinger. 1977. Conditioned taste aversion: A field application to coyote predation on sheep. *Behav. Biol.* 20:91–95.

Fletcher, H. J. 1965. The delayed-response problem. In A. M. Schrier, H. F. Harlow, and F. Stollnitz (eds.), *Behavior of Nonhuman Primates,* vol. 1. New York: Academic Press.

Fouts, R. 1973. Acquisition and testing of gestural signs in four young chimpanzees. *Science* 180:978–980.

Frisch, K. von. 1967. *The Dance Language and Orientation of Bees.* Cambridge: Harvard University Press.

Galef, B. G. 1976. Social transmission of acquired behavior: A discussion of tradition and social learning in vertebrates. *Adv. Stud. Behav.* 6:77–100.

Garcia, J., F. R. Ervin, and R. A. Koelling. 1966. Learning with prolonged delay of reinforcement. *Psychonom. Sci.* 5:121–122.

Garcia, J., W. G. Hankins, and K. W. Rusiniak. 1976. Flavor aversion studies. *Science* 192:265–266.

Garcia, J., and R. A. Koelling. 1966. Relation of cue to consequence in avoidance learning. *Psychonom. Sci.* 4:123–124.

Gardner, R. A., and B. T. Gardner. 1969. Teaching sign language to a chimpanzee. *Science* 165:664–672.

Gelber, B. 1952. Investigations of the behavior of *Paramecium aurelia.* I. Modification of behavior after training with reinforcement. *J. Comp. Physiol. Psychol.* 45:58–65.

———. 1965. Studies of the behavior of *Paramecium aurelia. Anim. Behav. Suppl.* 1:21–29.

Gilhousen, H. C. 1927. The use of the vision and of the antennae in the learning of crayfish. *Univ. Calif. Publ. Physiol.* 7:73–89.

Gonzalez, R. C., and M. E. Bitterman. 1969. Spaced-trials partial reinforcement effect as a function of contrast. *J. Comp. Physiol. Psychol.* 67:94–103.

Gonzalez, R. C., A. Potts, K. Pitcoff, and M. E. Bitterman. 1972. Runway performance of goldfish as a function of complete and incomplete reduction in the amount of reward. *Psychonom. Sci.* 27:305–307.

Greer, M. A. 1955. Suggestive evidence of a primary "drinking center" in the hypothalamus of the rat. *Proc. Soc. Exp. Biol.* 89:59–62.

Gustavson, C. R., et al. 1976. Prey-lithium aversions. I. Coyotes and wolves. *Behav. Biol.* 17:61–72.

Harlow, H. F. 1949. The formation of learning sets. *Psychol. Rev.* 56:51–65.

———. 1951. Primate learning. In C. P. Stone (ed.), *Comparative Psychology.* Englewood Cliffs, N.J.: Prentice-Hall.

Hawkins, R. D., T. W. Abrams, T. J. Carew, and E. R. Kandel. 1983. A cellular mechanism of classical conditioning in *Aplysia:* Activity-dependent amplification of presynaptic facilitation. *Science* 219:400–405.

Hebb, D. O. 1958. *Textbook of Psychology.* Philadelphia: Saunders.

Hilgard, E. R., and G. H. Bower. 1975. *Theories of*

Learning, 4th ed. Englewood Cliffs, N. J.: Prentice-Hall.

Hinde, R. A. 1973. Constraints on learning—An introduction to the problems. In R. A. Hinde and J. Stevenson-Hinde (eds.), *Constraints on Learning: Limitations and Predispositions.* New York: Academic Press.

Hinde, R. A., and J. Stevenson-Hinde, eds. 1973. *Constraints on Learning: Limitations and Predispositions.* New York: Academic Press.

Hodos, W., and C. B. G. Campbell. 1969. *Scala naturae:* Why there is no theory in comparative psychology. *Psychol. Rev.* 76:337–350.

Itani, J. 1958. On the acquisition and propagation of a new food habit in the troop of Japanese monkeys at Takasakiyama. *Primates* 1:131–148.

Janowitz, H. D., and M. I. Grossman. 1949. Some factors affecting the food intake of normal dogs and dogs with esophagotomy and gastric fistules. *Amer. J. Physiol.* 159:143–148.

Jensen, D. D. 1957. Experiments on "learning" in paramecia. *Science* 125:191–192.

———. 1965. Paramecia, planaria and pseudo-learning. *Anim. Behav. Suppl.* 1:9–20.

John, E. R. 1967. *Mechanisms of Memory.* New York: Academic Press.

———. 1972. Switchboard versus statistical theories of learning and memory. *Science* 177:850–864.

Kawai, M. 1965. Newly acquired precultural behavior of the natural troop of Japanese monkeys on Koshima Island. *Primates* 6:1–30.

Kety, S. S. 1982. The evolution of concepts of memory: An overview. In A. L. Beckman (ed.), *Neural Basis of Behavior.* New York: SP Medical.

Klopfer, P. 1957. An experiment with empathic learning in ducks. *Amer. Nat.* 91:61–63.

Köhler, W. 1925. *The Mentality of Apes.* New York: Harcourt, Brace.

Krasne, F. B. 1973. Learning in Crustacea. In, W. C. Corning, J. A. Dyal and A. O. D. Willows (eds.), *Invertebrate Learning,* vol. 2. New York: Plenum.

Lashley, K. S. 1929. *Brain Mechanisms and Intelligence.* Chicago: University of Chicago Press.

Lawick-Goodall, J. van 1968. A preliminary report on expressive movements and communication in the Gombe Stream chimpanzees. In P. Jay, ed., *Primates: Studies in Adaptation and Variability.* New York: Holt, Rinehart and Winston.

Lorenz, K. 1950. The comparative method in studying innate behavior patterns. *Symp. Soc. Exp. Biol.* 4:221–268.

Masterton, R. B., M. E. Bitterman, C. B. G. Campbell, and N. Hotton. 1976. *Evolution of Brain and Behavior in Vertebrates.* Hillsdale, N.J.: Erlbaum.

McConnell, J. V., A. L. Jacobson, and D. P. Kimble. 1959. The effects of regeneration upon retention of a conditioned response in the planarian. *J. Comp. Physiol. Psychol.* 52:1–5.

Metzgar, L. H. 1967. An experimental comparison of screech owl predation on resident and transient white-footed mice *(Peromyscus leucopus). J. Mammal.* 48:387–391.

Meyer, D. R., and H. F. Harlow. 1952. Effects of multiple variables on delayed response performance by monkeys. *J. Genet. Psychol.* 81:53–61.

Meyer, D. R., F. R. Treichler, and P. M. Meyer. 1965. Discrete-trial training techniques and stimulus variables. In A. M. Schrier, H. F. Harlow, and F. Stollnitz (eds.), *Behavior of Nonhuman Primates* vol. 1. New York: Academic Press.

Miller, N. E. 1957. Experiments on motivation studies combining psychological, physiological and pharmacological techniques. *Science* 126:1271–1278.

Paton, J. A. and F. Nottebohm. 1984. Neurons generated in the adult brain are recruited into functional circuits. *Science* 225:1046–1048.

Patterson, D. J. 1973. Habituation in a protozoan *Vorticella convallaria. Behaviour* 45:304–311.

Pinsker, H., I. Kupfermann, V. Castellucci, and E. R. Kandel. 1970. Habituation and dishabituation of the gill-withdrawal reflex in *Aplysia. Science* 167:1740–1742.

Ratner, S. C., and A. R. Gilpin. 1974. Habituation and retention of habituation of responses to air puff of normal and decerebrate earthworms. *J. Comp. Physiol. Psychol.* 86:911–918.

Ratner, S. C., and K. R. Miller. 1959. Classical conditioning in earthworms, *Lumbricus terrestris. J. Comp. Physiol. Psychol.* 52:102–105.

Ross, D. M. 1965. Complex and modifiable behavior patterns in *Calliactis* and *Stomphia. Amer. Zool.* 5:573–580.

Rozin, P. 1968. Specific aversions and neophobia resulting from vitamin deficiency or poisoning in half wild and domestic rats. *J. Comp. Physiol. Psychol.* 66:82–88.

Rumbaugh, D. M. 1968. The learning and sensory capacities of the squirrel monkey in phylogenetic perspective. In L. A. Rosenblum and R. W. Cooper (eds.), *The Squirrel Monkey.* New York: Academic Press.

Rushforth, N. D. 1973. Behavioral modifications in coelenterates. In W. C. Corning, J. A. Dyal, and A. O. D. Willows (eds.), *Invertebrate Learning,* vol. 1. New York: Plenum.

Sanders, G. D. 1973. The cephalopods. In W. C. Corning, J. A. Dyal, and A. O. D. Willows (eds.), *Invertebrate Learning,* vol. 3. New York: Plenum.

Seligman, M. E. P. 1970. On the generality of the laws of learning. *Psychol. Rev.* 77:406–418.

Shell, W. F., and A. J. Riopelle. 1957. Multiple discrimination learning in raccoons. *J. Comp. Physiol. Psychol.* 50:585–587.

Shettlesworth, S. J. 1972. Constraints on learning. *Adv. Stud. Behav.* 4:1–68.

Sterner, R. T., and S. A. Shumake. 1978. Bait-induced prey aversions in predators: Some methodological issues. *Behav. Biol.* 22:565–566.

Thompson, R., and J. V. McConnell. 1955. Classical conditioning in the planarian, *Dugesia dorotocephala. J. Comp. Physiol. Psychol.* 48:65–68.

Thorpe, W. H. 1956. *Learning and Instinct in Animals.* Cambridge: Harvard University Press.

————. 1963. *Learning and Instinct in Animals,* 2nd ed. London: Methuen.

Tinbergen, N. 1958. *Curious Naturalists.* New York: American Museum of Natural History.

Tinbergen, N., and W. Kruyt . 1938. Uber die Orientierung des Bienenwolfes (*Philanthus triangulum* Fabr.): III Die Bevorzugung bestimmter Wegmarken. *Zeit. Vergl. Physiol.* 25:292–334.

Travis-Neideffer, M. N., J. D. Niedeffer, and S. F. Davis. 1982. Free operant single and double alternation in the albino rat: A demonstration. *Bull. Psychonom. Soc.* 19:287–290.

Walcott, C. 1969. A spider's vibration receptor: Its anatomy and physiology. *Amer. Zool.* 9:133–144.

Warren, J. M. 1973. Learning in vertebrates. In D. A. Dewsbury and D. A. Rethlingshafer (eds.), *Comparative Psychology: A Modern Survey.* New York: McGraw-Hill.

Warren, J. M., and A. Barron. 1956. The formation of learning sets by cats. *J. Comp. Physiol. Psychol.* 49:227–231.

Wells, P. H. 1973. Honey bees. In W. C. Corning, J. A. Dyal, and A. O. D. Willows (eds.), *Invertebrate Learning,* vol. 2. New York: Plenum.

Woody, C. D. 1982. *Memory, Learning and Higher Function.* New York: Springer-Verlag.

11

SEXUAL BEHAVIOR AND REPRODUCTION

Most behavior relates to obtaining energy (food) and shelter, avoiding predators, and reproducing. Evolutionarily, an individual is successful if the relative frequency of its genes increases in subsequent generations. Not surprisingly, then, reproduction dominates the activities of most adult organisms.

The Pacific salmon (*Oncorhynchus* spp.), after reaching maturity at sea, works its way up rapids to reach its birthplace, expends its remaining energy in shedding masses of eggs or sperm, then dies. We can ask a number of "why" questions about the evolutionary and long-term ecological determinants of the salmon's reproductive behavior. For example, why does the salmon return to its birthplace to reproduce, and not to some other stream? Why does it put all its effort into a single, explosive, reproductive episode rather than producing smaller numbers of young at intervals? In the first part of the chapter, we look at the ultimate evolutionary forces that determine such things as sex ratios, frequency of reproduction, types of mating systems, and the amount of care given the young.

We can also discuss reproduction from the point of view of "how" questions. How does the salmon locate its ancestral stream at a certain time of year and arrive at its birthplace? In the second part of the chapter, we examine the proximate, or short-term, mechanisms of sex, such as neural and endocrine control of behavior, and how they are linked to environmental fluctuations.

ULTIMATE FACTORS IN REPRODUCTIVE BEHAVIOR

DIVERSITY OF REPRODUCTION

Many unicellular, plant, and invertebrate organisms reproduce asexually during times of environmental stability and predictability and undergo some form of sexual recombination during less stable, less predictable times. During the sexual phase the organism will often produce seeds or eggs that are resistant to environmental stress and are capable of being dispersed. Some sort of genetic recombination occurs in virtually all organisms. Bacteria and blue-green algae, among the prokaryotes, recombine genes in a process called **transformation.** Single-celled algae, such as diatoms, have eggs and sperm that resemble those of vertebrates; however their courtship behavior is probably not particularly complex. Some organisms — for example, earthworms, snails, and fish — are hermaphrodites; they possess functional sexual organs of both males and females. Usually hermaphrodites do not self-fertilize, possibly because the new genetic combinations created would not be novel enough to outweigh the costs of breaking up previously successful combinations. Some species of fish and some invertebrates are sequentially hermaphroditic; they change from female to male (**protogyny**) or from male to female (**protandry**) usually once in their lives (Figure 11–1).

COSTS AND BENEFITS OF SEX

Why is sexual reproduction so widespread? An organism maximizes its genetic contribution to offspring by reproducing asexually — that is, by creating carbon copies of itself — but during sexual reproduction, genotypes are broken up at meiosis, and the genes combine with those of another individual during fertilization, resulting in a 50 percent reduction in transmission of genes to the next generation, which is referred to as the **cost of meiosis** (Williams 1975). The energetic cost of raising young is about the same whether the offspring is produced sexually or asexually, so a female would have to produce two sexual offspring for each asexual offspring in order to pass the same number of genes to the next generation. If a female can get a male to share in raising the offspring, the cost of meiosis is reduced. If they share equally, a female can rear two sexual offspring for the same cost as a single asexual offspring, and the cost of meiosis might be considered zero (Wittenberger 1981). A further cost of meiosis, however, is that the unique combination of traits in the parent is lost in sexually produced offspring. This aspect of the cost of meiosis can be reduced if individuals mate with relatives, since they are likely to share the same combination of genes (Shields 1982).

A second cost of sex is the energy males expend in the production of large numbers of sperm, most of which never fertilize eggs. In fact, many males fail to reproduce at all and waste resources that could be used for reproduction in asexual forms. A third cost of sexual reproduction is that of courtship and mating, since it takes energy and risk to secure a partner (Daly 1978).

Given these costs, one might expect **parthenogenetic** (asexual) individuals to "infect" sexual populations and replace the sexually reproducing organisms. What are the possible benefits of sex and how can it be maintained in a population? Fisher (1958) first pointed out that sexually reproducing populations can evolve faster in a changing environment. Mutations occurring at different loci in different individuals can, through sexual recombination, appear as novel combinations in the offspring. A second, related, advantage was noted by Muller (1964) and is referred to as **Muller's ratchet.** Suppose that a deleterious mutation occurs in an individual member of an asexual population. The gene will be passed on to all its offspring. The only way an individual can arise that is free of this mutation is via back (or reverse) mutation, an unlikely event. Thus the ratchet turns one notch each time a deleterious mutation occurs, and mutations accumulate in the population. However, if sexual reproduction occurs, recombination between two (diploid) individuals with *different* muta-

FIGURE 11-1 Sequentially hermaphroditic coral reef fish
One male in the species *Anthias squamipinnis* defends a territory containing several females. The dominant female in the group (top) changes into a male (bottom) upon the death or disappearance of the male. Protogyny is common in reef fish.
Source: Photo by Douglas Shapiro.

tions could produce offspring with neither trait. In this way, harmful mutations can be "edited out" of a population.

The preceding arguments have emphasized the advantages of sex for the population, or group: faster evolution and elimination of mutations from the gene pool. Williams (1975) has championed benefits that might accrue to individuals, rather than populations, since natural selection is thought to act primarily at the level of the individual (Chapter 4). He noted that in species with both sexual and asexual reproduction, the sexual phase usually occurs prior to the onset of unpredictably changing environmental conditions. Parents produce genetically variable offspring "in the hopes" that a few will be adapted to one of the new environments. Williams uses a raffle analogy: If there is only one survivor (winner) in each different habitat, reproducing asexually would be like having tickets that all have the same number printed on them. The chances of picking a winner are much less than if all the tickets have different numbers, the case with sexually produced offspring.

Although Williams emphasized the importance of variations in the physical environment, others have argued that biotic fluctuations make it even more necessary to produce genetically variable progeny (Van Valen 1973). For instance, as predators are selected to become more efficient at catching their prey, the prey must continually evolve new ways to avoid being eaten. Such an evolutionary "race" takes place among competing species and between pathogens and their hosts as well. Bacteria and viruses, because of their short life spans, rapidly evolve new strains. The host produces variable offspring "in the hopes" that some will be resistant.

SEX RATIOS

In most animal species there are two sexes that are anatomically different and that are produced in about equal numbers. The possible genetic combinations for just two sexes are astronomical, and three or more sexes would add little except confusion in the selection of a mate. Although the sexes are anatomically similar in most microorganisms, fungi, and algae, division of labor is the rule in animals: females produce large, sessile, and energetically expensive eggs, and males produce small, motile, energetically cheap sperm. This difference in gamete size, referred to as **anisogamy,** may have resulted from disruptive selection (Parker, Baker, and Smith 1972) and sets the stage for other differences in reproductive behavior.

The sex ratios of populations of most species tend to be about 50:50 at birth or hatching, but may deviate significantly from equality among adults. The **operational sex ratio** is based on only the reproductively active members of the population. Deviations in this ratio can have a large impact on the mating system, since members of the abundant sex will compete for access to the scarcer sex.

Why does the sex ratio tend to be 50 : 50 at birth? Let us look at a population containing more males than females. Some males will be unable to find mates. Females tending to produce more females than males will be favored through natural selection because all of their female offspring will probably find mates, and the sex ratio will converge toward 50 : 50. The symmetrical argument also holds when the ratio is skewed in favor of females. The argument has been developed further by Hamilton (1967) and others, who point out that equality of parental investment in offspring of each sex is the important factor in the ratio. If twice as much effort by the parents is required to raise male young to maturity, we would expect a 1 male : 2 female sex ratio among young that reach maturity.

Some animals have physiological and behavioral control over sex determination. In social insects such as bees and ants *(Hymenoptera)*, males are derived from unfertilized (haploid) eggs and females from fertilized (diploid) eggs. Such a mode of sex determination is called **haplodiploidy.** When laying eggs early in the season, the fertilized queen does not release stored sperm, and male offspring result. Later in the season she produces all-female broods that become workers. In many species of turtles sex is determined by the temperature at which the egg is incubated; those at a low temperature become males, while those at a higher temperature become females (Bull 1980).

Among organisms that undergo a sex change are protogynous fishes such as the sea bass *(Anthias squamipinnis)* studied by Shapiro (1979) on Aldabra Island off the east coast of Africa. The mechanism that controls the fish's change from female to male is incompletely understood; it is triggered, however, by the death or departure of the (usually) single resident territorial male. Shapiro observed that, upon the permanent absence of the male, the largest female undergoes a marked change in behavior, engaging in more aggressive displays, as she assumes the male color pattern and reproductive role (see Figure 11–1). In another species, the saddleback wrasse *(Thalassoma duperrey)*, sex change is a function of the relative size of neighbors. When few larger fish (usually males) are present, females change into males (Ross, Losey, and Diamond 1983).

SEXUAL SELECTION

The success of an individual is measured not only by the number of offspring it leaves but also by the quality or probable reproductive success of those offspring. Thus, it becomes important who its mate will be. Darwin (1871) introduced the concept of **sexual selection,** a special process that produces anatomical and behavioral traits that affect an individual's ability to aquire mates. Sexual selection can be divided into two types: one in which members of one sex choose certain mates of the other sex **(intersexual selection),** and a second in which individuals of one sex compete among themselves for access to the other sex **(intrasexual selection).** One result of either type is that the sexes come to be different—that is, **dimorphic.**

FIGURE 11-2   Male peacock in display
In this extreme example of sexual selection the courting male spreads his tail feathers and shakes them in front of the female. Such plumage would seem to confer little advantage to male survival, but it may indicate to the female that male's genetic superiority. *Source:* Photo by Terence A. Gili from Animals Animals.

INTERSEXUAL SELECTION

In intersexual selection individuals of one sex (usually the males) "advertise" that they are worthy of an investment; then members of the other sex (usually the females) choose among them. Most naturalists after Darwin discounted the importance of mate choice in evolution, but it has recently become a popular topic of study. Fisher (1958) used birds as an example. Suppose a plumage characteristic in males is linked to some survival or reproductive advantage for the bearer. If females prefer those males as mates, they will have higher reproductive success and will likely produce sons with the trait and daughters that prefer the trait when choosing a mate of their own. Further development of the trait will proceed in males, as will the preference for that trait in the females, resulting in a runaway process. A possible result of such a process is seen in the peacock (Figure 11–2). Sexually selected traits may become so exaggerated that survival of the males is reduced. Counterselection in favor of less ornamented males will occur — due, for example, to greater predation on the more ornamented males — and sexual selection is brought to a steady state.

An alternative to Fisher's runaway selection hypothesis is the handicap hypothesis proposed by Zahavi (1975). Agreeing that sexual selection can produce traits that are detrimental to survival, he adds that they are both costly to produce and linked to superior qualities in the males. Thus a bird that can survive in spite of conspicuous and costly-to-produce tail feathers, or a stag that can afford to spend energy on a huge bony growth of antlers must be genetically superior and a worthy mate. Important to this hypothesis is the notion of "truth in advertising": the male's handicap must be linked to overall genetic fitness. Only in this way will females benefit by picking a male with this handicap.

Although Darwin, Fisher, and Zahavi thought of sexual selection as a process distinct from natural selection, some biologists argue that the two are inseparable. Kodric-Brown and Brown (1984) claim that most sexually selected traits are aids, rather than handicaps, to survival. Thus a stag with a big set of antlers may be dominant over other males and may have better access to a food supply. Much further field work needs to be done to adequately test these alternative hypotheses.

As an example of intrasexual selection in action, consider the bowerbirds of southeast Asia. Bowerbirds are unique in that they build a structure whose only function seems to be to attract mates. The male satin bowerbird (*Ptilonorhynchus violaceus*) of Australia and New Guinea stations itself alone in the forest, clears a space, weaves the bower from twigs, and decorates it with brightly colored objects (Figure 11–3). Gilliard (1963, 1969) reasoned that bowers function in place of showy plumage, and noted that those species with the fanciest bowers had the drabest plumage. He argues that such behavior arose to synchronize mating between males and females.

Borgia (1985) assessed the relative contribution of the male satin bowerbird's decorations to his mating success. Human observers and cameras controlled by infrared detection devices monitored activities at the bower. Removing decorations reduced mating success. Aspects of the bower that were positively related to mating success were the number of blue feathers, snail shells, and yellow leaves. The general construction of the bower and the density of sticks in the walls were also important. Borgia (1985) suggests that bowers function as "markers," conveying

FIGURE 11-3   Satin bowerbird
The male satin bowerbird *(Ptilonorhynchus violaceus)* constructs the bower as part of his courtship ritual to attract a female. Males building structures with more sticks and brightly colored objects are more successful in attracting mates.
Source: Photo by Richard A. Forster, courtesy of the Massachusetts Audubon Society.

information to females about the male's genetic quality. Dominant males have high-quality bowers which attract the most females.

INTRASEXUAL SELECTION

Intrasexual selection involves competition within one sex (usually males), with the winner gaining access to the opposite sex. Competition may take place prior to mating, as with ungulates such as deer (family Cervidae) and antelope (family Bovidae). Typically, males live most of the year in all-male herds; as the breeding season approaches, males engage in highly ritualized battles, using their antlers or horns. The winners of these battles gain dominance and do most of the mating (Figure 11–4). Antlers are better developed in those cervid species where males compete strongly for large groups of females (Clutton-Brock, Guiness, and Albon 1982).

It is often difficult to determine which type of sexual selection is operating to produce an observed effect, since members of both sexes may be present during courtship. The "roughout display" of the male brown-headed cowbird (*Molothrus ater*) consists of a "glunk-chee" sound, made as the bird spreads its tail, lowers its wings, topples forward, and nearly falls over. This display seems to intimidate other males *and* to attract females. Similarly, the chirp of male field crickets (family Gryllidae) is used to exclude other males from an area *and* to attract females (Alexander 1961). As a final example of this problem, you may have noted that we used the antlers of deer to demonstrate the effects of both female choice and male-male competition.

Females may incite competition among males and thus may gain some con-

FIGURE 11–4 **Use of antlers in combat between male Pere David deer** Winners of contests become dominant and control access to resources such as mates.
Source: New York Zoological Society Photo.

trol over the choice of mate. For example, female elephant seals *(Mirounga angustir-ostris)* vocalize loudly whenever a male attempts to copulate. This behavior attracts other males and tests the dominance of the male attempting to mate (Cox and Le Boeuf 1977). In response to the female's sounds, the dominant harem master will drive off low-ranking, potentially inferior mating partners.

Males may fight directly at the time of copulation, as exemplified by the male of the yellow dung fly *(Scatophaga stercoraria)* (Parker 1970a). A male will sometimes attack a mated pair, displacing the male and mating with the female (Figure 11–5).

Nor does competition among males to sire offspring cease with the act of copulation. Females of many species store sperm and may remate before sperm from the previous mating are used up, creating the possibility of **sperm competition** (Parker 1970b). Sperm competition can be thought of as a selection pressure leading to two opposing types of adaptation in males: those that reduce the chances that a second male's sperm will be used (first male advantage), versus those that reduce the chances that the previous male's sperm will be used (remator male advantage) (Gromko, Gilbert, and Richmond 1984).

First male adaptations include mate-guarding behavior and the deposition of copulatory plugs, both of which reduce the chance of sperm displacement by a second male. In fruit flies *(Drosophila melanogaster)* first males may also transfer to females "anti-aphrodisiac" substances that inhibit courtship by other males (Jallon, Antony, and Benamar 1981). Female spiders store sperm for long periods and often mate with several males. In laboratory studies of the bowl and doily spider *(Frontinella pyramitela),* Austad (1982) found that the first male's sperm had priority in fertilizing eggs over sperm of subsequent males, as is the usual case in spiders.

Among insects, however, the advantage usually seems to accrue to the sperm of the remator male (Parker 1970b). In the yellow dung flies he studied, the last male

FIGURE 11–5 Male dung flies fighting over female
The attacking male (on the left) is attempting to push the paired male (on the right) away from the female (only her wings are visible). Sometimes the attacking male succeeds in taking over the female and mating with her; she then continues to oviposit eggs fertilized by the second attacking male. *Source:* Photo by G. A. Parker.

to mate sired 80 percent of the offspring subsequently produced. The ultimate remator male adaptation may be the penis of the damselfly *(Calopteryx maculata)* which has a dual function: it removes sperm deposited in the female by a previous male via a special "sperm scoop" and then replaces it with its own (Waage 1979).

Moths and butterflies (Lepidoptera) produce two kinds of sperm. One kind is the usual type **(eupyrene)** that fertilizes the eggs. The second type **(apyrene)** contains no nuclear material but may comprise more than half the sperm complement. Why should males waste energy on these "dud" sperm? Silberglied, Shepherd, and Dickinson (1984) suggest that they are the result of sperm competition, possibly displacing eupyrene sperm from first males or delaying remating by the female.

Following conception, male-male competition may take a different form. In mice the **Bruce effect** operates early in pregnancy: a strange male (or his odor) causes the female to abort and become receptive (Bruce 1966). Among langur monkeys *(Presbytis entellus)* strange males may take over a group, driving out the resident male. The new male may then kill the young sired by the previous male (Hrdy 1977). Females who have lost their young soon become sexually receptive, and the new resident male can inseminate them. Infanticide by adult males thus may be viewed as a "remator male" adaptation.

MATING SYSTEMS

Anisogamy prevails in nearly all animals, with females investing more in gametes than males. According to Trivers (1972), this difference causes males to compete for access to females and to try to mate with more than one female, a condition referred to as **polygyny.** This results in greater variation in the reproductive success for males than for females; for each male that fertilizes the eggs from a second female, another male is likely to fertilize none. We have already seen that sexual selection tends to act more strongly on males than on females. However, not all species are polygynous.

In trying to evaluate the adaptive significance of differences in mating systems, we must look at ecological factors as well as historical ones. For example, group size may be related to predator pressure and food distribution. In the open plains, where large predators are present and food is widely distributed, omnivorous primates and grazing mammals such as ungulates live in large groups in which mating with several members of the opposite sex is likely for both males and females. In densely forested areas, where communication over long distances is difficult, small family units and monogamy prevail.

Most classifications of mating systems are based on the extent to which males and females associate (bond) during mating. **Monogamy** refers to association between one male and one female at a time. **Polygyny** refers to association between one male and two or more females at a time. **Polyandry** refers to association between one female and two or more males at a time. **Promiscuity** refers to the absence of any prolonged association and to multiple mating by at least one sex. The term **polygamy** usually includes polygyny and polyandry, but may also include promiscuity and thereby incorporates all multiple-mating, nonmonogamous-mating systems. One problem with classification systems based on bonding is that a judgment is needed as to what constitutes an association. Does it include parental care? What about species, such as many nonhuman primates, that live in year-round social groups? A prolonged association exists; but during breeding, females mate with several males, and males may mate with many females. Is this polygyny, polyandry, or promiscuity?

Emlen and Oring (1977) have developed an ecological classification of mating systems that may eliminate the need to make these judgments because it is based on the ability of one sex to monopolize or accumulate mates and emphasizes the ecological and behavioral potential for monopolization. Although they developed it primarily for birds, the classification of mating systems seems applicable to most vertebrate and insect taxa.

- **Monogamy:** Neither sex is able to monopolize more than one member of the opposite sex.

- **Polygyny:** Males control access to more than one female.
  Resource defense polygyny: Males control access to females indirectly by monopolizing critical resources.
  Female defense polygyny: Males control access to females directly, usually because females are grouped for other reasons.
  Male dominance polygyny: Mates or resources are not monopolizable; females select mates from aggregations of males, as in leks.

- **Polyandry:** Females control access to more than one male.
  Resource defense polyandry: Females control access to males indirectly by monopolizing critical resources.
  Female access polyandry: Females do not defend resources essential to males, but they interact among themselves to limit access to males.

MONOGAMY

In monogamous systems, neither sex is able to monopolize more than one member of the opposite sex. When the habitat contains scattered, renewable resources or scarce nest sites, monogamy is the most likely strategy. If there is no opportunity to monopolize mates, an individual will benefit from remaining with its initial mate and helping to raise the offspring. The formation of long-term pair bonds also seems advantageous because less time need be spent in finding a mate during each reproductive cycle. In long-lived birds such as sea gulls, those that breed with former mates have higher reproductive success, probably because of less aggression between mates and greater synchronization of sexual behaviors (Coulson 1966). About 90 percent of all bird species are monogamous (Figure 11–6).

FIGURE 11-6   Pair of California gulls with young
There is little sexual dimorphism in these male and female gulls *(Larus californicus).* In this monogamous species both sexes care for the young.
Source: Photo by Bruce Pugesek.

POLYGYNY

In polygynous systems, individual males have access to more than one female. In resource defense polygyny, males defend areas containing the feeding or nesting sites critical for reproduction, and a female's choice of a mate is influenced by the quality of the male and of his territory. Territories that vary sufficiently in quality may attain the "polygyny threshold," the point at which a female may do better to join an already mated male possessing a good territory than an unmated male with a poor territory (Orians 1969). Thus some males may get two or more mates while others get none (Figure 11-7). Typically males do not provide parental care and usually delay breeding until their third year of life, at which time their ability to obtain a good territory has increased.

Many species are probably facultatively, or optionally, polygynous. In habitats where feeding or nesting resources cannot be monopolized by males, monogamy is likely, while habitats with clumped resources that are defendable would favor polygyny. In this way we may expect to find variation in the mating system of a single species.

Female defense polygyny may occur when females are gregarious for reasons unrelated to reproduction. Some males monopolize females and exclude other males from their harems. In many species of seals, the females haul out on land to give birth, and they mate soon after. The females are gregarious because there are a limited number of suitable sites, and the males monopolize the females for breeding. Intense competition among males results in marked sexual dimorphism and a large variance in male reproductive success.

If males are not involved in parental care and have little opportunity to control resources or mates, male dominance polygyny may occur. If female movements or concentration areas are predictable, the males may concentrate in such areas and pool their advertising and courtship signals. Females then select a mate from the group of males. These areas are called **leks,** in which males congregate and

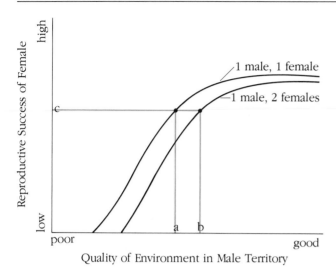

FIGURE 11-7   Reproductive success of female as a function of quality of male territory

One curve is for a monogamous pair of birds and another for a female who has to share the territory with another female. As territorial quality improves, so does the female's reproductive success. If territories vary enough from one male to the next, a female may have greater reproductive success by joining an already mated male with a good territory, a process that leads to polygyny. Here, a female mating with a male that is defending a territory of quality (a) would have the same reproductive success (c) as a female joining an already mated male in a superior territory of quality (b). For a test of this and several other models of polygyny in redwings see Lenington (1980).

Source: Data from G. H. Orians, "On the Evolution of Mating Systems in Birds and Mammals," *American Naturalist* 103 (1969):589–603.

defend small territories in order to attract and court females. Females select a mate, copulate, then leave the area and rear their young on their own. Older, more dominant males may occupy the central territories and do most of the copulating.

In the absence of territory or dominance, in some species a **scramble** takes place as the males try to mate. Female wood frogs *(Rana sylvatica)* congregate in a pond during a single night in early spring. Males rush about attempting to mate with fecund females (Berven 1981).

POLYANDRY

In polyandrous systems females control access to more than one male. Because female investment in eggs exceeds that of males in sperm, polyandry is rare. In most cases females provide parental care while males seek new mates. If food availability at the time of breeding is highly variable or if breeding success is very low due to high predation on the young or the eggs, females may have to produce many offspring. In birds, male incubation is common; a few cases of polyandry have been documented in which males do all the incubating and females lay multiple clutches.

Breeding sites of the American jacana *(Jacana spinosa)*, a large wading bird found in Central and South America, are limited and are divided into small territories by males (Jenni 1974). Female jacanas control superterritories that may encompass the nesting areas of several males. Frequently, several males incubate the clutches of one female, and she provides replacement clutches for them if, as often happens, nests are lost through predation. Breeding females are half again as big as the males, dominate the males, and provide little parental care. In this reversal of polygyny, the females specialize only in egg production.

Females of the migratory spotted sandpiper *(Actitis macularia)* compete for control of breeding territories and males provide most of the care of young (Oring and Lank 1982). Females arrive on the breeding grounds before the males and are **philopatric**—they return to the place where they were born. However, the vast majority of birds are either monogamous or polygynous and the reverse is generally true—males arrive first to establish territories and are more likely than females to return to the natal site to breed (Greenwood 1980).

ECOLOGY AND MATING SYSTEMS

One of the best examples of the way in which mating systems are related to resource distribution is provided by Orians's (1961) comparative study of blackbird social systems. The red-winged blackbird *(Agelaius phoeniceus)* is usually polygynous; a male defends a territory containing two or three females. The male returns three to four weeks ahead of the females, frequently to the same site as the previous year, and expends considerable energy defending his territory until the young are fledged. Males rarely help to raise the young; females do all of the nest building and incubating and nearly all of the feeding of the young, as is typical of polygynous species.

In parts of northern California, a very closely related species, the tricolored blackbird *(Agelaius tricolor)*, is **sympatric** (coexists) with the redwing (Orians 1961); its mating system, however, is quite different from that of the redwing. Male and female tricolored blackbirds pair off in a nomadic colony of anywhere from 100 to 200,000 birds. They establish territories, find mates, build nests, and lay eggs, all within one week. Activities are highly synchronous within each colony. As we can see from Figure 11–8, both sexes of the tricolored blackbird make a large investment but in a shorter time frame than that of the redwings.

Why do two very closely related species do things so differently in the same place and at the same time? The answer seems to be related to their energy source. Redwings have a relatively stable diet of seeds and insects, which is available for several months. A male redwing defends the same territory several years in a row, and each territory contains most of the food needed to support the females and the young. In contrast, tricolors go out from the colony on mass feeding fights, possibly

FIGURE 11-8   Time expenditure for pair of red-winged blackbirds and tricolored blackbirds during breeding season
The reproductive effort is spread out over more than four months in the redwing (a), but it is completed in about half that time by the tricolor (b). Tricolored blackbirds engage in mass feeding flights to concentrated food sources away from the breeding colony.
Source: G. H. Orians, "The Ecology of Blackbird *(Agelaius)* Social Systems," *Ecological Monographs* 31 (1961):285–312. Copyright 1961, the Ecological Society of America. Reprinted with permission.

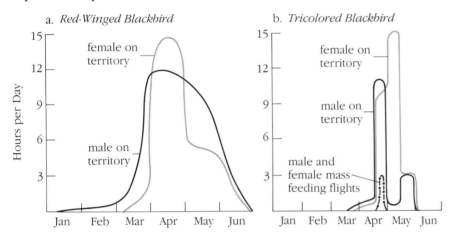

to assess concentrated food sources; and they attack rice and other grain fields at the time of seed maturity. Thus the tricolor mating system seems designed to take advantage of an ephemeral, but rich and concentrated, food source. When tricolors locate such an energy supply, they move in, establish a new colony, and reproduce before the source is gone. A similar strategy is used by African weaver finches *(Quelea)*, which move all over the continent, attacking rice and wheat fields, breeding quickly, then moving off to another ripening area (Ward 1971). Because of their nomadism these serious economic pests have been very difficult to control.

PARENTS AND OFFSPRING

PARENTAL INVESTMENT

Once eggs are laid or young are born, what may the parents do to improve the chances of their offspring to survive? We can define **parental investment** as any behavior toward offspring that increases the chances of the offspring's survival at the cost of the parent's ability to rear other offspring (Trivers 1972, 1974). Because an egg requires a greater investment of energy than does a sperm, male and female strategies differ. Since eggs are likely to be limited in number, we expect that males will compete for the opportunity to fertilize them and thus will be subject to sexual selection. A female is likely to mate, but, given her large investment, she will be particular about which male fertilizes her precious eggs. Males, on the other hand, will try to inseminate as many females as possible. Mating systems in which the male mates with more than one female should be the most common (Figure 11–9). Dewsbury (1982) has argued that although only one sperm fertilizes an egg, millions are generally required in each ejaculation to ensure fertilization of even a single egg. Also, there is a limit on the number of times most males can ejaculate within a certain period. Thus a male's investment in sperm is not necessarily trivial, and he, too, can be expected to be somewhat "choosey" about his mate. Nevertheless, it is generally assumed that for most species the female's investment in gametes is greater than the male's.

Competition by males that results in a greater variance in their copulatory success compared with that of females is referred to as the "Bateman effect." Bateman (1948) demonstrated that in laboratory populations of fruit flies *(Drosophila melanogaster)*, nearly all the females mated; however, many males failed to mate, and some males mated several times. In other words, the variance in copulatory success was higher for males than for females. The flies had chromosomal markers

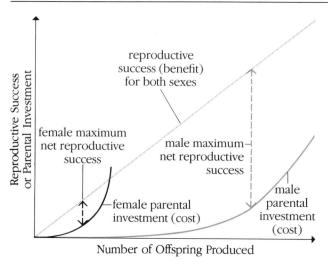

FIGURE 11-9 Parental investment and reproductive success as a function of number of offspring produced

Because the parental investment (cost) for females usually rises more steeply than for males, the optimum number of offspring (highest reproductive success at lowest cost) for females is less than for males. Males, consequently, may seek more than one mate to attain maximum net reproductive success. Broken lines indicate maximum male and female net reproductive success.

Source: Adapted with permission from R. L. Trivers, "Parental Investment and Sexual Selection," in Bernard Campbell, *Sexual Selection and the Descent of Man* (New York: Aldine Publishing Company). Copyright 1972 by Aldine Publishing Company.

FIGURE 11-10   Sexual dimorphism in elephant seals
Among a herd of females, two males fight to establish dominance. Males differ
strikingly from females, are about three times larger, possess an enlarged snout, or
proboscis, and have cornified skin around the neck. In this highly polygynous
species males invest nothing in their offspring other than DNA from sperm.
Source: Photo by Burney J. Le Boeuf.

so that parents of particular offspring could be identified. Males that copulated most
also sired the most offspring.

   In such polygynous systems, competition among males can be intense. For
example, mating of elephant seals takes place in colonies. Males establish a dominance hierarchy, and only the high-ranking males breed. Le Boeuf (1974) observed
that, typically, less than one-third of the males copulate at all, and the top five males
do at least 50 percent of the copulating (Figure 11–10 and Table 11–1). There is no
investment by the males in offspring beyond the sperm, as evidenced by the fact that
males may trample their own pups as they strive to inseminate females. The males

TABLE 11-1   Number and percent of male elephant seals copulating during consecutive breeding seasons

	1968	1969	1970	1971	1972	1973
Number of males present	103	120	125	136	146	180
Number of males copulating	14	17	32	41	51	62
Percent of males copulating	14	14	26	30	35	34
Percent copulations by the five most active males	83	92	69	65	53	48
Number of females present	193	243	311	352	408	470

Source: B. J. Le Boeuf, "Male-Male Competition and Reproductive Success in Elephant Seals," *American Zoologist* 14 (1974):163–176. Reprinted with permission.

have no way of knowing which young are their own, since pups are born a year after copulation.

In a few species the situation is reversed, and the male does most of the caring for young. Since the males' investment in offspring becomes larger than the females', we would expect females to compete for and try to attract males, rather than vice versa. Indeed, this reversal of the typical roles occurs in such fish as pipefishes and sea horses and in such polyandrous birds as phalaropes, jacanas, tinamous, and sandpipers.

Another type of investment occurs when the male provides nourishment to the female at the time of copulation, as in prenuptial feeding in birds. In some orthopteran insects, more than 25 percent of the male's weight may be transferred to the female in a spermatophore (Figure 11–11). This protein-rich meal increases the number and size of eggs produced by the female (Gwynne 1984). In cases where the male's investment in the spermatophore exceeds that of the female in eggs, males should become the choosey sex and females should compete for access to them. Gwynne (1981) found that male Mormon crickets (*Anabrus simplex*) mated more often with heavy females containing more eggs. Females, on the other hand, competed aggressively for access to singing males.

In many species both sexes care extensively for young, and the pair is monogamous; this is usually the case in birds. However, the investment in offspring is not equal at all times (Figure 11–12). It may be advantageous for one partner to desert the other and find a new mate if the remaining partner's investment is greater and the offspring are likely to survive anyway. Because of its large investment, the remaining partner gets stuck with caring for the young (Trivers 1972).

But should a parent continue to invest just because it has made a previous commitment in terms of time and energy? According to Dawkins and Carlisle (1976), such an individual would be committing the "Concorde fallacy," named in honor of the supersonic transport that was completed, even in the face of little likelihood of a profitable return, because of a large previous financial investment. Organisms, it is argued, should behave so as to increase future reproductive success regardless of prior investments in offspring. However, Trivers's concept of parental investment does consider future ability to invest. If one parent has made a large commitment,

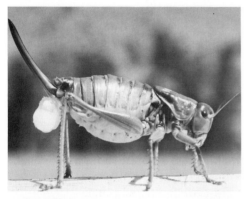

FIGURE 11-11   Mormon crickets with spermatophore

A female Mormon cricket is shown (a) mounting a male, (b) with attached spermatophore, and (c) consuming the spermatophore.

Source: "Sexual Differences Theory: Mormon Crickets Show Role Reversal in Mate Choice," Gwynne, D. T. *Science* Vol. 213, pp. 779–780, Fig. 1, 14 August 1981.

that parent may be physically unable to begin another breeding cycle; its reproductive success is thus increased most if it remains with its young.

Anisogamy may explain the propensity of females to raise their young while males compete for access to females and mate, if possible, with more than one female. Among birds, males can incubate eggs and feed young—even providing crop milk in the case of pigeons. However, in mammals, gestation and milk production are restricted to the female and there is relatively little the male can do to provide

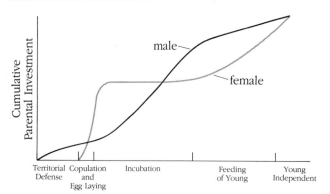

Territorial Defense — Copulation and Egg Laying — Incubation — Feeding of Young — Young Independent

FIGURE 11–12   Parental investment in offspring by male and female
In this hypothetical example we find that the male establishes and defends a territory, so his cost is initially higher; but the female invests heavily in eggs, and her cost soon exceeds that of the male. The male then incubates the eggs, and his cumulative investment exceeds hers until termination of parental care. Differences in investment may affect such behaviors as mate desertion.
Source: Adapted with permission from R. L. Trivers, "Parental Investment and Sexual Selection," in Bernard Campbell, *Sexual Selection and the Descent of Man* (New York: Aldine Publishing Company). Copyright 1972 by Aldine Publishing Company.

direct care for the young. In mammals whose young are relatively advanced (precocial) at birth, the opportunities for male investment are even lower, and males compete for multiple mates more than they do in species whose young are immature (altricial) at birth and in which the male can share more evenly the investment with the female (Zeveloff and Boyce 1980).

In polygynous species the Bateman effect sets the stage for a possible deviation from the usual 50:50 sex ratio, according to a model proposed by Trivers and Willard (1973). The model assumes that parental investment in male and female offspring is the same, and that mothers in the best physical condition produce healthier offspring that are better able to compete for mates or other resources. The reproductive success of males, which is highly variable, should be high if their mothers are in good condition, but low, perhaps zero, if their mothers are in poor shape; female offspring are likely to breed anyway, regardless of their mother's condition. Therefore, Trivers and Willard predicted that a female would produce male offspring if she were in good condition and female offspring if she were in poor condition; they cited data from mammals such as mink, deer, seals, sheep, and pigs in support of their model.

Meikle, Tilford, and Vessey (1984) found that high-ranking female rhesus monkeys (*Macaca mulatta*) in good physical condition produced significantly more male offspring than did low-ranking females. These sons did, in fact, have higher reproductive success than did sons of low-ranking females. Daughters, produced in excess by low-ranking mothers, reproduced almost as well as did daughters of high-ranking mothers. Producing daughters is a safe investment; producing sons is a gamble with the possibility of a big payoff in reproductive success. However, Williams (1979) reviewed sex ratio data for birds and mammals and concluded that the Trivers and Willard hypothesis is not supported. Additional data are needed to test this idea further.

The mechanism by which such an adjustment of sex ratio could occur is not understood. Intrauterine mortality rates are higher for male than for female embryos, and these rates are probably highest among mothers in poor condition.

PARENTAL CARE AND ECOLOGICAL FACTORS

The extent of parental care varies tremendously in the animal kingdom. The amount of care generally increases as the complexity of the organism increases. While many aquatic invertebrates simply shed eggs and sperm into the water, primate care may last for several years, which may amount to 25 percent of the offspring's life span. The kinds of parent-offspring relationships are varied and not simple to describe, however. Birds display a wide range of care techniques, from burying their eggs in rotting vegetation that provides heat for incubation and thus frees the parents, to sitting on the eggs without feeding until they are hatched.

The term **reproductive effort** is used to denote both the energy expended and the risk taken for breeding, which reduce reproductive success in the future. Finding mates and caring for young take extra energy, and parents may run greater risk of being consumed by predators. Individuals are faced with the decision, conscious or otherwise, of whether to breed now or wait until later. If the choice is to breed now, should some effort be spared for another attempt later?

What factors influence the investment parents make in their young after birth? Ecologists have attempted to relate environmental conditions to parental care. For example, species adapted to stable environments have a tendency toward larger body size, slower development, longer life span, and having young at intervals **(iteroparity)** rather than all at once **(semelparity).** Typically, individuals of these species will occupy a home range or territory (see Chapter 12). These stable conditions favor production of small numbers of young that receive extensive care and thus have a low mortality rate. In other words, the emphasis is on quality rather than quantity. Such species are said to be **"K" selected,** in reference to the fact that populations are usually at, or near, *K*, the carrying capacity of the environment.

Species that are adapted to fluctuating environments have high reproductive rates, rapid development, small body size, and little parental care. Their populations tend to be controlled by physical factors, and their mortality rate is high. Such species are said to be **"r" selected,** where *r* refers to the reproductive rate of the population. For example, Pacific salmon, by reproducing far upstream from feeding areas, must expend a great deal of energy before they can breed. Once they incur the great cost of migrating upstream, they breed explosively and die; in no other way can the benefit of reproduction become greater than the cost.

Other species, which do not have such a high initial cost before breeding, may defer reproduction or spread it out over time. In environments in which survival of offspring is low and unpredictable, parents may "hedge their bets" and put in a small reproductive effort each season. The California gull *(Larus californicus)* uses such tactics; the birds live fifteen years or more but rear only one or two chicks per year (Pugesek 1981, 1983). As the parents age, they increase their effort, laying more eggs, feeding the chicks more food, and defending them more vigorously, possibly because the parents' chances of surviving another year become smaller.

Note that the predictions of "bet-hedging" above contradict those of *r* and *K* selection. In unstable, unpredictable environments *r* selection should be important, favoring high reproductive rates. However, bet-hedging theory suggests low reproductive rates, and spreading reproductive effort across many breeding seasons. More data are needed from natural populations before we can resolve this apparent contradiction.

Prolonged dependency and extensive parental care are also favored when a species—for example, large mammalian carnivores such as the felids and canids—depends on food that is scarce and difficult to obtain. Much effort is spent searching for prey and, in some species, cooperation is needed for the kill. During the prolonged developmental period, the young benefit from considerable learning through observation of parents and through play.

The Old World monkeys and the great apes have the longest period of dependency. Typical of these species is an infancy of $1\frac{1}{2}$ to $3\frac{1}{3}$ years and a juvenile phase of 6 to 7 years, making up nearly one-third of the total life span. The reason for this prolonged dependency may be related to their flexibility of behavior that is shaped largely by learning. The complexity of monkey and ape social systems and the importance of kinship depend on a knowledge of individuals and an extensive behavioral repertoire.

PARENT-OFFSPRING CONFLICT

Anyone who has raised children or grown up with a sibling has observed frequent disagreement between parent and child. It is not unusual to see a mother rhesus monkey bat her ten-month-old infant or raise her hand over her head, thereby pulling her nipple from its mouth. Frequently the infant responds by throwing a "temper tantrum." We can view much of the conflict as a disagreement over the amount of time, attention, or energy the mother should give to the offspring (i.e., the infant wants more than the parent wants to give).

Although we often interpret such conflict in humans as being maladaptive and related to psychological problems of parent or child or to some negative cultural influence, work with nonhuman primates has led us to the idea that the conflict is part of the weaning process and is necessary for the infant to become an independent and functioning member of the social unit (Hansen 1966). It is not clear from this interpretation why the infant should protest so vigorously.

Another hypothesis, based on the coefficient of relationship (Chapter 4) and on parental investment, is that the conflict arises because natural selection operates differently on the two generations (Trivers 1974). From the mother's standpoint, she should invest a certain amount of time and energy in her offspring and then wean

FIGURE 11–13 Ratio of cost to mother and benefit to offspring as function of age of offspring When the offspring is young, the ratio is less than 1, and both the mother and offspring gain in terms of fitness from the relationship. Above 1, the fitness of the mother begins to decline, and so she should try to cease caring for this offspring and invest in new offspring. Between 1 and 2 is a zone of conflict (shaded area) because the fitness of the offspring is still increasing. Above 2, the fitness of the offspring also declines, and the offspring willingly becomes independent.
Source: Data from R. L. Trivers, "Parental Investment and Sexual Selection," in Bernard Campbell, *Sexual Selection and the Descent of Man* (New York: Aldine Publishing Company, 1972). Copyright 1972 by Aldine Publishing Company.

the young and invest in new young. When the cost to the mother in fitness exceeds the benefit, she should reject the young. From the offspring's standpoint, however, the offspring will profit from continued care until the cost to its mother is twice the benefit (since the offspring shares half of its mother's genes). When the cost is twice the benefit to the mother, then the offspring's own inclusive fitness will start to decline. At this point the offspring should leave the mother and allow her to increase the offspring's inclusive fitness by having more offspring (Figure 11–13).

Alexander (1974) challenged this idea; he argued that selection will work against behaviors of the offspring that allow them to cheat and to receive more than their share of care from the parent. Such offspring will pass those traits on to their own offspring, who will cheat them in turn, lowering the fitness of the parents. Thus, parents will win in cases of parent-offspring conflict.

PROXIMATE FACTORS IN REPRODUCTIVE BEHAVIOR

The remainder of this chapter emphasizes proximate causation and the more mechanistic aspects of reproduction.

CLIMATIC-HORMONAL INTERACTION

Most species live in seasonally fluctuating environments. The prime variables are temperature and rainfall, with temperature more variable in temperate and polar regions and rainfall more variable in tropical zones. Food supply is linked to these variables and may be the factor that actually triggers reproduction. Because of the delay between fertilization and birth, particularly long in mammals, environmental triggers must signal later conditions that will be optimal for birth.

Many species breed on an annual basis, with gonads regressed and sexual behavior absent during most of the year. In temperate and polar regions, change in photoperiod is most often the trigger for the onset of reproductive behavior. Species with a relatively short span between copulation and birth (e.g., most birds) cue on increasing daylength; species with long gestation periods (e.g., ungulate mammals) respond to decreasing daylength.

In his study on the white-crowned sparrow *(Zonotrichia leucophrys),* Farner (1964) worked out the details of the sparrow's response to changes in daylength. He manipulated daylength artificially to simulate spring conditions and found that it induced recrudescence of gonads and reproductive behavior out of season. He also found that the photoperiod response is regulated internally to some extent, since a refractory period that follows breeding must be completed before changes in daylength can induce another reproductive cycle.

Orians (1960) found that fine adjustment of the timing of reproduction can be made by weather. Once enlargement of gonads has been triggered by photoperiod, unusually warm weather may speed up the reproductive cycle, in some cases even causing breeding to occur in the fall. In the northeastern United States, we can hear song sparrows singing on warm days in December; and as the days get longer in February, bird singing increases regardless of weather (Nice 1941).

Some animals combine an annual cycle with lunar rhythmicity and breed at a particular phase of the moon, as exemplified by the palolo worm *(Leodice viridis)* of the South Pacific. Over a three-day period, six to eight days after the October-November full moon, the posterior halves of mature worms break off and swarm at the surface of coral reefs, where they release eggs and sperm (Smetzer 1969). In another example, peaks in conception rates occur among Malayan forest rats just before a full moon (Harrison 1952).

Much of the world is arid or semiarid. Many annual plants in these arid or semiarid environments respond rapidly to unpredictable rains by flowering and setting seeds that will germinate after the next rain. These seeds are a prime source of food for the fauna, and since their availability as food is unpredictable and brief, the animals must respond quickly in order to reproduce while an energy supply is plentiful. Some Australian birds remain in breeding condition with enlarged gonads for many months (Marshall 1960). The stimulus of green vegetation, or possibly the rain itself, triggers courtship and a rapid sequence of nest building, egg incubating, and feeding of young.

Hill kangaroos *(Macropus robustus)* are one of many Australian marsupials that practice a form of delayed implantation: courtship and copulation take place, but the fertilized egg remains unimplanted in the uterus (Ealey 1963). In response to rain or an abundance of food, implantation takes place, and within days the tiny young crawls into the pouch and attaches itself to a teat. Kangaroos, therefore, waste no time in getting hormonally primed, finding a mate, and copulating when food is abundant. During the weeks of development in the pouch, the mother can respond to unpredictable events, such as a resumption of dry conditions with little food, by simply throwing the young (joey) out of the pouch (Low 1978). Although such behavior might seem cruel and wasteful to us, it is not in the mother's interest to invest energy in young that probably wouldn't survive anyway; rather, natural selection has favored females who invest in young that have a better chance of survival.

SOCIAL-HORMONAL INTERACTION

Once animals have been brought into a general state of sexual readiness through climatic-hormonal interactions, shorter-term events coordinate the behavior of both sexes through social-hormonal interactions. One of the most detailed studies of the nature of these interactions is the work of Lehrman (1958a, 1958b) on ring doves *(Streptopelia risoria)*, which was also discussed in Chapter 7. In one series of experiments, Lehrman placed sexually experienced birds in a cage, presented them with a nest and eggs (Figure 11–14), and then looked for incubation behavior. Birds that he tested singly failed to incubate at all. The bird's habituation to the cage without contact with a mate did not increase its readiness to incubate; the bird's association with a mate for seven days prior to presentation of nest and eggs led it to incubate promptly. Lehrman concluded that association with a mate, involving the sounds and postures of courtship, and, to a lesser extent, the presence of nesting material brought about physiological changes leading to incubation. When he gave birds daily injections of progesterone, beginning seven days prior to presenting them with a mate, nest, and eggs, they incubated immediately. This behavior suggests that pairing and courtship cause production of progesterone, although other hormones, such as estrogen and prolactin, are also involved.

OVULATION, COPULATION, FERTILIZATION. In fish, amphibians, and birds, the female ovulates in response to both social and nonsocial stimuli and engages in sexual activity for a limited part of the reproductive season. In mammals the period of sexual receptivity at the time of ovulation is called **estrus,** and the interval from one estrus to the next is called the **estrous cycle.** Some mammals, like the fox *(Vulpes fulva),* are **monoestrous** — that is, the female is receptive for a few days once each year; others cycle regularly throughout much of the year. Variation also occurs in the stimulus necessary to cause ovulation. **Induced ovulators,** including rabbits, cats, and mustelids such as mink *(Mustela* spp.), require copulation for the release of the egg(s) from the ovary. Induced ovulation is characteristic of felids, with an extended estrous period of six to seven days and relatively brief courtship. With the exception of the lion *(Panthera leo),* felids are solitary with no lasting bond between the sexes (Kleiman and Eisenberg 1973). **Spontaneous ovulators,** such as canids and primates, release eggs whether or not the animals have copulated. Among canids courtship begins several weeks before the onset of estrus, and bonds between male

FIGURE 11-14   Experiments on pairs of ring doves with environmental and hormonal pretreatments

In all of these experiments, a nest with eggs was introduced on Day 0. If a pair of ring doves is put in a cage with a nest and eggs, they begin incubating 5 to 7 days later (a). Putting the pair into the cage with a partition between male and female at Day −7 has no effect, and it still takes 5 to 7 days for incubation to begin after introduction of nest and eggs and removal of the partition (b). If the pair is introduced with nesting material at Day −7, incubation begins immediately upon introduction of a nest with eggs (c). The absence of a nest in the pretreatment causes about a one-day delay in the onset of incubation (d). Pretreatment with progesterone causes immediate incubation when pair, nest, and eggs are introduced on Day 0 (e). *Source:* D. S. Lehrman, "Induction of Broodiness by Participation in Courtship and Nest-Building in the Ring Dove *(Streptopelia risoria)," Journal of Comparative and Physiological Psychology* 51 (1958):32–36 and from D. S. Lehrman, "Effect of Female Sex Hormones on Incubation Behavior in the Ring Dove," *Journal of Comparative and Physiological Psychology* 51 (1958):142–145. Copyright 1958 by the American Psychological Association. Adapted by permission of the publisher.

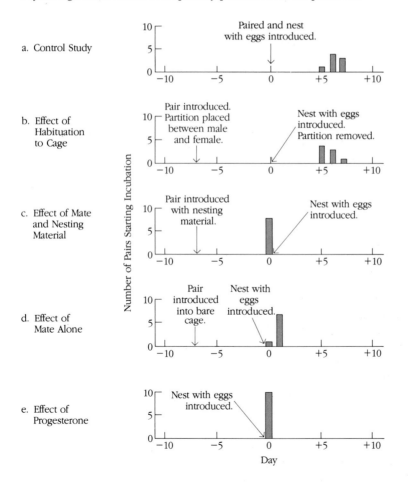

and female tend to last throughout the year (Kleiman and Eisenberg 1973). Primates undergo regular periods of **menses** during which the lining of the uterus is shed. The interval from one menses to the next is called the **menstrual cycle.** (For an outline of the hormonal and histological changes associated with estrus in mammals, see Figure 11–15).

Dewsbury (1975) has studied the mechanics of copulation in mammals, particularly rodents, in detail. Patterns within and between the individuals of a species are highly stereotyped and vary in the following four respects: (1) the presence or absence of a mechanical tie, or lock, between the penis and vagina, in which

FIGURE 11-15 Timing of events in estrous and menstrual cycles of mammals
In the closeup of part of the ovary (a), the growing follicle contains an egg that is shed into the oviduct after ovulation. The follicle then transforms itself into a corpus luteum that produces the hormone progesterone. Estrogen, produced by the ovary, and progesterone cause the thickening of the uterine lining. Part (b) shows a cross section of the uterine lining of a typical mammal through an estrous cycle and part (c) shows a cross section of the uterine lining of a primate, such as a monkey, ape, or human, through a menstrual cycle.
Source: L. B. Arey, *Developmental Anatomy* (Philadelphia: W. B. Saunders Company). Copyright 1954 by W. B. Saunders Company. Reprinted by permission of the publisher and author.

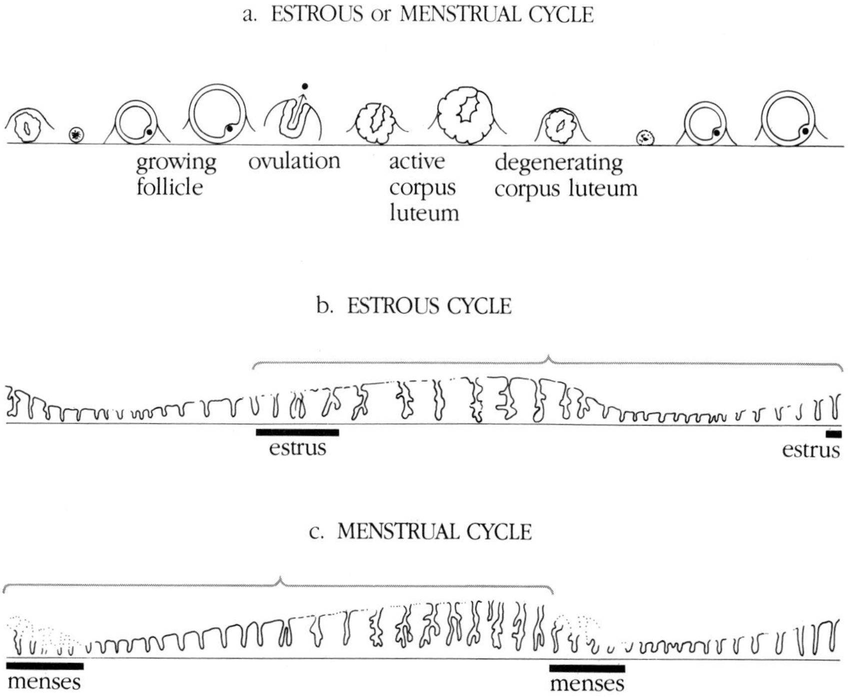

a. ESTROUS or MENSTRUAL CYCLE

growing follicle   ovulation   active corpus luteum   degenerating corpus luteum

b. ESTROUS CYCLE

estrus   estrus

c. MENSTRUAL CYCLE

menses   menses

the penis becomes engorged inside the vagina and the pair is temporarily unable to separate; (2) single or multiple thrusts during each insertion; (3) single or multiple intromissions preceding ejaculation; (4) single or multiple ejaculations in a single episode (Figure 11–16 and Table 11–2). The patterns are not restricted to large taxonomic units; for instance, carnivores such as wolves and dogs lock during copulation, but so do some rodents. We have only begun to explore the ecological and evolutionary significance of these differences. Rodents that lock, such as wood rats and grasshopper mice, are vulnerable to predation while the pairs are locked together. However, such species have elaborate nests or burrows, where predation is probably minimal. The lock may function to stimulate the female for pregnancy or it may be a product of sexual selection that reduces the possibility of another male's copulating with the female.

The stimuli that induce actual copulation vary tremendously, of course, but in many species there is a linear sequence of events by each partner in turn. If the sequence is broken at a certain point — for example, by an inappropriate response or an interruption — the female ceases to be receptive, and the pair either breaks up or starts over. Such complex reproductive patterns suggest a number of possible functions: (1) to coordinate physiological and behavioral events, such as nest building and egg formation in birds; (2) to provide an opportunity for assessment of certain features of the mate that might indicate potential reproductive success; or (3) more generally, to assure mating between conspecifics. Courtship patterns frequently serve as species-isolating mechanisms that prevent mating between closely related species.

Methods of fertilizing eggs vary among species. Frogs and many fish release

TABLE 11-2   Examples of different copulatory patterns in mammals

Common Name	Scientific Name	Lock?	Thrust?	Multiple Intromissions?	Multiple Ejaculations?	Pattern
Dog	*Canis familiaris*	yes	yes	no	yes	3
Wolf	*Canis lupus*	yes	yes	no	yes	3
Golden mouse	*Ochrotomys nuttalli*	yes	no	no	yes	7
House mouse	*Mus musculus*	no	yes	yes	yes	9
Montane vole	*Microtus montanus*	no	yes	yes	yes	9
Rhesus macaque	*Macaca mulatta*	no	yes	no	yes	11
Bonnet macaque	*Macaca rudiata*	no	yes	no	yes	11
Meadow vole	*Microtus pennsylvanicus*	no	yes	no	yes	11
Norway rat	*Rattus norvegicus*	no	no	yes	yes	13
Mongolian gerbil	*Meriones unguiculatus*	no	no	yes	yes	13
Bison	*Bison bison*	no	no	no	yes	15
Black-tailed deer	*Odocoileus hemionus*	no	no	no	no	16

Source: D. A. Dewsbury, "Patterns of Copulatory Behavior in Male Mammals," *Quarterly Review of Biology* 47 (1972):1–33. Reprinted with permission.

FIGURE 11-16 Classification scheme for male mammalian copulatory patterns
Species vary in the presence or absence of the following processes: copulatory lock, thrusting, multiple intromission, and multiple ejaculation. Varied combinations of these processes yield sixteen possible patterns of copulation.
Source: D. A. Dewsbury, "Patterns of Copulatory Behavior in Male Mammals, *Quarterly Review of Biology* 47 (1972):1–33. Reprinted with permission.

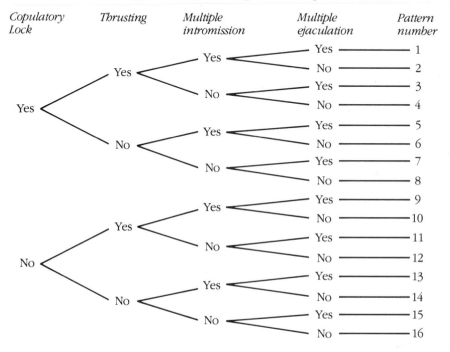

sperm over the eggs as the eggs are laid; some salamanders pick up a spermatophore that is deposited on the substrate by the male; all birds and mammals fertilize internally. In addition to ensuring that the eggs become fertilized, internal fertilization gives the female control over which male becomes the parent. Males, however, can be less certain of paternity with internal fertilization than with external fertilization. This difference may explain why parental care by males is more common in fish than in mammals.

TURKEY COURTSHIP. Hale, Schleidt, and Schein (1969) provide us with an example of a complex reproductive pattern in the courtship of the domestic turkey *(Meleagris gallopavo).* In response to the male display of strutting with his tail fanned, a receptive female crouches; as the male approaches and mounts her, she raises her head and he treads on her back, orienting himself to her head; she then raises her tail and, in response to pressure on her tail by the male, everts her oviduct. Up to this point, the sequence can be stopped and restarted at any point (see Figure 11–17). Once the hen everts her oviduct, the sequence must return to the beginning if it is interrupted, whether or not the female has come in contact with the male's everted cloaca and whether or not the male has inseminated her. (Most birds lack a penis; they transfer sperm via cloacal contact.) The female jumps up, ruffling her feathers, and may run away. The position of the hen's head during mounting and copulation is crucial to the male's orientation; if we present the male with a model whose head is at the side of its body, the male mounts sideways.

CAT SEXUAL BEHAVIOR.   Experimenters have extensively studied the interplay of neural and hormonal factors in the sexual behavior of domestic cats *(Felis domesticus)*. The timing of the peaks of estrus, which we can alter by modifying the photoperiod, are from mid-January through March and from May through June in northern latitudes. The rather violent courtship behavior for which cats are known and which is sometimes difficult to distinguish from fighting may promote follicle maturation in the female, and the small spines on the cat's penis provide stimulation to induce ovulation (Fox 1975).

By the fourth month of age, the cat secretes androgens, which stimulate development of the penile spines, but the male cat first exhibits mating behavior at eight to nine months of age. If we castrate the cat at four months, prior to puberty, it never exhibits mating behavior. If we inject the cat with testosterone, its mating behavior is restored. If we castrate the cat after puberty, the cat exhibits a variable decline in sexual behavior, depending on the amount of sexual experience the male had prior to castration; in males with extensive experience, frequency of copulation and latency to copulate decrease only slightly (Rosenblatt and Aronson 1958).

Aronson and Cooper (1966) investigated the role of sensory input in male feline sexual behavior. When they desensitized the glans penis by cutting the nerves, the cats did not decrease their sexual activity, but they were so disoriented they were unable to achieve intromission.

OCTOPUS SEXUAL BEHAVIOR.   Although much of our understanding of hormonal and neural correlates come from studies of vertebrates, experimenters have performed interesting studies with invertebrates. For example, in the octopus *(Octopus hummelincki)* the endocrine glands that control both reproduction and senescence are the optic glands, which are located in the orbital sinus. Normally the female octopus spawns once in her life; she eats little while caring for the eggs and dies shortly after the eggs hatch. When Wodinsky (1977) removed the optic glands after the octopus had laid her eggs, the female stopped brooding the eggs, ate, gained weight, and lived up to five months longer. Other semelparous invertebrates and fish exhibit such control, which demonstrates the relationship between short-term control of sex and the control of the entire life cycle. It would not be surprising to find that a single system is responsible for determining the age at which an organism reproduces, how often it does so, and when it dies. The timing of these and all other life history events is subject to natural selection, just as are morphological traits.

FIGURE 11-17 Sequential mating behavior in domestic turkeys
Arrows denote the path that the sequence follows. For instance sexual display by
the male is followed in the female by either avoidance (in an unreceptive female) or
sexual crouch. Sexual crouch is followed by approach of the male, and so on. Once
the female everts her oviduct, receptivity is terminated whether or not insemination
has taken place.
Source: Data from E. B. Hale, W. M. Schleidt, and M. W. Schein, "The Behavior of
Turkeys," in E. S. E. Hafez, ed., *The Behavior of Domestic Animals,* 2nd ed. (Baltimore, Md.: Williams & Wilkins, 1969), pp. 554–592.

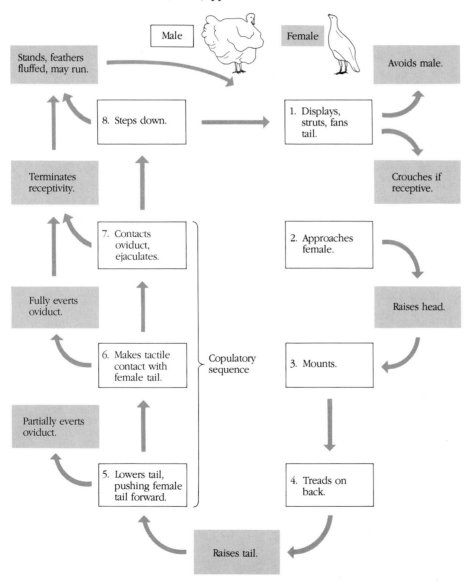

REPRODUCTION IN A PRIMATE — THE RHESUS MONKEY

In order to get a better feeling for the problems involved in research on reproduction, we look at the cycle of a well-known primate, the rhesus monkey *(Macaca mulatta).* We do not understand the reproductive events of this creature as well as we do those of the ring dove or the rat, and we will consider areas of needed research.

The rhesus monkey has been used widely as a biological model for increasing our understanding of human reproduction. Both rhesus monkeys and humans belong to the order Primates and the class Mammalia. The similarities that exist between the two species, such as the menstrual cycle, are no doubt due to common ancestry and are thus homologous. Other similarities, such as the striking resemblance in hands, may be due to convergent evolution, since both are highly adaptable "generalists" that evolved in similar tropical environments. But there are some important differences as well. Rhesus monkey females have a well-defined period of estrus, or sexual receptivity, around the time of ovulation, and they mate during a discrete five-to-six month breeding season. Human females are sexually receptive at all times of the cycle and at all times of the year; ovulation is not detectable by human males and seldom by the ovulating female herself. Rhesus monkeys are highly sexually dimorphic — the male has large canine teeth and weighs about 50 percent more than the female. As we might expect from this difference, rhesus monkey society is polygynous, with a small number of males monopolizing many females. Humans are less dimorphic and probably less polygynous (see Chapter 19).

MATING BEHAVIOR

The annual mating cycle of rhesus monkeys, which is very consistent and may be linked to photoperiod, begins in the fall in India, their native habitat. In these colonies, as in India, the monkeys live in stable social groups of from 20 to more than 100 males, females, and young. Rhesus monkeys were introduced on islands off the coast of Puerto Rico. One colony, Cayo Santiago, is off the east coast, and the other, La Parguera, is about 100 miles away from Cayo Santiago at about the same latitude. La Parguera monkeys begin mating in October, those in India begin in September, and those at Cayo Santiago begin in July. The two Puerto Rican colonies are on the same photoperiod, and temperatures differ little. The main difference, however, is in the timing of the rainy season. At Cayo Santiago rains begin in late spring, in India they start in early summer, and at La Parguera they begin in late summer (Figure 11–18). Following the onset of the rains, the vegetation becomes lush, and mating follows within a few weeks. Experimenters have not been able to determine whether the mere presence of rain and green vegetation is sufficient to trigger mating behavior or if there is some nutritive component, such as hormones produced by plants, that starts the cycle (Vandenbergh and Vessey 1968).

In the laboratory, under constant environmental conditions, rhesus monkeys breed year round; but when they are given access to outside runs, they typically do not mate during the hottest part of the summer. Females normally conceive once each year and give birth 5½ months later. Some degree of endogenous control is involved because females that lose infants from the previous year begin their estrous cycles earlier than do those with infants. Females with ligated oviducts, normal hormonally but unable to become pregnant, stop and start cycling along with the rest of the female population (Vessey and Marsden 1975).

An increase in redness of the skin around the perianal region in both sexes is evidence of the onset of mating. The redness pales in castrated adult males and the redness is restored when these males are treated with testosterone. If males are injected with estrogen, the sex color more closely resembles that of the female rhesus monkey (Vandenbergh 1965). Other hormones, such as progesterone, may be active in the sex skin color of the female rhesus monkey, and Phoenix (1977) reported that both testosterone and estrogen are essential for maximum sexual performance in females. Social factors are involved in the timing of mating; injecting castrated females with estrogen outside the regular breeding season brings intact males into breeding condition at the "wrong" time of year (Vandenbergh 1969). In one experiment Vandenbergh and Drickamer (1974) "seeded" a free-living group at la Parguera with two of these females four months before mating should have begun. As

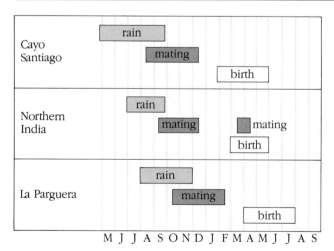

FIGURE 11-18   **Relationship between rainfall and reproduction in three populations of rhesus monkeys** Cayo Santiago and La Parguera, Puerto Rico, are only 100 miles apart and at the same latitude. The timing of the rainy season differs at Cayo Santiago, La Parguera, and in the rhesus monkey's native habitat in India. Mating begins about two months after the onset of the rainy season at each location. The small mating period that occured in spring in one localized area in India is atypical. *Source:* J. G. Vandenbergh and S. Vessey, "Seasonal Breeding of Free-Ranging Rhesus Monkeys and Related Ecological Factors," *Journal of Reproduction and Fertility* 15 (1968):71–79. Reprinted with permission.

evidenced by sexual behavior and births, the females in the seeded group bred early that year, compared with females in other, nonseeded groups.

As the mating season approaches, rhesus monkeys begin consort behavior —that is, a male and a female form a temporary pair bond and engage in much following and grooming. Males sometimes masturbate early in the breeding season; this behavior suggests that they come into breeding condition before females. Researchers have done little work to answer the important question of which sex initiates the consort behavior. Frequently the male begins to follow the female, but a female commonly approaches and "presents" her perianal region to the male as an invitation to mount. Rarely, an estrous female actually mounts the male before presenting.

Presumably, the color of the sex skin is a cue for sexual receptivity; trained observers can even document the color change within the female's cycle, and the male's sex skin reddens noticeably when he is involved in an intense consort. Michael and Bonsall (1977) have collected evidence that suggests the importance of odor as a sexual cue in rhesus monkeys. Under the influence of ovarian sex hormones, a bacterial system in the vagina produces an odoriferous substance that induces the male to mount. This substance, called copulin, is a mixture of short-chain fatty acids, such as acetic and butanoic. Males with plugged olfactory tracts will not mount estrous females; and a synthetic mixture of copulin, applied to the perianal region of a nonestrous female, causes intact males to mount. In this case they usually get slapped by the female since she does not react systemically to the locally applied copulin.

Goldfoot and his colleagues (1978) reported contradictory results. After testing three male-female pairs for several estrous cycles, they made the males permanently anosmic and found that the males' sexual behavior was not affected. Although we do not know the reasons for the contradiction in results, olfactory cues are apparently not absolutely necessary for the sexually active male.

There is some evidence that odors could influence reproductive behavior in humans. Acids similar to those present in monkeys are found in the human vagina, and human subjects could smell the differences among samples from different stages of the menstrual cycle (Michael and Bonsall 1977). In another study Doty et al. (1975) reported that human vaginal odors were scored by both men and women as least unpleasant at about the time the donors were ovulating.

The female rhesus monkey usually presents before the male mounts. Sometimes the male indicates his readiness to mount by placing a hand on her hip. When the male mounts, he places his hands on her lower back and grasps her ankles with his feet; she thus supports all his weight. Once mounted, the male usually attains intromission rapidly since the presence of a bone (baculum) in the penis reduces the time needed for erection. He thrusts rapidly for about three seconds, then dismounts. The pair may groom for a minute or two before he mounts again. These mounts tend to come closer and closer together until ejaculation, which is seen as a "freezing," occurs. Some males grimace or emit a bark vocalization during ejaculation. Usually at this point the female looks back, reaches out, and grabs one of the male's legs; this clutching reaction is thought to indicate orgasm in the female (Zumpe and Michael 1968). The existence of orgasm in nonhuman females is of interest since in most animals the female plays a rather passive role in copulation. Its existence is difficult to prove, but Goldfoot and his colleagues (1980) have reported that stump-tailed macaque *(Macaca arctoides)* females undergo uterine contractions and sudden increases in heart rate during the clutching behavior. (For a more detailed discussion, see Chevalier-Skolnikoff 1974.) (See Figure 11–19.)

After ejaculation, the male dismounts, and the pair may groom or rest. Mounting may resume in an hour or so, or the consort may break up. Rapid coagulation of the semen forms a plug in the vagina, perhaps reducing both loss of semen from the vagina and the likelihood that another male will copulate with the female.

Consorts persist only two or three days (the pair sleeps together, but little sexual activity takes place at night), and a female consorts with an average of three males during a nine-to-ten-day estrous period. Since the cycle length is about four weeks, the female will be receptive again in two to three weeks. About half the time the female does not conceive during the first cycle; most cycle about three times during the five-month breeding season. For some reason many females cycle once or twice after they have conceived; the hormonal and evolutionary causes of such behavior are puzzling.

Some females seem to improve their status in the group while they are consorting with a high-ranking male. Early experimental work with macaques and baboons emphasized the fluidity of female dominance, but we now know that the female hierarchy is much more stable than the male (Sade 1967). Consorting pairs often do team up against others, but the consort is short-lived and has no long-term effect on the female's status.

FIGURE 11–19 Copulation in rhesus monkeys
Males form temporary bonds with estrous females and follow, groom, and copulate exclusively with them for two or three days.
Source: Photo by Douglas B. Meikle.

BIRTH AND INFANCY

After a five-and-a-half-month gestation period, birth occurs at night, when the group is stationary. Although few normal births have been witnessed in natural populations of rhesus monkeys, there is no evidence that other members of the group assist in the delivery. From the time of birth, the infant clings unassisted to the belly of the mother. The mother expels the placenta within an hour and usually consumes it, as is the case in most mammals. Both nutritive and antipredator functions have been ascribed to this behavior.

The infant begins moving away from its mother within a day or two and by ten weeks of age is spending half its time away from her. Initially, the mother restrains the infant by pulling its leg as it struggles to escape; later, however, the mother initiates the separations as she begins to wean her infant (Vessey and Meikle 1984).

The infant's attachment to the mother is very strong; a ventrally clinging position, with a nipple in its mouth, is the infant's home base for the first year. A mother with a newborn youngster is a focus of attention for family members and other females, who gather around the mother, grooming her and making "girn" vocalizations. Clearly, their intent seems to be to touch and handle the infant. By far the most persistent are the two- to four-year-old nulliparous females, who may "steal" the infant during the first few weeks (Figure 11–20). A "kidnapped" infant emits a distinctive "lost" call that is recognized by its mother. Females that surgically have been made incapable of conception may refuse to surrender a kidnapped infant, but intact females usually surrender it within hours. After a few weeks the infant seems to lose some of its appeal and can usually escape under its own power (Vessey and Marsden 1975).

The amount and type of socialization that infants receive vary extensively among primates. Rhesus macaque mothers are relatively selfish with their infants; bonnet macaque (Macaca radiata) and langur (Presbytis entellus) mothers allow their infants to be passed around the first day. Such differences seem to be correlated with overall levels of aggression and dominance in the group. Simonds (1965) has noted that bonnet macaques are less aggressive and have more relaxed dominance interactions than do rhesus macaques. Rosenblum and Kaufman (1967) conducted a detailed laboratory study comparing bonnet with pigtailed (Macaca nemestrina) macaques. When the experimenters removed the mothers of pigtailed infants from small social groups, the infants showed acute separation distress. Infants of the less-restrictive bonnets tolerated the absence of their mothers with much less trauma. The relationship between early social experience and later social organization unfortunately tells us nothing about cause and effect. Does the high level of aggression in rhesus monkey groups result from the restrictive and punitive experience in early life, or does it cause it?

By the end of the first year, the infant rhesus monkey, which ranks just below its mother and siblings in the dominance hierarchy, has established a network of social relationships; it knows its own status and that of juveniles, adult females, and

FIGURE 11-20 Infant rhesus monkey handled by immature female "aunt"
Such attempts at care giving may prepare immature females to care for their first infant, but the infants seem to gain little.
Source: Photo by Dennis Ferguson.

its peers (Sade 1967). Male and female infants behave similarly in the first year, although male infants tend to engage in a bit more rough-and-tumble play than do females, and Mitchell and Brandt (1970) have provided some evidence that mothers hit their male infants and move away from them more than they do from their female offspring. One big difference is that males begin mounting other infants as early as eight weeks of age; this pattern is well developed by the end of the first year. Males deprived of social interactions during early life are incapable of properly mounting and inseminating a female later in adulthood (Harlow 1965; for contradictory evidence, see Meier 1965).

By the end of the first year, conflict between mother and infant is substantial. Although the infant can feed itself under optimal conditions by six months of age, it still nurses at the end of one year and still clings ventrally, with a nipple in its mouth, 5 to 10 percent of the time during the day and most of the time at night. The female gives birth once a year and rarely allows the infant, now a yearling, on the nipple once a new infant is born. She weans the yearling by withdrawing the nipple, pushing the yearling away, or hitting it. The yearling invariably jerks its head, makes a "geck" sound, and may throw a full-fledged temper tantrum.

BEHAVIOR DEVELOPMENT

After the first year the behaviors of the sexes become more and more different. Males spend increasing amounts of time with their male peers at the periphery of the group. One- and two-year-old males sometimes sleep together at night (Vessey 1973). But even as three- and four-year-olds, they often come back to groom their mothers and siblings. At about the age of three and a half, their testes descend, and they may visit the periphery of another social group in the breeding season. By seven years of age, they will have moved to a different social group from the one they were born into (Drickamer and Vessey 1973). Females never leave their natal group but spend increasing amounts of time grooming, handling, and carrying infants until their first young is born when they are about four years of age. As is typical of polygynous species, most males engage in little sexual behavior until well past (for rhesus monkeys, two or more years) maturity, although both sexes mature at about the same time (three and a half years).

SUMMARY

Asexual reproduction, by which an organism produces exact copies of itself in the next generation, might seem the most likely outcome of natural selection. Most organisms, however, reproduce sexually at some point in their life cycles, incurring the costs of meiosis and of finding a mate. The benefit of increased variability in an ever-changing environment apparently outweighs the costs. Many invertebrates have both sexual and asexual reproduction; some animals have functional male and female sex organs; and some fish and invertebrates change sexes during adult life.

Natural selection tends to maintain a 50:50 ratio of the two sexes. The gametes of the two sexes differ anatomically, with the female producing few, large, sessile eggs and the male producing many, small, motile sperm.

Sexual selection affects anatomy and behavior at the time of mating. Intersexual selection involves choices made between males and females, with the females usually choosing the males. In intrasexual selection, competition takes place within one sex (usually the male), with the winner gaining access to the opposite sex. Sexual selection may lead to the evolution of elaborate secondary sexual characteristics, particularly in males.

Ecological factors, such as food distribution and predator pressure, affect group size and thus the type of mating system. Monogamy, in which one male and one female form a pair bond for one or more breeding seasons, occurs when neither sex is able to monopolize more than one member of the opposite sex. In resource defense polygyny, males defend areas containing feeding or nesting sites critical for reproduction and thus gain access to more than one female. If females are gregarious, as in female defense polygyny, males may form harems. In male dominance polygyny, males may concentrate in an area and display to attract females for mating. Occasionally females monopolize males, as in resource defense polyandry or female access polyandry, but such cases of polyandry are restricted to a few species of birds.

Parental investment is any behavior toward offspring that increases the chances of the offspring's survival at the cost of the parents' ability to rear other offspring. Because the female's initial investment in eggs is greater than that of the male in sperm, the female is typically particular about the male she mates with; the male tries to mate with as many females as possible.

The more complex the organism is, the greater is the amount of parental care given to the young. As species adapt to stable environments, their body size tends to become larger, their life span increases, and they tend to have young at intervals (iteroparity) rather than all at once (semelparity). These conditions favor production of small numbers of young, which receive extensive care and thus are subject to a reduced mortality rate. An organism's dependence on food that is scarce and difficult to obtain also favors prolonged dependence of the young and extensive parental care. When survival of offspring is low and unpredictable, parents may bet-hedge, producing few offspring each year.

Parent-offspring conflict has traditionally been viewed as maladaptive or as a necessary part of the weaning process. More recent ideas point out that natural selection operates differently on the two generations, so that it is in the offspring's interest to receive more care than the parent is willing to give. Parents will win such conflicts and manipulate offspring to maximize their own reproductive success.

Many species breed on an annual basis, with recrudescence of gonads triggered by changes in photoperiod. Weather also affects the reproductive cycles of some species; for example, in dry environments rainfall may be the cue for the onset of the reproductive cycle in certain species. A few species are affected by lunar cycles. Around the time of fertilization, many sexually reproducing animals engage in behavior that insures that sperm come into contact with eggs (either internally or externally). The details of courtship and the mechanics of copulation are species-specific and function to coordinate the physiological and behavioral events leading to production of offspring, to assess the potential fitness of the mate, and to assure mating between conspecifics.

As shown in detail in the rhesus monkey, the physical and social environments interact with the neural-hormonal state at all levels to regulate reproduction.

Discussion Questions

1. Discuss the possible relevance to humans of Trivers's concept of parent-offspring conflict.

2. What sorts of ecological and phylogenetic circumstances might favor polyandry in animals?

3. The data in the table below relate to ring doves. Female ring doves were placed in a cage. Sounds from the breeding colony were played into some of the cages and not into others. Males were placed on the other side of a glass partition in the cages of the females so that the males and females could view each other but not make contact. Some of these males were castrated, others were intact. The numbers given are the median ranks of ovarian development in the females; the higher the number, the greater the development. All differences between groups were statistically significant except the 40.5 versus 44.5 rankings. Discuss these results in terms of the social-hormonal factors affecting reproduction and relate them to the data in Figure 11–14.

Median ranks of ovarian development

	With Castrated Male	With Intact Male	Combined Groups
With Colony Sound	40.5	55.5	46
Without Colony Sound	15.5	44.5	28
Combined Groups	24.0	49.5	

Source: D. Lott, S. D. Scholz, and D. S. Lehrman, "Exteroceptive Stimulation of the Reproductive System of the Female Ring Dove *(Streptopelia risoria)* by the Mate and by the Colony Milieu," *Animal Behaviour* 15 (1967):433–437. Reprinted by permission.

Suggested Readings

Daly, M., and M. Wilson. 1983. *Sex, evolution, and behavior,* 2nd ed. Boston: Willard Grant.
Basic text that covers the whole spectrum of sex, with emphasis on ecological and evolutionary aspects. Several chapters on evolution of human sexuality.

Maynard Smith, J. 1984. The ecology of sex. In J. R. Krebs and N. B. Davies (eds.), *Behavioural ecology: An evolutionary approach,* 2nd ed. Sunderland, Mass: Sinauer Associates.
Readable review of current theories on how sexual reproduction is maintained in populations.

Naftolin, F., et al. 1981. Sexual dimorphism. *Science* 211:1263–1324.
Series of articles reviewing the bases of sex differences. Deals with proximate mechanisms.

Thornhill, R., and J. Alcock. 1983. *The evolution of insect mating systems.* Cambridge: Harvard University Press.
Reviews insect reproductive behavior from an evolutionary perspective, with particular emphasis on sexual selection theory. Demonstrates the advantages of working with insects with their great diversity, short generation times, and high observability.

References

Alexander, R. D. 1961. Aggressiveness, territoriality, and sexual behavior in field crickets (Orthoptera: Gryllidae). *Behaviour* 17:130–223.

———. 1974. The evolution of social behavior. *Ann. Rev. Ecol. Syst.* 5:325–383.

Aronson, L. R., and M. L. Cooper. 1966. Seasonal variation in mating behavior in cats after desensitization of the glans penis. *Science* 152:226–230.

Austad, S. N. 1982. First male sperm priority in the bowl and doily spider, *Frontinella pyramitela* (Walckenaer). *Evolution* 36:777–785.

Bateman, A. J. 1948. Intra-sexual selection in *Drosophila*. *Heredity* 2:349–368.

Berven, K. A. 1981. Mate choice in the wood frog, *Rana sylvatica*. *Evolution* 35:707–722.

Borgia, G. 1985. Bower quality, number of decorations and mating success of male satin bowerbirds *(Ptilonorhynchus violaceus)*: An experimental analysis. *Anim. Behav.* 33:266–271.

Bruce, H. M. 1966. Smell as an exteroceptive factor. *J. Anim. Sci.,* Suppl. 25:83–89.

Bull, J. J. 1980. Sex determination in reptiles. *Quart. Rev. Biol.* 55:3–21.

Chevalier-Skolnikoff, S. 1974. Male-female, female-female, and male-male sexual behavior in the stumptail monkey, with special attention to the female orgasm. *Arch. Sex. Behav.*. 3:95–116.

Clutton-Brock, T. H., F. E. Guiness, and S. D. Albon. 1982. *Red deer: Behavior and ecology of two sexes.* Chicago: University of Chicago Press.

Coulson, J. C. 1966. The influence of the pair-bond and age on the breeding biology of the kittiwake gull, *Rissa tridactyla. J. Anim. Ecol.* 35:269–279.

Cox, C. R., and B. J. Le Boeuf. 1977. Female incitation of male competition: A mechanism in sexual selection. *Amer. Nat.* 111:317–335.

Daly, M. 1978. The cost of mating. *Amer. Nat.* 112:771–774.

Darwin, C. 1871. *The Descent of Man, and Selection in Relation to Sex.* New York: D. Appleton.

Dawkins, R., and T. R. Carlisle. 1976. Parental investment, mate desertion and a fallacy. *Nature* 262:131–133.

Dewsbury, D. A. 1975. Diversity and adaptation in rodent copulatory behavior. *Science* 190:947–954.

———. 1982. Ejaculate cost and male choice. *Amer. Nat.* 119:601–610.

Doty, R. L., et al. 1975. Changes in intensity and pleasantness of human vaginal odors during the menstrual cycle. *Science* 190:1316–1318.

Drickamer, L. C., and S. H. Vessey. 1973. Group changing in male free-ranging rhesus monkeys. *Primates* 14:359–368.

Ealey, E. H. M. 1963. The ecological significance of delayed implantation in a population of the hill kangaroo *(Macropus robustus).* In A. C. Enders (ed.), *Delayed Implantation.* Chicago: University of Chicago Press.

Emlen, S. T., and L. W. Oring. 1977. Ecology, sexual selection, and the evolution of mating systems. *Science* 197:215–223.

Farner, D. S. 1964. The photoperiodic control of reproductive cycles in birds. *Amer. Scientist* 52:137–156.

Fisher, R. A. 1958. *Genetical theory of natural selection.* New York: Dover Publications.

Fox, M. W. 1975. The behavior of cats. In E. S. E. Hafez (ed.), *Behavior of Domestic Animals.* 2nd ed. Baltimore: Williams & Wilkins.

Gilliard, E. T. 1963. The evolution of bower birds. *Sci. Amer.* 209:38–46.

———. 1969. *Birds of Paradise and Bower Birds.* Garden City, N.Y.: Natural History Press.

Goldfoot, D. A., S. M. Essock-Vitale, C. S. Asa, J. E. Thornton, and A. I. Leshner. 1978. Anosmia in male rhesus monkeys does not alter copulatory activity with cycling females. *Science* 199:1095–1096.

Goldfoot, D. A., H. Loon, W. Groeneveld, and A. K. Slob. 1980. Behavioral and physiological evidence of sexual climax in the female stumptailed macaque *(Macaca arctoides) Science* 208:1477–1479.

Greenwood, P. J. 1980. Mating systems, philopatry, and dispersal in birds and mammals. *Anim. Beh.* 28:1140–1162.

Gromko, M. H., D. G. Gilbert, and R. C. Richmond. 1984. Sperm transfer and use in the multiple mating system of *Drosophila*. In R. L. Smith (ed.), *Sperm competition and the evolution of animal mating systems.* New York: Academic Press.

Gwynne, D. T. 1981. Sexual difference theory: Mormon crickets show role reversal in mate choice. *Science* 213:779–780.

———. 1984. Courtship feeding increases female reproductive success in bushcrickets. *Nature* 307:361–363.

Hale, E. B., W. M. Schleidt, and M. W. Schein. 1969. The behavior of turkeys. In E. S. E. Hafez (ed.), *Behavior of Domestic Animals,* 2nd ed. Baltimore: Williams & Wilkins.

Hamilton, W. D. 1967. Extraordinary sex ratios. *Science* 156:477–488.

Hansen, E. W. 1966. The development of maternal and infant behavior in the rhesus monkey. *Behaviour* 27:107–149.

Harlow, H. F. 1965. Sexual behavior in the rhesus monkey. In F. Beach (ed.), *Sex and Behavior.* New York: Wiley.

Harrison, J. L. 1952. Moonlight and pregnancy of Malayan forest rats. *Nature* 170:73–74.

Hrdy, S. B. 1977. *Langurs of Abu : Female and Male Strategies of Reproduction.* Cambridge: Harvard University Press.

Jallon, J.-M., C. Antony, and O. Benamar. 1981. Un anti-aphrodisiaque produit par les mâles de *Drosophila melanogaster* et transfere aux femelles los de la copulation. *C. R. Acad. Sci. Paris,* 292:1147–1149.

Jenni, D. A. 1974. Evolution of polyandry in birds. *Amer. Zool.* 14:129–144.

Kleiman, D. G., and J. F. Eisenberg. 1973. Comparisons of canid and felid social systems from an evolutionary perspective. *Anim. Behav.* 21:637–659.

Kodric-Brown, A., and J. H. Brown. 1984. Truth in advertising: The kinds of traits favored by sexual selection. *Am. Nat.* 124:309–323.

Lack, D. 1968. *Ecological Adaptations for Breeding in Birds.* London: Methuen.

Le Boeuf, B. J. 1974. Male-male competition and reproductive success in elephant seals. *Amer. Zool.* 14:163–176.

Lehrman, D. S. 1958a. Induction of broodiness by participation in courtship and nest-building in the ring dove *(Streptopelia risoria). J. Comp. Physiol. Psychol.* 51:32–36.

———. 1958b. Effect of female sex hormones on incubation behavior in the ring dove *(Streptopelia risoria). J. Comp. Physiol. Psychol.* 51:142–145.

Lenington, S. 1980. Female choice and polygyny in redwing blackbirds. *Anim. Behav.* 28:347–361.

Low, B. S. 1978. Environmental uncertainty and the parental strategies of marsupials and placentals. *Am. Nat.* 112:197–213.

Marshall, A. J. 1960. Annual periodicity in the migration and reproduction of birds. *Cold Spr. Harb. Symp. Quant. Biol.* 25:499–505.

Meier, G. W. 1965. Other data on the effects of social isolation during rearing upon adult reproductive behavior in the rhesus monkey *(Macaca mulatta). Anim. Behav.* 13:228–231.

Meikle, D. B., B. L. Tilford, and S. H. Vessey. 1984. Dominance rank, secondary sex ratio and reproduction of offspring in polygynous primates. *Amer. Nat.* 124:173–188.

Michael, R. P., and R. W. Bonsall. 1977. Chemical signals and primate behavior. In D. Muller-Schwartze and M. M. Mozell (eds.), *Chemical Signals in Vertebrates.* New York: Plenum.

Mitchell, G., and E. M. Brandt. 1970. Behavioral differences related to experience of mother and sex of infant in the rhesus monkey. *Dev. Psychol.* 3:149.

Muller, H. J. 1964. The relation of recombination to mutational advance. *Mutation Res.* 1:2–9.

Nice, M. M. 1941. The role of territory in bird life. *Amer. Mid. Nat.* 26:441–487.

Orians, G. H. 1960. Autumnal breeding in the tricolored blackbird. *Auk* 77:379–398.

———. 1961. The ecology of blackbird *(Agelaius)* social systems. *Ecol. Monogr.* 31:285–312.

———. 1969. On the evolution of mating systems in birds and mammals. *Amer. Nat.* 103:589–603.

Oring, L. W., and D. B. Lank. 1982. Sexual selection, arrival times, philopatry and site fidelity in the polyandrous spotted sandpiper. *Behav. Ecol. Sociobiol.* 10:185–191.

Parker, G. A. 1970a. The reproductive behaviour and the nature of sexual selection in *Scatophaga stercoraria* L. (Diptera: Scatophagidae) IV. Epigamic recognition and competition between males for the possession of females. *Behaviour* 37:113–139.

Parker, G. A. 1970b. Sperm competition and its evolutionary consequences in the insects. *Biol. Rev.,* 45:525–568.

Parker, G. A., R. R. Baker, and V. G. F. Smith. 1972. The origin and evolution of gamete dimorphism and the male-female phenomenon. *J. Theoret. Biol.* 36:529–553.

Phoenix, C. H. 1977. Induction of sexual behavior in ovariectomized rhesus females with 19-hydroxytestosterone. *Horm. Behav.* 8:356–362.

Pugesek, B. 1981. Increased reproductive effort with age in the California gull *(Larus californicus). Science* 212:822–823.

———. 1983. The relationship between parental age and reproductive effort in the California gull *(Larus californicus). Behav. Ecol. Sociobiol.* 13:161–171.

Rosenblatt, J. S., and L. R. Aronson. 1958. The

decline in sexual behavior of male cats after castration with special reference to the role of prior sexual experience. *Behaviour* 12:285–338.

Rosenblum, L. A., and I. C. Kaufman. 1967. Laboratory observations of early mother-infant relations in pigtail and bonnet macaques. In S. A. Altmann (ed.), *Social Communication among Primates.* Chicago: University of Chicago Press.

Ross, R. M., G. S. Losey, and M. Diamond. 1983. Sex change in a coral reef fish: dependence of stimulation and inhibition on relative size. *Science* 221:574–575.

Sade, D. S. 1967. Determinants of dominance in a group of free-ranging rhesus monkeys. In S. A. Altmann (ed.), *Social Communication among Primates.* Chicago: University of Chicago Press.

Shapiro, D. Y. 1979. Social behavior, group structure, and the control of sex reversal in hermaphroditic fish. In J. S. Rosenblatt, R. A. Hinde, and C. Beer (eds.), *Advances in the Study of Behavior,* vol. 10. New York: Academic Press.

Shields, W. M. 1982. Philopatry, inbreeding, and the evolution of sex. Albany: State University of New York Press.

Silberglied, R. E., J. G. Shepherd, and J. L. Dickinson. 1984. Eunuchs: The role of apyrene sperm in Lepidoptera? *Am. Nat.* 123:255–265.

Simonds, P. E. 1965. The bonnet macaque in South India. In I. DeVore (ed.), *Primate Behavior: Field Studies of Monkeys and Apes.* New York: Holt, Rinehart and Winston.

Smetzer, B. 1969. Night of the palolo. *Nat. Hist.* 78:64–71.

Trivers, R. L. 1972. Parental investment and sexual selection. In B. Campbell (ed.), *Sexual Selection and the Descent of Man 1871–1971.* Chicago: Aldine.

———. 1974. Parent-offspring conflict. *Amer. Zool.* 14:249–264.

Trivers, R. L., and D. E. Willard. 1973. Natural selection of parental ability to vary the sex ratio of offspring. *Science* 179:90–92.

Van Valen, L. 1973. A new evolutionary law. *Evol. Theory,* 1:1–30.

Vandenbergh, J. G. 1965. Hormonal basis of sex skin in male rhesus monkeys. *Gen. Comp. Endocr.* 5:31–34.

———. 1969. Endocrine coordination in monkeys: Male sexual responses to the female. *Physiol. Behav.* 4:261–264.

Vandenbergh, J. G., and L. C. Drickamer. 1974. Reproductive coordination among free-ranging rhesus monkeys. *Physiol. Behav.* 13:373–376.

Vandenbergh, J. G., and S. Vessey. 1968. Seasonal breeding of free-ranging rhesus monkeys and related ecological factors. *J. Reprod. Fert.* 15:71–79.

Vessey, S. H. 1973. Night observations of free-ranging rhesus monkeys. *Amer. J. Phys. Anthrop.* 38:613–620.

Vessey, S. H., and H. M. Marsden. 1975. Oviduct ligation in rhesus monkeys causes maladaptive epimeletic (care-giving) behavior. In S. Kondo, M. Kawai, and A. Ehara (eds.), *Contemporary Primatology.* Basel: S. Karger.

Vessey, S. H., and D. B. Meikle, 1984. Free-living rhesus monkeys: Adult male interactions with infants and juveniles. In D. M. Taub (ed.), *Primate Paternalism.* New York: Van Nostrand Reinhold.

Waage, J. K. 1979. Dual function of the damselfly penis: Sperm removal and transfer. *Science* 203:916–918.

Ward, P. 1971. The migration patterns of *Quelea quelea* in Africa. *Ibis* 113:275–297.

Williams, G. C. 1975. *Sex and Evolution.* Princeton: Princeton University Press.

———. 1979. The question of adaptive sex ratio in outcrossed vertebrates. *Proc. Roy. Soc. Lond.* B 205:567–580.

Wittenberger, J. F. 1981. *Animal Social Behavior.* Boston: Duxbury.

———. 1979. The evolution of mating systems in birds and mammals. In P. Marler and J. G. Vandenbergh (eds.), *Handbook of Behavioral Neurobiology,* vol. 3, *Social Behavior and Communication.* New York: Plenum.

Wodinsky, J. 1977. Hormonal inhibition of feeding and death in *Octopus:* Control by optic gland secretion. *Science* 198:948–951.

Zahavi, A. 1975. Mate selection—A selection for a handicap. *J. Theor. Biol.* 53:205–214.

Zeveloff, S. I., and M. S. Boyce. 1980. Parental investment and mating systems in mammals. *Evolution* 34:973–982.

Zumpe, D., and R. P. Michael. 1968. The clutching reaction and orgasm in the female rhesus monkey (*Macaca mulatta*). *J. Endocr.* 40:117–123.

12

AGGRESSION

Konrad Lorenz (1966) states in his introduction that his book *On Aggression* is about "the fighting instinct in beast and man which is directed *against* members of the same species." His use of the word **instinct** points up one facet of the old nature-nurture controversy that has direct bearing on the behavior of humans. Is aggression a universal property of social animals, including humans, and is it an inborn trait whose expression is inevitable? How can we explain the widespread tendency of animals in a fight to use restraint and not to fight to the death even when the most aggressive individuals control resources such as food or mates?

In this chapter we define aggression, list some of its forms, and then consider two ways in which aggression is expressed: the social use of space and the dominance hierarchy. We look briefly at the internal causes of aggressive behavior—genetic, neural, and hormonal—and at external causes—the role of environment. In the concluding section we discuss control of aggression and implications for human behavior.

AGONISTIC BEHAVIOR

Aggression is a complex phenomenon with many functions and many causes, and it may include predatory behavior, in which the animal being attacked is eaten in the process. It can be defined as behavior that appears to be intended to inflict noxious stimulation or destruction upon another organism (Moyer 1976). The notion of intent is necessary to exclude such destructive behaviors as a person's accidentally stepping on an ant. Use of the word *aggression* emphasizes offensive behavior.

A term with a more precise definition is **agonistic** behavior, which is a system of behavior patterns that has the common function of adjustment to situations of conflict among conspecifics. The term includes all aspects of conflict, such as threats, submissions, chases, and physical combat, but it specifically excludes predatory aggression since, as Scott (1972) has argued, ingestive behavior is part of a separate behavioral system.

We can list the forms of aggressive behavior as follows (modified from Wilson 1975 and Moyer 1976):

- *Territorial:* exclusion of others from some physical space;
- *Dominance:* control, as a result of a previous encounter, of the behavior of a conspecific;
- *Sexual:* use of threats and physical punishment, usually by males, to obtain and retain mates;
- *Parental:* attacks on intruders when young are present;
- *Parent-offspring:* disciplinary action by parent against offspring (mostly in mammals, usually associated with weaning;
- *Predatory:* act of predation, possibly including cannibalism;
- *Antipredatory:* defensive attack by prey on predator, such as mobbing.

Most of these behaviors involve conflict among conspecifics and so would also be included under agonistic behavior. These forms of aggression serve very different functions within and between species, and they may even have evolved independent regulatory centers in the brain. Therefore, the term aggression is not a unitary concept. Most agonistic behavior involves competition for some resource — namely, food, water, access to a member of the opposite sex, or space, such as sites for nesting, wintering, or safety from predators. The resource at some point limits the survival and reproduction of individuals in the population. Most of our discussion concerns agonistic behavior, or aggressive behavior among conspecifics; we will consider competition between different species in Chapter 16.

TYPES OF COMPETITION

Competition sometimes is divided into two types: **scramble** and **contest.** If we throw a small piece of food between a couple of squirrel monkeys, they will probably both go after it, with the first one to reach the food taking it and running off (scramble). The second one may run after the first, grabbing for the piece of food, but no fighting is likely to result. If we do the same with two rhesus monkeys, the result will be different. One might stare at the other, bobbing its head up and down with its mouth open and making a "huh" sound; the other would probably make a grimace and face away. The first would then calmly walk up and take the food (contest). If the two were strangers, the confrontation could precipitate an overt fight. Most social species engage in contest competition, in which displays are used in place of combat and the limited resource is not fought over on each occasion.

Although aggression results from competition for resources, not all competition involves aggression. Organisms that use the same limited resource may never see or hear each other; individuals feeding on a plant by day compete but do not directly interact with those feeding on that plant by night. Ecologists use the term **interference competition** for situations in which organisms seeking a resource harm one another in the process, even if the resource is not in short supply.

EXTREME FORMS OF AGGRESSION: CANNIBALISM

The intensity of aggression varies widely, from actual physical contact and killing to a subtle threat, such as a direct stare or a raising of the eyebrows to expose the eyelids. So many cases of violent aggression and killing have been reported that some have questioned the idea that killing is always accidental, maladaptive, or otherwise anomalous. For example, the Indian langur monkey (*Presbytis entellus*) lives in a variety of habitats. In some areas social groups of langurs contain only one adult male, five to ten females, and their young. Several observers have reported instances in which a new male came into the group, chased out the old male, and killed some or all of the infants (Figure 12–1). One interpretation of such behavior is that the group is socially disorganized by the change in males, and the usual restraints on overt aggression are absent. Once the new male has established himself, he ceases his attack on the young. From the standpoint of the species or group, such behavior is maladaptive and should be rare, which, in fact, it is. A second interpretation of these events is that infanticide is adaptive, at least for the new male, since he removes the offspring of the presumably unrelated male and causes the females to come into estrus sooner to bear his own offspring (Hrdy 1977a, 1977b). Hrdy's view is consistent with the notion that social behavior results from the action of natural selection on the individual.

We can use the same arguments to explain acts of cannibalism performed by a variety of animals. Observers note that, particularly under crowded conditions, rodents often eat their young, usually those already dead. We could interpret such behavior as social pathology, which results from crowded conditions and leads to maladaptive behavior. We should note, however, that under crowded conditions or when food is scarce, the young would probably not survive anyway, so the parents save their reproductive investment for more propitious times by consuming their young. Rodents and lagomorphs frequently resorb embryos into their own bodies during the development of the embryos in the uterus, with the same effect as cannibalism. In kangaroos most of the development of the young takes place in the

FIGURE 12-1 Two female langur monkeys attacking male
The male langur monkey has stolen the infant of a third female (not visible). Males
frequently wound or kill infants sired by other males.
Source: Photo by Sarah Blaffer-Hrdy from Anthro-Photo.

pouch, and females can terminate "pregnancy" by simply throwing the young out of
the pouch.

Such killing of young is not limited to adults. Among some species of birds
older, and larger, young sometimes kill their younger, and smaller, siblings (Mock
1984). Such behavior might be expected only when food is so scarce that the survival
of the entire brood is threatened. Cannibalism, whether infanticide or siblicide,
seems to fit our definition of aggression; but such acts are relatively infrequent
compared to the ritualized combat associated with contest competition. (For recent
papers on this topic, see the book edited by Hausfater and Hrdy 1984.)

SOCIAL USE OF SPACE

The defense of fixed space against members of the same species accounts for much
of the agonistic behavior in a host of animals, ranging from limpets to long-billed
marsh wrens. This is not to say that all animals defend territories, as might be
assumed from reading such books as Robert Ardrey's *The Territorial Imperative*
(1966); in fact, the majority of species probably do not. Animals use space in a
number of different ways.

DEFINITIONS OF THE USE OF SPACE

The area used habitually by an animal or group, and in which the animal spends most of its time, is the **home range** (Figure 12–2). Most organisms spend their lives in a relatively restricted part of the available habitat and learn the locations of food, water, and shelter in this area. We may have difficulty determining the actual boundaries of the home range since an animal or group may occasionally wander some distance away to a place it will never revisit. To permit comparisons with other studies, we must specify what criteria are being used. For example, the white-footed mouse *(Peromyscus leucopus)* spends most of its time at the nest near the center of its

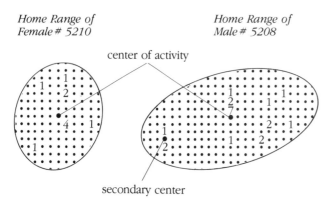

Home Range of
Female # 5210

center of activity

Home Range of
Male # 5208

secondary center

FIGURE 12–2   Capture locations of white-footed mice

Figures denote number of captures at live-trap stations spaced 7 meters apart, and dots represent traps where the mouse was not caught. The center of activity indicates the average of the coordinates. Most captures are close to the center of activity; but some mice, such as mouse #5208, seem to use secondary nest sites. Males usually have larger and less-exclusive home ranges than females. Ellipses include 95 percent of expected captures.

FIGURE 12–3   White-footed mouse with numbered ear tag in nest box
Nest boxes (left) can supplement live trapping to provide data on behavior and population dynamics. Note individually numbered metal tag on mouse's left ear (right). *Source:* Photos by Stephen H. Vessey.

home range. The likelihood of finding it away from that spot decreases according to a normal distribution. Thus we can arbitrarily propose that its home range be defined as that area in which the mouse can be found (captured, radio-tracked, and so on) 95 percent (or some other percentage) of the time. We can compute the home range in this case by calculating the area within which we expect 95 percent of the captures, assuming a normal distribution about the center of activity (95 percent confidence ellipse) (Figure 12–3).

The area of heaviest use within the home range is the **core area.** This location may contain a nest, sleeping trees, water source, or a feeding tree. As with home range, the designation of a core area is somewhat arbitrary but useful in understanding the behavior and ecology of different species or the same species in different habitats or at different population densities. Figure 12–4 illustrates home ranges and core areas of baboons in Africa.

The minimum distance that an animal normally keeps between itself and other members of the same species is its **individual distance.** In birds it is frequently the distance that one individual can reach to peck another; we see this spacing in

---

FIGURE 12–4 Home ranges and core areas of nine groups of baboons in Nairobi Park, Kenya
Although home ranges overlap extensively among groups of baboons *(Papio anubis),* core areas overlap little. Shading indicates core areas.
Source: From *Primate Behavior: Field Studies of Monkeys and Apes* by Irven DeVore. Copyright © 1965 by Holt, Rinehart and Winston. Reprinted with permission of CBS College Publishing.

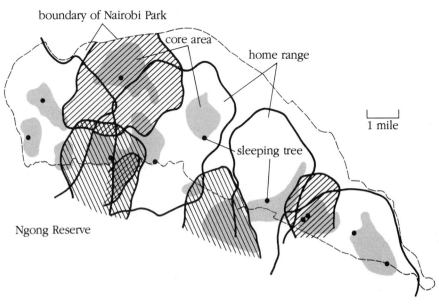

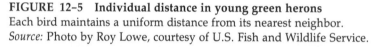

FIGURE 12–5   Individual distance in young green herons
Each bird maintains a uniform distance from its nearest neighbor.
Source: Photo by Roy Lowe, courtesy of U.S. Fish and Wildlife Service.

starlings that are lined up on telephone wires, in sea gull nests in a colony, or in green heron *(Butorides virescens)* fledglings on a tree branch (Figure 12–5). Individual distance, or **personal space,** exists for humans as well. For example, when talking to another person, most of us become uncomfortable and back off if the other person comes too close. The amount of personal space required varies from culture to culture and from situation to situation (Hall 1966).

An area occupied more or less exclusively by an animal or group and defended by overt aggression or advertisement is a **territory.** Although he did not discover the concept, Howard (1920) presented the first modern study of the phenomenon. Working mainly with aquatic birds, Howard described the defense of an area by a mated pair, pointed out the birds' need to defend a resource in order to breed, and noted the role of territory in population regulation. We treat this last point in Chapter 17 on population regulation.

To demonstrate territory we must show that an individual, mated pair, or group has exclusive use of some space, and we must also observe defense of that area. We can easily make these observations of conspicuous, diurnal species such as sea gulls (Figure 12–6), but how do we directly observe defense by a small, nocturnal, secretive, cryptically colored creature such as a white-footed mouse? Not surprisingly, even though millions of these mice have been snap-trapped, live-trapped, and radio-tagged, we still know very little about their social organization.

SIZE AND BOUNDARIES OF TERRITORY

The size of the territory or home range depends on the size and diet of the animal as well as on the particular resource the animal is defending. One type of territory or home range, most common in small mammals and insect-eating birds, is relatively large and includes the food supply, courtship, and nesting areas. Its size is an approximate function of the weight of the animal and of its metabolic rate (McNab 1963). For mammals *Area* = 6.76 $W^{0.63}$, where *Area* equals the expected home range or territory in acres, and *W* equals the body weight in kg. Thus a 20-gram mouse should have an area of 0.57 acres; this figure is not far off the mark. The productivity of the habitat is important, and white-footed mice range farther in less-productive habitats. If the area is defended against conspecifics, we would expect a smaller value. This relationship between home range or territory size and body weight suggests that ultimately what most vertebrates defend is the food source.

Territorial animals spend much time patrolling the boundaries of their space, singing, visiting scent posts, and making other displays. Such contest competition would seem to take more time and energy than would scramble competition. In fact, however, these displays have evolved so as to require relatively little energy and, once the territory has been established, the neighbors have been conditioned and need only occasional reminders to keep out. The cost of defense, then, could be less than the benefit of having exclusive use of a resource.

Ecologists are beginning to understand the factors that promote territoriality, including such things as the distribution of the limited resource in space and whether or not the availability of the resource fluctuates seasonally. A limited resource — say, food — that is uniformly distributed in time and space is most efficiently utilized if members of the population spread themselves out through the habitat, possibly

FIGURE 12-6   California gulls defending territory
The bird in the left foreground attacks an intruder. Both parents defend the nest site in this monogamous species. *Source:* Photo by Bruce Pugesek.

defending areas. A resource that is clumped in space and that is unpredictable might favor colonial living or possibly nomadism.

TERRITORIAL MODEL.   The relationship between the social use of space and the environment is clearly demonstrated by Orians's (1961) work with red-winged and tricolored blackbirds, which are sympatric in northern California. The tricolor, responding to a food source that is concentrated and unpredictable, is colonial, nomadic, and monogamous, and the males defend small territories; the redwing is polygynous, and the males defend large areas and space out regularly, responding to a food source that is less concentrated and more predictable.

Carpenter and MacMillen (1976), in their studies of nectar-feeding birds, in particular a Hawaiian honeycreeper *(Vestiaria coccinea),* have demonstrated the possibility of constructing a model to predict territorial behavior. They argue that in order for territorial behavior to occur, E, the basic cost of living, plus T, the added cost of defending a territory, must be less than the yield to the individual if not territorial (fraction a of productivity P) plus the extra yield gained by the reduced competition if territorial (fraction b of productivity P). In other words, $E + T < aP + bP$. Further development of the model leads to the prediction that at very high levels of food productivity, the birds can get enough food without excluding other birds, so territoriality disappears. Below a certain level of food availability, territorial behavior also should disappear since the resource is no longer worth defending. The birds then should switch to other foods or leave the area. Carpenter and MacMillen were able to estimate the parameters needed to test their model, and the honeycreepers behaved as predicted.

Not all territories include all the resources that the animal needs. Male bullfrogs *(Rana catesbeiana)* defend sites that are most suitable for development of the larvae (Howard 1978) (Figure 12–7). Nomadic species may exhibit some territoriality; for example, the tricolor blackbird defends only a small area around the nest. Some species have overlapping home ranges but defend a core area.

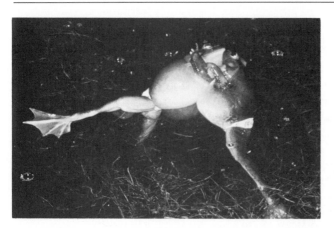

FIGURE 12–7   Male bullfrogs fighting
Males establish territories through contests, with larger males usually winning and controlling the better sites. Females preferentially mate with these males and lay eggs in their territories. Thus outcomes of male aggression are access to more females and higher reproductive success.
Source: Photo by Richard D. Howard.

FIGURE 12-8 Sage grouse lek in Montana
Males occupy small territories in one of the mating centers of the lek as females aggregate there. Males with more peripheral territories, away from a mating center, seldom have the opportunity to mate.
Source: Photo by R. Haven Wiley.

LEK. A mating system involving a peculiar type of territory is the **lek** (see Chapter 11). In this case the only resource that the organism defends is the space where mating takes place. Feeding and nesting occur away from the site. Lek, or arena, systems are characterized by promiscuous, communal mating; the males are likely to have evolved elaborate ornaments of plumages, as exemplified by the sage grouse (*Centrocerus urophasianus*) studied by Wiley (1973) (Figure 12–8). The same area, typically, is used year after year. Males arrive early in the breeding season and, through highly ritualized agonistic behavior, stake out their plots. Certain territories appear to be more attractive than neighboring ones, in the sense that males that control them do more breeding than others. Females move through the areas while the males display, mate with one or more males, then leave. While on the lek the males do little or no feeding; they spend all their time and energy patrolling the boundaries, displaying to other males, and attempting to attract females into their area. Although lek mating systems are rare among animals, they do occur in a wide range of species, from Hawaiian fruit flies (family Drosophilidae) to African hammerheaded bats (*Hypsignathus monstrosus*).

DOMINANCE

When two adult male laboratory mice of an aggressive inbred strain, such as CF1 or BALB/C, are socially isolated for a few weeks and then put together, they begin to fight. One or both may assume a hunched posture and advance with mincing steps; they may also vibrate their tails, producing a rattling sound, and one may roughly groom the other on the lower back. They may then start directing bites at each other's face, shoulders, and lower back, wrestling vigorously (see Figure 3–6). Before long one begins to get the best of things, and the other may try to escape. Once cornered, the newly submissive mouse will rear up, squeak, and box with its fore-paws. If the two are separated for a day and then reintroduced, the previous day's victor will establish his superiority more quickly and the submissive mouse will retire sooner. By including other mice, we can demonstrate that mice individually recognize each other; they remember which mice beat them up previously and which ones they were able to dominate. (Some of this information is communicated via urinary odors.) The establishment of dominance hierarchies requires individual recognition and learning based on previous encounters among the individuals.

DOMINANCE HIERARCHIES

We say that an animal is **dominant** if it controls the behavior of another (Scott 1966). In another sense, a prediction is being made about the outcome of future competitive interactions (Rowell 1974). If four or five mice that are strangers to each other are put together, several outcomes are possible. Most likely a despot will take over, and all the subordinates will be more or less equal. Another possibility is a linear hierarchy, where A dominates B, B dominates C, and so on (i.e., $A \rightarrow B \rightarrow C \rightarrow D$). Sometimes triangular relationships form, where

In other species coalitions may affect dominance, such that

but

$$C \; + \; B$$
$$\downarrow$$
$$A$$

The idea of dominance was first promulgated by Schjelderup-Ebbe (1922), who worked out the peck order in domestic chickens. Dominance hierarchies occur most often in arthropods and vertebrates that live in permanent or semipermanent social groups. There is much variation in the intensity of dominance and in the frequency of reversals. **Peck right** hierarchies are formed when all the aggression goes from dominant to subordinate (Table 12–1). In **peck dominance** hierarchies only a majority of agonistic acts go from dominant to subordinate. Among closely related species, one may see a clear-cut hierarchy in one species and not in the other. For example, African green monkeys *(Cercopithecus sabaeus)*, commonly kept in zoos, show little or no dominance hierarchy, even when access to some highly prized food is limited; the vervet *(Cercopithecus aethiops)*, a close relative, has a pronounced linear hierarchy. Species that are territorial in the wild may, when crowded in captivity, change over to a dominance hierarchy. Sometimes the dominance rank order is a function of the resource for which the animals are competing; for example, one individual may have first access to water, another to a favored breeding site.

Rowell (1974) and others have argued that the whole concept of dominance should be reassessed because it is largely a product of stress induced by captivity. Nevertheless, the phenomenon of social control is widespread, particularly among primates, including humans. We saw in Chapter 4 that differences in competitive ability provide the means by which natural selection acts. Dominance hierarchies seem to be a common result when socially living organisms engage in contest competition.

TABLE 12-1 Method for ranking individuals based on outcomes of dyadic (two-way) interactions in hypothetical group of five individuals identified as *A, B, C, D, E*
(a) Matrix of wins and losses in which individuals are arbitrarily arranged alphabetically. For instance, *B* was the winner and *A* was the loser in 8 contests; *A* never defeated *B*. (b) The order has been rearranged to maximize the values in the upper right half of the matrix. Thus the correct order is *C, D, B, E, A.* Some reversals did occur, as seen by numbers in the lower left half of the matrix. Thus *E* is dominant to *A* because *E* defeated *A* 4 times and *A* defeated *E* only once. This type of rearrangement works only if the hierarchy is linear.

a.

		Loser				
		A	B	C	D	E
	A		0	0	0	1
	B	8		0	1	4
Winner	C	7	7		6	9
	D	12	8	0		5
	E	4	0	0	1	

b.

		Loser				
		C	D	B	E	A
	C		6	7	9	7
	D	0		8	5	12
Winner	B	0	1		4	8
	E	0	1	0		4
	A	0	0	0	1	

DOMINANCE HIERARCHY IN THE RHESUS MONKEY

To understand the nuances of social control, let us examine in some detail the dominance relationships in a relatively aggressive primate with a well-developed linear hierarchy. By the time infant rhesus monkeys *(Macaca mulatta)* are three to four months old, they have developed most of the threat and submission postures and vocalizations of the adult (Figure 12–9). Although fights are not common among infants, by the end of the first year the infants in a group have established a linear peck order (Vessey and Meikle 1984).

DOMINANCE IN FEMALES.   Position in the hierarchy is determined by the mother's rank (Sade 1967; Marsden 1968). Adult females are linearly ranked, with sons and daughters generally ranked just below their mothers. The infants get support from mothers and siblings; offspring of lower-ranking mothers soon learn to submit. Sometimes a one-year-old infant chases a two- or three-year-old juvenile, but only if the latter comes from a lower-ranking family. Hierarchies among families are very stable; those observed at La Parguera rhesus colony in Puerto Rico did not change in the eighteen years since the groups formed. Because female rhesus monkeys never change groups, extensive genealogies develop over the years. As the heads of gene-alogies die off, the highest-ranking daughters take over. At about puberty (three years), each female becomes dominant to all her older sisters for reasons that are not clear; perhaps the greater support received from her mother gives the young female an advantage in agonistic encounters. If the mother should die, the new dominant female could be as young as three years old.

   Dominance in these families manifests itself subtly but constantly at a food or water source or in access to resting spots, with the dominants supplanting subordi-nates. Fights break out sporadically in the group for no apparent reason, and family members are quick to support each other. In some groups the highest-ranking female and some of her daughters collaborate with high-ranking males to break up fights among other families and to oppose threats from outside the group.

DOMINANCE IN MALES.   Dominance among males is more complex because males leave the natal group shortly after puberty (three and a half years). While in the natal group, they rank below their mothers. In a new group, of course, family membership is meaningless, and the males must form new alliances. As males move into new groups, dominance relationships are quickly formed. Surprisingly, size and previous fighting experiences seem not to influence rank directly; as long as they stay in the new group, their rank is positively correlated with seniority: those in the group the longest, rank the highest (Drickamer and Vessey 1973). Establishing ties with the central, high-ranking females of the group may be of importance in this respect. Males generally move into a group that contains an older brother (half-sib), and brothers may collaborate to stay in the group and attain high rank, thereby increas-ing their inclusive fitness (Meikle and Vessey 1981).

   Although there is no proof that attaining high rank affects mating success in rhesus monkeys, studies suggest that such a relationship exists (Smith 1981) and,

FIGURE 12–9   Threat and submission in rhesus monkeys
The open-mouthed threat (top) by an adult female rhesus monkey is directed at a monkey out of view to the right. The submissive grimace (bottom) by the young male on the left is in response to the alert posture of the dominant male on the right.
Source: Photos (top) by Douglas B. Meikle (bottom) by Stephen H. Vessey.

therefore, that there should be some selective advantage to attaining high rank. Furthermore, in times of food shortage, when access to food is restricted, high-ranking individuals have a distinct advantage. All group members, even the lowest ranking, probably benefit from a dominance hierarchy because there is less energy spent in scrambling for resources. Also, as we shall see later, the advantages of group living, even as a subordinate, more than compensate for the stress endured.

Alpha males.   Early monkey watchers were much impressed with dominant (alpha) males and wrote about their confident walk, upright tail carriage, and sexual exploits and about how they led the group from place to place and how they determined the rank of their group in the intergroup dominance hierarchy. The usual method of collecting field data was *ad libitum* note-taking: the observer simply wrote down behaviors that caught his or her attention. This technique led to biased conclusions because the most conspicuous individuals were the large, high-ranking males and the most conspicuous behavior was aggression. More recent field studies use bias-free techniques such as focal animal observations, in which the observer singles out and observes certain individuals exclusively for a fixed time interval (Altmann 1974).

Much of the mystique of the dominant male evaporated when observers noted that these males are usually the last to get involved in a skirmish, situate themselves safely in the center of the group, and let the lower-ranking, peripheral males do most of the fighting. The alpha male has little to do with the group's rank, does not actually lead the group, and may not be the most sexually active. The ascription to alpha males of the control role of breaking up fights by charging one or both the combatants and protecting the group against serious extragroup threats does seem to have substance (Bernstein 1966) (Table 12–2).

ADVANTAGES OF DOMINANCE

The advantages of being dominant might require little comment if firm data were available for many species. Studies of group-living birds and mammals indicate that the dominant animals are well fed and healthy. Subordinates may be malnourished or diseased and thus suffer higher mortality. Christian and Davis (1964) have reviewed data for mammals showing that low-ranking individuals, frequent losers in fights, have higher levels of adrenal cortical hormones than do dominants. These hormones elevate blood sugar and prepare the animal for "fight or flight." The cost is a reduction in antigen-antibody and inflammatory responses — the body's defense mechanisms — and a reduction in levels of reproductive hormones.

TABLE 12–2   Change in spatial relations and agonistic behavior of group's second-ranking male after removal of alpha male

Days after Removal of Alpha Male	Hours Observed	Percentage of Time Seen at			Fights Broken Up/hr	Attacks on Females/hr
		Edge	Inside	Midst		
−30 to 0	4.2	90	10	0	0.5	0.2
0 to 8	2.2	12	38	50	0.3	2.7
8 to 34	6.0	13	9	78	1.3	1.0
34 to 90	5.8	5	10	85	0.9	0.3

Source: S. H. Vessey, "Free-Ranging Rhesus Monkeys: Behavioural Effects of Removal, Separation, and Reintroduction of Group Members," *Behaviour* 40 (1971):216–227. Reprinted with permission of the publisher and the author.

Higher rank is often directly correlated with reproductive success, as Le Boeuf (1974) noted in his study on elephant seals (see Table 11–1). But in some cases the highest-ranking male may be so busy displaying, that lower-ranking males copulate with the females. In many species the male dominance hierarchy is age graded, with younger, lower-ranking males working their way up the hierarchy as older males die off or leave the group. Thus, low rank does not necessarily mean low reproductive success. We need studies that document the lifetime reproductive success of members of natural populations.

Many popular treatments have emphasized the importance of dominance and aggression in "keeping the species fit," since the survivors of fierce battles are the most vigorous and therefore "improve" the species by passing those traits on. The simplest and most likely explanation of aggression and dominance is that individuals benefit. As with other traits, there is an optimum level of aggression, depending on the individual's particular social and physical environment. Individuals that are too aggressive are selected against, as are those that are too passive.

INTERNAL FACTORS IN AGGRESSION

LIMBIC SYSTEM

Most of what we understand about central nervous system control of agonistic behavior comes from studies on mammals, particularly the cat (Flynn 1967). The brain structures involved are part of the limbic system (see Figure 6–17). The hypothalamus is involved in defense and escape behavior in animals as diverse as pigeons, cats, monkeys, and opossums. Using lesions, electrical stimulation, and single neuron recordings from specific areas of the brain, researchers have found that different brain sites are responsible for different types of aggression. For example, electrical stimulation, in cats, of the ventromedial nucleus of the hypothalamus produces growling, hissing, and attacking with claws (defensive attack). Stimulation of the lateral hypothalamic area produces a biting attack with no defensive elements. Thus thinking of agonistic behavior as involving a single neural system may be an oversimplification. Other areas of the brain that are involved in aggression are the amygdala of the forebrain and the central gray in the midbrain. These areas are connected by nerve pathways, and they interact; for example, electrical stimulation of certain areas in the thalamus causes cats to attack rats (Bandler and Flynn 1974). Using special staining techniques, axons were traced to areas in the central gray. Stimulation of those latter sites elicited similar attacks.

We can study the effects of electrical stimulation of specific areas of the brain in seminatural social groups by the use of radio transmitters (Delgado 1967). Monkeys with electrodes implanted in certain parts of the thalamus, hypothalamus, or

FIGURE 12–10   Radio-stimulated attack in rhesus monkey
Social aggression in rhesus monkeys can be elicited by telestimulation of hypothalamic structures.
Source: J. G. Herndon, A. A. Perachio, and M. McCoy, "Orthogonal Relationship between Electrically Elicited Social Aggression and Self-Stimulation from the Same Brain Sites," *Brain Research* 171 (1979):374–380. Reproduced courtesy of Yerkes Regional Primate Research Center.

central gray became aggressive when the electrodes were activated (Figure 12–10). When stimulation was applied to the hypothalamus, in some cases lower-ranking monkeys became dominant as a result of this manipulation (Robinson, Alexander, and Bowne 1969).

HORMONES

Interacting with the limbic system are neurosecretions and hormones. Epinephrine (adrenaline) and norepinephrine (noradrenaline) are related to physiological arousal. These and other substances, such as dopamine and serotonin, may act as neurotransmitters and may affect aggressiveness, but much further work needs to be done.

The effect of castration on sexual and aggressive behavior in males has been known since the time of Aristotle. Early work on chickens demonstrated that testosterone made them more aggressive and increased their dominance rank. Some of the hormone changes are rapid; male rhesus monkeys that have been defeated in fights show declines in circulating testosterone within hours (Rose, Bernstein, and Gordon 1975). In many species of mammals, exposure to sex steroids early in ontogeny is a primary factor in the appearance of aggressive behavior in adulthood. Female rhesus monkeys that received testosterone before birth and that were tested as juveniles threatened more and played more roughly than did untreated females (Phoenix 1974). In species of mammals where multiple embryos are present in the uterus, hormones produced by one sex may affect the other (vom Saal 1983). When tested as adults, males that were between males in utero were found to be more aggressive than males that were between females. Similarly, females that were between males as embryos fought more than females that were between other females.

Reproductive hormones other than testosterone also affect aggression. Luteinizing hormone (LH), rather than testosterone, increases aggression in some species of birds (Davis 1963). Castrated zebra finches *(Poephila guttata)* receiving estrogens become just as aggressive as those receiving testosterone. Once injected, hormones undergo chemical changes as they are broken down by the body. It appears that the metabolites of these hormones are the substances actually causing the increase in aggression (Harding 1983).

A study by Reinisch (1981) has shown that pregnant women who were treated with synthetic progesterone produced offspring who, when they were tested as adolescents, were found to be more aggressive than a similar group of adolescents not treated prenatally. See chapters 6 and 7 for a discussion of neural and hormonal mechanisms.

GENETICS

The synthesis of nerve structures, neurosecretions, and other compounds is under genetic control, and agonistic behavior is shaped by natural selection, as is any other behavior. Artificial selection can lead to significant changes in levels of aggression within just a few generations. For example, Ebert and Hyde (1976) tested wild female house mice for aggressiveness and, by selecting those with high scores and those with low scores, produced two lines, one with highly aggressive females and one with passive females. The unselected control lines were intermediate, as expected. Domestic laboratory strains of mice differ widely in aggressiveness (Southwick and Clark 1968), as do dog breeds. Siamese fighting fish *(Betta splendens)*, fighting cocks, and even crickets have been artificially selected over the years for performance in contests, with large sums of money riding on the outcome.

A genetic link to aggression in humans has been postulated to explain the fact that males with two Y chromosomes (XYY males) are found more frequently than expected in maximum security mental institutions, given their frequency in the population at large. However, it turns out that XYY males also tend toward below-average intelligence, and thus merely may be more likely to get caught! Also, there is no indication that they are more violent than are normal males, nor do they have elevated levels of testosterone (Mazur 1983).

EXTERNAL FACTORS IN AGGRESSION

LEARNING AND EXPERIENCE

Although genes control production of the hormones, neurosecretions, and neurons that produce aggressive behavior, researchers generally accept the fact that some factor outside the animal triggers the response. Previous experience can produce semipermanent changes in the expression of agonistic behavior, and animals can be conditioned to win or lose. If we give a naive male laboratory mouse that has been socially isolated for a few weeks a few easy victories over less aggressive mice, he will turn into a fighter mouse and will quickly attack any other male in sight. Likewise, we can "create" a loser by repeatedly exposing a naive male laboratory mouse to highly aggressive mice (Scott and Fredericson 1951).

Interest in primates has centered on the importance of early experience on later levels of aggression. For instance, we know that monkey societies differ widely in the frequency and intensity of aggression. Bonnet macaques *(Macaca radiata)* have a rather loose dominance hierarchy, with little fighting and much friendly interaction. Pigtail macaques *(Macaca nemestrina)* have a more rigid hierarchy, with more fighting and a low tolerance of strangers. Treatment of infants also differs among the two species. Pigtail mothers are very restrictive and do not let others handle or carry the infant as much as do bonnet mothers. The latter tolerate "aunting" by young females. Bonnet infants are cared for by others if the mother is removed; in contrast, pigtail infants become very depressed and huddle in the corner (Rosenblum and Kauffman 1968). Whether this difference in early experience is responsible for the different social structures that characterize the adults of the two species remains to be demonstrated.

Scott (1975, 1976) has further emphasized the role of experience and learning in the development of aggression. Although he acknowledges the genetic and physiological bases of aggression, Scott in no way believes that the expression of hostility is inevitable. He emphasizes the importance of early experience and learning in the development of aggression and suggests that one of the most important controls of agonistic behavior is passive inhibition. Animals form the habit of not fighting, particularly during critical periods of development. Some evidence in support of this view comes from studies of rats and mice (reviewed by Scott 1966) and monkeys (Rosenblum and Kauffman 1968). Although it is clear that animals can be trained in the laboratory to be passive and nonaggressive, researchers have not clarified the role of early experience in the restraint of aggression in natural populations.

PAIN AND FRUSTRATION

More direct causes of aggression are pain and frustration. Researchers have shown that noxious stimuli, such as loud noises, foot shocks, tail pinches, and intense heat, cause different lab species to attack a wide variety of objects, from conspecific cage mates to tennis balls (Ulrich, Hutchinson, and Azrin 1965). Although referred to as attack behavior, the response often seems to be defensive in nature; for instance, shocked rats assume an upright "boxing" posture that rats losing a fight normally exhibit. These defensive attacks occur in a laboratory situation in which the victim is prevented from escaping (i.e., the victim is cornered). The full range of attack behaviors seen in the wild is not elicited by painful stimuli.

Researchers have explored the frustration-aggression connection in humans.

In some early experiments researchers allowed children to view a room full of toys but denied them access to it; when they finally let them into the room, the children frequently smashed toys and fought more than did a control group that was allowed immediate access to the room (reviewed in Berkowitz 1969 and Johnson 1972). Frustration is now considered to be only one of several causes of aggression. Restricted access to resources triggers aggression in most animals. Rhesus monkeys on a reduced diet fought less and became lethargic, but when limited food was placed in a small number of feeders, fighting became intense (Southwick 1967).

SOCIAL FACTORS

Another strong stimulus for aggression is the presence of strangers. In most species, introducing a strange animal into an established social group produces the violent reaction of **xenophobia,** usually among members of the same sex as the stranger (Southwick et al. 1974). The amount of fighting tends to decrease with time, but acceptance of strangers is usually a slow process. A related phenomenon is **isolation-induced aggression,** in which socially isolated animals become hyperactive and prone to attack other animals (Scott 1966).

Crowding per se tends to increase interaction rates but does not always increase aggression. The overall group size, presence of strangers, or restricted access to resources has a much more powerful effect on aggression than does reducing the space available to an already established social unit.

The relationship between sex hormones, particularly testosterone, and aggression is striking in seasonally breeding species. As the gonads increase in size in response to environmental changes in photoperiod, rainfall, vegetation, and so forth, fighting and wounding increase as well. Most of this increase is related to competition for breeding territories, social rank, or access to females. In some monogamous species (e.g., seagulls) fighting occurs during the early phases of pair formation, before a sexual bond develops. In others, courtship and copulation themselves seem to have aggressive components, as evidenced by the appalling racket mating cats make, which makes the listener wonder whether it is a fight to the death or courtship. Other carnivores also have rather violent mating behavior. When polecats mate, the male grabs the female by the scruff of the neck and drags her back to his nest. Some of this aggression at the time of mating probably stems from the fact that males tend to court females rather indiscriminately, often making advances when the females are not sexually receptive. Some mammals, mainly carnivores and lagomorphs, are induced ovulators, and vigorous courtship plus copulation may be needed to trigger the release of an egg into the oviduct as copulation occurs.

Some evidence has shown that the initiation of feeding in groups of fish increases intragroup agonistic behavior. Albrecht (1966) proposed that predatory (feeding) and agonistic behaviors, although functionally distinct, are motivationally linked such that motivational summation occurs. However, Poulsen and Chiszar (1975) found that receipt of aggression inhibited feeding in submissive bluegill sunfish *(Lepomis macrochirus)*. Feeding was unaffected by levels of aggression in dominant bluegills. They conclude that motivational summation does not occur, and that feeding and aggression are independently motivated.

RESTRAINT OF AGGRESSION

DISPLAYS

Displays, discussed more fully in Chapter 13, are behaviors that have evolved to communicate information. Many displays communicate agonistic behavior, such as a threat or a submission. For instance, Figure 12–11 shows a threatening and a submissive dog; the postures are in many ways opposites of one another (Darwin's

FIGURE 12–11   Threatening and submissive postures in the dog
Note the features that are opposites, such as ear and tail positions, shape of spine, and general posture.
Source: C. Darwin, *Expression of the Emotions in Man and Animals* (London: John Murray, 1872).

theory of **antithesis**). Displays that communicate an animal's intentions and mood may inhibit attack by another. Lorenz (1966), Scott (1966), and others have argued that such behaviors are necessary to avoid large-scale killing and wounding, which would be to the detriment of the social group or species. Species with effective weapons like large canine teeth or horns are most likely to have evolved such displays since the potential for inflicting severe damage to opponents is so great.

GAMES THEORY MODEL

The species preservation argument implies that natural selection operates at the group or species level — how else could individuals be selected that yield to dominants without a struggle or that fail to go all out for the kill when given the opportunity? In contrast, Tinbergen (1951) pointed out that ritualized contests would reduce the risk of injury to the aggressor and that threats, which take less energy than physical combat, are of advantage to individuals as well as to groups.

Maynard Smith and Price (1973) independently developed more detailed models of how such displays could be selected for at the level of the individual. One such model is the Hawks-Doves game (Parker 1984). In a hypothetical population all individuals, whom we will call "doves," use only ritualized, (i.e., conventional) signals; they display, but always retreat at the first sign of serious conflict and thereby avoid injury. If a mutant fighter, a "hawk," appears and fights vigorously enough to win each contest, the hawk will be highly successful. If hawk behavior is heritable, hawks will become more common and doves less so in subsequent generations. But as the hawks increase in number, the likelihood that contests will involve two hawks rather than a hawk and a dove also increases. The average benefit of winning (controlling the resource) may be outweighed by the cost of losing (being injured or killed); the doves will then do better. The result is that either group does well when it is rare but is outdone by the other when it is common (Table 12–3).

The situation is similar to the one on sex ratio selection (Chapter 11) (Maynard Smith 1976). In both situations the best strategy is a function of what all the others in the population are doing. The relative benefits of winning and costs of losing bring about a stable mixture of hawks and doves, which we refer to as an evolutionary stable strategy (ESS). In any population we can expect either a mixture of those that are always highly aggressive and those that always display without combat *or* individuals that vary their behavior and follow a mixed strategy of being either aggressive or noncombative a certain proportion of the time. One conclusion we may draw from this model is that noncombative individuals that control their aggression by displaying and retreating rather than escalating conflicts are following an adaptive strategy; they are not simply the losers or "less fit" victims of the aggressive dominants.

Whatever its evolutionary origins, the ritualization of agonistic displays into

TABLE 12-3   Payoff matrix for hawks-doves game

- H = hawk: tries to damage opponent, retreats only if injured
- D = dove: tries to settle amicably, retreats if opponent escalates
- B = benefit of winning (control of some resource)
- C = cost of injury

Against:

		H	D
Payoff to:	H	$\dfrac{B-C}{2}$	B
	D	0	$\dfrac{B}{2}$

Hawks share equal chances of victory against other hawks and win against doves. Doves lose against hawks but are not harmed, while against other doves they have equal chances of victory.

If $B > C$, pure hawk would be an evolutionary stable strategy (ESS), while it would never pay to be a dove. However, if $C > B$, the ESS is a mixed strategy such that p, the probability of playing a hawk, $= B/C$, and we expect both hawks and doves in the population. As the costs of injury increase relative to the benefits of winning the contest, the fewer hawks we would expect to see.

contests rather than struggles has occurred in practically all species with social behavior. As we have seen, killing and wounding do occur; however, given the weapons available and the competition for resources that probably occurs at some point in the lives of all organisms, serious injury is much rarer than is potentially possible.

RELEVANCE FOR HUMANS

Control of aggression in non-human animals may have special relevance for us. Lorenz (1966) has argued that humans are aggressive for the same reasons as are other animals. However, humans have only recently acquired the means of inflicting serious injury through the use of weapons; we have not evolved the ritualized displays typical of species with such natural weapons as canine teeth or horns. Thus aggression in humans escalates, with violent consequences.

Humans also become violent, according to Lorenz, because we have few harmless outlets for aggression — a particular problem in modern society, where so little of our time and energy is expended in subsistence activities. Lorenz, in his hydraulic model of behavior, demonstrates his theory of how action-specific energy (in this case for aggression) builds up until it is in some way released. Although other research discredits the idea that specific "energies" are stored in the brain, the discovery that different chemicals are involved in transmitting information to specific parts of the brain could give credence to his theory. If aggression is inevitable, Lorenz argues, then we must find harmless ways to vent it. This cathartic approach suggested to him the importance of ritualized tournaments — for example, sports events — in which participants and spectators alike can work out their hostilities. But what about the riots triggered by the sports events themselves?

The idea that we are more aggressive if we are deprived of aggression is not supported by experiments on laboratory animals. For instance, rats trained to kill mice showed no tendency to increase their rate of killing after being deprived of mice for a few days (Van Hemel and Myer 1970). A number of studies of humans have tried to evaluate the effect of the observation of aggression, mainly in movies and on television, on subsequent aggressive behavior. Possible changes after viewing violence, if any, would be either a decrease in aggressiveness through catharsis or an increase in aggressiveness through some type of learning, through a generalized arousal, or through a reduction in inhibitions. The results of television and movie viewing are not clear-cut — some studies have shown increases, and some have shown decreases (Feshbach and Singer 1971) — but a 1982 report by the National Institute of Mental Health concluded that violence on television *does* lead to aggressive behavior by children and teenagers who watch the programs. Not surprisingly, television networks have challenged the findings (Walsh 1983). Controlled laboratory experiments have indicated that various visual stimuli (for example, comic book violence) do increase the subjects' aggressive behavior (Berkowitz 1969).

Conditioning to situational stimuli may be important in humans, as Berkowitz and LePage (1967) have shown. The researchers instructed a subject to play the role of an experimenter trying to teach others a task. When the "subjects" made a mistake, the "experimenter" (i.e., the real subject) was supposed to administer a punishment in the form of an electric shock. The researchers could then measure the aggressiveness of the real subject by the number of shocks he or she delivered. In this study the variable was the presence or absence of a gun on a table near the "experimenter." The real subject administered more shocks when the weapon was present than when it was absent. Presumably the gun facilitated aggression because of its association with violence.

SOCIAL CONTROL AND SOCIAL DISORGANIZATION

Scott (1975) has developed the idea of the social control of aggression. Observation of agonistic behavior in cichlid fish (*Cichlasoma biosallatum*) (aptly named the Jack

Dempsey fish, after the heavyweight boxer) has shown that when strange fish of the same species are continually introduced into the group, fighting remains at a high level; if group membership is allowed to stabilize, fighting declines to a relatively low level. One cause of aggression, then, seems to be social disorganization, as seen in newly formed groups or those that have been disturbed. When a high-ranking male or female rhesus monkey dies or leaves the group, aggression within the group may increase. New males moving into the central hierarchy attack females and are sometimes mobbed by them in return. The absence of a high-ranking control animal contributes to the increase of aggression. As new relationships are established within the group and as control animals emerge and begin to break up fights, group aggression decreases (Vessey 1971). The following tactics can be used by any social animal to avoid or deescalate aggression (modified from Marler 1976):

- Keep away,
- Evoke behavior that requires proximity, is physically incompatible with aggression, and reduces arousal (e.g., grooming),
- Avoid provoking extreme arousal of frustration,
- Use submissive displays (antithesis of aggressive displays),
- Behave predictably,
- Divert attack elsewhere.

Maintenance of a territory or a position in a dominance hierarchy requires constant reinforcement by songs, threats, or other displays; individual recognition and memory of the results of previous encounters are always involved. Although male macaques and baboons use their enormous canines in fights against other males, once they have formed a stable hierarchy, they can maintain rank without them, as happened in a captive colony of Japanese monkeys in Oregon after experimenters removed the males' canines (Alexander and Hughes 1971).

Aggression in human societies, particularly noninstitutionalized crimes of violence (i.e., excluding war), can be traced, in part, to social disorganization and xenophobia. Sociologists use the term **community** to refer to a social system with stable membership whose members interact extensively among themselves. When group membership is unstable and when relationships and roles are uncertain, alienation and aggressive interactions are frequent (Scott 1976). The high rates of crimes of violence in inner cities may be spurred, in part, by the unstable family structure.

SUMMARY

Agonistic behavior, defined as social fighting among conspecifics, includes all aspects of conflict, such as threat, submission, chasing, and physical combat, but excludes predation. Aggression emphasizes overt acts intended to inflict damage on another, and may include predation, defensive attacks on predators by prey, and attacks on inanimate objects. All social arthropods and vertebrates engage in some form of agonistic behavior, which may take the form of dominance hierarchy, territoriality, combat for mates, or parent-offspring conflict.

Most agonistic behavior relates to competition and may be a scramble, in which the competitors struggle directly for the mutually sought-after resource, or a contest, in which conspecifics fight first and the winner then gets the resource. Agonistic behavior is usually highly ritualized and communicative in nature; killing or wounding is infrequent. Recent observations interpret overt violence as an expression of selfish behavior on the part of an individual rather than as a maladaptive response to abnormal conditions.

Animal behaviorists have long held that the widespread use of restraint and display in social species has evolved for the preservation of the group or the species. Models based on game theory suggest that natural selection, acting at the level of the individual, can produce an evolutionary stable strategy in which a certain proportion of individuals in a population are aggressive and a certain proportion are passive.

Much conflict behavior involves the social use of space. Home range is the area used habitually by an individual or group; core area is the zone of heaviest use within the home range. Many social animals maintain a personal space around themselves, also referred to as individual distance. Territory is the space that is used exclusively by an individual or group and that is defended. Individual territories occur most often when a needed resource is predictable and evenly distributed in space. Some territories include food and water supply, nest site, and mates; other territories contain only one resource that is defended — for example, a place where only mating takes place, as in the lek.

Species that live in more or less permanent groups usually develop dominance hierarchies, in which individuals control the behavior of conspecifics on the basis of the results of previous encounters. Dominance hierarchies are not always determined by fighting ability; age, seniority, maternal lineage, and formation of alliances with friends have been shown to be important in some mammals.

Researchers have studied the causes of aggression from both an internal and external perspective. Internal factors include the subcortical limbic system, in particular the thalamus, hypothalamus, amygdala, and central gray. Researchers have linked various neurotransmitters and hormones with the exhibition of different forms of aggression. For example, testosterone is most often associated with aggression, but other hormones are important in producing aggression in some species. Artificial selection in the laboratory for high and low aggression demonstrates a genetic component of aggression upon which natural selection can act. In laboratory experiments, researchers have conditioned animals to be dominant or subordinate, which leads us to conclude that previous experience and learning affect later aggressive behavior.

External factors that influence aggression include early learning and experience, pain and frustration due to restricted access to resources, xenophobia, crowding, and environmental factors that trigger the onset of reproduction. Control of aggression is of prime interest because it relates to noninstitutionalized violence in humans. Some researchers have emphasized the inevitability of aggression in humans and have sought ways to divert it into harmless channels. Others have suggested that the modification of the social environment through early experience and avoidance of social disorganization can help to control aggression. Encouraging noncompetitive habits during early experience and creating stable communities may help reduce violent agonistic behavior.

Discussion Questions

1. Distinguish between agonism and aggression.

2. Discuss ways aggression is restrained in animals. Evaluate the arguments put forth to explain the evolution of such restraint.

3. What, if anything, can we learn about aggression in humans by studying aggression in nonhumans?

4. The data below present ranks and maternal lineages among rhesus monkeys at La Parguera, Puerto Rico in June 1969. Part of A Group, which was formed in 1962 when unrelated adult females were placed together in a large cage, is shown. Females 34, 227, 184, and 236 established a linear dominance hierarchy in the order listed; they were then released. The offspring of those females are shown connected to the females with a line; their ages are given in parentheses. Most male offspring left the group at three or four years of age and are, therefore, not shown. The bottom row of numbers and letters shows the monkeys ranked on the basis of agonistic encounters, with those having won the greatest number of encounters on the left, and the least, on the right. Males are in squares. What kinds of rules seem to apply in determining dominance orders in rhesus monkeys? Can you explain how or why such rules come about?

Ranks and maternal lineages among rhesus monkeys

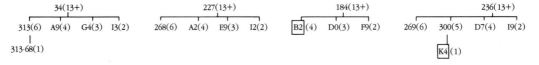

34, G4, A9, 313, 313-68, I3, 227, E9, A2, 268, I2, 184, |B2|, D0, F9, 236, D7, 300, |K4|, 269, I9

Suggested Readings

Johnson, R. N. 1972. *Aggression in Man and Animals.* Philadelphia: Saunders.
Basic, introductory text with emphasis on humans.

Lorenz, K. 1966. *On Aggression.* New York: Harcourt, Brace & World.
The ethologist's perspective. Emphasizes the adaptive
significance of aggression.

Moyer, K. E. 1976. *The Psychobiology of Aggression.* New York: Harper & Row.
The psychologist's perspective deals with neural and
hormonal factors.

Southwick, C. H. 1970. *Animal Aggression: Selected Readings.* New York: Van Nostrand Reinhold.
Balanced collection of authoritative review papers on
agonistic behavior of nonhumans.

Svare, B. B., ed. 1983. *Hormones and Aggressive Behavior.* New York: Plenum Press.
Series of review articles synthesizing research in the last
fifteen years.

References

Albrecht, H. 1966. Zur Stammesgeschichte einiger Bewegungsweisen bei Fischen untersucht am Verhalten von *Haplochromis* (Pisces, Cichlidae). *Zeit. Tierpsychol.* 23:270–301.

Alexander, B. K., and J. Hughes. 1971. Canine teeth and rank in Japanese monkeys (*Macaca fuscata*). *Primates* 12:91–93.

Altmann, J. 1974. Observational study of behavior: Sampling methods. *Behaviour* 49:227–267.

Ardrey, R. 1966. *The Territorial Imperative.* New York: Atheneum.

Bandler, R. J., Jr., and J. P. Flynn. 1974. Nerve pathways from the thalamus associated with regulation of aggressive behavior. *Science* 183:96–99.

Berkowitz, L. 1969. *The Roots of Aggression.* New York: Atherton.

Berkowitz, L., and A. LePage. 1967. Weapons as aggression-eliciting stimuli. *J. Per. Soc. Psychol.* 7:202–207.

Bernstein, I. S. 1966. Analysis of a key role in a capuchin (*Cebus albifrons*) group. *Tulane Stud. Zool.* 13:49–54.

Carpenter, F. L., and R. E. MacMillen. 1976. Threshold model of feeding territoriality and test with a Hawaiian honeycreeper. *Science* 194:639–642.

Christian, J. J., and D. E. Davis. 1964. Endocrines, behavior and population. *Science* 146:1550–1560.

Darwin, C. 1873. *On the Expression of the Emotions in Man and Animals.* New York: D. Appleton.

Davis, D. E. 1963. The physiological analysis of aggressive behavior. In W. Etkin (ed.), *Social Behavior and Organization among Vertebrates.* Chicago: University of Chicago Press.

Dawkins, R., and J. R. Krebs. 1978. Animal signals: Information or manipulation? In J. R. Krebs and N. B. Davies (eds.), *Behavioural Ecology: An Evolutionary Approach.* Sunderland, Mass.: Sinauer Associates.

Delgado, J. M. R. 1967. Social rank and radio-stimulated aggressiveness in monkeys. *J. Nerv. Ment. Disease* 144:383–390.

Drickamer, L. C., and S. H. Vessey. 1973. Group changing in free-ranging male rhesus monkeys. *Primates* 14:359–368.

Ebert, P. D., and J. S. Hyde. 1976. Selection for agonistic behavior in wild female *Mus musculus. Behav. Genet.* 6:291–304.

Eleftheriou, B. E., and J. P. Scott. 1971. *The Physiology of Aggression and Defeat.* New York: Plenum.

Feshbach, S., and R. D. Singer. 1971. *Television and Aggression.* San Francisco: Jossey-Bass.

Flynn, J. P. 1967. The neural basis of aggression in cats. In D. C. Glass (ed.), *Neurophysiology and Emotion.* New York: Rockefeller University Press and Russell Sage Foundation.

Hall, E. T. 1966. *The Hidden Dimension.* Garden City, N.Y.: Doubleday.

Harding, C. F. 1983. Hormonal influences on avian aggressive behavior. In B. B. Svare (ed.), *Hormones and Aggressive Behavior.* New York: Plenum.

Hausfater, G., and S. B. Hrdy, eds. 1984. *Infanticide: Comparative and Evolutionary Perspectives.* New York: Aldine.

Howard, H. E. 1920. *Territory in Bird Life.* New York: Atheneum.

Howard, R. D. 1978. The evolution of mating strategies in bullfrogs, *Rana catesbeiana. Evolution* 32:850–871.

Hrdy, S. B. 1977a. *Langurs of Abu: Female and Male Strategies of Reproduction.* Cambridge: Harvard University Press.

———. 1977b. Infanticide as a primate reproductive strategy. *Amer. Scientist* 65:40–49.

Johnson, R. N. 1972. *Aggression in Man and Animals.* Philadelphia: Saunders.

Le Boeuf, B. J. 1974. Male-male competition and reproductive success in elephant seals. *Amer. Zool.* 14:163–176.

Lorenz, K. 1966. *On Aggression.* New York: Harcourt, Brace & World.

McNab, B. K. 1963. Bioenergetics and the determination of home range size. *Amer. Nat.* 97:133–140.

Marler, P. 1976. On animal aggression: The roles of strangeness and familiarity. *Amer. Psychol.* 31:239–246.

Marsden, H. M. 1968. Agonistic behavior of young rhesus monkeys after changes induced in social rank of their mothers. *Anim. Behav.* 16:38–44.

Maynard Smith, J. 1976. Evolution and the theory of games. *Amer. Scientist* 64:41–45.

Maynard Smith, J., and G. R. Price. 1973. The logic of animal conflict. *Nature* 246:15–18.

Mazur, A. 1983. Hormones, aggression, and dominance in humans. In B. B. Svare (ed.), *Hormones and Aggressive Behavior.* New York: Plenum.

Meikle, D. B., and S. H. Vessey. 1981. Nepotism among rhesus monkey brothers. *Nature* 294:160–161.

Mock, D. W. 1984. Siblicidal aggression and resource monopolization in birds. *Science* 225: 731–733.

Moyer, K. E. 1976. *Psychobiology of Aggression.* New York: Harper & Row.

Orians, G. H. 1961. The ecology of blackbird (*Agelaius*) social systems. *Ecol. Monogr.* 31:285–312.

Parker, G. A. 1984. Evolutionarily stable strategies. In J. R. Krebs and N. B. Davies (eds.), *Behavioural Ecology: An evolutionary approach,* 2nd ed. pp. 30–61. Sunderland, Mass.: Sinauer Associates.

Phoenix, C. H. 1974. Prenatal testosterone in the nonhuman primate and its consequences for behavior. In R. C. Friedman, R. N. Richart, and R. L. Van de Wiele (eds.), *Sex Differences in Behavior.* New York: Wiley.

Poulsen, H. R., and D. Chiszar. 1975. Interaction of predation and intraspecific aggression in bluegill sunfish *Lepomis macrochirus. Behaviour* 55:268–286.

Reinisch, J. M. 1981. Prenatal exposure to synthetic progestins increases potential for aggression in humans. *Science* 211:1171–1173.

Robinson, B. W., M. Alexander, and G. Bowne. 1969. Dominance reversal resulting from aggressive responses evoked by brain telestimulation. *Physiol. Behav.* 4:749–752.

Rose, R. N., I. S. Bernstein, and T. P. Gordon. 1975. Consequences of social conflict on plasma testosterone levels in rhesus monkeys. *Psychosomatic Med.* 37:50–61.

Rosenblum, L. A., and I. C. Kauffman. 1968. Variations in infant development and response to maternal loss in monkeys. *Amer. J. Orthopsych.* 38:418–426.

Rowell, T. E. 1974. The concept of social dominance. *Behav. Biol.* 11:131–154.

Sade, D. S. 1967. Determinants of dominance in a group of free-ranging rhesus monkeys. In S. Altmann (ed.), *Social Communication among Primates.* Chicago: University of Chicago Press.

Schjelderup-Ebbe, T. 1922. Beiträge zur Sozialpsychologie des Haushuhns. *Zeit. Psychol.* 88:225–252.

Scott, J. P. 1966. Agonistic behavior of mice and rats: A review. *Amer. Zool.* 6:683–701.

———. 1972. *Animal Behavior,* 2nd ed. Chicago: University of Chicago Press.

————. 1975. Violence and the disaggregated society. *Aggressive Behav.* 1:235–260.

————. 1976. The control of violence: Human and nonhuman societies compared. In A. Neal (ed.), *Violence in Animal and Human Societies.* Chicago: Nelson-Hall.

Scott, J. P., and E. Fredericson. 1951. The causes of fighting in mice and rats. *Physiol. Zool.* 24:273–309.

Smith, D. G. 1981. The association between rank and reproductive success of male rhesus monkeys. *Amer. J. Primat.* 1:83–90.

Southwick, C. H. 1967. An experimental study of intragroup agonistic behavior in rhesus monkeys, *Macaca mulatta. Behaviour* 28:182–209.

————. 1970. *Animal Aggression: Selected Readings.* New York: Van Nostrand Reinhold.

Southwick, C. H., and L. H. Clark. 1968. Interstrain differences in aggressive and exploratory activity of inbred mice. *Comm. Behav. Biol.,* Part A 1:49–59.

Southwick, C. H., M. F. Siddiqi, M. Y. Farooqui, and B. C. Pal. 1974. Xenophobia among free-ranging rhesus groups in India. In R. L. Halloway (ed.), *Primate Aggression, Territoriality, and Xenophobia.* New York: Academic Press.

Tinbergen, N. 1951. *The Study of Instinct.* Oxford: Clarenden Press.

Ulrich, R. E., R. R. Hutchinson, and N. H. Azrin. 1965. Pain-elicited aggression. *Psych. Record.* 15:111–126.

Van Hemel, P. E., and J. S. Myer. 1970. Satiation of mouse killing by rats in an operant situation. *Psychon. Sci.* 21:129–130.

Vessey, S. H. 1971. Free-ranging rhesus monkeys: Behavioural effects of removal, separation and reintroduction of group members. *Behaviour* 40:216–227.

Vessey, S. H., and D. B. Meikle. 1984. Free-ranging rhesus monkeys: Adult male interactions with infants and juveniles. In D. Taub (ed.), *Paternal Behavior.* New York: Van Nostrand Reinhold.

vom Saal, F. S. 1983. Models of early hormonal effects on intrasex aggression in mice. In B. B. Svare (ed.), *Hormones and Aggressive Behavior.* New York: Plenum.

Walsh, J. 1983. Wide world of reports. *Science* 220:804–805.

Wiley, R. H. 1973. Territoriality and nonrandom mating in sage grouse, *Centrocercus urophasianus. Anim. Behav. Monogr.* 6:85–169.

Wilson, E. O. 1975. *Sociobiology: The New Synthesis.* Cambridge: Harvard University Press.

Wilson, A. P., and C. Boelkins. 1970. Evidence for seasonal variation in aggressive behavior by *Macaca mulatta. Animal Behav.* 18:719–724.

13

COMMUNICATION

Animals convey information to members of their own species and to other species as well through an incredible diversity of sounds, colors, flashing lights, smells, and postures. The song of a male white-crowned sparrow *(Zonotrichia leucophrys)*, for example, is specific not only to the species but to an area and to the individual. By singing, the male may be warning neighboring males away, warning neighbors of danger, or may be attracting females.

Wilson (1975) defined **biological communication** as an action on the part of one organism (or cell) that alters the probability pattern of behavior in another organism (or cell) in a fashion adaptive to either one or both of the participants. The word **adaptive** implies that the signal or response is to some extent genetically controlled and under the influence of natural selection. But this definition presents some difficulty. As Marler (1967) pointed out, what about the mouse that rustles in the grass, making sound that enables the owl to catch it? This case fits Wilson's definition, but would we really say that the mouse is communicating with the owl? Another difficulty centers on who benefits from communication. Although both sender and receiver frequently benefit, sometimes the sender benefits at the expense of the receiver, as when a parent bird uses a broken wing distraction display to lead a predator away from her young. For this reason Slater (1983) defines communication as "the transmission of a signal from one animal to another such that the sender benefits, on average, from the response of the recipient."

Through extensive field observations ethologists have found that communication behavior occurs in regular patterns, recurring nearly unchanged from event to event; these sequences are called **fixed action patterns (FAPs).** The most striking are the stereotyped sequences that occur as animals interact with conspecifics—courtship, territory defense, or dominance encounters. The primary function of these behaviors is to signal. We begin by examining how signals are coded to convey information and how we can measure the information transfer that is taking place. In the second half of the chapter we consider the kinds of information conveyed and the sensory channels through which the signals are received. In addition, we look at two examples of symbolic "language" in nonhuman animals, in which multiple sensory channels transfer several pieces of information. In his study of great crested grebes, Julian Huxley called their strange postures **displays;** Moynihan (1956) defines display as any behavior patterns especially adapted in physical form or frequency to function as social signals. Ethological interest has focused on the ways in which displays have evolved from other, noncommunicative behavior. Near the end of the chapter we discuss the evolution of conspicuous body structures and the ritualization of behavior, and finally, we return to the problem of who benefits from communication and consider honesty in signalling.

HOW SIGNALS CONVEY INFORMATION

DISCRETE AND GRADED SIGNALS

Some signals are **discrete** (digital) and others are **graded** (analog). The alarm calls of many species, which are typically given at the same intensity each time and are relatively constant across species, permit communication among different species. These discrete signals are of a frequency and duration that make them difficult for predators to pinpoint (Marler 1957).

Graded signals may vary in intensity as a function of the strength of the stimulus. The waggle dance of the honeybee illustrates both the complexity and the graded properties of signals. The number of turns per minute is inversely proportional to the distance from the food source, and the duration and liveliness of the dance increase with the quality of the food. Signals that might at first seem discrete often, upon closer study, turn out to be graded. Although the bursts of light emitted by fireflies seem to be discrete and species-specific, they vary in intensity and duration under different conditions.

DISTANCE AND DURATION

Although most species are limited to about twenty to forty different displays, signals can vary in several other ways that increase their information content. The distance a signal travels may vary. In some cases the detection threshold of the receiver is no more than several molecules of a substance, as with sex attractants of many insects.

Thus a small amount of material produced by a female can be detected by a male several kilometers downwind. Visual displays, at the other extreme, usually operate over much shorter distances. The duration of a signal may also vary. Alarm signals, such as chemicals produced by many invertebrates, may have a localized, short-term effect and thus a rapid fade-out time (Wilson 1975). Signals such as male secondary sexual characteristics in polygynous species can last for long periods of time. Information about reproductive status, like the antlers of male deer or the bright red perineal region of female rhesus monkeys in estrus, is conveyed during the breeding season. In contrast, the brightly colored epaulets on the red-winged blackbird, although present during the entire breeding season, are conspicuous only when the male exposes and erects these feathers during the song spread.

COMPOSITE SIGNALS, SYNTAX, AND CONTEXT

Two or more signals can be combined to form a composite signal with a new meaning. For example, equids such as zebras communicate hostile behavior by flattening their ears and communicate friendliness by raising their ears (discrete signals). They indicate the intensity of either emotion by the degree to which the mouth opens (graded signal). The mouth-opening pattern is the same for both hostile and friendly behavior (Figure 13–1).

Animals can convey additional information with a limited number of displays through changing the **syntax** — the sequence of displays. For example, the two composite signals A and B would have different meanings depending on whether A

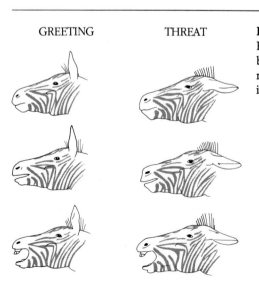

GREETING   THREAT

FIGURE 13–1   Composite facial signals in zebras
Ears convey a discrete signal. They are either laid back as a threat or pointed upward as a greeting. The mouth conveys a graded signal and opens variably to indicate the degree of hostility or friendliness.

FIGURE 13-2   Metacommunication in dogs
The play bow performed by the dog on the right communicates that behaviors that
follow are play.
Source: Photo by Marc Bekoff.

or B came first. No evidence exists that nonhuman animals use syntax naturally.
However, in the rather special case of language learning in chimpanzees, chimps
assemble words in novel ways as they communicate with their human companions
(Rumbaugh and Gill 1976).

The same signals can have different meanings depending on the **context**—
that is, on what other stimuli are impinging on the receiver. For example, the lion's
roar can function as a spacing device for neighboring prides, as an aggressive display
in fights between males, or as a means of maintaining contact among pride
members. The "ruff-out" display of the male cowbird serves in courtship as well as
in conflict with other males.

METACOMMUNICATION

Increasing the information content of displays by **metacommunication**—
communication about communication—is theoretically possible: one display
changes the meaning of those that follow. We can see good examples in play
behavior: animals use aggressive, sexual, and other displays in play, but they precede such behavior by an act that communicates the message that "what follows is
play, join in" (Bekoff 1977). Canids such as dogs and wolves precede play with the
play bow (Figure 13–2). Monkeys communicate play behavior through a relaxed,
open-mouthed face.

Disagreement exists about the importance of metacommunication (Smith

1984), and some observers have used the word a bit loosely. For example, male rhesus monkeys *(Macaca mulatta)* sometimes communicate dominance by carrying the tail elevated in an "S" shape over the back. While this posture conveys status and mood, it does not really change the meaning of behaviors that follow; rather, it communicates that aggressive behavior is likely to follow.

MEASUREMENT OF COMMUNICATION

Suppose that we wish to learn something about the way crayfish *(Oronectes rusticus)* communicate. If we place crayfish that are strangers together in a tank in the lab, they assume various postures with the large claws and a dominance hierarchy, which is positively related to the size of the animal, develops (Bovjberg 1956).

OBSERVATION

Our first step is to identify motor patterns that are relatively constant by observing the animals and describing the movements they make. We might divide the behavior patterns into the following acts: **retreat**—a rapid, backward swimming motion; **cheliped presentation**—the movement of the large claw from a downward-facing position to one that is horizontal to the substrate; **cheliped extension**—the rapid movement of the opened claw toward the other crayfish; **forward locomotion**—movement toward the other crayfish; and **fighting**—striking and pinching the other crayfish (Figure 13–3).

Simple observation might suggest that communication is occurring since certain behaviors occur only in the presence of other animals. Perhaps a behavior performed by one individual —say, cheliped extension—is usually followed by a particular behavior performed by the other animal—say, retreat. In that case we suspect that communication has occurred. But the occurrence of any behavior is probabilistic (stochastic). How can we be sure that what one crayfish does affects the behavior of the other?

QUANTIFICATION

OBSERVED AND EXPECTED FREQUENCIES. For a more thorough analysis, we can compare the frequency of each behavioral response against the expected frequency if the response were random, and without regard to the behavior of the other crayfish. Table 13–1 shows observed and expected frequencies. The results of the

a.

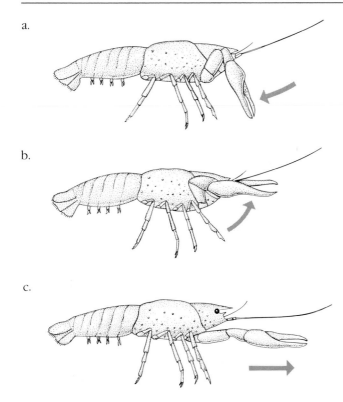

b.

c.

FIGURE 13-3 Cheliped (large claw) presentation and extension in the crayfish

In the first phase of cheliped presentation (a), the cheliped tip is lowered and the claw is opened, and in the second phase (b), the cheliped is raised. Part (c) shows cheliped extension. Arrows denote direction of movement. Analysis of these signals (Tables 13–1 and 13–2) shows that cheliped presentation by one crayfish increases the chance of a fight with a second crayfish, whereas cheliped extension increases the likelihood that a second crayfish will retreat.

Source: S. de Roth, "Communication in the Crayfish *(Orconectes rusticus),"* Master's thesis, Bowling Green State University, 1974. Reprinted with permission.

TABLE 13-1 Frequency distributions of agonistic behavior patterns in crayfish
The frequencies expected, if random, are given in parentheses.

Signaler's Signal	Receiver's Response						
	No Change	Retreat	Cheliped Presentation	Cheliped Extension	Forward Locomotion	Fighting	Total Observed
no change	4 (0.9)	1 (5.6)	0 (0.4)	4 (1.2)	0 (0.1)	0 (0.8)	9
retreat	6 (4.0)	14 (24.3)	0 (1.8)	7 (5.2)	3 (0.3)	9 (2.7)	39
cheliped presentation	6 (2.7)	6 (16.2)	6 (1.2)	0 (3.5)	0 (0.2)	8 (2.3)	26
cheliped extension	8 (17.5)	126 (106.1)	3 (7.7)	24 (22.6)	0 (1.3)	9 (14.9)	170
forward locomotion	17 (8.5)	39 (51.8)	6 (3.7)	15 (11.0)	0 (0.6)	6 (7.3)	83
fighting	0 (7.4)	63 (44.9)	3 (3.2)	3 (9.6)	0 (0.5)	3 (6.3)	72
total observed	41	249	18	53	3	35	399

Source: S. de Roth, "Communication in the Crayfish *(Oronectes rusticus)."* Master's thesis, Bowling Green State University, 1974. Reprinted with permission.

TABLE 13-2   Outcome of statistical analysis of frequency distributions of agonistic behavior patterns in crayfish

| | Receiver's Response | |
Signal	Facilitates	Inhibits
no change		
retreat	fighting	retreat
cheliped presentation	fighting	retreat, cheliped extension
cheliped extension	retreat	no change, fighting
forward locomotion		
fighting	retreat	no change, cheliped extension

Source: S. de Roth, "Communication in the Crayfish *(Oronectes rusticus)."* Master's thesis, Bowling Green State University, 1974. Reprinted with permisson.

chi-square test are presented in Table 13–2. Note that some behaviors facilitate response and others inhibit them. Some, such as forward locomotion, seem to have little effect on other behaviors.

INFORMATION THEORY.   Another approach is to quantify the amount of information transmitted from one individual to another, a technique referred to as **information analysis.** Here information is measured as the reduction of the uncertainty of the behavioral response of an individual. The data show that when one crayfish presents its cheliped, the other is less likely to retreat and more likely to fight; there is a reduction in the uncertainty as to whether fighting or retreating will occur. How much information has been transmitted?

We can first compute the information contained in the behavioral repertoire of one individual according to a method developed independently by communication theorists Shannon and Weiner and usually referred to as the **Shannon-Weiner index** (Shannon and Weaver 1949).

The formula is

$$H = \Sigma\, p_i \log_2 p_i,$$

where H is the information content, i is the occurrence of a particular signal and p_i is the probability of occurrence of that signal. Each signal is considered to be discrete and to contain one message, yes or no, and is thus part of a binary system. For this reason, we use logarithms to the base two, and our unit of information is the "bit." To measure the information in a two-animal system, we measure information at the source, the signaler, and at the receiver. In our example, the signaler information for one crayfish is calculated from data in Table 13–1 and appears in Table 13–3. The first signal, no change, occurred 9 times out of a total of 399 acts, so $p_i = 9/399$ or

TABLE 13-3 Computation of signaler information for crayfish data in Table 13-1 using Shannon-Weiner formula

Signal (i)	Proportion of Total Acts (p_i)	$p_i \log_2 p_i$
no change	9/399 = .022	−.1211
retreat	39/399 = .098	−.3284
cheliped presentation	26/399 = .065	−.2563
cheliped extension	170/399 = .426	−.5244
forward locomotion	83/399 = .209	−.4720
fighting	72/399 = .180	−.4474
Total	1.000	$\Sigma = -2.1497$
		= −H (signaler information)

Note: $\log_2 X = \dfrac{\log_{10} X}{\log_{10} 2}$

.022. $\text{Log}_2 p_i = -5.504$ and p_i times $\log_2 p_i = -.1211$ (Table 13–3). The rest of the signals are treated the same way, and then summed to give H. We can calculate the receiver information in the same way. To measure the information actually transmitted, which will be less than either of these values, we must find the conditional probabilities of the signals that evoke each response; Wilson (1975, pp. 194–198) and Attneave (1959) outline the complete procedure, which is not shown here.

The signaler crayfish actually transmitted 1.11 bits of information per behavior pattern, resulting in a 59 percent restriction of the second animal's behavior. One advantage of this type of analysis is that it permits comparison among different organisms with regard to the diversity and efficiency of their communication systems. Hazlett and Bossert (1965) compared several species of crustaceans and concluded that the amount of information transmitted per signal remains relatively constant across species. We might expect natural selection to maximize the sender's ability to restrict or control the receiver's behavior, if Slater's definition of communication is correct.

FUNCTIONS OF COMMUNICATION

Another approach to the study of communication is to classify signals by function. The ultimate function of any communication is increased fitness, and we will examine this function more closely when we consider the evolution of displays. Although we can list proximate functions of communication according to different criteria, the classifications tend to be artificial and arbitrary and are made mainly for researchers' convenience. The following functions of communication are modified from Wilson (1975) and Smith (1984).

GROUP SPACING AND COORDINATION

Group-living animals use a variety of signals that seem to function to keep members in touch. The highly arboreal *Cebus* monkeys of the South American rain forest forage in dense vegetation. A group of fifteen may spread out over an area 100 meters in diameter as they search the treetops for fruits. In addition to the sound of moving branches, an observer hears a continual series of "contact" calls from the different members of the group. An individual that becomes isolated utters a "lost" call, which is much louder than the contact calls. Marler (1968) has suggested that primates use the following types of spacing signals: (1) distance-increasing signals, such as branch shaking, which may result in another group's moving away; (2) distance-maintaining signals, such as the dawn chorus by howler monkeys (*Alouatta* spp.) that regulates the use of overlapping home ranges; (3) distance-reducing signals, for example, the contact or lost calls of *Cebus* monkeys; and (4) proximity-maintaining signals, such as occur during social grooming within groups.

Among fishes, conspicuous color patterns, especially in coral reef fish, are species-specific and attract conspecifics and hold them together in schools. Social insects emit a variety of chemicals that result in assembly of conspecifics, as exemplified by termites, ants, and bees (Wilson 1971).

RECOGNITION

SPECIES RECOGNITION.   **Species recognition** is crucial prior to mating to avoid infertile matings between members of closely related species. Animals generally communicate messages that are much more individualized than those needed merely for species recognition, however.

INDIVIDUAL RECOGNITION.   The **playback** experiment has been a critical tool in the analysis of communications used in individual recognition, particularly for birdsong. Researchers record vocalizations with a special directional microphone and a high-quality tape recorder and can loop the tape for repeated playbacks, with the loudspeaker placed directly in the field. Observers can score the responses of resident birds according to singing frequency and approaches to the speaker before and after playback. Indigo buntings (*Passerina cyanea*) produce a complex song that is quite variable. Most of the phrases are in twos (sweet-sweet, chew-chew, and so forth). Emlen (1972) analyzed tape recordings of the song and spliced pieces of the tapes in varying order. When he played them back in the field and observed the responses of territorial males, he discovered the significance of much of the song. Part of the sequence was species-specific, communicating the message, "I am an indigo bunting," and part varied from individual to individual. Thus territorial males can identify particular males (Figure 13–4). Changing the ordering of notes (syntax) was not important in species recognition, but changing the rhythm or the frequency

FIGURE 13-4 Audiospectrograms of indigo bunting
Tapes of a normal song were cut and spliced together to test the function of various
attributes of the song. The normal song and the nonpaired song elicited typical
responses, but temporal changes, such as reducing the time between notes by one
half or increasing it twofold, interfered with song recognition.
Source: S. T. Emlen, "An Experimental Analysis of the Parameters of Bird Song
Eliciting Species Recognition," *Behaviour* 41 (1972):130–171. Reprinted with
permission.

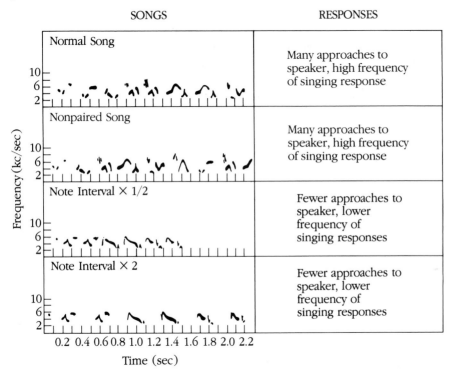

of the notes themselves reduced the agonistic responses of resident birds because the
sounds were no longer recognized as those of other indigo buntings.

NEIGHBOR RECOGNITION. Falls and Brooks (1975), to test whether white-throated
sparrow males recognize neighbors individually or as a class, studied the effect of
playback location within the resident male's territory. When they placed the speaker
at a neighbor's territorial boundary, the resident responded more strongly to a
stranger's song than to the neighbor's, as expected. But when they placed the
speaker on the boundary opposite the neighbor's boundary, the resident treated the

neighbor's song as a stranger's and responded vigorously. When they placed the speaker in the center of the territory, responses to the neighbor's song were intermediate. These results demonstrate that a territorial male can discriminate among the songs of its neighbors and can also associate each song with the appropriate location.

CLASS RECOGNITION. Class recognition occurs mainly in social insects in which castes are treated differentially. Nest queens receive preferential treatment in terms of food and care from workers because of pheromones produced by the queens. Males are discriminated against as a group; they receive less food from the workers and are even driven from the colony in times of food scarcity (Wilson 1971).

DEME RECOGNITION. Local dialects in bird song have been demonstrated for a number of geographically separated populations of white-crowned sparrows. Baker (1983) showed that females from one population in Colorado perform copulation solicitation displays when hearing songs of males from their own population but rarely when hearing songs from another population (Figure 13–5). This tendency may restrict mating to members of locally adapted demes. Whether or not the senders, in this case the males, benefit from such choosiness by females remains to be seen.

KIN RECOGNITION. Communication may be involved in the differential responses of many organisms to their close relatives. For example, tadpoles of the American toad *(Bufo americanus)* prefer to associate with siblings over nonsiblings, even after being reared in isolation (Waldman 1982). Waldman hypothesized that some substance in the egg jelly, contributed by the mother, was used as a cue. Other data from insects, amphibians, birds, and mammals demonstrate kin recognition even in the absence of interactions with kin early in life (Holmes and Sherman 1983). Chemical, auditory, and visual cues all have been implicated.

How can kin identify each other even if they have never interacted? One possibility is **phenotype matching,** where the individual uses kin whose phenotypes are learned by association as a referent. The referent is then compared with the stranger (Holmes and Sherman 1983). For example, female sweat bees *(Lasioglossum zephyrum)* guard the nest entrance and admit only nestmates (usually their sisters). In a laboratory experiment Buckle and Greenberg (1981) reared some guards with sisters and some with nonsisters. In later tests the guard bees admitted strangers, but only if they were sisters of the guards' nestmates. The evolutionary significance of kin recognition is discussed in chapters 15 and 19.

REPRODUCTION

The most striking displays involve reproduction. Receptive females or males may advertise their condition, court a member of the opposite sex, form a bond, copulate, or perform postcopulatory displays. As discussed in chapter 11, these behaviors involve species identification, assessment of individual condition, and coordination of the neuroendocrine systems.

FIGURE 13-5

(A) Sound spectrograms of home-dialect tutor songs and stimulus songs in white crowned sparrows. The spectrograms lie in the frequency range 3 to 6 kHz (vertical axis) and are about 2 seconds long (horizontal axis). (B) Postures showing the response of each female subject to the two stimulus song dialects. Juveniles are unshaded and adults are shaded. The lordotic posture is the copulation solicitation display. Above each female is indicated the number of displays elicited from the subject during a 21-minute test. Juvenile and adult females responded more to the home dialect than to the alien dialect.

Source: Early experience determines song dialect responsiveness of female sparrows. M. C. Baker et al. *Science* Vol. 214, pp.819–821, 13 November 1981. Copyright 1981 by the American Association for the Advancement of Science.

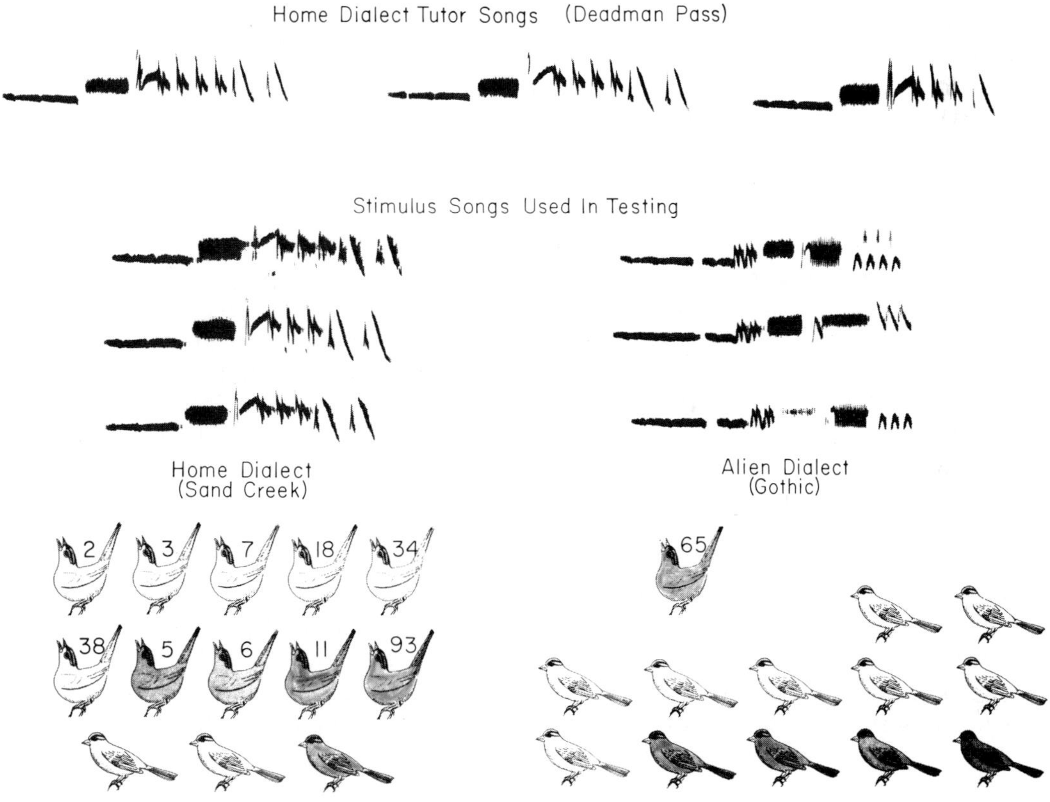

Home Dialect Tutor Songs   (Deadman Pass)

Stimulus Songs Used In Testing

Home Dialect
(Sand Creek)

Alien Dialect
(Gothic)

AGONISM AND SOCIAL STATUS

Particularly in social groups in which members of a species are in close proximity, it is sometimes in an individual's interest to compete with and fight with others for possession of a resource — be it food, space, or access to another individual. Physical combat is expensive in terms of energy and increases the risk of death or injury, even for winners. Social species have evolved displays that communicate information about an individual's mood (recall the tail of the rhesus monkey) and the way it is likely to behave in the near future. As a result of previous encounters, the animal may be dominant or submissive, and its behavior is thus predictable. Presented with a limiting resource, the submissive individual will yield to the dominant without an overt fight. Many species compete for resources only on infrequent occasions; thus agonistic interactions make up only a small proportion of their behavioral repertoire. Dominant, aggressive displays tend to be opposites of submissive displays, exemplifying the principle of antithesis first mentioned by Darwin (1872) (see Figure 12–11).

ALARM

Animals use vocalizations and chemicals to alert group members to danger. Male song, which functions in reproductive isolation and in species-specific territorial defense, varies greatly within and among closely related species. Sympatric species, those with overlapping ranges, are likely to have simple, hard-to-locate alarm calls that differ little among species. Species that live together and are endangered by the same predators benefit mutually by minimizing divergence among alarm vocalizations (Marler 1973). For example, Marler was unable to tell the difference between the "chirp" alarm calls of African blue monkeys *(Cercopithecus mitis)* and red-tailed monkeys *(Cercopithecus ascanius)*, whereas the male "songs" differed greatly between the two monkey species. Vervet monkeys *(Cercopithecus aethiops)* communicate **semantically** in that they use different signals to warn about different objects in their environment. Group members climb trees when they hear alarm calls given in response to leopards, they look up when they hear eagle alarms, and they look down when they hear snake alarms (Struhsaker 1967). Young vervets give alarm calls in response to a variety of animals, and their ability to classify predators and give appropriate alarm calls improves with age (Seyfarth, Cheney, and Marler 1980).

Some invertebrates, such as starfish and earthworms, produce chemical alarm substances. Mice release a substance in the urine when they are electrically shocked or when they are beaten up by another mouse. This substance acts as an alarm and causes other mice to avoid the area.

Tracing the evolution of alarm calls presents a challenge to biologists because such signals seem unlikely to benefit the caller. Sherman (1977) studied individually marked Belding's ground squirrels *(Spermophilus beldingi)* (Figure 13–6) and found that whenever a terrestrial predator, such as a weasel or a coyote, was spotted, the calling squirrel stared directly at the predator while sounding the alarm. Sherman suggested many hypotheses, two of which we discuss, to explain this behavior. First, the predator may abandon the hunt once it is spotted by the potential prey — in this hypothesis, the caller is behaving selfishly. Second, others in the area may benefit from the warning even though the caller may be harmed — in this hypothesis, the caller is behaving altruistically. Sherman demonstrated that callers attract predators and are more likely to be attacked after calling, so they are not behaving selfishly.

Because he kept records on mothers and offspring, Sherman knew that the males leave the area several months after birth and that the females are sedentary, with daughters breeding near their birthplaces. He also found that adult and yearling females are much more likely to call than would be expected by chance, and males are less likely to do so. Furthermore, females with living female relatives such as mothers or sisters in the area, but not necessarily with offspring, call more frequently in the presence of a predator than those with no living female relatives. Sherman therefore concluded that the most likely function of the alarm call is to warn family members. The behavior is phenotypically altruistic (reducing direct fitness) but genotypically selfish (increasing indirect fitness), and evolution by kin selection is indicated (see Chapter 4).

Models have been powerful tools in the study of the function of alarm calls, as seen in the study of geese postures (Inglis and Isaacson 1978). When alarmed by sudden auditory or visual stimuli, geese adopt the extreme head-up posture (Figure 13–7). When "flocks" of decoys were placed in grainfields in which geese had been causing damage, the response of the real flocks depended on the postures of the models. In general, fields containing model flocks with a high proportion of extreme head-up decoys were avoided, while those with mostly head-down decoys attracted the real flocks.

HUNTING FOR FOOD

One of the selection pressures in favor of group living is a group's increased efficiency in finding food, which involves both communication about its location and cooperation in securing it. This is exemplified in the African wild dog *(Lycaon pictus)*, a canid only distantly related to domestic dogs and wolves. Members of the pack

FIGURE 13-6 Belding's ground squirrel giving alarm call
This conspicuously calling squirrel, a lactating female, is more likely to be attacked by a predator than is a noncaller. Nearby squirrels benefit since they can remain hidden or take cover. Females with mothers, sisters, or offspring in the vicinity are most likely to call.
Source: Photo by George D. Lepp, courtesy of Paul Sherman.

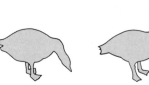 

Head Down: attractive

Head Up: aversive only in presence of extreme head-up shapes

Extreme Head Up: aversive

FIGURE 13-7   Models of Brent goose postures
When "flocks" of geese models are placed in grainfields, real geese avoid fields containing models in the extreme head-up (alarm) posture. *Source:* Data from I. R. Inglis and A. J. Isaacson, "The Responses of Dark-Bellied Brent Geese to Models of Geese in Various Postures," *Animal Behaviour* 26 (1978):953–958.

engage in a frenzy of nosing, lip-licking, tail-wagging, and circling before they run off to hunt prey many times larger than an individual dog (Lawick and Lawick-Goodall 1971). Such behavior seems important in coordinating activities. Social carnivores may even communicate information about what type of prey they are about to hunt. Kruuk (1972) has noted that hyenas hunting zebra were observed on some occasions to pass by prey they had hunted on other occasions.

Chimpanzees *(Pan troglodytes)* communicate the location of food and may actually lead others to it (Menzel 1971). Menzel removed captive chimps from their home pen. He then showed the location of food hidden in the home pen to one chimp and allowed that chimp to lead the others to it. No special signals were used; rather, the leader moved toward the hidden food purposively, looking back at the others periodically. The information transferred seemed to be "something in those bushes ahead has aroused his expectancy of edibles."

In an advancing wave of army ants, movements of the swarm are influenced by tactile stimulation with antennae and by the laying down of a trail pheromone. A scouting forager that has discovered a food source dashes back and forth between the nearest raiding column and the food; 50 to 100 ants are recruited to the food source within the first minute (Chadab and Rettenmeyer 1975).

GIVING AND SOLICITING CARE

A wide variety of signals is used between parent and offspring or among other relatives in the begging and offering of food. As Tinbergen (1951) pointed out, the red spot on the herring gull's lower bill stimulates and directs a pecking response by the chick; and the chick's contact with the parent's beak in turn stimulates the parent to regurgitate the food. Distress calls by the young are individually recognizable by the parent once the young are capable of leaving the birth site under their own power. When they are chilled, baby mice produce high-frequency sounds that are inaudible to humans but audible to adult mice, who can assist them. Bird species in which both parents care for the young use complex displays. When one parent arrives at the nest to relieve the other, recognition is first established and a gradual transfer takes place as the initial aggressive responses of the resident parent are overcome.

SOLICITING PLAY

As we discussed earlier, play consists of behavior patterns that may have many different functions in the adult: sex, aggression, exploration, and so forth. The play bow in canids (Figure 13–2) is communication about play and informs others that the motor patterns that follow are not the real thing. The function of play itself is a subject of debate. Observers usually agree that they can recognize play but have had great difficulty ascribing definitions or functions to it. They most often suggest that the function of play is to help develop motor skills and behavior patterns used later in adult life (Fagen 1981).

SYNCHRONIZATION OF HATCHING

Precocial birds, such as pheasants and ducks, lay large clutches of eggs. There is a premium on synchronous hatching because the mother, with the young following as a group, leaves the area to feed or to get into the water to reduce predation by terrestrial vertebrates. Species nesting in tree holes leave the area permanently on the day the chicks hatch. Late-hatching chicks are vulnerable to predation. A few days prior to hatching, the chicks begin to vocalize. This communication is involved in hatching synchrony, since similar-aged eggs incubated separately hatch over a period of several days (Vince 1969).

CHANNELS OF COMMUNICATION

ODOR

From an evolutionary standpoint, the earliest type of communication was chemical, for odor is used throughout the animal kingdom, with the exception of most bird species. Most pheromones are involved in mate identification and attraction, spacing mechanisms, or alarm. The greatest amount of research has been done on insects and mammals. There are several probable reasons why the widespread use of chemical signals has evolved: such signals can transmit information in the dark, can travel around solid objects, can last for hours or days, and are efficient in terms of the cost of production (Wilson 1975). However, because these pheromones must diffuse through air or water, they are slow to act and have a long "fade-out" time.

We have a relatively good understanding of insect pheromones. For example, the sex attractant bombykol, produced by the female silk moth, has been isolated, and we are reasonably familiar with the male's perceptual system (see Figure 6–7). A single molecule of bombykol triggers a nerve impulse in a receptor cell in an antenna of a male. About 200 receptor-cell firings in one second lead to a behavioral response (Schneider 1974). The male responds by flying upwind, equalizing the pheromone concentration on both antennae, until he reaches the female. With the commercial synthesis of insect sex attractants, traps baited with pheromone to lure males are used to control such pests as the gypsy moth in the northeastern United States.

Many of the recent studies of pheromones in mammals have been done on rodents. Two general classes of substances, which differ in effect, have been identified: **priming pheromones,** which produce a generalized response, such as triggering estrogen and progesterone production that leads to estrus; and **signaling pheromones,** which produce an immediate motor response, such as the initiation of a mounting sequence. Bronson (1971) suggested that pheromones in mice have the following functions:

- *Signaling pheromones*
    Fear substance
    Male sex attractant
    Female sex attractant
    Aggression inducer
    Aggression inhibitor

- *Priming pheromones*
    Estrus inducer
    Estrus inhibitor
    Adrenocortical activator

In addition, substances produced in male mouse urine speed up maturation in young females (Lombardi and Vandenbergh 1977). Other urinary products inhibit aggression, increase aggression, stimulate the adrenal cortex, block implantation of embryos, and so on. At this point it is not clear how many different chemicals are involved; different functions may be served by the same pheromone. The sources of these products include the sexual accessory glands and even the plantar tubercles on the mouse's feet.

Many of the substances produced by mammals function as a means of staking out territories or home ranges, much as does birdsong. The advantage of pheromones, as mentioned previously, is that the odor may last for many days and nights, a significant asset since many animals are nocturnal. Since these substances are often associated with the urinary and digestive systems, eliminative behavior is often highly specialized. Hyena clans mark the boundaries of their territories by establishing latrine areas (Figure 13–8). Clan members defecate simultaneously in an area, then paw the ground. The feces turn white and become quite conspicuous. In the

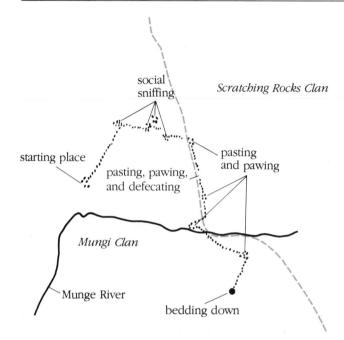

FIGURE 13-8   Activities of the Mungi hyena clan along its boundary
On a particular evening six members of the clan left the starting place or "club" where they had been sleeping and moved toward the boundary with the Scratching Rocks clan, sniffing along the way. Once at the boundary they sought out tall grass stems for pasting and pawing. Two members of the clan also defecated at a site that many others had used previously. After more pasting and pawing, the clan bedded down, having covered about two kilometers in one hour's time. Members of one clan rarely trespass into the territory of another clan.
Source: H. Kruuk, *The Spotted Hyena* (Chicago: University of Chicago Press, 1972), p. 259. Reprinted with permission.

same or in another area, hyenas engage in "pasting." Both sexes possess two anal glands that open into the rectum, just inside the anal opening. When pasting, the hyena straddles long stalks of grass; as the stems pass underneath, the animal everts its rectum and deposits a strong smelling whitish substance on the grass stems (Kruuk 1972).

SOUND

Much more information about immediate conditions can be transmitted faster by sound than by chemicals. Sound can be produced by a single organ, and it can also travel around objects, through dense vegetation, and can be used in the dark. Information can be conveyed by both frequency and amplitude modulation. As we can see in sound spectrographs, a unit of birdsong contains sounds of various frequencies and amplitudes. Low-frequency sounds travel great distances and are used by animals with large home ranges. For example, howler monkeys *(Alouatta)* in the Neotropical rain forest signal to other groups with low-frequency calls. Animals with smaller home ranges—for example, squirrel monkeys *(Saimiri sciureus)*—use higher-frequency sounds, which dissipate rapidly. Such calls serve to maintain contact among group members.

Aside from human speech, birdsong, whose development has been analyzed by Thorpe (1958), Marler and Tamura (1962), and others, is probably the most complex auditory communication. Birds raised in isolation develop abnormal songs; but if they are exposed to normal song prior to the first breeding season, they modify their songs to match. Such imitation occurs under natural conditions, and young males develop songs resembling those of their neighbors.

Although part of the song develops without any experience, normal, species-specific song must be learned from others, most likely the father (Figure 13–9). The learning that takes place is highly **selective**: hand-reared swamp sparrows failed to learn the song of the closely related song sparrow (both in the genus *Melospiza*) but readily learned the song of other swamp sparrows (Marler and Peters 1977). Birds may modify their song to match the local dialect if they move into a new area at the start of a breeding season. Females may not mate as readily with strange-sounding males (Baker 1983).

FIGURE 13-9   Chaffinch song experiments

A male chaffinch *(Fringilla coelebs)* reared in isolation has a simpler song (a) than one that hears other male chaffinches singing (b). If a tree pipit *(Anthus trivialis)* song (c) is played to a male chaffinch while it is reared in isolation, some elements of the tree pipit song are incorporated into the chaffinch song (d).
Source: W. H. Thorpe, "The Learning of Song Patterns by Birds, with Special Reference to the Song of the Chaffinch, *Fringilla coelebs*," *Ibis* 100 (1958):535–570. Reprinted with permission.

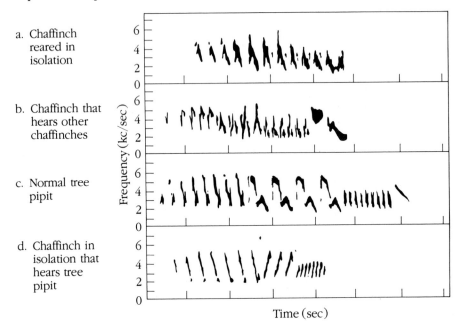

High-frequency sounds are used by a variety of animals, particularly mammals. The distress calls of young rodents and some of the vocalizations of dogs and wolves are well above the range of human hearing, as are the echolocation sounds of bats. Although bat sounds are used mainly to locate food objects, communication also occurs between predator and prey. Noctuid moths, for example, do not produce sounds themselves, but they possess tympanic membranes on each side of the body that receive sonar pulses from bats (Roeder and Treat 1961). Depending on the location and intensity of sound stimulation, the moth may fly away in the opposite direction, dive, or desynchronize its wingbeat to produce erratic flight.

Underwater sound has properties somewhat different from sound in air. Fish and invertebrates produce a variety of sounds, some of which have only recently been investigated. Marine mammals produce clicks, squeals, and longer, more-complex sounds incorporating many frequencies; most of these are short-duration sounds, which are thought to function in echolocation. Baleen whales (family Mysticeti) produce lower and longer sounds than do the toothed whales, such as dolphins. Payne and McVay (1971) analyzed the sounds of the humpback whale *(Megaptera novaneangliae)*, which are varied and occur in sequences of seven to thirty minutes duration and are then repeated. The songs have a great deal of individuality, and each whale adheres to its own for many months before developing a new one (Figure 13–10). Researchers have as yet ascribed no clear function to these sounds, but they serve to maintain group cohesion across thousands of miles because of the low attenuation of low-frequency sounds in water.

FIGURE 13-10   Song of humpback whale
The song of a humpback whale can be broken up into units, phrases, themes, songs, and song sessions. Each whale sings its own variation of the song, which may last up to a half hour.
Source: Data from R. S. Payne and S. McVay, "Songs of Humpback Whales," *Science* 173 (1971):585–597. Data, 13 August 1971.

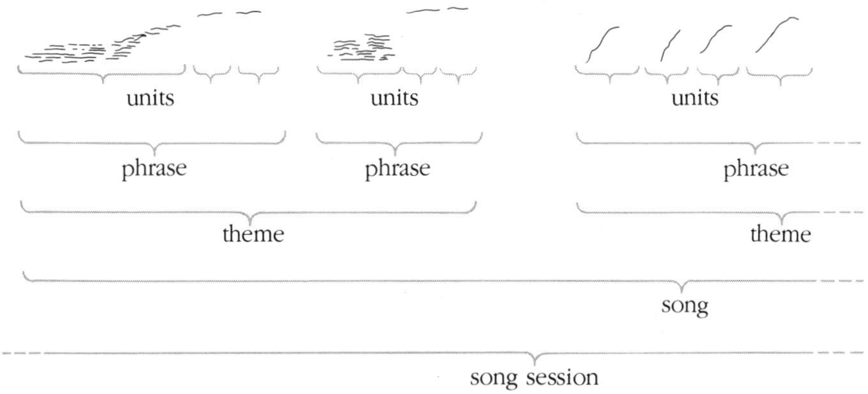

TOUCH

Short-range communication in the form of physical contact is used by many invertebrates whose antennae, covered with receptors, are the first part of the body to contact other objects and organisms. Antennae are used by nonsocial insects, such as cockroaches, and by social insects, such as bees. The honeybee often performs the waggle dance in a dark hive; therefore, much of the information about the type and location of food comes from tactile communication as the workers' antennae contact the dancer and pick up taste cues.

Perhaps the most widespread use of tactile displays occurs during copulation. In many rodents stimulation of the back end of an estrous female produces concave arching of the back and immobility (lordosis). In some mammals vaginal stimulation induces ovulation.

In most primates grooming is an important social activity (Figure 13–11) and seems to have a function not only in the removal of ectoparasites but also as a "social cement" in the reaffirmation of social bonds. Most grooming takes place between close relatives, but long-term relationships, indicated by grooming patterns, may exist between nonrelatives (Sade 1965). In the large, multimale groups characteristic of macaques (*Macaca* spp.) and baboons (*Papio* spp.), grooming between the sexes is confined largely to the mating season. South American titi monkeys (*Callicebus*), which live in monogamous groups, entwine their tails when resting. These signals may not be complex, but they are no less important than other signals.

SURFACE VIBRATION

Information can be conveyed by patterns of surface vibrations. Males of one species of water strider (*Gerris remigis*) send out ripples of a certain frequency, and receptive females respond by moving toward the source. When a female gets within a certain distance, the male switches to courtship waves (Wilcox 1972). Wilcox (1979) also observed that males generate high-frequency (HF) waves when they are close to another water strider. If return HF waves are not picked up from the second strider, the first attempts copulation. In order to demonstrate the function of HF, Wilcox used an ingenious playback method to program *females* to send out HF. He glued a magnet to the female's foreleg and allowed her to move freely inside an electrical coil. When Wilcox played an electrical copy of the male HF signal through the coil, the magnet moved the female's leg and she involuntarily sent out the HF signal. When Wilcox placed a rubber mask over the male's eyes to eliminate possible visual cues, he found that when the male approached a female, it always attempted to copulate when the female was not sending an HF signal and never did when the female was sending an HF signal. We can apply this type of playback experiment to studies of other types of substrate-transmitted signals, such as web communication among spiders.

FIGURE 13-11   Female rhesus monkey grooming offspring
In addition to removing ectoparasites and other foreign matter from the skin and hair, grooming acts as "social cement," solidifying social bonds. Most grooming occurs between close relatives.
Source: Photo by Douglas B. Meikle.

ELECTRIC FIELD

Some sharks and electric fish have electroreceptors that they use both passively and actively in detecting objects and in communicating socially. Sharks *(Scyliorhinus caniculus)* detect the electric field produced by prey flatfish that are buried in the sand (Kalmijn 1971). In addition to electrolocating objects, electric fish of the African family Mormyridae communicate information about species identity (Hopkins and Bass 1981), individual identity, and sex by modulating the shape of the electric organ discharge (see Figure 6–5). Moller (1976) demonstrated that members of this family also use electric organ discharges to maintain group coordination in schools. By altering either wavelength or pulse duration, they can communicate threat, warning, submission, and so on (Bullock 1973). The advantages of this sensory mode are that it is useful in dark, murky waters, it can travel around and even through certain objects, and it provides precise information on location.

VISION

The need for a direct line of sight and ambient light limit their use, but within social groups, visual displays enable the receiver to locate the signaler precisely in space and time. Monkeys and apes, with a few exceptions, are social, diurnal primates that rely extensively on visual displays. Primates ourselves, we human observers have studied visual systems more than other systems.

If you live east of the Rocky Mountains, you have probably seen fields and lawns sparkle with flashes of fireflies. These flashes are emitted by beetles of the family Lampyridae that have specialized photogenic tissue in the abdomen. Such behavior is known to be related in some way to mate attraction, with each species having its own flash code and flashes of the males varying in intensity, duration, and interval in a species-specific way, as do the females' responses (Carlson and Copeland 1978). Within a species the flash interval varies, depending on whether the

---

FIGURE 13-12   Responses of female *Photuris versicolor* to flashes of males
Unmated virgin fireflies of the same species (top row) usually answer only the triple flash of males of their own species. Mated females become *femmes fatales,* answering flashes of different species (middle row). Females mated to sterile males do not become *femmes fatales* (bottom row) and respond significantly less.
Source: Data from J. E. Lloyd, Aggressive mimicry in *Photuris:* Firefly femmes fatales, *Science* 149 (1965):653–654.

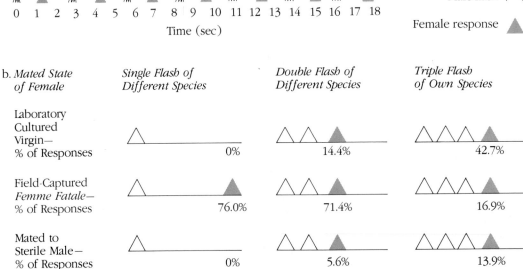

a. *Flash Codes for* Photuris versicolor

Time (sec)

Male flash

Female response

b. *Mated State of Female* — Single Flash of Different Species — Double Flash of Different Species — Triple Flash of Own Species

Laboratory Cultured Virgin— % of Responses: 0%, 14.4%, 42.7%

Field-Captured *Femme Fatale*— % of Responses: 76.0%, 71.4%, 16.9%

Mated to Sterile Male— % of Responses: 0%, 5.6%, 13.9%

male is searching for a female or courting one he has found. In one particular species *(Photuris versicolor)*, the female, once she has mated, may mimic the flash response of females of closely related species, thereby luring males to her and then devouring them. Such females are aptly termed *femmes fatales* by Lloyd (1965) (Figure 13–12). To make matters even more complicated, males of some species of the genus *Photuris* mimic males of other species in order to lure hunting *femmes fatales* of their *own* species into a second mating (Lloyd 1980). In other words, a mated female of species X who is mimicking the female of species Y in order to lure and eat a male of species Y is herself lured into another mating by a male of species X who is mimicking a male of species Y! Copeland (1983) argues that more data are needed before male mimicry can be assumed.

Fish and some invertebrates are able to change color within seconds by expanding and contracting chromatophores beneath the skin (Figure 13–13). The most spectacular in this regard is the octopus; waves of color advance and recede according to the animal's mood. Because of the limitations of visual displays, these displays are usually coupled with other modes of communication, such as audition. For instance, in the song spread, the redwing spreads its tail, lowers its wings, and raises its epaulets at the same time as it renders the song.

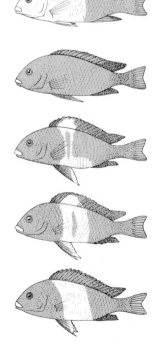

a. Frightened or submissive fish

b. Fish in neutral state

c. Beginning appearance of yellow band used in courtship

d. Increasing appearance of yellow band used in courtship

e. Maximum appearance of yellow band used in courtship

FIGURE 13-13   Graded visual signals in the mouthbrooder cichlid fish
An increasing expression of yellow band used in courtship shows in parts (c) through (e).
Source: Data from W. Wickler, "Zur Sociologie des Brabantbuntbarsches, *Tropheus moorei* (Pisces, Cichlidae)," *Zeitschrift für Tierpsychologie* 26 (1969):967–987.

"LANGUAGE"

FOOD LOCATION IN HONEYBEES

Hours or days may go by before a foraging honeybee *(Apis mellifera)* discovers a sugar or honey solution placed outdoors, but then new bees arrive within minutes. Aristotle thought that the other bees simply followed the forager to the food. However, von Frisch (1967) demonstrated that recruitment takes place even when the forager is not allowed to return from the hive to the food source. He hypothesized that bees obtain information on food location through odor communicated in the hive by the forager's "dance." Hive members maintain antennal contact with the dancer's body and taste samples of regurgitated food. But von Frisch's observations later suggested to him that the bees were getting more specific information on location of food sources, and he went on to develop his famous dance-language hypothesis.

By placing food sources at varying distances and angles from enclosed observation hives, von Frisch found that foragers perform a dance on the comb. With food at short distances (from 20 to 200 meters, depending on the strain of bees used), returning foragers perform the round dance, a series of circles with reversals in direction every second or so. With food at greater distances, they perform the waggle dance, a figure eight with a straight portion in the middle of the figure (Figure 13–14). The forager waggles its body and emits sound bursts during the straight run. The direction of the food relative to the sun is the same as the direction of the straight run relative to gravity. The duration of the straight run increases with distance at the rate of about one complete waggle per 25 meters. The area that the dance occupies on the comb, the duration of each complete figure eight cycle, and the duration of sound bursts are all also correlated with distance to the food source. Von Frisch argued that this symbolic "language" is used to communicate information about the location of food.

Although these behaviors are highly correlated with food location, how do we prove that the bees actually use this information? Some have argued that, since correlations exist, the dance must have a purpose, and that purpose must be communication. Such teleological reasoning has been unacceptable to others, however. First, many species of insects have the ability to transpose an angle flown or walked with respect to the sun into an angle with respect to gravity (Gould 1976). Correlations between waggling or buzzing and distance to food sources exist in species of insects besides honeybees, yet there is no indication that these correlations are part of a language. Second, von Frisch's studies had not eliminated the possibility that odor was solely responsible for the target accuracy of the recruits since odors of specific locales could be transmitted during the dance. Wenner (1967) and others set out to test the olfactory hypothesis further. After many experiments they concluded that the following foraging rule could account for what was observed: After the dance recruits leave the hive, they drop downwind and pick up the odors to which

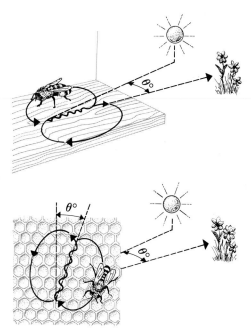

FIGURE 13-14   Waggle dance of the honeybee
A foraging honeybee returns after discovering a food source. If the bee dances outside the hive, it waggles or vibrates its body as it passes through the straight run, which points directly toward the source. If it dances on an enclosed vertical comb, it orients itself by gravity and substitutes a point directly overhead for the sun. The angle θ between the sun and the food source is the same as that between a point directly overhead and the food source.
Source: Data from K. von Frisch, *The Dance Language and Orientation of Bees* (Cambridge: Harvard University Press, 1967).

they have been recruited. There were a number of differences between von Frisch's and Wenner's experiments that could have accounted for the discrepancies. In general, Wenner used concentrated sucrose foods with strong odors at short distances from the hive. Von Frisch used low concentrations with weak odors at much greater distances. Dancing would be of less utility in the former situation than in the latter.

We could demonstrate the use of the dance with an experimental design similar in principle to Wilcox's study of the role of high-frequency vibrations in water-striders, which we discussed earlier. If the forager gave wrong information about food location during the dance, but gave correct information about odor, we would be able to see where the recruits ended up. Gould (1976) placed a light in the hive, thereby causing the foragers to orient their dance to it rather than to gravity. When the ocelli (the three simple eyes between the compound eyes) were painted over, the bees foraged and danced normally but were less sensitive to light. When they returned to the hive after discovering a food source, their waggle dance was oriented to gravity even in the presence of a light in the hive. But the untreated recruits oriented to the light. If recruits used the dance information, they would be misinformed and would go to a place that corresponded to the angle of the dance and the light. If the food source was at an angle of 20 degrees to the right of the sun, the partially blind forager, using gravity, would dance at an angle of 20 degrees to

FIGURE 13–15   Effect of artificial light in the beehive on recruitment direction
The presence of a light in a beehive causes the dancing foragers to use the light
rather than the vertical axis as the reference point. When foragers are partially
blinded so that they can see the sun but not the light in the hive, they use the
vertical referent, opposite gravity, whereas recruits use the light. Bars denote the
number of recruits to six stations within 30 minutes, and arrows denote the angle
indicated by the dance in the three different experiments (a, b, and c). In all cases the
food was located at 0 degrees. As the angle of the light was shifted, the dances
indicated a new direction, and the distribution of recruits shifted accordingly. These
misdirection experiments prove that bees use the direction information in the
waggle dance.
Source: J. L. Gould, "The Dance-Language Controversy," *Quarterly Review of Biology*
51 (1976):211–244. Reprinted with permission.

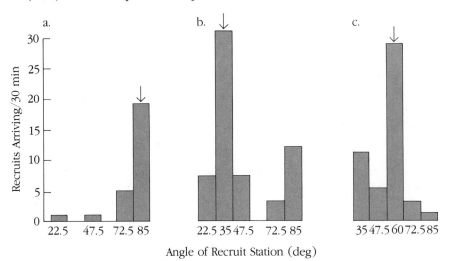

the right of the vertical. But if a light was placed in the hive at an angle of 40 degrees
to the right of the vertical, the untreated recruits should end up at a place 60 degrees
to the right of the sun, off by 40 degrees. In fact, Gould's bees did just that, demonstrating that they used direction information from the dance (Figure 13–15). In other
experiments (reviewed in Gould 1976) researchers fed foragers a poison, which
caused them to waggle at a rate lower than normal. Recruits wound up short of the
food supply by the predicted amount, demonstrating that bees also use distance
information in the waggle dance.

    One result of these studies has been to demonstrate redundancy in communi-
cation systems. Both odors of specific locales and symbolic dance are used in many
cases. The communication is multichanneled as well; touch, smell, vision, and sound
all play important roles.

LANGUAGE IN CHIMPANZEES

Communication using true language has traditionally provided a clear-cut separation of humans from other animals. By **true language** we mean both the use of symbols for abstract ideas and an understanding of syntax, so that symbols convey different messages depending on their relative positions.

Numerous investigators have explored the potential of chimpanzees *(Pan troglodytes)* to learn language. An early subject was the Hayes's (1951) home-reared chimp, which was taught to use human language. Its vocabulary consisted of only a few simple words, such as "cup" and "mama." Chimps seem to lack the motor ability to pronounce human sounds, but different approaches have demonstrated that they do not lack the ability to deal with the other aspects of learning complex language. The Gardners (Gardner and Gardner 1971) used the American Sign Language for the Deaf (Ameslan), and their chimp, Washoe, learned over 100 words. Another chimp, Sarah, learned to use plastic pieces as word symbols and could communicate with them (Premak 1971).

Using computer-age technology, Rumbaugh and Gill (1976) taught their chimp, Lana, to press buttons with symbols embossed on them in order to gain access to food and drink from vending machines or to human companionship (Figure 13–16). In addition to a vocabulary of several hundred words, she used verbs and pronouns and could assemble words in novel ways. Table 13–4 presents a conversation between Lana and an experimenter.

Sometimes Lana posed questions for herself, answered them, and then carried out the activity questioned. Once Gill observed her pose the question, by way of the keyboard, "? Lana groom," to which Lana gave the answer, "Yes," and followed it by grooming herself.

FIGURE 13–16 Chimpanzees at the computer keyboard
Each key of the keyboard has a different symbol that represents a word. The positions of the keys are scrambled frequently to avoid the use of position cues. In order to obtain goals such as food, water, and companionship, chimps must press the keys in the correct order.
Source: Photo by Elizabeth Rubert, courtesy of Yerkes Regional Primate Research Center.

TABLE 13-4   Conversation between Tim Gill and the chimpanzee Lana (June 11, 1975)

Conditions: Formal Test. Tim had entered the anteroom with a bowl of monkey chow. Lana had asked that it be loaded into the machine; however, the conditions of the test called for Tim *not* to comply, to load cabbage for vending instead, and to declare that chow (which she had requested) was in the machine. Although Lana might have asked the machine to vend "chow," she did not—[which was] appropriate to the fact that cabbage, and not monkey chow, was in the vendor. She said:

Lana:	Please machine give piece of cabbage.	16:53
	? You *(Tim)* put chow in machine. *(5 times)*	16:54 & 16:55
Tim:	*(lying)* Chow in machine. *(in response to each of the 5 requests)*	
Lana:	? Chow in machine.	16:57
Tim:	*(still lying)* Yes.	16:57
Lana:	No chow in machine. *(which was true)*	16:57
Tim:	? What in machine. *(repeat once)*	16:57 & 16:58
Lana:	Cabbage in machine. *(which was true)*	16:59
Tim:	Yes cabbage in machine.	16:59
Lana:	? You move cabbage out-of machine.	17:00
Tim:	Yes. *(whereupon he removed the cabbage and put in the monkey chow)*	17:01
Lana:	Please machine give piece of chow. *(repeatedly until all was obtained)*	17:01

Conclusion: Lana discerned what had, in fact, been loaded in the machine, did not concur with Tim's assertion that it was "chow," asked that he remove it, and then asked for "chow" when it was loaded for vending.

Source: D. M. Rumbaugh and T. V. Gill, "The Mastery of Language-Type Skills by the Chimpanzee (Pan)." In S. R. Harnad, H. D. Steklis, and J. Lancaster, eds., *Origins and Evolution of Language and Speech.* Annals of the New York Academy of Sciences 280:562–578. Reprinted with permission of the publisher and the author.

Lana's conversations were pragmatic; once she obtained her immediate goal of food, drink, or companionship, the conversation ended. Her curiosity was related to her immediate needs, and she showed no interest in extending her knowledge of the world or how things in it work. This pragmatism contrasts with language development in humans, who use language at an early age to gain information about all aspects of the environment.

A controversy perhaps more intense than that over bee dance-language resulted when Terrace and his collaborators (1979) asserted that chimps could not really create sentences. They worked with their own chimp, Neam Chimpsky (Nim for short, named after the famous linguist Noam Chomsky), and also re-analyzed the videotapes and films made by other investigators. Nim mastered a respectable vocabulary of sign language words and, like other chimps, used them to convey information to another individual. Nim did use two-sign combinations that were syntactically consistent; he signed the correct "eat banana," for example, more often than "banana eat." But he was apparently imitating his teachers' previous utterances or responding to other cues, rather than creating sentences on his own. The Gardners said in response that Nim was trained in an environment unlikely to produce spontaneous behavior and that the film segments of Washoe that Terrace analyzed were too short to demonstrate the complexity of communication (Marx 1980).

Recently the Rumbaughs have questioned whether Lana and her successors, Austin and Sherman, were using symbolic representation in the same way that humans do (Savage-Rumbaugh, Rumbaugh, and Boysem 1980). **Symbolization** means the use of arbitrary symbols to *refer* to objects and events that are removed in time and space. Although chimps learn to string words together in social interaction routines to attain goals, this accomplishment is not proof that they can do more than *associate* a word with an object — that is, their language learning does not demonstrate a referential relationship.

ORIGINS OF DISPLAYS

One of the primary goals of ethologists has been to demonstrate that displays have evolved from noncommunicative behaviors. The evolutionary process is referred to as **ritualization.** A behavior pattern may undergo the following changes during ritualization (from Eibl-Eibesfeldt 1975):

1. Change in function;
2. Change in motivation;
3. Exaggeration of movements in frequency and amplitude, but concurrent simplification;
4. "Freezing" of movements into postures;
5. Stereotyping of the behavior, while keeping frequency and amplitude relatively constant even if motivation varies;
6. Development of conspicuous body structures, such as ornamental feathers, enlarged claws, manes, sailfins.

One of the behaviors thought to have given rise to displays is **intention movement** — low-intensity, incipient movement. More often, displays seem to have evolved from **displacement activities,** which sometimes occur in conflict situations when an animal is undecided as to the appropriate response to a stimulus. For instance, male ducks sometimes preen their wings during courtship, or they may just touch their feathers rather than actively groom them. Possibly this behavior is a displacement activity resulting from a conflict between sexual and agonistic behaviors. In mandarin ducks this behavior has become a display involved in courtship. Several conspicuous feathers, or sails, have evolved and are exposed during this sham preening (Figure 13–17). Other suggested origins of displays are food exchange, comfort movements, and thermoregulatory patterns; the latter refers to displays that involve feather erection, the original function of which was the regulation of body temperature in birds (Morris 1956).

Comparisons of closely related species can provide further information about the evolution of displays (see also Chapter 18 for a discussion of the comparative method). A common feature of the courtship of pheasants and their relatives, including the domestic chicken, is food-enticing. The male chicken *(Gallus gallus)* scratches several times with its feet and pecks at the ground while calling; if no food

FIGURE 13–17   Male mandarin duck with modified primary feather or sail
The mandarin duck provides an example of the evolution of conspicuous morphological traits in the ritualization of courtship display. The male points to the sail with its beak during courtship.
Source: Photo by Michael Hopiak, courtesy of Cornell University Laboratory of Ornithology.

is present, he picks up stones as if they were food objects. The hen usually comes running, and the male can then attempt to copulate with her. The ring-necked pheasant *(Phasianus colchicus)* performs the same display. The male impeyan pheasant *(Lophorus impejanus)* bows low, with a slightly spread tail, and pecks the ground. When the hen approaches and searches for the food, he spreads his wings and tail feathers. The peacock pheasant *(Polyplectron bicalcaratum)* scratches the ground and then bows, with wings and tail spread. If he is given food, he will offer it to the female. Finally, the peacock *(Pavo)* male spreads his tail, shakes it, and moves back several steps, then points downward with his beak. The fanned tail arched over his head seems to focus the attention of the hen on the ground in front of him. Young male peacocks food-entice in the "original" form, with scratching and pecking, and develop the ritualized form as they mature (Figure 13–18).

This series of behavior patterns shows that the more ancestral, or primitive, form of courtship involves the actual searching for food by the male and the offering of it to the female in order to attract her to him for possible copulation. The most advanced form, which is the furthest evolutionarily from the ancestral pattern, is shown by the peacock (Figure 13–18), whose search for food has become a highly ritualized display. The movements are highly exaggerated, and special plumage has evolved.

HONESTY IN COMMUNICATION

Wilson's definition of communication, with which we began this chapter, did not specify which organism—sender or receiver—benefits from the behavior. Many definitions imply that both sender and receiver must benefit, and that there is evolution toward maximization of information transfer. It is but a short step to state that communication promotes the welfare of the species.

FIGURE 13-18 Food calling and courtship in phasianid birds
The rooster shows food calling, or "tidbitting," the original or primitive behavior in
(a). In parts (b) through (e), the food is no longer there, but the female is attracted to
a spot on the ground by displays of the male. Such comparative series are taken as
evidence for the evolution of displays.
Source: Data from R. Schenkel, "Zur Deutung der Balzleistungen einiger Phasianiden
und Tetraoniden," *Ornithologische Beobachter* 53 (1956):182–201.

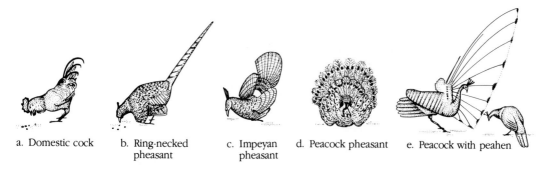

a. Domestic cock   b. Ring-necked   c. Impeyan   d. Peacock pheasant   e. Peacock with peahen
                       pheasant   pheasant

An alternative, sociobiological view of communication starts with the assumption that natural selection acts primarily at the level of the individual; thus
communication is a means by which the sender *manipulates* others for his or her own
benefit. The receiver may benefit or may be harmed. Signals become ritualized so
that the sender can control the behavior of the receiver with a minimum of wasted
energy (Dawkins and Krebs 1978; Krebs and Dawkins 1984). In some cases information is transmitted to the benefit of both. The honeybee dancer could be thought
of as manipulating her sisters. The forager reenacts the flight from the hive to the
food through the dance; the recruits amplify the dance into a real flight, and they
return with much more food than the forager could have brought back alone. In
other cases the receiver clearly does not benefit, as is the case with the male firefly
that is lured to his death by the *femme fatale.*

This account suggests that deceit should be widespread both between and
within different species. Interspecific deceit is rather common, particularly in predator-prey relationships (see Chapter 15), but intraspecific deceit seems to be much less
common, possibly because the sender and receiver belong to a common gene pool
(Dawkins and Krebs 1978). If a mutant "liar" with gene *A* appears in a population of
nonliars (gene *a*), it may increase its fitness by deceiving conspecifics. As gene *A*
spreads, the likelihood of deceiving another bearer of gene *A* increases. If bearers of
gene *A* can recognize and lie only to bearers of gene *a*, gene *A* will become fixed in the
population. The lying habit would eventually cease, however, because there would
be no one left to deceive. Lying is an advantage only when most others tell the truth.
Selection also would favor the ability to detect deceit by others and thus keep lying
relatively rare.

Dawkins and Krebs emphasized the advertising aspects of communication.
The purpose of the display is not to inform but to persuade. Exaggeration and

redundancy are the rule; and ritualization occurs to increase the persuasive power of the behavior, not to maximize information transfer.

SUMMARY

Communication may be defined as an action on the part of one organism that alters the probability pattern of behavior in another organism in a fashion adaptive to either one or both of the participants. Behavior patterns that are specially adapted to serve as social signals are termed displays.

Social signals, which vary in fade-out time, effective distance, and duration, convey information by being discrete or graded. They may be combined to form composite signals, and the order in which they appear may affect the information transmitted (syntax). Metacommunication, communication that alters the meaning of the communication that is to follow, is seen among nonhuman animals mainly in conjunction with play behavior.

Communication can be measured by comparing the frequencies of occurrence of each behavioral response against the expected frequencies if the responses were occurring randomly. If the response frequencies are significantly altered, information transfer is taking place. The use of information theory allows us to quantify the amount of communication in the case of discrete signals.

Communication functions in group spacing and coordination; individual, species, and class recognition; reproduction; agonism and social status; alarm; hunting for food; giving and soliciting care; soliciting play; and synchronization of hatching. Channels of communication include odor (mainly via pheromones), sound, touch, surface vibration, electric field, and vision. An example of complex visual communication is sign language learning in the chimpanzee. Although they lack the motor ability to produce the sounds of human language, chimps can acquire vocabularies of several hundred words by means of hand signs or substitute symbols. Chimps can use nouns, pronouns, and verbs to converse about their immediate needs, but their ability to create sentences and use true symbolism has been questioned.

By way of a dance on the comb, honeybees communicate to fellow workers information about the distance and direction of food. Odor cues also seem to be important, but several experiments have demonstrated that the dance information is of primary importance in locating the food.

Displays are thought to have evolved from such noncommunicative behaviors as intention movements and displacement acts. As behaviors change in function and motivation, natural selection produces exaggerated movements and postures. Conspicuous body structures, such as ornamental plumes on birds or claws on crabs, may have evolved to reduce ambiguity of the message and hence uncertainty of the response. By comparing differences in displays among closely related species, ethologists infer the evolutionary pathways of this process of ritualization.

Some definitions of communication imply that both sender and receiver must benefit and that there is evolution toward maximization of information transfer. An alternate, sociobiological, view argues that communication is a means by which the sender manipulates the receiver, who may benefit or may be harmed; the purpose of a display is to persuade, not to inform.

Discussion Questions

1. Although we have discussed how communication operates in several discrete sensory channels, most displays involve more than one channel. Discuss examples in which two or more channels are used. Why have such multimedia displays evolved? What methods would you use to demonstrate the function of such displays?

2. Using the crayfish data in Table 13–1 and the Shannon-Weiner formula, compute H for the receiver. Using Wilson (1975), Attneave (1959), or some other source, compute the information transmitted.

3. The table below shows the mean number of scent markings made during a ten-minute period by Maxwell's duikers — small, forest-dwelling antelopes — which were in three groups (I, II, III). What do these data tell us about the function of marking and the social organization of this species? What can you say about the Type A and Type B females?

Group Membership	Marking Activity When with Own Group	Marking Activity after Presence of Additional	
		Male	Female
	Males		
I	6.6	15.2	6.1
II	5.8	10.7	6.2
III	4.4	8.6	4.1
	Type A Females		
I	3.5	3.7	18.6
II	3.4	3.1	12.2
III	1.5	0	1.7
	Type B Females		
I	0.06	0	0.09
II	0	0.1	0
III	0.04	0	0.03

Source: K. Ralls, "Mammalian Scent Marking," *Science* 171 (1971):443–449. Copyright 1971 by the American Association for the Advancement of Science. Reprinted with permission.

4. Discuss the significance of play behavior in animals. Include the phyletic distribution of play and its occurrence in relation to social organization and ecology. Discuss the concept of metacommunication as it relates to play behavior. How would you define play operationally? For a discussion of play behavior, see Fagen (1981).

Suggested Readings

Halliday, T. R., and P. J. B. Slater, eds. 1983. *Animal Behavior,* vol. 2: *Communication.* N.Y.: Freeman.
A readable and introductory book, emphasizing the sociobiological approach. Chapters on sensory mechanisms, environment, and evolution.

Sebeok, T. A., ed. 1984. *How Animals Communicate,* 2nd ed. Bloomington: Indiana University Press.
A massive tome with chapters by specialists on different taxonomic groups.

Smith, W. J. 1984. *Behavior of Communicating,* 2nd ed. Cambridge: Harvard University Press.
An in-depth text that covers the entire subject, emphasizing the ethological approach.

Wilson, E. O. 1975. *Sociobiology: The New Synthesis.* Cambridge: Harvard University Press.
Chapters 8, 9, and 10 provide a concise overview of communication, with an explanation of information theory.

References

Attneave, F. 1959. *Applications of Information Theory to Psychology:* New York: Holt, Rinehart and Winston.

Baker, M. C. 1983. The behavioral response of female Nuttall's white-crowned sparrows to male song of natal and alien dialects. *Behav. Ecol. Sociobiol.* 12:309–315.

Bekoff, M. 1977. Social communication in canids: Evidence for the evolution of a stereotyped mammalian display. *Science* 197:1097–1099.

Bovjberg, R. V. 1956. Some factors affecting aggressive behavior in crayfish. *Physiol. Zool.* 29:127–136.

Bronson, F. 1971. Rodent pheromones, *Biol. Reprod.* 4:344–357.

Buckle, G. R., and L. Greenberg. 1981. Nestmate recognition in sweat bees *(Lasioglossum zephyrum):* Does an individual recognize its own odour or only odours of its nestmates? *Anim. Behav.* 29:802–809.

Bullock, T. H. 1973. Seeing the world through a new sense: Electroreception in fish. *Amer. Scientist* 61:316–325.

Carlson, A. D., and J. Copeland. 1978. Behavioral plasticity in the flash communication systems of fireflies. *Amer. Scientist* 66:340–346.

Chadab, R., and C. W. Rettenmeyer. 1975. Mass recruitment by army ants. *Science* 188:1124–1125.

Copeland, J. 1983. Male firefly mimicry. *Science* 221:484–485.

Darwin, C. 1872. *Expression of the Emotions in*

Man and Animals. Chicago: University of Chicago Press, 1965.

Dawkins, R., and J. R. Krebs. 1978. Animal signals: Information or manipulation? In J. R. Krebs and N. B. Davies (eds.), *Behavioural Ecology: An Evolutionary Approach.* Sunderland, Mass.: Sinauer Associates.

de Roth, S. 1974. Communication in the crayfish *(Orconectes rusticus).* Master's thesis, Bowling Green State University.

Eibl-Eibesfeldt, I. 1975. *Ethology: The Biology of Behavior.* New York: Holt, Rinehart and Winston.

Emlen, S. T. 1972. An experimental analysis of the parameters of bird song eliciting species recognition. *Behaviour* 41:130–171.

Fagen, R. 1981. *Animal Play Behavior.* N.Y.: Oxford University Press.

Falls, J. B. 1969. Functions of territorial song in the white-throated sparrow. In R. A. Hinde (ed.), *Bird Vocalizations: Their Relations to Current Problems in Biology and Psychology: Essays Presented to W. H. Thorpe.* Cambridge: Cambridge University Press.

Falls, J. B., and R. J. Brooks. 1975. Individual recognition by song in white-throated sparrows. II. Effects of location. *Can. J. Zool.* 53:1412–1420.

Frisch, K. von. 1967. *Dance Language and Orientation of Bees.* Cambridge: Harvard University Press.

Gardner, B. T., and R. A. Gardner. 1971. Two-way communication with an infant chimpanzee. In A. M. Schrier and F. Stollnitz (eds.), *Behavior of Nonhuman Primates,* vol. 1. New York: Academic Press.

Gardner, R. A., and B. T. Gardner. 1969. Teaching sign language to a chimpanzee. *Science* 165:664–672.

Gould, J. L. 1976. The dance-language controversy. *Quart. Rev. Biol.* 51:211–244.

Hayes, J. H. 1951. *The Ape in Our House.* New York: Harper & Brothers.

Hazlett, B. A., and W. H. Bossert. 1965. A statistical analysis of the aggressive communications systems of some hermit crabs. *Anim. Behav.* 13:357–373.

Holmes, W. G., and P. W. Sherman. 1983. Kin recognition in animals. *Amer. Scientist* 71:46–55.

Hopkins, C. D., and A. H. Bass. 1981. Temporal coding of species recognition signals in an electric fish. *Science* 212:85–87.

Huxley, J. S. 1914. The courtship-habits of the great crested grebe *(Podiceps cristatus);* with an addition to the theory of sexual selection. *Proc. Zool. Soc. Lond.* 35:491–562.

Inglis, I. R., and A. J. Isaacson. 1978. The responses of dark-bellied brent geese to models of geese in various postures. *Anim. Behav.* 26:953–958.

Kalmijn, A. J. 1971. The electric sense of sharks and rays. *J. Exp. Biol.* 55:371–383.

Krebs, J. R., and R. Dawkins. 1984. Animal signals: Mind-reading and manipulation. In J. R. Krebs and N. B. Davies (eds.), *Behavioural Ecology: An Evolutionary Approach,* 2nd ed. Sunderland, Mass.: Sinauer Associates.

Kruuk, H. 1972. *The Spotted Hyena: A Study of Predation and Social Behavior.* Chicago: University of Chicago Press.

Lawick, H. van, and J. van Lawick-Goodall. 1971. *Innocent Killers.* Boston: Houghton Mifflin.

Lloyd, J. E. 1965. Aggressive mimicry in *Photuris:* Firefly femmes fatales. *Science* 149:653–654.

———. 1980. Male *Photuris* fireflies mimic sexual signals of their females' prey. *Science* 210:669–671.

Lombardi, J. R., and J. G. Vandenbergh. 1977. Pheromonally induced sexual maturation in females: Regulation by the social environment of the male. *Science* 196:545–546.

Marler, P. 1957. Specific distinctiveness in the communication signals of birds. *Behaviour* 11:13–39.

———. 1967. Animal communication signals. *Science* 157:769–774.

———. 1968. Aggregation and dispersal: Two functions in primate communication. In P. C. Jay (ed.), *Primates: Studies in Adaptation and Variability.* New York: Holt, Rinehart and Winston.

———. 1973. A comparison of vocalizations of red-tailed monkeys and blue monkeys, *Cercopithecus ascanius* and *C. mitis,* in Uganda. *Zeit. Tierpsychol.* 33:223–247.

Marler, P., and W. J. Hamilton. 1966. *Mechanisms of Animal Behavior.* New York: Wiley.

Marler, P., and S. Peters. 1977. Selective vocal learning in a sparrow. *Science* 198:519–521.

Marler, P., and M. Tamura. 1962. Song "dialects" in three populations of white-crowned sparrows. *Condor* 64:368–377.

Marx, J. L. 1980. Ape-language controversy flares up. *Science* 207:1330–1333.

Menzel, E. W. 1971. Communication about the environment in a group of young chimpanzees. *Folia Primat.* 15:220–232.

Moller, P. 1976. Electric signals and schooling behavior in a weakly electric fish, *Marcusenius cyprinoides* L. (Mormyriformes). *Science* 193:697–699.

Morris, D. 1956. The feather postures of birds and the problem of the origin of social signals. *Behaviour* 9:75–113.

Moynihan, M. 1956. Notes on the behaviour of some North American gulls. I. Aerial hostile behaviour. *Behaviour* 10:126–178.

Payne, R. S., and S. McVay. 1971. Songs of humpback whales. *Science* 173:585–597.

Premak, D. 1971. Language in chimpanzee? *Science* 172:808–822.

Ralls, K. 1971. Mammalian scent marking. *Science* 171:443–449.

Roeder, K. D., and A. E. Treat. 1961. The detection and evasion of bats by moths. *Amer. Scientist* 49:135–148.

Rumbaugh, D. M., and T. V. Gill. 1976. The mastery of language-type skills by the chimpanzee (Pan). In S. R. Harnad, H. D. Steklis, and J. Lancaster (eds.), *Origins and Evolution of Language and Speech. Ann. N.Y. Acad. Sci.* 280:562–578.

Sade, D. S. 1965. Some aspects of parent-offspring and sibling relations in a group of rhesus monkeys, with a discussion of grooming. *Amer. J. Phys. Anthrop.* 23:1–17.

Savage-Rumbaugh, E. S., D. M. Rumbaugh, and S. Boysem. 1980. Do apes use language? *Amer. Scientist* 68:49–61.

Schneider, D. 1974. The sex-attractant receptor of moths. *Sci. Amer.* 231:28–35.

Sebeok, T. A. 1965. Animal communication. *Science* 147:1006–1014.

———. ed. 1984. *How Animals Communicate,* 2nd ed. Bloomington: Indiana University Press.

Seyfarth, R. M., D. L. Cheyney, and P. Marler. 1980. Monkey responses to three different alarm calls: Evidence of predator classification and semantic communication. *Science* 210: 801–803.

Shannon, C. E., and W. Weaver. 1949. *Mathematical Theory of Communication.* Urbana: University of Illinois Press.

Sherman, P. W. 1977. Nepotism and the evolution of alarm calls. *Science* 197:1246–1253.

Slater, P. J. B. 1983. The study of communication. In T. R. Halliday and P. J. B. Slater (eds.), *Animal Behaviour,* vol. 2: *Communication.* New York: Freeman.

Smith, W. J. 1984. *Behavior of Communicating,* 2nd ed. Cambridge: Harvard University Press.

Struhsaker, T. T. 1967. Auditory communication among vervet monkeys *(Cercopithecus aethiops).* In S. A. Altmann (ed.), *Social Communication among Primates.* Chicago: University of Chicago Press.

Terrace, H. S., L. A. Petitto, R. J. Sanders, and T. G. Bever. 1979. Can an ape create a sentence? *Science* 206:891–902.

Thorpe, W. H. 1958. The learning of song patterns by birds, with special reference to the song of the chaffinch, *Fringilla coelebs. Ibis* 100:535–570.

Tinbergen, N. 1951. *The Study of Instinct.* Oxford: Clarendon Press.

Vince, M. A. 1969. Embryonic communication, respiration and the synchronization of hatching. In R. A. Hinde (ed.), *Bird Vocalizations: Their Relations to Current Problems in Biology and Psychology: Essays Presented to W. H. Thorpe.* Cambridge: Cambridge University Press.

Weldman, B. 1982. Sibling association among schooling toad tadpoles: Field evidence and implications. *Anim. Beh.* 30:700–713.

Wenner, A. M. 1967. Honeybees: Do they use the distance information contained in their dance maneuver? *Science* 155:847–849.

Wilcox, R. S. 1972. Communication by surface waves: Mating behavior of a water strider (Gerridae). *J. Comp. Physiol.* 80:255–266.

———. 1979. Sex discrimination in *Gerris remigis:* Role of a surface wave signal. *Science* 206:1325–1327.

Wilson, E. O. 1971. *Insect Societies.* Cambridge: Harvard University Press.

———. 1975. *Sociobiology: The New Synthesis.* Cambridge: Harvard University Press.

14

MIGRATION, ORIENTATION, AND NAVIGATION

Many species of birds, from tiny hummingbirds and warblers to cranes and hawks, migrate southward from temperate regions over water and land, often to the same location each winter and back to the same tree each spring. Other animals undertake seasonal movements, many disperse from natal sites, and most animals move about every day as they feed, nest, and search for mates. How do these animals find their way?

In the first section of the chapter, we look at **migration**—a periodic movement from one location and climate to another location and climate—and its evolutionary history. In the second part of the chapter, we examine the means of orientation and navigation that have evolved and the cues animals use to find their way not only in long-distance migrations, but also in their home areas. We can define **orientation** as the way in which an organism positions itself in relation to external cues. **Navigation** is the process by which an animal uses various cues to determine its position in reference to a goal as it moves about from place to place. **Homing,** which involves the use of navigation, is the ability of an animal to return to its home site or locale after being displaced. **Piloting** can be defined as the animal's use of familiar landmarks to find a goal or direction.

MIGRATION

BIRDS

Most of us are familiar with the fall and spring migrations of various bird species. The honking of geese and their V-shaped formations winging overhead herald the true arrival of fall each year, just as the appearance of certain warbler species signals the onset of spring (Figure 14–1). We can ask four major questions about migratory flights: (1) What species migrate, from where to where, and when? (2) What external cues and internal physiological events trigger migration? (3) What is the evolutionary history of migratory behavior? and (4) What types of cues do birds use to guide them in their long-distance flights? We examine the first three questions now and treat the last question in the second half of the chapter.

DESCRIPTION OF MIGRATORY BEHAVIOR.   Information about which bird species migrate, where their summer and winter ranges are located, what route they follow in flight, and at what speed they travel over long distances is gathered through several techniques. A common method is banding, in which researchers fit small, colored or numbered cylinders of metal or plastic on birds' legs; the toes prevent the bands from slipping off. Researchers capture the birds in special nets or in traps, place bands on the legs, and make careful records of such data as the species, sex, age, location of capture. When the same bird is later caught, shot, or found dead in another location, researchers hope the band will be noticed and the information reported. In the United States the U.S. Fish and Wildlife Service is the clearinghouse for securing permission to band birds and for reporting all information pertaining to banding. From these accumulated records of bands and recovered bands, we can draw a picture of migratory species and collect vital facts about their migrations.

   For example, from banding data we have learned that many duck species,

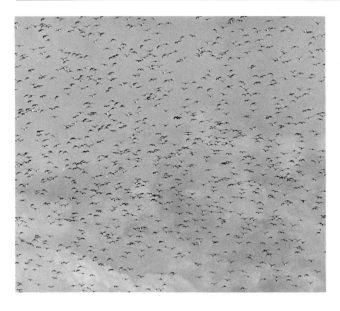

FIGURE 14–1   Migrating knots
Each fall many thousands of birds of many species desert their summer residences in the north temperate zones to fly southward. Some fly in small groups; others, such as knots (*Calidris canutus*), fly in large flocks numbering in the thousands.
Source: Photo by Eric Hosking.

including teal *(Anas discors)* and mallards *(Anas platyrhynchos)*, spend the summer months nesting in marshes and lakes in the northern United States or Canada and fly to the southern states for the winter. Examples of spectacular migration speeds have also come from banding data (see Griffin 1974); for instance, a sandpiper flew 2,300 miles from Massachusetts to the Panama Canal Zone in nineteen days (an average speed of 125 miles per day), and a very small (10 grams) lesser yellowlegs *(Tringa flavipes)* traveled 1,930 miles from Massachusetts to the island of Martinique in the Caribbean in just six days (a rate of more than 320 miles per day).

We can also use radar in the study of bird migration to ascertain the approximate number of birds in a flock, the speed of movement, the altitude at which the flock is moving, and the flight path of the flock. We can also focus a telescope on the moon; the birds' passage across the lighted background allows us to estimate the numbers of birds, and through the silhouettes, to identify some of the species.

TRIGGERS. The timing of migration is probably under the control of endogenous biological clock mechanisms that are set by and adjusted to external stimuli. Two processes appear to be involved: a phase of preparation for migration and the actual triggers that initiate migration. Among the stimuli that are part of the birds' environment, the most prevalent and consistent is daylength. The annual cycle of increasing daylength in the spring and declining daylength in the fall is consistent from year to year. Thus, the amount of daylength is the birds' best cue for initiating preparations for a fall migration.

The preparation phase is characterized by two major features: increasing fat deposition and migratory restlessness *(Zugenruhe)*. The metabolic system of most migratory birds goes through two cycles each year—one in the fall and another in spring—before each migratory flight. At these times the birds add large amounts of fat, the food reserves that they need for energy during flight. For the bird species whose pre- and postmigratory body weights have been determined, the average weight loss is 30 to 40 percent. Species that make long-distance flights over water or that fly nonstop from the summer residence to the winter resting ground store virtually all of the necessary food energy for the flight before starting. Among the birds that employ this strategy are some warbler species (family Parulidae) that migrate to the Caribbean islands, flying nonstop from Cape May, New Jersey, and similar locations. Other species, like the white-throated sparrow *(Zonotrichia albicollis)* (Figure 14–2), store some fat but make frequent stops to replenish their energy supply.

Control of metabolism is centered, in part, in the hypothalamus and pituitary gland. Hormones secreted by the master gland affect the way in which foods are metabolized and the accumulation of fat reserves (Meier 1973). After the fall migration the pituitary apparently enters a refractory phase but "reawakens" in the spring in time to trigger both the deposition of fat for migration and the production and secretion of sex hormones, the latter signal preparation for breeding after the birds return to their summer home. Earlier, observers suggested that because changing daylengths stimulated both migration and changes in sexual activity, these two activities were interrelated. Experiments with golden-crowned sparrows *(Zonotri-*

FIGURE 14-2   White-throated sparrow
The white-throated sparrow provides an example of a migratory species that only stores part of its needed food reserves before initiating its flight. These sparrows make several stops along the route for additional food.
Source: Photo by G. Ronald Austing.

chia atricapilla) (Morton and Mewaldt 1962) have indicated that castrated birds still develop fat deposits; the two processes may involve some of the same hormones and triggers, but they are not directly causally linked.

A second characteristic of the preparation phase is migratory restlessness, or **Zugunruhe.** As birds approach the time of migration, they show increased activity levels; automated perches, placed in cages with captive birds, measure the birds' movements, as shown in Figure 14–3 (Farner 1955). Restlessness is related to day-length and may also be affected by other weather conditions, such as storms; birds exhibit more restlessness under conditions that are favorable for migratory flights.

Thus some type of endogenous rhythm, which involves neuroendocrine mechanisms, appears to regulate the processes of fat deposition and the onset of migratory restlessness (see Meier 1973). The initiation of migration is triggered by the accumulation of sufficient fat deposits, possibly monitored by neural and hor-monal systems, and by the appearance of favorable weather conditions. Once initi-ated, some migrations — for example, the nonstop flights of birds such as golden plovers *(Pluvialis dominica)* — are all-or-none; there is no turning back. For other species, like robins *(Turdus migratorius)*, blackbirds *(Agelaius phoeniceus)*, and those that migrate primarily over land, the migratory flock may stop temporarily or even reverse direction toward the south for a short while if, as often happens in the spring, it encounters unfavorable weather (cold, storms).

EVOLUTION OF MIGRATION.   We should consider two aspects of the function of migratory behavior: the advantages and how the behavior evolved. One clear ad-vantage for birds that migrate is that they have an adequate food supply and favorable climatic conditions for all months of the year. By leaving the north temper-ate zones during the harsh portion of the year, the birds avoid the perils of cold or stormy weather and the constant risk of running short of food. Some bird species

FIGURE 14-3   *Zugunruhe* in male white-crowned sparrow
During the molting period, prior to the onset of *Zugunruhe*, or migratory restlessness
(a), the sparrow exhibits a different pattern of activity than it does after the onset of
Zugunruhe (b). Lined bars represent the twilight, shaded bars indicate periods of
darkness, and the open bars are daylight hours.
Source: D. S. Farner. "The Annual Stimulus for Migration: Experimental and
Physiologic Aspects," in A. Wolson (ed.), *Recent Studies in Avian Biology* (Urbana:
University of Illinois Press, 1955). Copyright 1955 University of Illinois Press.
Reprinted with permission.

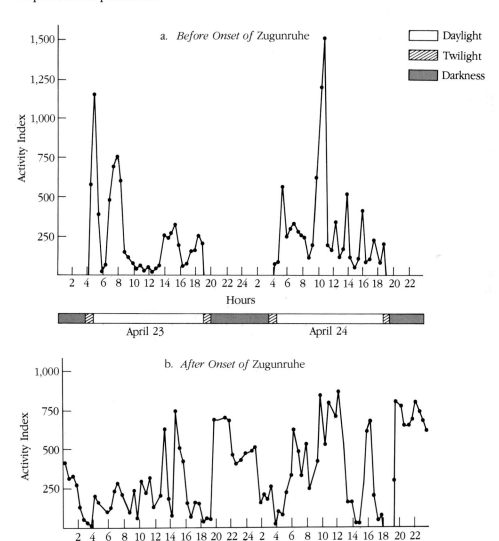

have adapted their diets and other behaviors — for example, huddling for warmth at night — to permit them to remain in northern latitudes throughout the year. Indeed, as more sources of artificial food, such as garbage or food supplies placed in feeders, have become available, a few members of some bird species — robins, for example — have ceased to migrate. However, if many species remained in their summer locations throughout the year, the demands for food would far outweigh available supplies.

Other advantages for migratory birds involve reproduction. Although the breeding season is shorter in the north, daylength is increased, and individual birds are able to concentrate their reproductive activities into fewer days. They can also take advantage of seasonally rich food supplies at the northern latitudes. An additional advantage is that by concentrating the time of mating and rearing of young, a period when both adults (of many species) and their offspring, either eggs or pre-fledging young, are most vulnerable to predation, birds may reduce the risk of death or unsuccessful reproduction. Also, because the breeding season is short, the probability of reproductive synchrony is enhanced.

Migration also has at least two significant consequences for the genetics and evolution of birds. Geographic dispersal is facilitated because of migration. Individuals of a species may establish themselves in new areas as the result of migratory flights, either by settling down in a new location along the migration route or possibly by being forced off course by winds or inclement weather during migration. Because individuals of migratory species are subjected to at least two different environments as well as the interim environment during migration, they may be subjected to rigorous natural selection pressures. However, nonmigratory species are also subject to rigorous selection due to conditions they face by remaining in the same geographical area throughout the harsh seasons of the year.

Three theories have been advanced to account for the evolution of migratory behavior in birds. One theory claims that certain bird species evolved in southern latitudes, near the equator, and, as time passed, filled new niches and expanded their ranges northward where food supplies were abundant and climatic conditions were acceptable, but only for a portion of the year (Lincoln 1950; Graber 1968). These species may have found it necessary to return to more favorable conditions nearer their original tropical and subtropical homelands for the winter season each year.

An alternate theory holds that some migratory bird species may have evolved in northern latitudes at a time when conditions were such that seasonal movements to avoid harsh conditions were not necessary. Subsequent periods of glaciation — for example, in the Pleistocene — forced these birds to fly south each year to seek better conditions (Sauer 1963). Their present annual migrations northward can be interpreted as attempts to return to their ancestral homelands in northern climes to breed each year.

Still another theory relates bird migration to continental drift (Wolfson 1948). The subtropical or tropical evolutionary origins of many bird species in Gondwonaland was followed by migratory movements out to the newer land masses as continental drift occurred.

MAMMALS

Among small mammals, the best-known migrants are bats of the order Chiroptera. Much of what we know about bat migrations has been derived from tagging bats, a technique similar to bird banding (see Griffin 1970). Some bats that live in more temperate climates hibernate during the winter months; others, including some species that also hibernate, migrate southward to warmer climates. The distances of these migratory movements vary from a few hundred to over one thousand kilometers. For instance, little brown bats *(Myotis lucifugus)* move from the New England states in a southwesterly direction to areas of Pennsylvania and adjacent states (Davis and Hitchcock 1965); and in western North America, hoary bats *(Lasiurus cinerea)* move from summer ranges in northern coniferous forest into central California and Mexico.

Among the larger mammals, some ungulates migrate; perhaps the most spectacular of these migrations is that of the caribou *(Rangifer tarandus)* of the Arctic. Enormous herds of caribou migrate between winter ranges in Canada's central latitudes and summer ranges in the northern Arctic (Figure 14–4) (Murie 1935; Banfield 1954; Lent 1966). The summer ranges are north of the timberline, where the caribou subsist on grasses and lichens. With the fall migration southward, they return to the shelter and food supplies provided by the trees.

FIGURE 14–4 Migration routes of caribou
Source: H. L. Gunderson, *Mammalogy* (New York: McGraw-Hill, 1976), p. 343. Copyright © 1976. Used with the permission of McGraw-Hill Book Company.

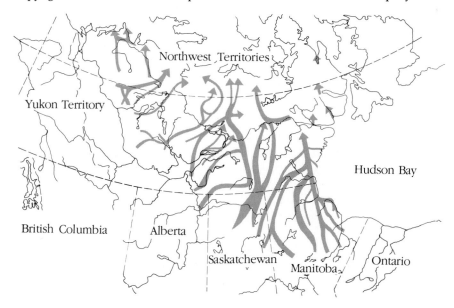

Certain marine mammals — for example, gray whales *(Eschrichtius gibbosus)* (Figure 14–5) and harbor seals *(Phoca vitulina)* — also exhibit patterns of seasonal migration. Gray whales spend summers in the Arctic Ocean and North Pacific Ocean; in the winter they inhabit shallow waters off the coast of Mexico. Beginning in late January, gray whales calve in the warm waters off the coasts of Baja California and Mexico. Interestingly, the adult whales do not appear to feed either during the migration south or while they are in the breeding area. Instead, they utilize the fat deposits accumulated the previous summer in the food-rich Arctic waters. In March or early April, the whales begin the long trip back north.

Orr (1970) has stated that three categories of factors — alimental, climatic, and gametic — help to explain the evolution of migration in mammals (and in many other classes of organisms, too). Alimental factors pertain to diet and available food supply; the need to obtain sufficient food energy underlies the evolution of migratory behavior in species like the caribou. Climatic factors are those that concern the ambient conditions, including temperature, rainfall, and so forth. Gametic factors appear to be important in some species, for example, the gray whales just described and the Pacific salmon *(Oncorhynchus* spp.). Food and environmental conditions do not appear to explain their migratory behavior. Rather, the migration seems to bring the sexes together for the purpose of mating, or it provides a place for birth and early development of young that will enhance successful reproduction (reviewed by Dingle 1980).

INVERTEBRATES

Monarch butterflies *(Danaus plexippus)* of the eastern United States fly south to sites in Mexico for the winter; and monarchs from the west coast move to locations along the southern California coast (Figure 14–6). Unlike birds and mammals, monarchs

FIGURE 14-5   Gray whale migrating up the California coast
Gray whales spend the winter months in warm lagoons off the coast of Baja California and other parts of the Mexican coast. They migrate northward each spring and summer to the Arctic and North Pacific oceans. *Source:* Photo by Steven Katona.

FIGURE 14-6 Monarch butterflies (Danaus plexippus) at overwintering site
The monarch butterfly is a seasonal migrant, flying south to Mexico or southern California for the winter. The butterflies that return north the following spring are the result of several generations of breeding en route.
Source: Photo by L. Brower.

do not restrict their breeding activities to their summer residences in northern locations. Adult butterflies migrate southward in the fall and overwinter in large aggregations in trees (Zahl 1963; Brower et al. 1977; Calvert and Brower 1981). Those that survive until the following spring breed at or near the overwintering site. Then as the monarchs move northward, they complete up to three additional generations. Thus the residents of some summer locations are the third or fourth generation descendants of the butterflies that migrated south the previous fall.

The locust (family Acrididae) illustrates a periodic movement type of insect migration. Young locusts, or hoppers, that develop in low-density populations exhibit a generally low level of activity, and the adult locusts after metamorphosis are normally rather solitary and do not engage in sustained flights. When young hoppers develop in dense populations, they tend to be more active and are gregarious; the adults, too, are more gregarious and exhibit longer flights (Johnson 1969). Populations of locust species vary between sedentary and migratory phases, depending on the interaction of social density and environmental triggers such as food availability and climate that serve as migratory stimuli. When food supplies diminish, changes in the physiology of the developing insects and increased population densities interact to produce migratory adult forms; large swarms, which may number in the millions, set forth on long flights and consume prodigious quantities of vegetation in their path.

OTHERS

Other organisms, representing all groups in the animal kingdom, exhibit migratory behavior (Dingle 1980). Pacific Ocean salmon migrate upstream to the place of their birth to spawn. Certain eels *(Anguilla rostrata)* live much of their lives in fresh waters in North America and Europe and migrate to the Sargasso Sea region of the Atlantic each year for breeding. Many frogs from North America's fast-flowing freshwater streams migrate to lakes or ponds to breed in an environment in which egg deposition and fertilization are more practicable. Among reptiles the desert tortoise migrates between summer and winter habitats.

ORIENTATION AND NAVIGATION

EARLY WORK

One of the early investigators of animal orientation, Jacques Loeb (1918), theorized that asymmetrical stimulation of an animal's sensory organs results in differential contraction of muscles on the opposite side of the animal until the symmetry is restored for both sense organs and muscle actions. Other investigators (e.g., Mast 1938) reached the conclusion that not all animal orientation could be fitted to Loeb's sensory organ-muscle scheme and that different organisms have different systems of orientation.

Later, Fraenkel and Gunn (1940) summarized the work on animal orientation up to that date and defined general classes of orienting reactions. **Kineses** are random locomotion patterns in which there is no orientation of the organism's body axis in relation to the source of stimulation. **Taxes** are directed reactions involving (in a single stimulus situation) an orientation of the long axis of the body in line with the stimulus source. Movements toward the stimulus are positive taxes; movements away from the source of stimulation are negative taxes. For instance, planaria exhibit a negative phototaxis — movement away from a light source (Figure 14–7). Extensive studies have revealed other, often complex, systems of orientation and navigation in different organisms, only a few of which we examine here.

BIRDS

In the last three decades, a variety of studies have been conducted on orientation in and navigation by birds. Many of the hypotheses that have been advanced for birds are also potentially applicable to other animals, such as insects and mammals (see reviews by Mathews 1968; Griffin 1974; Emlen 1975a; Able 1980).

Navigation requires both a compass to provide directional information and a map to provide the animal with information on its position relative to home or some other goal. As the following summaries of evidence for different cues illustrate, considerable evidence exists for the compass, but we still know very little about the "map" (Keeton 1969, 1970).

TOPOGRAPHIC FEATURES. Use of familiar topographic features is certainly a factor in navigation, but it is probably of secondary importance compared with other cues. Nonetheless, many diurnal and nocturnal migratory birds are influenced by and may utilize topographic features. For example, birds congregate near and fly along coastlines or river valleys, and they may also use landmarks for piloting, particularly in areas at both ends of the migration, where they are more familiar with specific topographic features. However, the use of topography as a guidance system has inherent drawbacks. How do first-time migrants who have never learned the landmarks find their way? What if a storm blows birds off the normal route into areas they have never traversed before? Also, visual landmarks alone cannot provide the necessary compass direction for proper orientation over long distances. Both landmarks and certain topographic features could be used in combination with some type of compass mechanism to provide sustained orientation during longer movements. Also, in familiar terrain, landmarks may suffice without any compass (Able 1980).

Familiar landmarks may play a critical role in finding particular stopover locations, as in the case in migrating waterfowl (Bellrose 1964, 1971). Evidence from radar studies of bird migration, however, has shown that nocturnal migrants in different parts of the country ignore most topographic features (Gauthreaux 1971; Richardson 1972; Drury and Nisbet 1964). Studies in which homing pigeons *(Columba livia)* have been fitted with frosted lenses over their eyes (Schmidt-Koenig and

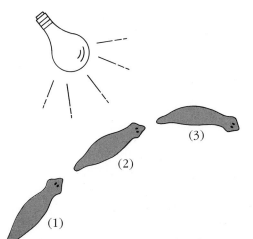

FIGURE 14–7 Planaria exhibiting negative phototaxis
Planaria turn away from a source of stimulation, in this case a light source. In this instance the light was turned on when the planarian reached position (2).

Schlichte 1972; Schmidt-Koenig and Walcott 1973) have revealed that when the birds can locate the sun, even if they see clearly for only three meters upon release in a new location, they return to the loft or to within a short distance of home. When investigators used airplane tracking and ground radar to follow the flight paths of homing pigeons with frosted lenses, they found that the paths were generally in the direction of home. The pigeons' navigation system does not require detailed vision and can lead birds that have been released 13 to 20 kilometers from home to within a short distance of the loft.

SUN. The most prominent regular cue for diurnal migratory birds, and possibly also for those that migrate at night, is the sun. The latter take a bearing at sunset and use that bearing to fly at night (Able 1982). The early studies of bird navigation devoted considerable attention to the sun (see review by Keeton 1974). Kramer (1950, 1951) first showed that European starlings *(Sturnus vulgaris)*, placed in an outdoor cage with the sun visible, exhibited migratory restlessness in the appropriate direction in the spring and fall (Figure 14–8a,b). When Kramer used mirrors to alter the apparent position of the sun, the starlings' migratory restlessness shifted direction in a predict-

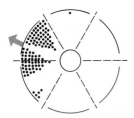

a. Clear skies

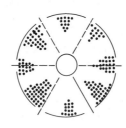

b. Overcast skies

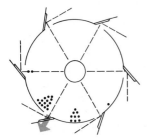

c. Sun's image
   deflected 90°
   counterclockwise
   by mirrors

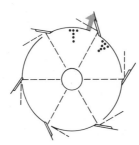

d. Sun's image
   deflected 90°
   clockwise
   by mirrors

FIGURE 14–8   Starling orientation experiment
Each circular diagram illustrates the orientation of diurnal spontaneous migratory activity in a caged European starling. Experimenters caged the bird in an outdoor pavilion with six windows and conducted tests during the migratory season (a) under clear skies, (b) under overcast skies, (c) with the sun's image deflected 90° counterclockwise by mirrors, and (d) with the sun's image deflected 90° clockwise by mirrors. Arrows denote mean direction of activity, and each dot within the circles represents ten seconds of fluttering activity. The dotted lines indicate the direction of light coming from the sky.
Source: S. T. Emlen, "Migration, Orientation and Navigation." In D. S. Farner and J. R. King, eds., *Avian Biology*, vol. V (New York: Academic Press, 1975), p. 152. Reprinted with permission of the publisher and the author.

able manner (Figure 14–8c,d). As we discuss later, other organisms (e.g., insects, turtles) appear to use some form of sun-compass navigation system.

After the initial discovery of sun-compass orientation in birds, theories were put forth to explain how birds use the sun cue (see Mathews 1968). Some investigators theorized that birds use the sun only to gain a compass bearing to head in a particular direction. Others theorized they use true navigation — the ability to orient toward a goal regardless of its direction and without the use of familiar landmarks (Griffin 1955).

One way to demonstrate the use of a sun compass is to examine whether birds maintain a constant orientation at different times of the day even though the sun's position varies as it traverses the sky. Data from some diurnal migrants provides support for this alternative. Clock-shift experiments have provided a second, more rigorous test of sun-compass orientation. First, clock shifts of six hours, brought about by artificially shifting the birds' internal clock mechanism, resulted in predictable alterations of 90 degrees in the initial bearing of birds heading for the home site. This finding is consistent with the notion that birds use the sun as a simple compass. As a further test birds were clock-shifted six hours slow and released at a site 100 miles south of the home loft (Figure 14–9). If the birds were using true sun navigation, they should have decided that they were 4,000 miles east of home and headed directly west. However, the birds headed directly east, as would be predicted if they were using the sun *only as a compass* and not as both a map and compass (Keeton 1974).

The sun, then, is an important directional cue for birds that are diurnal migrants and for homing pigeons. Evidence in support of the hypothesis that the sun can be used as both compass and map is lacking.

STELLAR CUES. When nocturnally migrating birds are placed in outdoor cages with a view of the clear night sky, they exhibit migratory restlessness in a direction appropriate to the seasonal migration (Kramer 1949, 1951; Sauer 1957). Sauer exposed

---

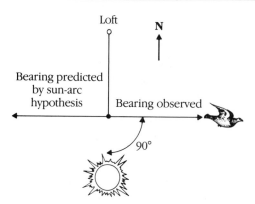

FIGURE 14-9 Clock-shifted pigeon experiment
Experimenters released six-hour-slow clock-shifted pigeons at noon at a sight 100 miles south of their home loft. The birds' internal clock indicates it is 6:00 a.m., so, if they are using sun-arc navigation, sight of a noon sun should lead them to determine their location as thousands of miles east of home, and they should depart westward. They actually depart eastward. This experiment and others support the contention that pigeons use the sun as a directional compass, but not as a "map."
Source: Data from W. T. Keeton, "The Orientation and Navigational Basis of Homing in Birds." *Advances in the Study of Behavior* 5 (1974):47–132.

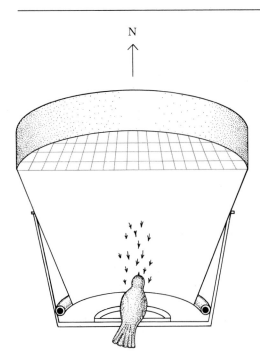

N

FIGURE 14–10   Circular cage for orientation experiment
A special cage enables researchers to study the pattern or direction of restlessness in migrating birds. A bird stands on the ink pad at the bottom of the funnel-shaped cage where it can view the stars overhead through a wire mesh top. Each jump it makes as if to fly is recorded on paper on the sides of the funnel. (The cage is depicted here in cross-section.)
Source: S. T. Emlen, "Migrating Orientation in the Indigo Bunting. I. Evidence for Use of Celestial Cues," *Auk* 84 (1967):309–342.

birds to planetarium skies, which permitted him to manipulate star patterns experimentally. Not only were the results of outdoor studies confirmed, but when he shifted star patterns 180 degrees in the planetarium, the direction of the *Zugunruhe* exhibited by the birds also shifted. Under cloudy skies or with no stars visible and with only diffuse, dim illumination in the planetarium, the birds exhibited random orientation. Sauer's studies on warblers (Sylviidae) have been confirmed in both field and laboratory experiments by Emlen (1967a, 1967b) on indigo buntings *(Passerina cyanea)*. Emlen's technique is innovative; the bird stands on an ink pad at the bottom of a funnel-shaped cage (Figure 14–10). The sides of the funnel are covered with paper, and the top is a wire mesh screen that permits the bird to see the overhead sky. When the bird jumps up against the sides of the funnel in its restlessness, it leaves marks on the paper. Investigators can turn the record of these marks into a vector diagram for purposes of analysis.

In additional studies, Emlen (1970, 1975b) rotated the night sky in the planetarium. He conducted an experiment with three groups of young indigo buntings. Individuals in group 1 were raised in a windowless room with only diffuse light. Individuals in group 2 were allowed to see the normal night sky in the planetarium, with a normal rotation of the heavenly bodies around the pole star once every other day. Birds in group 3 were raised the same way as those in group 2, except that the heavenly bodies were rotated around Betelgeuse, a bright star in the constellation Orion. The indigo buntings were later measured in the funnel apparatus for their migratory orientation under planetarium skies.

Two major conclusions resulted from this experiment. First, exposure to stellar sky patterns is necessary for normal southward migratory orientation in young buntings. Birds in group 1 exhibited random patterns of orientation when placed under the normal night sky. Second, birds in group 3 oriented 180 degrees away from Betelgeuse, as if headed south, using that star to define the southerly direction. Early experience, thus, plays a critical role in determining the migratory orientation of buntings, and the sky pattern they learn at this time may be used throughout life. These findings also help to account for the evolution of stellar cue orientation in spite of the fact that the earth's magnetic poles change location approximately every 13,000 years and star patterns therefore change their positions in the sky. Birds do not inherit a star map or knowledge of a specific star pattern. Rather, they inherit a predisposition to learn the sky pattern they see when they are very young.

As was the case in our discussion of the compass/map role of the sun, we are again faced with the question of whether birds are using the stars merely as a compass or whether they are capable of true navigation using stellar configurations (Figure 14–11). Use of the stars differs from use of the sun because the night sky

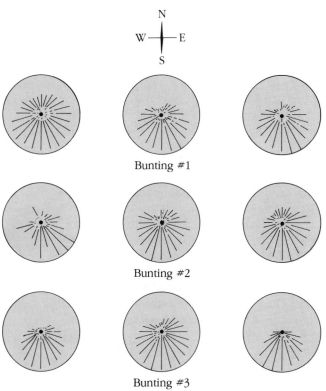

N

W ── E

S

Bunting #1

Bunting #2

Bunting #3

8PM–10PM   11PM–1AM   2AM–4AM

FIGURE 14–11 Measurement of migratory restlessness with vectors Vector diagrams depict the results of studies in which migratory restlessness is measured either by an observer, with the bird in a circular cage, or automatically via the use of the funnel apparatus shown in Figure 14–10. Within each circle the length of a vector (directional line) indicates the amount of migratory restlessness activity oriented in that particular compass direction. Here data for three birds studied by Emlen illustrate that during three two-hour time blocks at night the orientation of indigo buntings does not shift. This indicates that the birds are capable of time compensation to account for the movement of the pattern of stars in the sky in an apparent circle.
Source: S. Emlen, "Migratory Orientation in the Indigo Bunting, *Passerina cayanea*," *Auk* 84 (1967):463–489. Reprinted with permission.

contains many more potential cues. Because the stars shift (and do so at varying speeds, depending on their position), birds must and some can compensate, as has been shown by Emlen (1967b). To use true star navigation, the birds would have to use several star patterns and would need several compensation rates for these different groupings. Emlen (1975a) and Able (1980, 1982) have suggested that although no data as yet demonstrate star navigation convincingly, data do exist to support the hypothesis that birds use the stars as a directional or compass mechanism.

METEORLOGICAL CUES.   Migratory activity in a particular region tends to be concentrated on a few days or nights. Birds are capable of responding to favorable weather conditions (Drury and Keith 1962; Kreithen and Keeton 1974a). Data support the conclusion that birds often fly downward on their migratory flights (Bellrose 1967; Richardson 1971; Bruderer and Steidinger 1972). Do birds select favorable winds *after* they have oriented themselves using other types of cues or, as some (e.g., Able 1974) have suggested, might the birds be using the winds themselves as a directional cue?

Further experiments (Able 1982; Able et al. 1982) monitored the natural nocturnal migratory flights of individuals of several *Passerina* species under various weather and wind conditions. When migrants could observe the sun near the time of sunset or the stars, they flew in the appropriate migratory direction regardless of wind direction. If observations were made under totally overcast skies with no view of sun or stars possible, birds often flew downwind and thus sometimes in seasonally inappropriate directions. When migrant white-crowned sparrows were fitted with frosted lenses and released from balloons aloft, they headed downwind, even though that direction was sometimes seasonally inappropriate. These results indicate the primacy of visual cues in birds for determining appropriate migratory direction and that, when deprived of visual cues, some species can determine wind direction.

OLFACTORY CUES.   Several investigators have suggested that olfactory cues, primarily odors carried on the wind, may serve as an aid in orientation, particularly in homing pigeons (Papi et al. 1972, 1973, 1974). The most direct test of such a hypothesis would involve the removal of the olfactory capacity and the testing for expected orientation and homing behavior. When the olfactory sense is blocked or when nerves are transected, behavioral effects that are not related to the homing ability being tested may result; and, therefore, unequivocal evidence of the possible role of olfaction in pigeon homing is difficult to provide. Indeed, studies by Keeton and Brown (1976), Keeton, Kreithen, and Hermayer (1976), Schmidt-Koenig and Phillips (1978) and Hartwick, Kiepenheuer, and Schmidt-Koenig (1978) have all supplied data that demonstrate little or no effect on homing behavior when the nasal passages are blocked or a local anesthetic is used on the nasal epethelium.

GEOMAGNETIC CUES.   Evidence from several sources has accumulated that supports a role for geomagnetic cues in orientation and navigation in some birds. After many failures by experimenters in attempting to determine whether birds could even sense the .05 Gauss geomagnetic field of the earth (e.g., Emlen 1970; Kreithen and Keeton, 1974b), Brookman (1978), after a successful laboratory experiment, reported conditioning pigeons to magnetic fields. Measurement of *Zugunruhe* in European robins *(Erithacus rubecula)* has provided a second line of evidence. Birds were placed in a cage around which a set of Helmholtz coils, which provide an artificial magnetic field, was constructed. Wiltschko and Wiltschko (1972; and Wiltschko 1972) have used this apparatus to show that the birds were not responding to the polarity of the horizontal component of the magnetic field; they did not shift the direction of their migratory restlessness when horizontal polarity was shifted. However, they did reverse direction when the vertical component was reversed; and they reversed the direction of orientation in a predictable manner. In an artificial magnetic field with a zero vertical component and strong horizontal component, the activity was random.

Other evidence for the role of magnetic fields in orientation comes from the work of Southern (1969, 1972) on ring-billed gulls *(Larus delawarensis)*. Gulls in cages exhibited a strong tendency to walk in a southerly direction, except when storms produced temporary aberrations in the earth's magnetic field. Also when gulls with small magnets affixed to their backs were taken some distance from home and then released, they displayed a random pattern of dispersion. Control gulls without the magnets exhibited normal and consistent directional headings to the south. Additional confirmation has been reported by Keeton (1971) and Larkin and Keeton (1976), who used bar magnets attached to the wings of homing pigeons.

Investigators have also utilized radar and airplane tracking to monitor the behavior of homing pigeons fitted with Helmholtz coils on their heads (Walcott 1972, 1977). Further, Walcott, Gould, and Kirschvink (1979) have reported the existence of an organ, located along the midline of the pigeon's brain, that contains magnetite granules. This organ may prove to be the source of sensitivity to geomagnetism in the pigeon.

As both Emlen (1975a) and Able (1978) have noted in their reviews of bird orientation and navigation, tremendous progress has been made since the 1960s in new field and laboratory techniques for studying the question and also in overall knowledge of what cues may be important.

It is clear from much of the foregoing discussion that many bird species possess multiple systems for orientation; the use of available cues apparently follows some type of hierarchical scheme. Which mechanism the organism employs depends on its preferences for using certain types of cues and the prevailing weather conditions at that location.

MAMMALS

In their studies on orientation and navigation in mammals, researchers have devoted particular attention to small rodents and bats. In many of the studies on rodents, experimenters have displaced the animal for a distance and observed the rate of return in terms of the number of animals that successfully returned home, the speed with which they returned, and aspects of the rodent's orientation as they left the displacement site. In some studies experimenters have attached reflector tape or radio transmitters to the animals to plot the rodents' course during homing.

Rodents have an area, called the home range, in which they carry out their daily functions. When experimenters have determined the home range, they can displace the rodent and monitor its homing behavior until the rodent's return and recapture within its home range (see review by Joslin 1977a, which includes a number of excellent comments on methodological problems encountered in attempting such homing experiments). Data from several species of mice indicate that, when displaced, these animals can find their home range again, some from a considerable distance (up to 30 kilometers) and with speeds ranging up to 300 meters/hour. In most studies the rodents are generally displaced for distances of only up to 500 meters and may take several days to return to the home range and reenter a trap (see Joslin 1977a). The return rate, or the percentage of mice that successfully return home, varies with the species and the nature of the habitat. Overall figures have indicated return rates ranging from 3 to 5 percent to over 80 percent.

The most plausible explanation of homing in rodents, based on present data, appears to be **piloting**—that is, location by using familiar landmarks or terrain—combined with random search when the displacement is for greater distances (Joslin 1977a, 1977b). An animal learns about its habitat in several ways, including exploration of the home range. Some exploratory forays may take the rodent a distance beyond the home range. Rodents also learn about the habitat during dispersal when the young move away from the natal site to establish home ranges of their own. Home ranges differ greatly in size among species, and even within the same species, from areas of only several hundred square meters in vole species to several square kilometers or more in deermice and squirrels. For example, a rodent whose home range has an approximate diameter of 200 meters may also know something about the terrain and its landmarks for 200 meters or more in all directions around the perimeter of that home range. Thus when a rodent is displaced for a short distance, it may be in an area about which it already knows something—this larger area is sometimes referred to as the **life range.** Alternatively, when it is displaced for a greater distance, the animal may exhibit random, or possibly systematic, searching behavior until it encounters familiar terrain and can again rely on piloting to find the home area.

Little work has been done on how rodents orient within their home ranges; however, Drickamer and Stuart (1984) investigated the movements of deermice *(Peromyscus).* In the winter the patterns of tracks left by the deermice on fresh snow can be mapped (Figure 14–12). Drickamer and Stuart used traverses—each segment of the path from tree to tree or from tree to log—to examine the potential use of cues. They found that 86 percent of all traverses were oriented from one tree to another tree and covered distances ranging from one to thirty meters. The concentration of tracks suggests that the vertical tree trunks are used for orienting. A significant positive relationship ($r = 0.51$) between the length of the segment traveled by the mouse and the diameter of the tree to which it was headed further suggests that

FIGURE 14-12 Deermice navigating forest floor
Deermice (left) utilize trees in their environment as cues for piloting their way
around the top of the snow cover on the forest floor. Many of their movements are
from one tree to another (right).
Source: Photo by Lee C. Drickamer.

 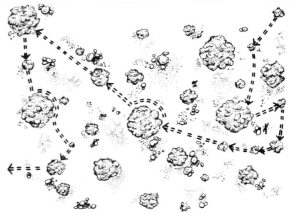

larger trees serve as more conspicuous objects on the horizon for orientation. Laboratory investigations (Joslin 1977b) have supported the hypothesis that deermice may use conspicuous vertical cues on the horizon for orienting to goals.

The navigation and orientation systems of bats have received considerable attention. As we noted in Chapter 3, bats use an echolocation system for orienting with respect to objects in the environment and in capturing prey. Bats send out high-frequency sounds that bounce off objects in the environment, return as echoes, and provide information the bat utilizes to control its flight pattern (Griffin 1958, 1974).

Griffin and Galambos (1941) investigated whether bats use the high-frequency sounds to avoid objects and navigate through their environment. They tested bats in a large room in which a barrier of wires was set up. The observers scored "hits" and "misses" as bats flew back and forth past the wires. Data from these experiments (Table 14–1) show clearly that any treatments that impaired the bat's ability to send out high-energy sounds or receive the return echoes were associated with a lower percentage of misses. In other words, when the experimenter hindered its capability for using echolocation, the bat was more likely to touch or strike the wires.

To test whether bats use visual cues, experimenters blindfolded bats and then displaced them for a distance of five miles from the cave in which they were roosting (Mueller and Emlen 1957). The number of bats returning and the speed of their return to the cave (where they were caught in a large net set up across the cave's entrance) were the same for blindfolded bats and for bats with unimpaired vision.

TABLE 14-1   Bat navigation by echolocation

Bats (mostly little brown bats, *Myotis lucifugus)* were tested for their ability to navigate after various impairments of hearing, sound production, and vision. The test room contained 16-gauge wires hung vertically with a 1-foot (0.3 m) spacing interval. When bats flew near the wall, floor, or ceiling, their flights were not counted. The scores in the table represent percentages of flights through the wires during which the bats did not touch the wires. Specific controls used for each experiment are indicated in the "experimental treatment" column. A chance score was calculated as about 35 percent misses.

Number of Bats Used	Experimental Treatment	Experimentals		Controls	
		Number of Flights	Average % Misses	Number of Flights	Average % Misses
28	Both eyes covered (controls untreated)	2,016	76	3,201	70
12	Both ears covered (controls untreated)	1,047	35	1,297	66
9	Ears and eyes covered (controls with only the eyes covered)	654	31	832	75
8	Glass tubes in ears and closed (controls with the same tubes in ears but open)	580	36	636	66
12	Both ears covered (controls with one ear covered)	853	29	560	38
6	Eyes and one ear covered (controls with only the eyes covered)	390	41	590	70
7	Mouth covered (controls with eyes covered or intact)	549	35	442	62

Source: D. A. Griffin and R. Galambos, "The Sensory Basis of Obstacle Avoidance by Flying Bats," *Journal of Experiment Zoology* 86 (1941):481–506.

We should note, however, that in experiments with blindfolds there have been varying techniques (for example, blindfolds or goggles), varying results from different experimenters (Mueller 1966; Barbour et al. 1966), and varying interpretations of the results. At longer distances, vision appears to play a role in the orientation and navigation of bats (Williams and Williams 1967, 1970; Williams, Williams, and Griffin 1966). Bats can see with their eyes; studies of optomotor responses and experimental training of bats to respond to visual patterns have revealed visual acuities of several degrees (Suthers 1966; Chase and Suthers 1969). We will need improved experimental techniques and designs to learn more about navigation in bats.

INVERTEBRATES

Migrating locusts, flying in swarms, give the appearance of oriented directional movement. Studies of locust swarms have revealed that wind exerts a displacing effect on individuals and thus on the swarm over time. Within the swarm there are groups of locusts that are flying in directions different from that of the swarm as a whole. When these subgroups reach the periphery of the swarm, they turn back inwards toward the main flow of the swarm (Rainey 1959, 1962; Waloff 1958).

The means by which honey bees *(Apis mellifera)* communicate the location of food sources to one another in the hive and then make foraging flights to locate the food have been studied by a number of investigators (von Frisch 1967; Lindauer 1961; Gould 1975, 1976; Wenner 1971, 1974). Chapter 13 on communication covered this topic in some detail.

Bees and some other insects, including ants, can perceive polarized light (von Frisch 1967). The plane of polarization can provide an axis for orientation for these animals. It is also possible that in some animals the perception of polarized light may enable them to locate the position of the sun when it is obscured by partially overcast skies. The use of polarized light for orientation has also been suggested for a variety of other organisms, including fish (Waterman and Hashimoto 1974), octopodes (Waterman 1966), and salamanders (Adler and Taylor 1973).

Many of the cues discussed in this chapter with regard to orientation in birds and mammals have also been demonstrated in some invertebrates. Among these are sun-compass orientation in spiders, beetles, and ants; geomagnetic force cues in bees; and landmarks in a variety of species (see Able 1980). In addition, data have been reported which suggest beachhoppers *(Talitrus)* use the moon (lunar orientation) to guide their nocturnal movements.

FISH

We have defined **homing** behavior as the use of landmarks and navigation to return to a homesite. An animal may become displaced from its home by weather, or, as in the case of the Pacific salmon, it may return to a home location in the course of the life cycle. Alternatively, researchers may artificially move an organism from its homesite to another location in order to investigate the cues and means of navigation it employs to return home.

Female Pacific salmon lay eggs in stream beds among the cracks and crevices in the rocks, where male salmon fertilize them by depositing sperm over the eggs. After hatching, the fry develop in the home stream during the summer and then migrate to the ocean, where they mature. Some three to five years later (depending on the species), most of the surviving adults return to spawn and die in the stream where they hatched (Figure 14–13).

How do the salmon find the stream where they hatched? The results of numerous experiments conducted over many years (see Hasler 1960) indicate that

FIGURE 14-13   Pacific salmon migrating upstream
After spending much of their adult life in the ocean, Pacific salmon migrate upstream to breed in the location where they were hatched. The mechanism by which they relocate their natal stream appears to involve chemical cues in the water.
Source: Photo by Ronald Thompson from Frank W. Lane.

young salmon fry are somehow imprinted with the odor of the natal stream. When adult salmon migrate upstream, they follow the odor or pattern of odors from the home stream, making correct turns at each junction until they arrive back at the natal stream. The olfactory stimuli involved in this process have not yet been clearly identified.

AMPHIBIANS AND REPTILES

Orientation has been studied experimentally in southern cricket frogs (*Acris gryllus*) and northern cricket frogs (*Acris crepitans*) (Ferguson, Landreth, and Turnispeed 1965; Ferguson, Landreth, and McKeown, 1967). Experimenters placed captured frogs in a large, circular plastic pen in the water, where the frogs could see no landmarks on the horizon. When southern cricket frogs were released in the pen under sunlit skies, they swam in the direction that would have taken them to land had they been released off the shore at their home pond. The results were the same when the experimenters carried the frogs to the pen with or without giving the frogs a view of the sky overhead, when they released the frogs under starlit skies, and when they released the frogs under moonlit skies. When researchers placed the frogs in the pen on a moonless night, with only stars visible in the sky, the frogs oriented themselves in two opposite directions; some headed toward the home shore and others headed directly away from home. When experimenters released the frogs during the twilight period, after sundown but before the stars or moon were visible, or under cloudy, overcast conditions, the frogs oriented in a random fashion.

Cricket frogs held in darkness for periods of thirty hours to seven days exhibited a decline in accuracy of orientation, which culminated in random orientation after seven days. "Dephased" frogs reestablished proper orientation when experimenters exposed them to normal dark/light cycles or to daily fluctuations of temperature and humidity. These frogs appeared to use a learned shore position, celestial cues, and a timing mechanism to orient themselves properly to the home shore. Similar results have been obtained for northern cricket frogs.

Carr (1967) has summarized a number of investigations on orientation by green turtles (Figure 14–14). Female turtles *(Chelonia mydas)* deposit their eggs on sandy beaches on certain islands; they lay about 100 eggs on each trip to the beach, and make from three to seven trips spaced twelve days apart. In nonreproductive periods the turtles live in areas with abundant turtle grass, generally along the coasts of continental land masses. In a specific study (Carr 1967), turtles were banded at Ascension Island (Figure 14–15). Young turtles that hatched on the island floated on currents that took them to the coast of South America. Adults had to swim against the same current to reach Ascension Island to reproduce. How did they find their way? The current may have certain features that allow the turtles to distinguish it.

FIGURE 14-14 Female green turtle on nesting beach
Sea turtles have been studied extensively by tagging. The turtles return periodically to the same beach to lay their eggs.
Source: Photo by Archie Carr.

FIGURE 14–15   Turtle tagging project on Ascension Island
The number of turtles tagged at each nesting beach location is shown by a figure.
Solid triangles indicate mainland Brazil locations where researchers recovered
turtles. Open triangles on nesting beach sites show turtles that returned after three
years. Solid hexagons denote turtles that returned in the fourth year to nest,
possibly having made two round trips to Brazil. Arrows indicate prevailing ocean
currents.
Source: A. Carr, "Adaptive Aspects of the Scheduled Travel of Chelonia," in R. M.
Storm, ed., *Animal Orientation and Navigation* (Corvallis: Oregon State University
Press, 1967), p. 43. Reprinted with permission.

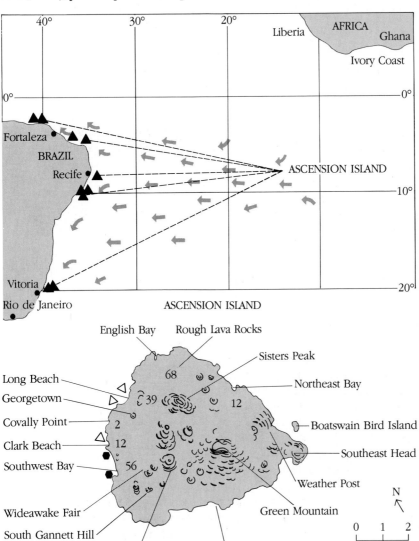

This current could provide a general orientation, but not the precise information needed to target a small land mass in the midst of many square miles of ocean. Studies using green turtles have provided tentative data to support a hypothesis of sun-compass orientation (Carr 1965, 1967; Ehrenfeld and Carr 1967).

SUMMARY

Many species of animals, including birds, mammals, and insects, migrate seasonally to areas with more favorable conditions in order to avoid the harsh weather conditions and diminished food supplies of the winter months.

The routes and dynamics of bird migration can be studied by direct observation, use of radar, and observation of the moon through a telescope. The triggers for migration include factors that stimulate fat deposition and migratory restlessness, or *Zugunruhe,* and the actual signals that initiate migratory movements.

The advantages of migration for birds include having an adequate food supply, the concentration of activities related to reproduction during the longer summer days in northern latitudes, and the synchrony between sexes that may ensure a higher probability of successful reproduction. Migration may affect geographical dispersal in birds and may have evolved as a seasonal movement northward and southward from sites of evolutionary origin in or near the tropics. Alternatively, some migratory species may have evolved in northern climates and have adopted seasonal patterns of movement in response to shifts in climate.

Some bats and caribou migrate for climatic or alimental reasons; gray whales apparently migrate for gametic reasons. Monarch butterflies exhibit a unique long-distance migratory pattern in that the returning migrants in the spring are fourth-generation descendents of those that flew south. Locusts exhibit periodic movements involving sedentary and migratory phases.

Investigators have studied homing pigeons and other bird species to determine the mechanisms of orientation and navigation. They have found that some bird species are capable of using the sun or stars as a compass for orientation and that birds appear, in many instances, to use topographic landmarks for piloting. Investigations have shown that meteorological cues can influence migratory flight. Most recently, a number of reports have provided evidence for the role of geomagnetism in bird orientation. Birds appear to have several orientation systems at their disposal; the choice of system depends on prevailing conditions.

Homing experiments with small rodents have illustrated their use of piloting as a means of navigation. Among mammals, extensive studies have been carried out on the echolocation system used for navigation in bats, which is supplemented by vision. The sun, landmarks, and geometric forces are also used by some invertebrates for orientation. The life cycle of the Pacific salmon demonstrates homing behavior based on chemical cues, and studies of cricket frogs and sea turtles have suggested that the sun or stellar cues are used as a "compass" for orientation.

Discussion Questions

1. Much of the work on orientation and migration by birds has produced a number of new techniques and methods. Develop an experiment to investigate orientation in a nonavian animal and apply some of the techniques used with birds.

2. In addition to those discussed in the chapter, what are the advantages and disadvantages of migration for organisms of different species?

3. What comparisons or generalizations can you make about the cues used for orientation and navigation by the animals discussed in this chapter?

4. Recent studies support the hypothesis that birds can use geomagnetic cues for orientation. Data from Baker (1980) suggest that the same may be true for humans. The data presented below are taken from a hypothetical study similar to a study conducted by Baker. Individuals were blindfolded and taken by car to a location 10 kilometers from home and released one by one (just like homing pigeons). The initial orientation bearing for these individuals at two different sites is recorded here (each dot represents the initial bearing for one individual). Fifteen people were tested for each site. What conclusions can you draw about geomagnetic cues and orientation by humans? What methodological problems might you encounter in conducting such an experiment?

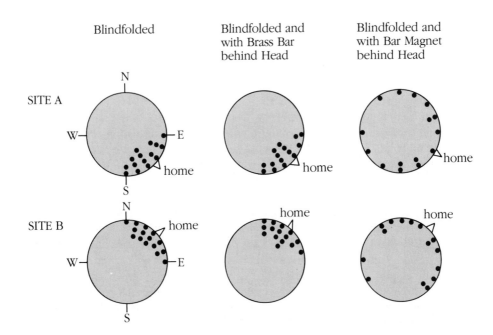

Suggested Readings

Emlen, S. T. 1975. Migration, orientation and navigation. In D. S. Farner and J. R. King, eds., *Avian Biology*, vol. 5, New York: Academic Press.
Summary with historical development of bird orientation theories. Solid presentation of experimental approaches and techniques by a respected authority in this field.

Gauthreaux, S. A. 1980. *Animal Migration, Orientation and Navigation.* New York: Academic Press.
An edited volume of review articles covering many aspects of these topics. Good mix of mechanistic and evolutionary perspectives between and among the chapters.

Gould, J. L. 1980. The case for magnetic sensitivity in birds and bees (such as it is). *Amer. Sci.* 68:256–267.
Very readable. Touches on information relating magnetic cues to the orientation of animals.

Orr, R. T. 1970. *Animals in Migration.* London: Macmillan.
Engaging summary of migratory phenomena in all forms of animal life. Well illustrated and readable.

Schmidt-Koenig, K., and W. T. Keeton, eds. 1978. *Animal Migration, Navigation, and Homing.* New York: Springer-Verlag.
Edited by two major contributors in this research area. Experts treat all theories of orientation and navigation. Best source of references on this topic for the beginning student. Most papers are readable without extensive background.

References

Able, K. P. 1974. Environmental influences on the orientation of nocturnal bird migrants. *Anim. Behav.* 22:225–239.

———. 1978. Field studies of the orientation cue hierarchy of nocturnal songbird migrants. In K. Schmidt-Koenig and W. T. Keeton (eds.), *Animal Migration, Navigation, and Homing.* New York: Springer-Verlag.

———. 1980. Mechanisms of orientation, navigation and homing. In S. A. Gauthreaux (ed.), *Animal Migration, Orientation, and Navigation.* New York: Academic Press.

———. 1982. Field studies of avian nocturnal migration orientation. I. Interaction of sun, wind and stars as directional cues. *Anim. Behav.* 30:761–767.

Able, K. P., V. P. Bingman, P. Kerlinger, and W. Gergits. 1982. Field studies of avian nocturnal migratory orientation. II. Experimental manipulation in white-throated sparrows (*Zonotrichia albicollis*) released aloft. *Anim. Behav.* 30:768–773.

Adler, K., and D. H. Taylor. 1973. Extraocular perception of polarized light by orienting salamanders. *J. Comp. Physiol.* 87:203–212.

Baker, R. 1980. Goal orientation by blindfolded

humans after long-distance displacement: Possible involvement of a magnetic sense. *Science* 210:555–557.

Banfield, A. W. 1954. Preliminary investigation of the barren ground caribou. II. Life history, ecology and utilization. *Can. Wild. Serv. Wild. Mgmt. Bull.* No. 10B.

Barbour, R. W., W. H. Davis, and M. D. Hassell. 1966. The need of vision in homing by *Myotis sodalis. J. Mammal.* 47:356–357.

Bellrose, F. C. 1964. Radar studies of waterfowl migration. *Trans. N. Amer. Wild. Nat. Conf.* 29:128–143.

———. 1967. Orientation in waterfowl migration. *Proc. Ann. Biol. Colloq.,* Oreg. State Univ. 27:73–99.

———. 1971. The distribution of nocturnal migrants in the air space. *Auk* 88:397–424.

Brookman, M. A. 1978. Sensitivity of the homing pigeon to an earth-strength magnetic field. In K. Schmidt-Koenig and W. T. Keeton (eds.), *Animal Migration, Navigation, and Homing.* New York: Springer-Verlag.

Brower, L. P., et al. 1977. Biological observations on an overwintering colony of monarch butterflies (*Danas plexippus* Danaidae) in Mexico. *J. Lepid. Soc.* 31:232–242.

Bruderer, B., and P. Steidinger. 1972. Methods of quantitative and qualitative analysis of bird migration with a tracking radar. *NASA Spec. Publ.* NASA SP-262:151–167.

Calvert, W. H. and L. P. Brower. 1981. The importance of forest cover for the survival of overwintering monarch butterflies (*Danaus plexippus* Danidae). *J. Lepid. Soc.* 35:216–225.

Carr, A. 1965. The navigation of the green turtle. *Sci. Amer.* 212:78–86.

———. 1967. Adaptive aspects of the scheduled travel of Chelonia. *Proc. Ann. Biol. Colloq.,* Oreg. State Univ. 27:35–36.

Chase, J., and R. A. Suthers. 1969. Visual obstacle avoidance by echolocating bats. *Anim. Behav.* 17:201–207.

Davis, W. H., and H. B. Hitchcock. 1965. Biology and migration of the bat, *Myotis lucifungus,* in New England. *J. Mammal.* 46:296–313.

Dingle, H. 1980. Ecology and evolution of migration. In, S. A. Gauthreaux (ed.), *Animal Migration, Orientation and Navigation.* New York: Academic Press.

Drickamer, L. C., and J. Stuart. 1984. Peromyscus: Snow tracking and possible cues used for navigation. *Amer. Mid. Nat.* 111:202–204.

Drury, W. H., and J. A. Keith. 1962. Radar studies of songbird migration in coastal New England. *Ibis* 104:449–489.

Drury, W. H., and I. C. T. Nisbet. 1964. Radar studies of orientation of songbird migrants in southeastern New England. *Bird Banding* 35:69–119.

Ehrenfeld, D. W., and A. Carr. 1967. The role of vision in the sea-finding orientation of the green turtle (*Chelonia mydas*). *Anim. Behav.* 15:25–36.

Emlen, S. T. 1967a. Migratory orientation in the indigo bunting. *Passerina cyanea. Auk* 84:463–489.

———. 1967b. Migratory orientation in the indigo bunting. I. Evidence for use of celestial cues. *Auk* 84:309–342.

———. 1970. Celestial rotation: Its importance in the development of migratory orientation. *Science* 170:1198–1201.

———. 1975a. Migration, orientation and navigation. In D. S. Farner and J. R. King (eds.), *Avian Biology,* vol. 5. New York: Academic Press.

———. 1975b. The stellar-orientation system of a migratory bird. *Sci. Amer.* 233(2):102–111.

Farner, D. S. 1955. The annual stimulus for migration: Experimental and physiologic aspects. In A. Wolfson (ed.), *Recent Studies in Avian Biology.* Urbana: University of Illinois Press.

Ferguson, D. E., H. F. Landreth, and M. R. Turnispeed. 1965. Astronomical orientation of the southern cricket frog, *Acris gryllus. Copeia* 1965:58–66.

Ferguson, D. E., H. F. Landreth, and J. P. McKeown. 1967. Sun compass orientation of the northern cricket frog, *Acris crepitans. Anim. Behav.* 15:45–53.

Fraenkel, G. S., and D. L. Gunn. 1940. *The Orientation of Animals.* London: Oxford University Press.

Frisch, K. von. 1967. *Bees, Their Vision, Chemical Senses and Language.* Ithaca: Cornell University Press.

Gauthreaux, S. A., Jr. 1971. A radar and direct visual study of passerine spring migration in southern Louisiana. *Auk* 88:343–365.

Gould, J. L. 1975. Honeybee recruitment: The dance-language controversy. *Science* 189:685–693.

———. 1976. The dance-language controversy. *Quart. Rev. Biol.* 51:211–244.

Graber, R. R. 1968. Nocturnal migration in Illinois: Different points of view. *Wilson Bull.* 80:36–71.

Griffin, D. A. 1955. Bird navigation. In A. Wolfson (ed.), *Recent Studies in Avian Biology*. Urbana: University of Illinois Press.

———. 1958. *Listening in the Dark*. New Haven: Yale University Press.

———. 1970. Migrations and homing of bats. In W. A. Wimsatt (ed.), *Biology of Bats*, vol. 1. New York: Academic Press.

———. 1974. *Bird Migration*. New York: Dover.

Griffin, D. A., and R. Galambos. 1941. The sensory basis of obstacle avoidance by flying bats. *J. Exp. Zool.* 86:481–506.

Hartwick, P., J. Kiepenheuer, and K. Schmidt-Koenig. 1978. Further experiments on the olfactory hypothesis of pigeon homing. In K. Schmidt-Koenig and W. T. Keeton (eds.), *Animal Migration, Navigation, and Homing*. New York: Springer-Verlag.

Hasler, A. D. 1960. Guideposts of migrating fishes. *Science* 132:785–792.

Johnson, C. G. 1969. *Migration and Dispersal of Insects by Flight*. London: Methuen.

Joslin, J. K. 1977a. Rodent long distance orientation ("homing"). *Adv. Ecol. Res.* 10:63–90.

———. 1977b. Visual cues used in orientation in white-footed mice, *Peromyscus leucopus*: A laboratory study. *Amer. Mid. Nat.* 98:303–318.

Keeton, W. T. 1969. Orientation by pigeons: Is the sun necessary? *Science* 165:922–928.

———. 1970. Orientation by pigeons. *Science* 168:153.

———. 1971. Magnets interfere with pigeon homing. *Proc. Nat. Acad. Sci. USA* 68:102–106.

———. 1974. The orientation and navigational basis of homing in birds. *Adv. Stud. Behav.* 5:47–132.

Keeton, W. T., and A. I. Brown. 1976. Homing behavior of pigeons not disturbed by application of an olfactory stimulus. *J. Comp. Physiol.* 105:259–266.

Keeton, W. T., M. L. Kreithen, and K. L. Hermayer. 1976. Orientation by pigeons deprived of olfaction by nasal tubes. *J. Comp. Physiol.* 114:289–299.

Kramer, G. 1949. Über Richtungstendenzen bei der nächtlichen Zugunruhe gekäfigten Vögel. In E. Mayr and E. Schüz (eds.), *Ornithologie als biologische Wissenschaft*, Heidelberg: Winter.

———. 1950. Orientierte Zugaktivitätgekäfigter Singvögel. *Naturwissenschaften* 37:188.

———. 1951. Eine neue Methode zur Erforschung der Zugorientierung und die bisher damit erzielten Ergebnisse. *Proc. Xth Inter. Ornithol. Congr.*, Uppsala. 271–280.

Kreithen, M. L., and W. T. Keeton. 1974a. Detection of changes in atmospheric pressure by the homing pigeon, *Columbia livia. J. Comp. Physiol.* 89:73–82.

———. 1974b. Attempts to condition homing pigeons to magnetic stimuli. *J. Comp. Physiol.* 91:355–362.

Larkin, T. S., and W. T. Keeton. 1976. Bar magnets mask the effect of normal magnetic disturbances on pigeon orientation. *J. Comp. Physiol.* 110:227–231.

Lent, P. C. 1966. Calving and related social behavior in the barren-ground caribou. *Zeit. Tierpsychol.* 23:701–756.

Lincoln, F. C. 1950. Migration of birds. *U.S. Fish and Wild. Serv. Washington, D.C. Circ.* 16:1–102.

Lindauer, M. 1961. *Communication among Social Bees*. Cambridge: Harvard University Press.

Loeb, J. 1918. *Forced Movements, Tropisms and Animal Conduct*. Philadelphia: Lippincott.

Mast, S. O. 1938. Factors involved in the process of orientation of lower organisms in light. *Biol. Rev.* 17:68–90.

Mathews, G. V. T. 1968. *Bird Navigation*, 2nd ed. Cambridge: Cambridge University Press.

Meier, A. H. 1973. Daily hormone rhythms in the white-throated sparrow. *Amer. Sci.* 61:184–187.

Morton, M. L., and L. R. Mewaldt. 1962. Some effects of castration on migratory sparrows (*Zonotrichia atricapilla*). *Physiol. Zool.* 35:237–247.

Mueller, H. 1966. Homing and distance-orientation in bats. *Zeit. Tierpsychol.* 23:403–421.

Mueller, H., and J. T. Emlen. 1957. Homing in bats. *Science* 126:307–308.

Murie, O. J. 1935. Alaska-Yukon caribou. *N. Amer. Fauna* (U.S. Dept. Agric.) 54:1–93.

Orr, R. T. 1970. *Animals in Migration*. New York: Macmillan.

Papi, F., et al. 1972. Olfaction and homing in pigeons. *Monit. Zool. Ital.* (n.s.) 6:85–95.

———. 1973. An experiment for testing the hypothesis of olfactory navigation of homing pigeons. *J. Comp. Physiol.* 83:93–102.

———. 1974. Olfactory navigation of pigeons: The effect of treatment with odorous air currents. *J. Comp. Physiol.* 94:187–193.

Rainey, R. C. 1959. Some new methods for the study of flight and migration. *Proc. XVth Inter. Congr. Zool.*, London. 866–870.

———. 1962. The mechanisms of desert locust swarm movements and the migration of insects. *Proc. XIth Inter. Congr. Ent.* 3:47–49.

Richardson, W. J. 1971. Spring migration and weather in eastern Canada: A radar study. *Amer. Birds* 25:684–690.

———. 1972. Autumn migration and weather in eastern Canada. *Amer. Birds* 26:10–17.

Sauer, E. G. F. 1957. Die Sternenorientierung nächtlich ziehender Grasmücken *(Sylvia atricapilla, borin und curruca). Zeit. Tierpsychol.* 14:29–70.

———. 1963. Migration habits of golden plovers. *Proc. Inter. Ornithol. Congr., 13th.* 454–467.

Schmidt-Koenig, K., and J. B. Phillips. 1978. Local anesthesia of the olfactory membrane and homing in pigeons. In K. Schmidt-Koenig and W. T. Keeton (eds.), *Animal Migration, Navigation, and Homing.* New York: Springer-Verlag.

Schmidt-Koening, K., and H. J. Schlichte. 1972. Homing in pigeons with reduced vision. *Proc. Nat. Acad. Sci. USA* 69:2446–2447.

Schmidt-Koenig, K., and C. Walcott. 1973. Flugwege und Verbleib von Brieftauben mit getrübten Haftschalen. *Naturwissenschaften* 60:108–109.

Southern, W. E. 1969. Orientation behavior of ring-billed gull chicks and fledglings. *Condor* 71:418–425.

———. 1972. Magnets disrupt the orientation of juvenile ring-billed gulls. *BioScience* 22:476–479.

Suthers, R. A. 1966. Optomotor responses by echolocating bats. *Science* 152:1102–1104.

Walcott, C. H. 1972. Bird navigation. *Natural History* 81(June):32–43.

———. 1977. Magnetic fields and orientation of homing pigeons under sun. *J. Exp. Biol.* 70:105–123.

Walcott, C., J. L. Gould, and J. L. Kirschvink. 1979. Pigeons have magnets. *Science* 205:1027–1029.

Waloff, Z. 1958. The behaviour of locusts in migrating swarms. *Proc. 10th Inter. Congr. Ent.,* Montreal 2:567–570.

Waterman, T. H. 1966. Systems analysis and the visual orientation of animals. *Amer. Sci.* 54:15–45.

Waterman, T. H., and H. Hashimoto. 1974. E-vector discrimination by the goldfish optic tectum. *J. Comp. Physiol.* 95:1–12.

Wenner, A. M. 1971. *Bee Language Controversy.* Boulder, Colo.: Educational Programs Improvement Corp.

———. 1974. Information transfer in honeybees: A population approach. In L. Krames, P. Pliner, and T. Alloway (eds.), *Nonverbal Communication,* vol. 1. New York: Plenum.

Williams, T. C., and J. M. Williams. 1967. Radio tracking of homing bats. *Science* 155:1435–1436.

———. 1970. Radio tracking of homing and feeding flights of a neotropical bat, *Phyllostomus hastatus. Anim. Behav.* 18:302–309.

Williams, T. C., J. M. Williams, and D. R. Griffin. 1966. The homing ability of the neotropical bats, *Phyllostomus hastatus,* with evidence for visual orientation. *Anim. Behav.* 14:468–473.

Wiltschko, W. 1972. The influence of magnetic total intensity and inclination on directions preferred by migrating European robins *(Erithacus rubecula). NASA Spec. Publ.* NASA SP-262:569–578.

Wiltschko, W., and R. Wiltschko. 1972. The magnetic compass of European robins, *Erithacus rubecula. Science* 176:62–64.

Wolfson, A. 1948. Bird migration and the concept of continental drift. *Science* 108:23–30.

Zahl, P. A. 1963. The mystery of the monarch butterfly. *Nat. Geogr.* 123:588–598.

PART FOUR

BEHAVIORAL ECOLOGY AND ANIMAL POPULATIONS

15 *Habitat Selection*

16 *Feeding Relationships*

17 *Behavior and Population Regulation*

15

HABITAT SELECTION

In Part Four we consider, from an environmental perspective, how animals solve three basic problems: finding a suitable habitat (Chapter 15), obtaining food, and avoiding becoming the food of others (Chapter 16). A final chapter deals with behavioral aspects of abundance — the problem of population regulation (Chapter 17).

One of the big differences between plants and animals is that plants are generally dependent on natural agents, such as currents of air or water, for dispersal. The result is an essentially random dissemination of plant individuals; few ever reach environments conducive to survival and reproduction. Animals, in contrast, have well-developed locomotive abilities, at least during some point in the life cycle; and they play more active roles in finding places to live. **Habitat selection** can be defined as the choosing of a place in which to live. This definition does not imply that the choice is necessarily a conscious one or that individuals make a critical evaluation of the entire constellation of factors confronting them. More often the choice is an "automatic" reaction to certain key aspects of the environment.

This chapter concerns the **distribution** of a species — that is, the presence of a species in, or its absence from, a particular habitat, and the dispersal in a habitat of individuals within a species. We first consider how dispersal ability, other organisms, and physical and chemical factors restrict habitat use. We then look at the role of proximate factors in habitat selection — the environmental and social cues that influence the choice of a place to live. Finally, the roles of genetics, early experience, and tradition in the development of habitat preferences are examined. The question of the abundance of a species is considered in Chapter 17. When examining the distribution and abundance of most species, areas of higher population are often found near the geographic center of a species' range, with abundance decreasing outward (Figure 15–1). The **fundamental niche** — the multidimensional space that a species occupies under ideal conditions with no competition — is often much larger than the **realized niche** — the space occupied under real world conditions that involve competitors, predators, disease, and so forth.

FACTORS RESTRICTING HABITAT USE

If a species occupies an area and reproduces there, we know that all its needs are met and that it can compete with other species successfully. A useful way to identify the factors affecting the distribution of a species is to determine why it is *absent* from a place. We should examine possible reasons one by one, by using transplant experiments, by giving individuals a choice of artificial habitats in the laboratory, or by building enclosures in the field that encompass different habitats.

In a transplant experiment organisms are moved to a new environment, and their survival and reproduction are monitored. Because organisms sometimes survive but fail to reproduce, a long-term project covering several generations is necessary. In a study of the distribution of heathland ants in England, Elmes (1971) dug up colonies and moved them to sites that differed in temperature and in the amount of moisture. He moved eighteen colonies of one species *(Lasius niger)*, which normally inhabit low, wet heathland, to higher, drier sites and moved six to other low, wet

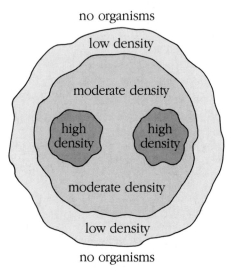

no organisms

low density

moderate density

high density   high density

moderate density

low density

no organisms

FIGURE 15–1   Typical relationship between distribution in space and density for a species
Areas of optimal habitat support the highest densities and are surrounded by suboptimal habitats with lower densities. This distribution pattern contrasts with that shown in Figure 15–4.

TABLE 15-1 Transplants of heathland ant colonies in England
Ant colonies of *Lasius niger*, which naturally occur in low, wet areas, were moved either to a similar habitat as a control or to higher, drier sites. Each transplanted colony was observed for five years.

Location of Transplant	Years Survived						Total Number of Colonies Transplanted
	0	1	2	3	4	5+	
Control							
Lasius niger zone 11 m level	1		1			4	6
Dry Heath							
13 m level			1		1	3	5
17–20 m level	6	3				1	10
25 m level	1	1	1				3

Source: Data from G. W. Elmes, "An Experimental Study on the Distribution of Heathland Ants," *Journal of Animal Ecology* (1971) 40:495–499.

areas as controls. Elmes monitored their survival over a five-year period and found that the higher the colony was moved, the less likely it was to survive (Table 15-1). Those at the 13 m level did as well as the controls at 11 m. One of the main factors affecting survival was competition with other species that normally inhabit the higher, drier part of the heath. This experiment demonstrates the need for long-term monitoring of transplants.

If a transplant is successful, two possible factors may explain why a species is not found naturally in the transplant area: (1) the area is inaccessible because the dispersal ability of the organism is limited or (2) the organism fails to recognize the area as a suitable habitat [referred to by Lack (1933) as the "psychological factor"]. If a transplant fails, the causal factors may be the presence of other species (competitors, parasites, pathogens) or physical and chemical factors (temperature, pH, and so on). In trying to understand why a species is absent from an area, we can thus proceed through a series of steps (Krebs 1985), as shown in Figure 15-2.

DISPERSAL

Is a species absent from a place because it cannot get there? Many cases of successful introductions by humans have demonstrated that locomotive abilities adequately explain a species' absence. The European starling *(Sturnus vulgaris)* originally was found in most of Europe and Asia. After several unsuccessful introductions of small numbers of starlings into the United States, eighty pairs were released in New York City's Central Park in 1890 by Eugene Scheifflin of the Acclimatization Society. The "goal" was to familiarize Americans with all the birds in Shakespeare's plays (Miller 1975). Within fifty years starlings had reached the West Coast, and today they are

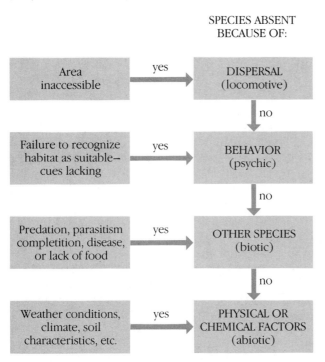

SPECIES ABSENT
BECAUSE OF:

Area inaccessible →yes→ DISPERSAL (locomotive) →no→

Failure to recognize habitat as suitable— cues lacking →yes→ BEHAVIOR (psychic) →no→

Predation, parasitism completition, disease, or lack of food →yes→ OTHER SPECIES (biotic) →no→

Weather conditions, climate, soil characteristics, etc. →yes→ PHYSICAL OR CHEMICAL FACTORS (abiotic)

FIGURE 15–2   Methodological approach for studying the geographical distribution of a species
A useful way to proceed in the analysis of a species' distribution is to determine why that species is *not* in a particular place. Four basic factors are listed; behavior may be involved in each and the factors may interact. *Source:* Data from C. J. Krebs, *Ecology: The Experimental Analysis of Distribution and Abundance,* 3rd ed. (New York: Harper & Row, 1985), p. 39.

probably the most numerous bird species in the country. Starlings nest in tree cavities and are more aggressive than most native species; they even evict other species from nest holes. The eastern bluebird *(Sialia sialis)* is one native species that now occupies only a fraction of its former range partly because of its unsuccessful competition with the starling. Although they are insectivorous during the summer, starlings are generalists and switch to seeds in the winter; many native insectivorous species have to head south to find insects. Other aspects of the starlings' habits have led to their success in modern industrialized countries: a tolerance of loud noises and air pollution, a willingness to roost and perch in various places, from trees to bridge supports, and a preference for feeding in grassy areas.

Distributions on islands offer a further test of dispersal powers. MacArthur and Wilson (1967) developed a model that predicts a dynamic equilibrium of the number of species on islands. Although the species change through time, the total number of species remains constant. The immigration rates of new species are affected by island size; smaller islands are smaller targets and therefore have lower rates. Also important is distance from the colonizing pool of species on the mainland; islands farther away have lower rates (Figure 15–3). Thus small islands equilibrate at fewer species than large islands, and distant islands equilibrate at fewer species than near islands.

Ecologists sometimes characterize species as being **r-** or **K-selected;** *r* refers to the rate of population increase, and *K* means the number in the population at the upper limit, or carrying capacity, of the environment. Species that are *r*-selected have high reproductive rates, rapid development, and great powers of dispersal (Pianka 1970); they live in unstable environments in which recolonization of areas is necessary. Diamond (1974) studied the birds of New Guinea and nearby islands and found that certain species were always the first to recolonize islands that had lost their fauna due to volcanic explosions or tidal waves; he referred to these *r*-selected species as "supertramps." *K*-selected species have low reproductive rates, slow development, and limited powers of dispersal; characteristically inhabitants of stable environments, these species may fail to colonize new areas separated by relatively small barriers. One island Diamond studied was separated from New

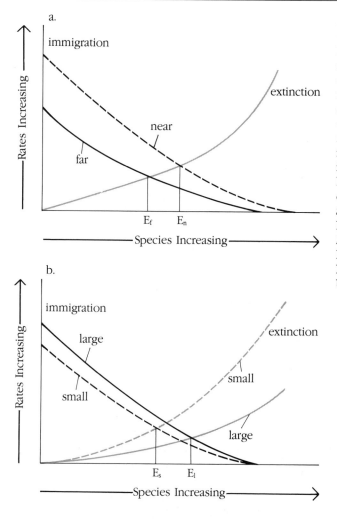

FIGURE 15-3   Immigration and extinction rates of species on islands
Researchers hypothesize that the number of species on islands reaches a dynamic equilibrium as a function of immigration and extinction rates. In (a) an island far from a colonizing source or mainland should equilibrate with fewer species, E_f, than an otherwise identical near island, E_n. In (b) a large island equilibrates with more species, E_l, than a small one, E_s, at the same distance from the mainland.
Source: Robert H. MacArthur and Edward O. Wilson, *The Theory of Island Biogeography.* Copyright © 1967 by Princeton University Press, Fig. 16–3. Reprinted by permission of Princeton University Press.

Guinea by only 10 m of water, yet had only half the New Guinea species expected on the basis of the availability of comparable habitats in the two areas.

BEHAVIOR

Sometimes behavior patterns keep species from occupying apparently suitable habitats. Two closely related species of British birds, the tree pipit *(Anthus trivialis)* and the meadow pipit *(Anthus pratensis),* are both ground nesters and eat similar types of food, but the tree pipit is absent from many treeless areas that the meadow pipit inhabits. The two also have similar songs, but the meadow pipit ends its aerial song on the ground, and the tree pipit always ends its aerial song on a perch, such as a tree or pole. Thus the tree pipit is excluded from areas that it could otherwise occupy because of a specific behavior pattern (Lack 1933).

Female mosquitoes of the genus *Anopheles,* many species of which transmit such diseases as malaria, are very particular about where they lay their eggs. In the southern part of India, *Anopheles culifacies* eggs and larvae are found only in new rice fields, where the plants are less than twelve inches high; however, eggs transplanted to old fields have yielded normal numbers of larvae and adults, and other species of *Anopheles* lay their eggs in these mature fields. Russell and Rao (1942) demonstrated that the mechanical obstruction of the rice plants inhibited *culifacies* females from laying eggs. Glass rods or bamboo strips "planted" in the water had the same effect; shade was not a factor. *Culifacies* females oviposit while they are on the wing, performing a hovering dance two to four inches above the water; possibly the obstructions interfered with this oviposition dance. However, in one experiment researchers partially submerged a box with no lid in the shallow water; although the box did not directly interfere with oviposition, few eggs were laid.

Why should a species not take advantage of a suitable habitat? One possibility is that the habitat is not actually suitable, perhaps because of competition, predation, or other factors the scientist may have failed to detect. Or such habitats may not have been suitable in the past: if organisms responding to certain environmental cues in previous optimal habitats left more offspring, their genetically influenced behaviors would become widespread and persist. New environments, although suitable, may not contain those cues and therefore are not utilized.

OTHER SPECIES

Do other species keep a particular species out? Even if an organism can and "wants" to get to a place, other factors may prevent its becoming established. These factors could be predators, parasites, disease agents, allelopathic agents (plant poisons or antibiotics), or interspecific competitors. Demonstrating conclusively that one species prevents an area from being colonized by another is difficult, but experimental and observational data do point in that direction. For instance, an elaborate series of experiments by Kitching and Ebling (1967) has shown how mussels *(Mytilus edulis)* were kept out of protected bays along the coast of Ireland by three species of crabs and one species of starfish. Where the coast was unprotected, crabs were restricted by wave action, and small mussels could survive. In sheltered waters, mussels survived only in areas, such as steep rock faces, that the predators could not reach. Kitching and Ebling proposed that four criteria must be met before we can conclude that a predator restricts the habitat of its prey: (1) if they are protected from predators, prey will survive when transplanted to a site where they normally do not occur; (2) the distributions of prey and predator are negatively correlated; (3) the predator is observed eating the prey; and (4) the predator can be shown to destroy prey in transplant experiments.

The more similar two species are, the more likely they are to compete intensely and thereby restrict each other's distribution, but once again positive proof is difficult to obtain. Most often we rely on presumptive evidence in comparisons of the distributions of closely related species in several habitats. Although a species is usually less numerous at the edge of its distribution than at the center (as shown in Figure 15–1), sometimes a species is abruptly replaced by a close relative, with both species at maximum density just above and just below the interface (Figure 15–4). If

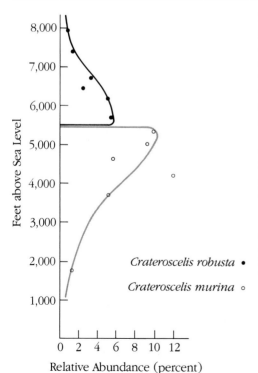

FIGURE 15–4   Distribution of two species of warblers
Abundance is measured as a percentage of all bird individuals observed. As one sample from the side of a mountain in New Guinea shows, *Crateroscelis murina* reaches maximum abundance at 5,400 feet and is abruptly replaced by *C. robusta*. This distribution can be compared with that in Figure 15–1.
Source: J. M. Diamond, "Distributional Ecology of New Guinea Birds," *Science* 179 (1973):759–769. Copyright © 1973 by the American Association of the Advancement of Science. Reprinted with permission of the publisher and the author.

interspecific competition restricts distribution, we would expect one species to extend its range in the absence of the other. This phenomenon has been demonstrated by Diamond (1978) with bird distributions on mountain tops in New Guinea, where a second, closely related, species may be absent due to inability to disperse (Figure 15–5). In these cases the single species occupies a much larger altitude range than it does on islands where both are present.

Competitive exclusion was actually witnessed by Orians and Collier (1963) when colonial tricolored blackbirds (*Agelaius tricolor*) moved into a marsh already occupied by red-winged blackbirds (*Agelaius phoeniceus*). After the invasion the redwing territories were restricted to the periphery. Competition can be studied experimentally by removing one species and noting changes in nearby competitors or by introducing closely related species and monitoring the success of each, as Vaughan and Hansen (1964) did with two species of pocket gophers (*Thomomys bottae* and *Thomomys talpoides*). Slight differences between species in dispersal powers and environmental tolerances led to one or the other species' winning out.

PHYSICAL AND CHEMICAL FACTORS

If a transplant experiment fails, and no evidence exists that biotic factors have eliminated the species, some combination of physical and chemical factors may be involved. Each organism has a range of tolerances for these factors, and much of its behavior is directed toward staying within these limits. Temperature and moisture are the master factors that limit the distribution of life on earth, but physical factors, such as light, soil structure, or fire, and chemical factors, such as oxygen, soil nutrients, salts, or pH, may be important as well. Most research on the effects of physical and chemical factors on organisms is done by physiological ecologists.

CHOICE OF BREEDING SITES

DISPERSAL OR PHILOPATRY

The discussion thus far has dealt with the problem of habitat selection at the species-distribution level. Animals may also make choices about whether to remain at (or return to) the natal site or to disperse to other breeding locations. In most species of birds and mammals members of one sex tend to disperse, while members of the other sex are **philopatric,** breeding near the place where they were born (Greenwood 1980). Among mammals it is usually the males that disperse, while among birds the opposite is true. The reason for this difference may be that most bird species are monogamous, with the male defending a territory that contains resources vital to him and his mate. It is probably easier for a male to establish such a territory and

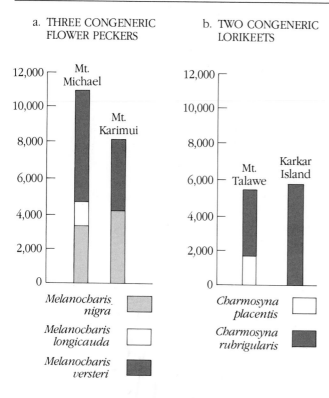

a. THREE CONGENERIC
   FLOWER PECKERS

b. TWO CONGENERIC
   LORIKEETS

Melanocharis nigra

Melanocharis longicauda

Melanocharis versteri

Charmosyna placentis

Charmosyna rubrigularis

FIGURE 15-5   Altitudinal ranges of species of birds on New Guinea mountains and surrounding islands
In (a) three similar congeneric flower peckers, *Melanocharis nigra, M. longicauda,* and *M. versteri,* occupy nonoverlapping areas up to about 11,000 feet on Mt. Michael. On Mt. Karimui, which is smaller and more isolated than Mt. Michael, *M. longicauda* is absent. In (b) two congeneric lorikeets. *Charmosyna placentis* and *C. rubrigularis,* occupy an area on Mt. Talawe, but only *C. rubrigularis* colonized Karkar Island. Altitudinal ranges in all cases are nonoverlapping and are larger in the absence of other species.
Source: J. M. Diamond, "Ecological Consequences of Island Colonization by Southwest Pacific Birds I: Types of Niche Shifts," *Proceedings of the National Academy of Sciences USA 67* (1970):529–536. Reprinted with permission of the publisher and the author.

attract a mate in, or close to, his natal site, where he is familiar with the location of resources and/or predators. On the other hand, many species of mammals are polygynous. The females form the stable nucleus and the males attempt to maximize their access to them, frequently moving from one group to another (Greenwood 1980).

The ultimate cause of dispersal from the natal site is assumed to be inbreeding avoidance. If one or the other sex disperses, there will be less chance of matings between related individuals. Among black-tailed prairie dogs *(Cynomys ludovicianus),* young males leave the family group before breeding, while females remain. Also, adult males usually leave groups before their daughters mature (Hoogland 1982). Among primates such as vervet monkeys *(Cercopithecus aethiops),* males leave the natal group at, or shortly after, sexual maturation. They usually transfer to a neighboring group with age peers or brothers (Figure 15–6). Several years later they may again transfer alone to a third group. Cheney and Seyfarth (1983) argue that this pattern of non-random followed by random movement minimizes the chances of mating with close kin. Packer (1979) reported that a male baboon *(Papio anubis)* that failed to disperse at sexual maturity and mated with relatives sired offspring that

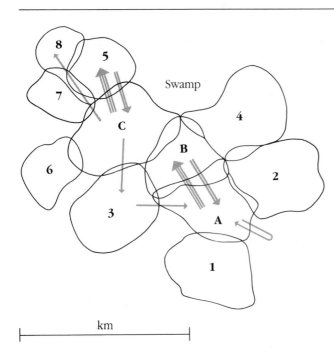

FIGURE 15-6 Distribution of migration by natal and young adult male vervet monkeys into and from the three study groups between March 1977 and July 1982.
Arrows indicate direction of movement; each line indicates one male. Letters indicate ranges of the study groups; numbers indicate ranges of regularly censused groups. Only groups with ranges adjacent to the study group are shown.
Source: D. L. Cheney and R. M. Seyfarth, "Nonrandom Dispersal in Free-ranging Vervet Monkeys: Social and Genetic Consequences." *American Naturalist* 122 (1983):392–412.

had low survival compared to the offspring of outbred males. Thus, in this case there seems to be a real cost associated with inbreeding.

If inbreeding depression were the only factor involved, we might expect individuals to disperse as far as possible from relatives. However, such is not the case. According to Shields (1982) most species that have been adequately studied are philopatric. He cites cases of **outbreeding depression,** in which matings between members of different populations within a species yield less fit offspring. Members of a population may possess adaptations to local conditions which are lost with outbreeding. Perhaps that is why white-crowned sparrow (*Zonotrichia leucophrys*) females respond sexually to the songs of males from their natal area but not to males from other areas (Baker 1983; see also Chapter 13).

A certain degree of inbreeding may be advantageous. Recall our discussion in Chapter 11 about the costs of meiosis: in sexually reproducing organisms the loss of genes in the offspring can be reduced by one half if the parents are related. Furthermore, according to Shields (1982) gene complexes are less likely to be disrupted in matings between relatives. Finally, kin selection (Chapter 4), which results in the evolution of cooperative behaviors, can operate only when relatives are in proximity. We predict more cooperative behavior within philopatric species and within the philopatric sex. Indeed, among Belding's ground squirrels (*Spermophilus beldingi*) females are philopatric and engage in altruistic alarm calling (Sherman 1981; see also Chapter 13).

From the preceding arguments we might predict some sort of "optimal" inbreeding strategy, in which matings between very close relatives (siblings, or parents and offspring) are avoided but matings with more distant relatives are favored. Bateson (1982) has found that female Japanese quail *(Coturnix coturnix)* spend more time in proximity to first cousins than in proximity to siblings or more distant relatives. If these tendencies reflect later sexual preferences, an optimal level of inbreeding might be among first cousins.

MICROHABITAT CHOICE AND REPRODUCTIVE SUCCESS

Although it is assumed that an individual that makes a "correct" choice of habitat has higher reproductive success than one that makes an "incorrect" choice, few studies have actually measured this relationship. Witham (1980) studied the life history of the aphid *Phemphigus betae*, a plant parasite about 0.6 mm long that feeds on leaves of the cottonwood tree. In the spring, after hatching from eggs laid the previous fall in the bark of the tree, females called stem mothers move up the trunk and select a leaf on which to feed. This activity triggers the formation of a hollow gall on the leaf, within which the female produces offspring parthenogenetically.

Witham found that females settling on large leaves have higher reproductive success than females on small leaves. Not surprisingly, aphids select the largest leaves, leaving small ones vacant. However, latecomers may have to choose whether to take an already occupied large leaf or an unoccupied small one. If she takes an occupied leaf, she will have to settle further from the base, where there is less food. Stem mothers farther from the base were found to be smaller in size and produce fewer young than those closer to the base. Stem mothers may engage in shoving and kicking contests that last for days, with the largest aphid usually getting the basal position. Choice of habitat in these insects is nonrandom and results in higher average fitness than would random selection of leaves.

ENVIRONMENTAL CUES

Organisms may respond directly to critical environmental factors. Speed or frequency of locomotion may be dependent on the intensity of stimulation. Isopods, such as *Porcellio*, are found in moist environments; when placed in a humidity gradient, they move faster in drier air and eventually wind up at the moist end of the gradient (Fraenkel and Gunn 1940). Likewise, the turning rates of protozoans are dependent on intensity of stimulation. (Types of orientation behavior involved in habitat selection were discussed in Chapter 14.)

Organisms may integrate more than one environmental variable—for instance, temperature and humidity. Several species of fruit flies *(Drosophila)* prefer warm temperatures in a laboratory gradient apparatus (Prince and Parsons 1977), but only if the humidity is high. At low humidity they move to the cooler area, thereby reducing water loss and thus increasing the probability of survival. This adaptive pattern is a response not only to external cues but also to the individual's physiological state.

Sale (1970) hypothesized the existence of a simple mechanism of habitat selection in fish that is based on levels of exploratory behavior. Sense organs monitor specific stimuli in the environment and send a summation of pertinent stimuli back to central nervous system centers, which regulate the amount of exploration. As the constellation of cues approaches some optimum level, exploratory behavior ceases and the animal stays where it is.

An alternative hypothesis is that an animal has a cognitive map of the ideal habitat and that its behavior is goal directed. Working with a species of surgeon fish, the Hawaiian manini (*Acanthurus triostegus*), Sale (1970) tested juvenile fish in laboratory tanks with various water depths and bottom covers (Figure 15–7). Exploration time was least in the tank with shallow water and bottom cover and highest in the tank with shallow water and no bottom cover. In choice tests and field observations, most fish preferred shallow areas with bottom cover. Thus there is no need to suggest the inheritance of complex cognitive maps and goal-directed behaviors; rather, the animal simply moves more in an unsuitable habitat and less in a suitable one.

Sale's model still does not explain how the animal "knows" what is suitable and what is not, or how stimuli from multiple cues are integrated. Nor does it explain the role of photoperiod in the response of dark-eyed juncos (*Junco hyemalis*) to photographs of their natural habitat. Birds kept in the lab under a winter photoperiod (9L:15D—9 hours of light and 15 hours of darkness) preferred slides of their southern, winter habitat (pine and hardwood forest). After daylength was increased to 15L:9D, the birds' preferences shifted to the northern, summer habitat (grassland and conifer forest) (Roberts and Weigl 1984).

Social cues may also affect choice of habitat. Large sized juncos (usually males) dominate smaller individuals (usually females and juveniles) in wintering flocks. Ketterson (1979) explained the finding that females usually migrate farther south than males by hypothesizing that subordinate birds are forced to migrate farther to avoid competing with dominants. In their lab study Roberts and Weigl (1984) found that during the short days (simulating winter) small, subordinate juncos showed the strongest preference for winter scenes.

The proximate cues to which animals respond when selecting a habitat may not be the same as the ultimate factors that have brought about the evolution of the response. The ultimate reason a mouse chooses to live in a wooded area rather than a field might be an inability to tolerate the higher temperatures in the field. But rather than responding directly to temperature, it might cue on the geometric shapes of trees.

The blue tit *(Parus caeruleus),* a European relative of the chickadee, lives in oak woodlands, where most of its preferred food is found (Partridge 1978). But the blue

FIGURE 15-7   Laboratory environments for testing habitat selection in fish
In all four test environments (a) small, flat rocks were present bearing a film of algae as a food source. Variables were water level (deep or shallow) and covers under which the fish could hide (present or absent). Exploration time, as shown in (b), was least in the environment with shallow water and cover present. This habitat is the one most similar to the preferred habitat in the field.
Source: P. F. Sale, "A Suggested Mechanism for Habitat Selection by the Juvenile Manini *Acanthurus triostegus sandvicensis* Streets," *Behaviour* 35 (1970):27–44. Reprinted with permission.

a. TEST ENVIRONMENTS

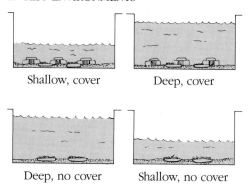

b. EXPLORATION TIME IN TEST ENVIRONMENTS

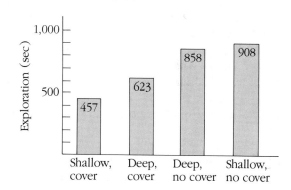

tit establishes its breeding territory each year before leaves and caterpillars (the staple food) have even appeared, so it must be using other features, such as shape of the trees, as cues to the habitat. Although birds have been studied intensively, we know little about the signals to which they respond in choosing a particular habitat. Possible cues are stimuli from landscape; sites for nesting, singing, or feeding; food itself; or other animals. Social cues may be important, particularly among colonially nesting birds. Laughing gulls *(Larus atricilla)*, studied by Hailman, nested on only one of two similar islands. When a storm destroyed the nesting island, they moved over to the other one. Thus a "suitable" habitat cannot be defined by its structural features alone (Klopfer and Hailman 1965).

Wood frogs *(Rana sylvatica)* lay their eggs during a brief period each spring in temporary ponds, those that dry up in the summer. All the egg masses are deposited in one place in the pond. The physical features of this location seem less important than does the presence of an egg mass, which triggers other females to lay their eggs there (Howard 1980). By placing an egg mass in one part of the pond before any other eggs had been laid, Howard was able to induce all the females to lay their eggs there. Eggs in the center of such a mass may be protected from predators and from temperature fluctuations (Berven 1981).

DETERMINANTS OF HABITAT PREFERENCE

HEREDITY

How can we sort out the roles that genetics, early learning, and tradition play in habitat choice? If two animals reared from birth in identical environments are found to differ in habitat preference when they are tested as adults, we can conclude that those differences must be due to hereditary factors. When coal tits *(Parus ater)* and blue tits were reared in aviaries with no vegetation and then presented with a choice between oak and pine branches, coal tits preferred pine and blue tits preferred oak (Figure 15–8). The differences correspond to the distribution of these birds in nature and to the response of wild titmice in aviaries (Partridge 1974, 1978). It is perhaps not surprising that blue tits feed more efficiently in oak than do coal tits (Partridge 1976).

EARLY EXPERIENCE

Some experimenters have modified the environments of young birds to test whether their genetic predisposition to respond to certain stimuli can be altered by early experience. For example, when Klopfer (1963) placed wild-caught chipping sparrows *(Spizella passerina)* in a room containing both pine and oak branches, they preferred the pine, as they usually do in nature (Table 15–2). Klopfer reared a group of eight nestlings in a covered cage that prevented their seeing any foliage. When he tested them at two months of age, these birds, like their wild-caught counterparts,

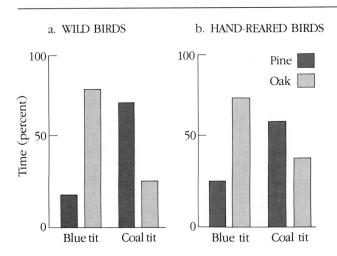

FIGURE 15–8   Time tits spent in oak and pine habitats in the laboratory Both wild and hand-reared blue tits preferred the oak habitat to the pine habitat. Wild and hand-reared coal tits, however, preferred the pine habitat. These preferences reflect those of each species in the wild.
Source: Data from L. Partridge, "Habitat Selection" in J. R. Krebs and N. B. Davies (eds.), *Behavioural Ecology: An Evolutionary Approach* (Sunderland, Mass.: Sinauer Associates, 1978).

CHAPTER 15 Habitat Selection

TABLE 15-2 Results of laboratory experiment to test vegetation preference of chipping sparrows

One side of the test chamber contained oak branches; the other, pine. Findings indicate that the preference for pine could be modified to some extent by early experience.

Chipping Sparrows	Percent Time Spent in Pine	Percent Time Spent in Oak
Wild-caught adults	71	29
Laboratory-reared, no foliage exposure	67	33
Laboratory-reared, oak foliage exposure only	46	54

Source: Data from P. H. Klopfer, "Behavioral Aspects of Habitat Selection: The Role of Early Experience," *Wilson Bulletin* 75 (1963):15–24.

preferred pine to oak. Klopfer reared a third group of ten birds in the presence of oak leaves to see if their preference could be shifted. When he tested them at two months, they were ambivalent, spending about equal time in each habitat. This result suggests that their inherited predisposition for pine could be affected to some extent by early experience.

After these tests the oak-reared birds were removed to a large aviary located in a pine-hardwood forest. Within the aviary itself, however, were only some small, broadleaf shrubs. When Klopfer retested the birds their preference for pine had increased. After eight months in an aviary containing pine, the birds' preference for pine was even stronger. These results suggest that the birds retain the ability to shift to the naturally preferred vegetation type.

Wecker (1963), in his study of deermice (*Peromyscus maniculatus*), did an extremely thorough study of habitat selection in mammals. This species, one of the more common North American rodents, is divided into many geographically variable subspecies of two general types: the long-eared, long-tailed forest form and the smaller, short-eared, short-tailed grassland form. In the laboratory the grassland form (*Peromyscus maniculatus bairdi*) does well in forest conditions, where its food preference and temperature tolerance are similar to those of the forest subspecies; thus experimenters assume that the avoidance of forests in the grassland deermouse is a behavioral response (Harris 1952).

Wecker's objective was to assess the genetic basis of this behavior and to test the idea that "habitat imprinting" (Thorpe 1945) is important. He constructed an enclosure halfway in a forest and halfway in a grassland, released the mice in the middle, and recorded their locations. The animals he tested were of the grassland subspecies (*bairdi*) and were of three basic types: (1) wild-caught in grassland; (2) offspring, reared in laboratory, of wild-caught; (3) reared in laboratory for 20 generations. Both wild-caught mice and their offspring selected the grassland half of the enclosure, regardless of previous experience. Laboratory stock and their offspring showed no preference, whether or not they had been raised in forest conditions. However, laboratory stock reared in a grassland enclosure until after weaning

TABLE 15-3   Habitat selection by deermice *(Peromyscus maniculatus bairdi)* as a function of hereditary background and early experience
The outdoor test enclosure was forest on one side and grassland on the other. Preference was measured by the percentage of time, amount of activity, and depth of penetration by mice in each side of the enclosure.

Number of Mice Tested	Hereditary Background	Early Experience	Habitat Preference
12	Grassland	Grassland	Grassland
13	Laboratory	Grassland	Grassland
12	Grassland	Laboratory	Grassland
7	Grassland	Forest	Grassland
13	Laboratory	Laboratory	None
9	Laboratory	Forest	None

Source: Data from S. C. Wecker, "The Role of Early Experience in Habitat Selection by the Prairie Deer Mouse, *Peromyscus maniculatus bairdi,"* *Ecological Monographs* 33 (1963):307–325.

showed a strong preference for the grassland when tested later (Table 15–3). Wecker reached the following conclusions:

1. The choice of grassland environment by grassland deermice is predetermined genetically.
2. Early grassland experience can reinforce this innate preference but is not a necessary prerequisite for subsequent habitat selection.
3. Early experience in forest or laboratory is not sufficient to reverse the affinity of this subspecies for the grassland habitat.
4. Confinement of these deermice in the laboratory for twelve to twenty generations results in a reduction of the hereditary control over the habitat selection response.
5. Laboratory stock retain the capacity to "imprint" on early grassland experience but not on forest.

Wecker also suggested that learned responses, such as habitat imprinting, are the original basis for the restriction of this subspecies to grassland environments; genetic control of this preference is secondary.

In a series of laboratory experiments designed to explore the importance of early experience on bedding preference in inbred mice *(Mus musculus),* Anderson (1973) raised animals either on cedar shavings or on a commercial cellulose material. When he tested them later, he found that the mice preferred the bedding on which they had been raised, although females raised on cellulose "drifted" toward cedar shavings in subsequent tests. Naive mice preferred cedar shavings (Figure 15–9).

Habitat imprinting occurs among migratory vertebrates, which commonly tend to return to the vicinity of their birth to breed. In the case of **iteroparous** organisms (those that have their young at intervals), such as birds, the adults of most species return to the same area to nest each year. **Semelparous** breeders (those that

FIGURE 15-9 Sleeping site selection of mice as a function of prior bedding experience and test day

Mice were born and raised on either cellulose or cedar bedding and were then given a choice of bedding types at either 30 or 60 days of age. Mice generally preferred the bedding type on which they had been raised. There was, however, a "drift" toward cedar preference across test days in females. Older females raised on cellulose initially showed no preference but by test day 3 they chose cedar.
Source: Data from L. T. Anderson, "An Analysis of Habitat Preferences in Mice as a Function of Prior Experience," *Behaviour* 47 (1973):302–339.

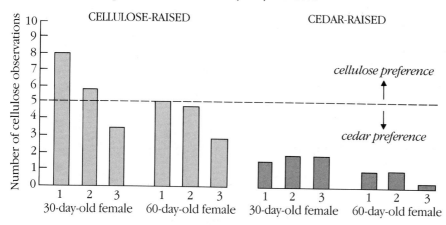

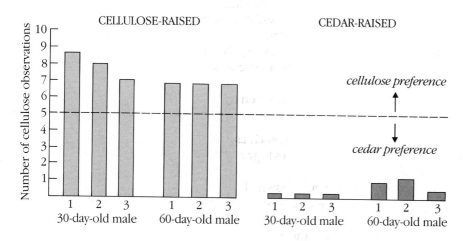

have their young all at once), such as salmon, breed only once. After feeding in the open ocean for several years, salmon return to the same upstream spawning bed where they were hatched. The salmon's olfactory system is programmed to respond to unique odors of the home stream during a critical period in the first few weeks of life (Hasler 1966).

TRADITION

Inherited tendencies and imprinting may be involved in restricting habitat choice to a small part of the potential range, but **tradition** — behavior passed from one generation to the next through the process of learning — may also be an important factor for such species as sheep (Geist 1971).

Mountain sheep *(Ovis canadensis)* live in unisexual groups. Females are likely to stay in the natal group but may switch to another female group when they are between one and two years of age. Young rams desert the natal group after the second year of life and join all-ram bands. Mothers do not tend to chase their young away at weaning, as most other mammals do. Females follow an older, lamb-leading female, males follow the largest-horned ram in the band. When rams mature, they are followed by younger rams and pass their habitat preferences on to them.

Until the last century mountain sheep occupied a much larger range in North America and Asia than they do today. Measures enacted to protect sheep, mainly hunting regulations, have done little to increase their numbers. A formerly inhabited part of the range that appears intact is not colonized, and transplants to suitable areas are often unsuccessful. In contrast, deer *(Odocoileus* spp.) and moose *(Alces alces)* have recolonized areas rapidly and have reached population densities higher than ever (Geist 1971).

Why have sheep failed to extend their range while moose and deer have done so? Geist has pointed out that deer and moose, which are relatively solitary beasts, establish ranges by individual exploration after being driven out of the mother's range; sheep, in contrast, transmit home-range knowledge from generation to generation and often associate with group members for life.

Understanding the niches of these species can help explain the sheep system, which may seem a rather poor adaptation to the environment. The moose is associated over much of its range with short-lived plant communities that follow in the wake of forest fires. Moose in these environments are r-selected "pioneer species" and disappear as the climax coniferous forest regenerates. Moose habitats are subject to rapid expansion after fires, and moose must continually colonize new habitats. Each spring when her new calf is born, the cow drives away her yearling, which may wander some distance before establishing its new range.

Sheep habitats are formed by stable, long-lasting, climax grass communities, which exist in small patches. Geist has argued that, given the distance between patches and the ease with which wolves can pick off sheep, the best strategy for the sheep is to stay on familiar ground. Sheep also have at least two and as many as seven seasonal home ranges, which may be separated by twenty miles or more. These areas are visited regularly by the same sheep year after year at the same time; knowledge of the location of these ranges and the best times to visit them is transmitted from one generation to the next. Because new habitats rarely become available, there is no advantage to an individual's dispersing and attempting to colonize other areas.

SUMMARY

Habitat selection refers to the choice of a place in which to live. Factors affecting this choice can be best understood by a stepwise approach to the question of why a species is absent from a particular place. The transplant experiment, in which organisms are relocated outside their natural ranges and their survival and reproduction are monitored, is an important tool. First, if the transplant is a success, we can assume that the organism's dispersal powers could not overcome geographical barriers. The success of introductions of foreign species by humans and data on colonization of islands suggest that this inability to disperse often explains a species' absence. Second, in a more complex case, organisms may fail to colonize an otherwise suitable area because of behavioral responses to specific features of the habitat. Third, the presence of such other factors as competitors, parasites, predators, or diseases may also explain a species' absence. Evidence for exclusion due to competition comes in large part from range expansion by one species in the absence of another. Experimental demonstration of exclusion by predators requires the species to survive in the absence of the predator. Fourth, physical and chemical factors that are beyond the range of tolerance of organisms can restrict a species' distribution. Temperature and moisture are very important and may interact with other factors.

Behavior patterns may restrict organisms to a fraction of the habitat that they seem, to us, to be equipped to occupy. However, apparently suitable areas may in fact not be so in the long run. Behavior patterns that restrict a species' distribution may have evolved because of negative factors associated with other habitats in the evolutionary past.

Within a species' range individuals may disperse or may remain in the natal area to breed. In mammals males typically disperse while females are philopatric. In birds the opposite is generally true. By dispersing, the chances of matings between close relatives are reduced, minimizing inbreeding. However, there may also be costs to outbreeding, leading to philopatry and a preference for mates from the same local population. A few studies have measured the impact of habitat selection on reproductive success. Among insects such as aphids information about habitat quality and the density of competitors is used to make choices that will maximize reproductive success.

The proximate cues animals use in habitat selection include direct locomotive responses to such environmental variables as temperature and humidity. Little is known about the cues used by vertebrates; fish may use olfaction, and birds may respond to vegetation type. The cues may be only indirectly related to the ultimate factors that determine survival; for example, birds select breeding habitats before the leaves and staple insect foods have emerged.

Several experiments with birds and mammals have been conducted to determine the roles of genes and experience in habitat selection. Selection of the "correct" habitat is under some degree of genetic control, as has been demonstrated by studies in which animals have been reared in isolation and later tested in various habitats. However, early experience can modify later choices.

Tradition, the transmission of knowledge of habitats from one generation to the next, is thought to be important in mountain sheep. These animals live in close social groups in stable habitats but seasonally occupy two or more ranges. Other ungulates — for example, moose — often live in unstable habitats and gain knowledge of suitable places to live through individual exploration.

Discussion Questions

1. Suppose that you wanted to understand why species X is present in a particular habitat and species Y, closely related to X, is absent. Outline the steps you would follow to explain the distribution.

2. Review the life history of an organism of your choice. Try to explain the dispersal patterns of each sex in light of our discussion of the costs and benefits of inbreeding and outbreeding. Include ecological as well as genetic factors in your answer.

3. Although tradition is claimed to play a role in the habitat preference of some animals, firm data are lacking. Design an experiment to demonstrate the role of tradition in habitat choice.

4. The following figure shows the distribution of two species of chipmunk, *Eutamias dorsalis* and *Eutamias umbrinus*, in the Great Basin area of Nevada. What conclusions about animal distribution can we draw from these data?

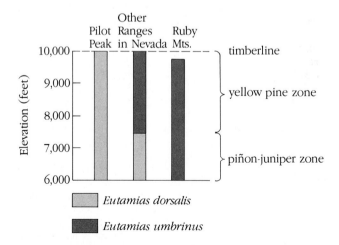

Source: Data from E. R. Hall, *Mammals of Nevada* (Berkeley: University of California Press, 1946).

Suggested Readings

Krebs, C. J. 1985. *Ecology: The Experimental Analysis of Distribution and Abundance,* 3rd ed. New York: Harper & Row.
Part 2 of this basic text treats the problem of species distribution. Chapter 5 specifically treats behavioral processes in habitat selection.

Partridge, L. 1978. Habitat selection. In J. R. Krebs and N. B. Davies (eds.), *Behavioural Ecology: An Evolutionary Approach.* Sunderland, Mass.: Sinauer Associates.
One of the few attempts to treat the entire topic. Emphasis is on birds.

Shields, W. M. 1982. *Philopatry, Inbreeding and the Evolution of Sex.* Albany: State University of New York Press.
A provocative essay arguing that most species of plants and animals that have low fecundity are philopatric, that philopatry promotes inbreeding, and that the benefits of all but extreme inbreeding outweigh the costs.

Wecker, S. C. 1964. Habitat Selection. *Sci. Amer.* 211(4):109–116.
Describes Wecker's classic experiments with deermice. Serves as a model for experiments on habitat selection.

References

Anderson, L. T. 1973. An analysis of habitat preference in mice as a function of prior experience. *Behaviour* 47:302–339.

Baker, M. C. 1983. The behavioral response of female Nuttal's white-crowned sparrows to male song of natal and alien dialects. *Behav. Ecol. Sociobiol.* 12:309–315.

Bateson, P. P. G. 1982. Preference for cousins in Japanese quail. *Nature* 295:236–237.

Berven, K. A. 1981. Mate choice in the wood frog, *Rana sylvatica. Evolution* 35:707–722.

Cheney, D. L., and R. M. Seyfarth. 1983. Nonrandom dispersal in free-ranging vervet monkeys: Social and genetic consequences. *Am. Nat.* 122:392–412.

Diamond, J. M. 1974. Colonization of exploded volcanic islands by birds: The supertramp strategy. *Science* 184:803–806.

———. 1978. Niche shifts and the rediscovery of interspecific competition. *Amer. Scientist* 66: 322–331.

Elmes, G. W. 1971. An experimental study on the distribution of heathland ants. *J. Anim. Ecol.* 40:495–499.

Fraenkel, G., and D. L. Gunn. 1940. *Orientation of Animals: Kineses, Taxes, and Compass Reactions.* New York: Oxford University Press.

Geist, V. 1971. *Mountain Sheep: A Study in Behavior and Evolution.* Chicago: University of Chicago Press.

Greenwood, P. J. 1980. Mating systems, philopatry, and dispersal in birds and mammals. *Anim. Behav.* 28:1140–1162.

Harris, V. T. 1952. An experimental study of habitat selection by prairie and forest races of the deer mouse, *Peromyscus maniculatus. Contrib. Lab. Vert. Biol., Univ. Mich.* 56:1–53.

Hasler, A. D. 1966. *Underwater Guideposts: Homing of Salmon.* Madison: University of Wisconsin Press.

Hoogland, J. L. 1982. Prairie dogs avoid extreme inbreeding. *Science* 215:1639–1641.

Howard, R. D. 1980. Mating behaviour and mating success in wood frogs. *Anim. Behav.* 28:705–716.

Ketterson, E. D. 1979. Aggressive behavior in wintering dark-eyed juncos: Determinants of dominance and their possible relation to geographic variation in sex ratio. *Wilson Bull.* 91:371–383.

Kitching, J. A., and F. J. Ebling. 1967. Ecological studies at Lough Ine. *Adv. Ecol. Res.* 4:197–291.

Klopfer, P. H. 1963. Behavioral aspects of habitat selection: The role of early experience. *Wilson Bull.* 75:15–22.

Klopfer, P. H., and J. P. Hailman. 1965. Habitat selection in birds. *Adv. Stud. Behav.* 1:279–303.

Krebs, C. J. 1985. *Ecology: The Experimental Analysis of Distribution and Abundance,* 3rd ed. New York: Harper & Row.

Lack, D. 1933. Habitat selection in birds with special reference to the effects of afforestation on the Breckland avifauna. *J. Anim. Ecol.* 2:239–262.

MacArthur, R. H., and E. O. Wilson. 1967. *Theory of Island Biogeography.* Princeton: Princeton University Press.

Miller, J. W. 1975. Much ado about starlings. *Natur. Hist.* 84(7):38–45.

Orians, G. H., and G. Collier. 1963. Competition and blackbird social systems. *Evolution* 17:449–459.

Packer, C. 1979. Inter-troop transfer and inbreeding avoidance in *Papio anubis. Anim. Behav.* 27:1–36.

Partridge, L. 1974. Habitat selection in titmice. *Nature* 247:573–574.

———. 1976. Field and laboratory observations on the foraging and feeding techniques of blue tits *(Parus caeruleus)* and coal tits *(Parus ater)* in relation to their habitats. *Anim. Behav.* 24:534–544.

———. 1978. Habitat selection. In J. R. Krebs and N. B. Davies, eds., *Behavioural Ecology: An Evolutionary Approach.* Sunderland, Mass.: Sinauer Associates.

Pianka, E. R. 1970. On *r*- and *K*-selection. *Amer. Nat.* 104:592–597.

Prince, G. J., and P. A. Parsons. 1977. Adaptive behaviour of *Drosophila* adults in relation to temperature and humidity. *Aust. J. Zool.* 25:285–290.

Roberts, E. P., Jr., and P. D. Weigl. 1984. Habitat preference in the dark-eyed junco *(Junco hyemalis):* The role of photoperiod and dominance. *Anim. Behav.* 32:709–714.

Russell, P. F., and T. R. Rao. 1942. On relation of mechanical obstruction and shade to ovipositing of *Anopheles culifacies. J. Exp. Zool.* 91:303–329.

Sale, P. F. 1970. A suggested mechanism for habitat selection by the juvenile manini *Acanthurus triostegus sandvicensis* Streets. *Behaviour* 35:27–44.

Sherman, P. W. 1981. Kinship, demography, and Belding's ground squirrel nepotism. *Behav. Ecol. Sociobiol.* 8:251–259.

Shields, W. M. 1982. *Philopatry, Inbreeding, and the Evolution of Sex.* Albany: State University of New York Press.

Thorpe, W. H. 1945. The evolutionary significance of habitat selection. *J. Anim. Ecol.* 14:67–70.

Vaughan, T. A., and R. M. Hansen. 1964. Experiments on interspecific competition between two species of pocket gophers. *Amer. Mid. Nat.* 72:444–452.

Wecker, S. C. 1963. The role of early experience in habitat selection by the prairie deer mouse, *Peromyscus maniculatus bairdi. Ecol. Monogr.* 33:307–325.

Witham, T. G. 1980. The theory of habitat selection: Examined and extended using *Pemphigus* aphids. *Am. Nat.* 115:449–466.

16

FEEDING RELATIONSHIPS

The food supply, as the energy source, usually sets the upper limit of the population of a species in an area and is thus the resource for which individual organisms most often compete. In fact, life itself can be viewed as a process whereby organisms, under control of DNA and RNA, channel the sun's energy to build and maintain complex organic compounds and to replicate. Dawkins (1976) has presented the argument in its extreme form by stating that organisms are simply the vehicles used by genes to obtain energy and replicate themselves.

In this chapter we briefly examine feeding relationships as they relate to ecosystems and consider optimal foraging strategies. We then take a look at food-catching techniques and the relationship between social organization and food resources, concluding with a discussion of antipredator techniques. In some cases morphological as well as behavioral adaptations are considered, since the two are closely linked.

ECOSYSTEMS AND TROPHIC LEVELS

The concept of the trophic, or feeding, level and the flow of energy up the food chain from one level to another is basic to the study of ecology. **Energy** can be defined as the ability to do work. The first law of thermodynamics states that energy can be neither created nor destroyed in any system but can be transferred from one form (such as light) to another (such as electricity). The second law of thermodynamics, in turn, states that when such transfer occurs, the process is not 100 percent efficient; much energy is lost, usually as heat, and is no longer available to the system. Because of this loss, the number of trophic levels in an ecosystem is limited in most cases to only three or four. The first step—the producer level—in any food chain is the conversion of the electromagnetic energy of sunlight into the chemical energy of sugar by photosynthetic plants. About 98 percent of the available energy is lost here: some is reflected and some is of unusable wavelengths. Of the consumers, herbivores that feed on plants may utilize 10 percent of the remaining available energy; carnivores may capture as much as 15 percent of the energy available in the herbivores, and the rest is lost as heat or utilized by bacteria and fungi. With such low efficiencies, three or four steps will obviously exhaust the energy supply. Although the total energy available is less at higher trophic levels, the efficiencies of energy capture tend to increase at these higher trophic levels. When a weasel catches a mouse, it gets a concentrated packet of energy; the same cannot be said of a mouse when it eats grass.

Several generalities emerge from this understanding of energy flow. Because less energy is available at higher trophic levels, organisms that feed at high levels are less common than those that feed at lower levels. They also have large home ranges and are active, with high metabolic rates, since they have to search for a small number of concentrated energy sources. The flow of energy through a system of four trophic levels is shown in Figure 16–1.

We can conveniently arrange ecosystems in linearly ascending trophic levels, and organisms can be classified according to diet, or position in the food chain—(herbivores, carnivores, insectivores, granivores, and so on). However, closer study shows that such chains are rarely linear. Food chains are really food webs with interacting levels. Many species, such as some birds, are seed-eating herbivores in the nonbreeding season and insectivores in the breeding season; others feed at different trophic levels during different stages of development.

Species also vary in size and in the kinds of food they eat. Some are specialists throughout their life cycles, and others are generalists. Competing organisms tend to become very efficient at harvesting one or a few resources, and a division of labor among specialists results. As MacArthur (1972) has pointed out, this tendency to specialize is the ultimate reason why there are so many species. For example, Bernon (1981) studied the insect community inhabiting dung pads of herbivores in southern Africa. Among the predators were seven species of robber flies (of the family Asilidae), each specializing on different prey types. One species of robber fly (*Hoplisto-*

FIGURE 16-1 Energy flow through a generalized community
Inputs are sunlight and imported organic matter. Outputs are heat (respiration) and
exported organic matter. Because so much energy is not utilized but lost as heat at
each trophic level, few communities have more than four trophic levels.

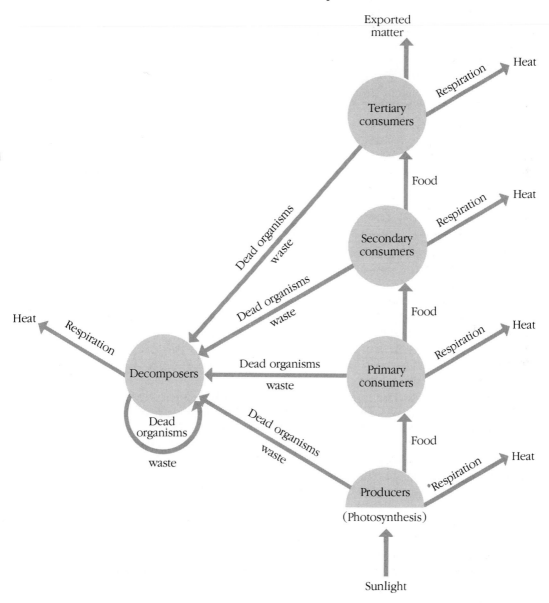

FIGURE 16-2   Robber fly specialist with a just-captured dung beetle
This African species of robber fly *(Hoplistomerus nobilis)* perches on a fresh dung pad and intercepts dung beetles in midair as the latter fly in to colonize the pad. Ninety-six percent of the robber fly's prey are dung beetles.
Source: Photo by Gary Bernon.

merus nobilis) is a beetle specialist; it perches on a dung pad minutes after the dung has been deposited and captures dung beetles as they fly in to colonize the pad (Figure 16–2).

FORAGING STRATEGIES

Studies of foraging behavior have focused on four basic problems: (1) If food occurs in clumps, or **patches,** what path should an animal follow within a patch to encounter the most food? (2) What items should it eat? (3) How long should it stay in a patch? (4) Which patch should it visit next? These are some of the decisions an animal must make; of course they are not necessarily conscious decisions. The problems are basically the same whether we speak of birds as "preying" on seeds, or of lions on wildebeests. We can predict that natural selection favors organisms whose decisions maximize energy intake. One of the goals of research in behavioral ecology is to find out whether animals do, in fact, follow decision rules that maximize foraging efficiency — that is, that they forage optimally.

Researchers usually assume that what is being maximized is the net rate of energy intake, which is intake per unit of time. There are certain **energy costs** in obtaining food — mainly search, pursuit, handling, and eating — that must be subtracted from the **benefit** — in this case, the energy in the food. A rate is obtained by dividing the net energy gain by the time it takes to do all of the above.

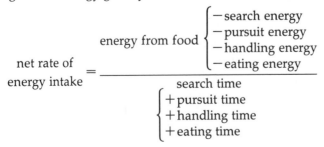

$$\text{net rate of energy intake} = \frac{\text{energy from food} \begin{cases} -\text{search energy} \\ -\text{pursuit energy} \\ -\text{handling energy} \\ -\text{eating energy} \end{cases}}{\begin{cases} \text{search time} \\ +\text{pursuit time} \\ +\text{handling time} \\ +\text{eating time} \end{cases}}$$

MOVEMENT RULES WITHIN A PATCH

How can we tell whether or not animals maximize the rate of energy intake? Pyke (1979a) studied a relatively simple system in which bumblebees fed on nectar from monkshood flowers and transported the food to the colony for storage (Figure 16–3). After they arrived at a plant, the bees usually started to forage on the lower flowers of an inflorescence, a cluster of flowers. What rule did they follow to determine their movements to subsequent flowers? After recording thousands of actual movements in the field, Pyke came up with the following rule: the bees moved to the closest flower not most recently visited unless the last movement was downward and was not the first movement on an inflorescence. In the latter case, the bees moved to the closest flower not most recently visited. The bee was most likely to depart from an inflorescence after reaching either the top or, less often, the bottom flower. Departures under other circumstances were most likely after a downward movement.

Pyke's next step was to try to figure out what the bees should be doing to maximize the net rate of energy intake and then to compare observed and expected behavior. He found that flowers at the bottom of the plant usually had more nectar than did those at the top, so the observed tendency for bees to start foraging on a low flower and work their way up makes sense. However, a tradeoff was necessary because the cost of flying from the top of one plant to the bottom of the next was greater than the cost of going to the middle of the next plant; perhaps this is why some bees started in the middle. In developing expected movement rules, Pyke had

FIGURE 16-3 Foraging bumblebee Bumblebee worker *(Bombus terricola)* is collecting nectar from flower of fireweed *(Epilobium angustifolium)*. *Source:* Photo by Candace Galen, from C. Galen and R. C. Plowright. Contrasting movement patterns of nectar-collecting and pollen-collecting bumblebees *(Bombus terricola)* on fireweed *(Chamaenerion angustifolium)* inflorescences. *Ecological Entomology* (1985) 10:9–17.

to calculate: (1) the time bees spent at each flower and the time they spent flying between flowers, inflorescences, and plants; (2) the energy gain, as determined from caloric analysis of the nectar; (3) the energy cost of foraging; and finally (4) the net rate of energy intake for each possible movement rule. Of a number of possible movement rules he considered, two provided equally high rates of energy intake, and one is similar to the observed rule: Always choose the closest flower not previously visited.

Pyke's approach to the analysis of bumblebee foraging has been criticized by Heinrich (1983), a long-time student of bumblebee physiology and behavior. Heinrich argues that the use of optimal foraging theory tells us little about the proximate mechanisms underlying the behavior and may lead us in the wrong direction. In trying to explain why bees usually start at the bottom flower in an inflorescence and work up, he points out that the flowers bees feed on usually hang down, and can be most easily entered from below. When artificial inflorescenses were constructed in which the topmost flowers had the most nectar rather than the bottommost, bees continued to visit the bottom flowers first.

Other constraints may keep an animal from foraging optimally. Predators or competitors may be in the area, forcing the forager to expend time and energy in addition to that indicated in the formula above. Nor can we always assume that energy gain is being maximized. For instance, the golden-winged sunbird *(Nectarinia reichenowi)*, an African nectar-feeder resembling a hummingbird, defends a feeding territory, even in the non-breeding season. Pyke (1979b), using data collected by Gill and Wolf (1975), tested the notion that these birds were maximizing the rate of energy intake. The birds spent far less time foraging and defending and more time just sitting than was predicted. What they seemed to be doing was minimizing costs, eating and defending just enough to maintain themselves. However, these data were obtained during the nonbreeding season. Presumably, foraging rates and territory defense would increase to meet the demands of reproduction in the breeding season.

CHOICE OF FOOD ITEMS

The barn owl *(Tyto alba)* is a highly efficient nocturnal predator on small mammals. In southwestern New Jersey these owls roost in tree cavities or silos, foraging in fields over a radius of several kilometers. Colvin (1984) found that although more than 90 percent of the available small mammals were white-footed mice *(Peromyscus leucopus)* and house mice *(Mus musculus)*, and less than 5 percent were meadow voles *(Microtus pennsylvanicus)*, 70 percent of the mammals eaten were meadow voles (Figure 16–4). Clearly, these owls are not simply taking prey in proportion to its abundance. What factors influence the choice of prey types?

Once an animal has located a food item *(j)*, it should pursue that item only if during the expected pursuit time it could not expect to locate and to catch a better food item (MacArthur 1972). If P_j is the time needed to pursue and to capture an item

FIGURE 16-4 The percentage of small mammals eaten (a) compared to the percentage available, based on trapping (b)
In this study from southwestern New Jersey barn owls are meadow vole specialists.
Source: B. A. Colvin. 1984. Barn owl foraging behavior and secondary poisoning hazard from rodenticide use on farms. Ph.D. dissertation, Bowling Green State University.

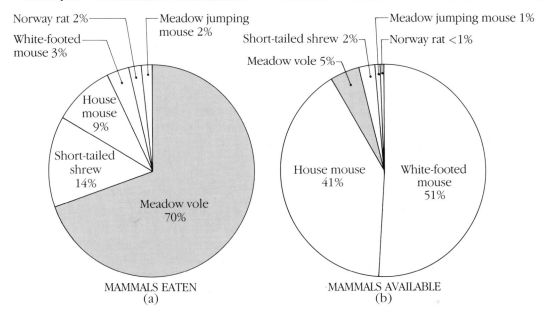

just located, and $\overline{P}$ and $\overline{S}$ are, respectively, the average pursuit and search times in the previous diet, item j should be added if $P_j < \overline{P} + \overline{S}$.

Suppose that an animal, such as a bird, eats small, stationary insects on leaves of plants. The bird must expend a lot of energy to find these prey, but they are easy to catch and eat; $\overline{S}$ is large compared to $\overline{P}$. If the bird encounters a new insect type on a leaf, its pursuit time probably will be small, too; P_j is likely to be less than $\overline{S}$ and $\overline{P}$, and the new item will be included. Such animals, including many species of insects, are likely to be generalists.

A different creature, such as a lion, may often have prey in sight, so that $\overline{S}$ is low. However, the lion has a difficult time catching anything, so $\overline{P}$ is high. The pursuit time of most new prey types the lion might sight would be greater than $\overline{P}$ and $\overline{S}$, that is, $P_j > \overline{P} + \overline{S}$, and so the lion would not chase the new prey. Such "pursuers" are thus likely to be specialists, rarely adding newly encountered species to their diet.

Another application of the relationship between food choice and pursuit and search times is to the productivity of the environment. In an unproductive environment, where food is scarce, $\overline{S}$ is large; therefore most new items will be included, and the animal will be a generalist. In more productive habitats, $\overline{S}$ will be smaller, and the same species will be more specialized. Thus great blue herons *(Ardea herodias)* eat a much wider range of foods in the unproductive lakes of the Adirondack Mountains in New York State, where they breed, than they do in the more productive waters of Florida, where they winter (MacArthur 1972).

Competition is likely to affect diet; its main result, referred to by MacArthur (1972) as the **compression hypothesis** (Figure 16–5), is to restrict the use of habitat. Unless the overall productivity of a feeding area is lowered by competitors, we would not expect much change in the number of different food items taken. The reason is that the rarity of an item does not affect the decision whether or not to pursue it once it has been found. However, if competition is intense enough to diminish the overall food supply, mean search times $(\bar{S})$ will increase and new items will be added to the diet.

Food preference has a heritable component, as Dix (1968) has demonstrated in garter snakes *(Thamnophis sirtalis)*. He compared isolation-reared offspring of garter snakes collected in Florida and Massachusetts. All the snakes accepted worms

---

FIGURE 16–5   MacArthur's compression hypothesis
As the number of species increases from (a) to (b), the habitat occupied in each species shrinks, but the range of acceptable food items in the diet changes little. Similarly, as species from a species-rich source (b) colonize an uninhabited island (a), the habitat occupied expands but the range of foods eaten remains the same. This hypothesis applies only to short-term, nonevolutionary changes. Note that species 1 and 2 in situation (a) occupy a large range of habitats. In (b) the habitat range is smaller but the range of foods eaten is about the same.
Source: Data from R. H. MacArthur, *Geographical Ecology: Pattern in the Distribution of Species* (New York: Harper & Row, 1972).

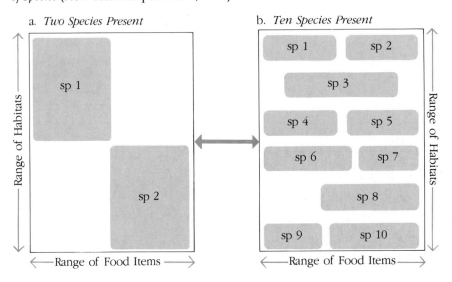

and frogs, but only the Florida stock took fish readily. Florida garter snakes are more aquatic in their habits than are northern forms and frequently feed on concentrated pools of fish during the dry season in the Everglades. On the other hand, experience also affects later food preference, as discussed in Chapter 9.

A limit to the number of different kinds of food an animal can search for efficiently at one time probably exists. Birds sometimes overlook palatable items to zero in on one or two prey types at a time. Such foragers have formed a **search image** and will continue to hunt those prey even when they are less common than other palatable foods. This phenomenon could be due to the forager's mental limitations; of many acceptable foods, it can remember only a few during any particular search. For example, if you were asked to search a telephone book for the occurrences of twenty different first names, you might pick five and search the list once, then repeat the search with five more, and so on, rather than trying to remember all twenty names.

An animal does not demonstrate a search image just because it overlooks potential prey and seems to choose one prey type more frequently than expected on the basis of the prey's relative density. Thus, in the case of the barn owl it would be premature to hypothesize a search image. We must first rule out simpler explanations, such as differences in palatability of prey types, ease of capture, or fear of a novel stimulus. Pietrewicz and Kamil (1979) tested whether blue jays (*Cyanocitta cristata*), after previous encounters with cryptic moths, improve their ability to detect that species and thus show a specific search image (Figure 16–6). The jays viewed slides of naturally occurring prey; the pictures were of either one of two species of

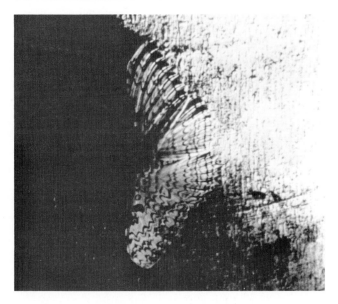

FIGURE 16-6 Cryptic moth
Source: Photo by Steve Pollick.

•

cryptic moths or of the substrate with no moth. When the experimenters projected a picture of a moth, the jays could get a food reward by pecking a key ten times. When no moth was present, pecking the food key not only resulted in no reward, but in a delay of the next slide as well; the correct response was to peck the advance key that started the next slide. In arranging the slides, the experimenters used either "runs," with only one species of moth shown, or "nonruns," with both species of moth shown. In both runs and nonruns, slides of no moths were mixed in to test the jays' ability to detect the presence or absence of the cryptic moth.

The result was that in "runs," with only one species, the jays improved their ability to detect the moth. In "nonruns," when both species were mixed in, the jays did not improve. The jays thus showed an increased ability to detect prey species to which they had been recently exposed. However, the search image in this case seems to be restricted to one species of prey at a time.

MOVING TO A NEW PATCH

Leaving the problem of choice of food items, let us assume that an animal has been feeding in a patch and is gradually depleting its food supply. When should it leave that patch and where should it go? If it is maximizing its new energy gain, it should leave when its expected net gain from staying declines to its expected net gain from travelling to and foraging in a new patch (Charnov 1976). This model, referred to as the **marginal value theorem,** can best be explained graphically, where energy gain is plotted against time in patch (Figure 16–7a). The travel time to the patch and the gain curve are fixed by the environment. Note that as the travel time increases, the optimum leaving time also increases (Figure 16–6b and c). This rather simple model makes a number of assumptions, including that the forager "knows" the travel time between patches and can compare the current rate of intake with that of the recent past. Despite this simplicity and the assumptions, studies of a variety of insects, birds, and mammals tend to support the model (McNair 1982).

The marginal value theorem has also been tested in **central place foragers,** animals that can carry more than one item at a time back to a central location for storage or feeding to offspring. The problem here is not only when to leave the patch, but the number of items to collect before returning to home base. Giraldeau and Kramer (1982) conducted a field experiment with chipmunks, which carry seeds back to a burrow for storage and later consumption. They placed "patches" of sunflower seeds in trays at different distances from burrows and recorded patch time, travel time, and the weight of seeds collected (load size). They found that the rate of seed collection declined as the cheek pouches filled. Also, as predicted by the model, load size and patch time increased as the travel distance increased. However, since the actual values were not as predicted, these researchers feel that more data are needed before the marginal value theorem can be accepted as a complete model.

The question of which patch to choose next has received less attention than

FIGURE 16-7 Predicting when to leave a patch: The marginal value theorem
As a forager stays in a patch and depletes the food resource, the rate of gain
decreases (graph a). Lines AB and CD (graph b) are tangents to the gain curve used
to determine the optimal patch resident time (T_{opt}). For a long travel time (Line AB)
the optimal patch time is greater than for a short travel time.
Source: Redrawn after J. R. Krebs and R. H. McCleery, 1984. Optimization in
ecology. In J. R. Krebs and N. B. Davies (eds.), *Behavioral Ecology: An Evolutionary
Approach,* 2nd ed. Sunderland, Mass.: Sinauer Associates.

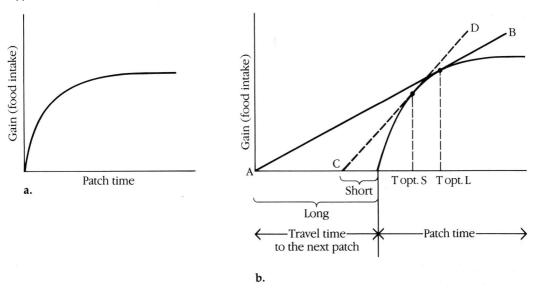

the "giving up time" described above. Animals faced with two patches of unknown
prey density might be expected to sample both, switching to the one having higher
prey density (Shettleworth 1984). Without checking the other patch, they might end
up in the poorer one. In a laboratory study using great tits *(Parus major)*, the birds
were given a choice of two feeders, one at each end of an aviary. Each feeder
required the bird to hop on a perch a certain number of times to get a mealworm.
Initially, the birds fed at both feeders, but they gradually shifted to the one with the
better reward ratio (Krebs, Kacelnik, and Taylor 1978).

FEEDING TECHNIQUES

Natural selection has resulted in a variety of techniques for maximizing the net rate
of energy intake. Some of these techniques are considered below.

TROPHIC LEVELS

As a consequence of the second law of thermodynamics, low trophic levels have more energy available, and, therefore, there is some advantage to feeding as low on the food chain as possible. One way is to feed directly on the producers, photosynthetic plants. Another way is to feed on dead organic material, as do many invertebrates, such as clams and earthworms. These animals, called **detritus feeders,** often filter food out of water or soil; they are usually sessile or slow moving. A majority of animals gather food through filter feeding and grazing, thereby taking advantage of the large supply of potential energy at low trophic levels.

Large size is often associated with grazing and filter feeding. Gorillas *(Gorilla gorilla)* and elephants *(Loxodonta africana),* feed on vegetable matter; and the largest of all animals, the blue whale *(Balaenoptera musculus),* uses its baleen mouth parts as a strainer to filter seawater for small, planktonic organisms such as krill. Animals feeding at higher trophic levels are generally smaller and more active than herbivores.

RESOURCE PARTITIONING

Many species divide up resources and are able to coexist only through differences in feeding behavior. MacArthur (1958) studied five species of warblers in Maine and Vermont. Each species of warbler, which is a small, insect-feeding bird that breeds in the boreal (mainly spruce) forest, feeds primarily in a certain part of a tree; thus the Cape May warbler *(Dendroica tigrina)* feeds near the top and outermost branches, while the myrtle, or yellow-rumped, warbler *(Dendroica coronata)* feeds in all parts of the lower branches (Figure 16–8). The types of movements of the birds differ also. The Cape May moves mostly in a vertical plane while the yellow-rumped moves mostly in a tangential plane (Figure 16–9).

The feeding action of the warblers can be classified as long flights, hawking, or hovering. Thus, although the warblers are morphologically similar and can eat similar kinds of food, by feeding in different places and in different ways, each is exposed to different kinds of food. Analysis of stomach contents has confirmed that different species of warblers do, in fact, eat different kinds of arthropods (MacArthur 1958).

MODIFYING FOOD SUPPLY

Some herbivores modify the food supply so as to increase it. The activities of browsers and grazers in the grasslands, for example, stimulate the growth of some species of grass and prevent succession to other community types such as forests.

FIGURE 16-8   Feeding positions within trees of Cape May warbler and yellow-rumped warbler
Shaded areas denote zones of most concentrated activity. The Cape May warbler feeds primarily in the outermost top branches of evergreen trees, the yellow-rumped warbler feeds on interior and bottom branches.
Source: Data from R. H. MacArthur, "Population Ecology of Some Warblers of Northeastern Coniferous Forests," *Ecology* 39 (1958):599–619.

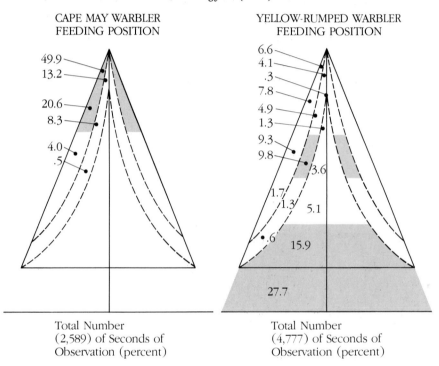

Some species of intertidal limpets increase the food supply with the mucus trail they secrete during locomotion (Connor and Quinn 1984). Two species *(Lottia gigantea* and *Collistella scabra),* which are solitary and return to a home scar on a rock, produce mucus that acts as an adhesive trap for algae. This substance also stimulates growth of the algae that the limpets feed on as they retrace their path home.

The ultimate in nonhuman manipulation of producers in an ecosystem is the "agriculture" practiced by certain fungus-growing ants, widespread in the New World tropics (Weber 1966). These ants cultivate yeast or fungi on organic material that they gather and carry into their nests. The organic material ranges from caterpillar feces used by ants of the genus *Cyphomyrmex* to the fresh leaves cut by the leaf

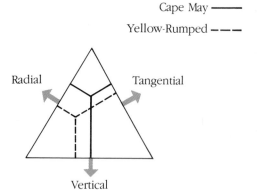

Cape May ———
Yellow-Rumped – – –

Radial

Tangential

Vertical

FIGURE 16-9   Components of flight motion of Cape May and yellow-rumped warblers while feeding on insects in trees

The lengths of the lines are proportional to the total distance each species moves in radial, tangential, and vertical directions. Vertical movements are up and down parallel to the trunk of the tree; tangential movements are circular at the same elevation; and radial movements are in and out along branches, perpendicular to the trunk. Cape May move mainly vertically while yellow-rumped move mainly tangentially. Such flight differences and the tendency to feed in different parts of the tree expose the birds to different species of insects.

Source: Data from R. H. MacArthur, "Population Ecology of Some Warblers of Northeastern Coniferous Forests," *Ecology* 39 (1958):599–619.

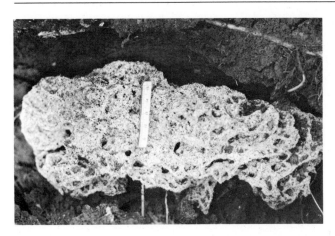

FIGURE 16-10   Fungus garden of ant

Ants of the species *Acromyrmex octospinosus* cultivate fungus as a food source. This garden, measuring approximately 31 by 11 centimeters, was found under a log in Trinidad.
Source: Photo by Neal A. Weber.

cutter ants of the genus *Acromyrmex* and is placed in nest cavities as much as six meters below ground (Figure 16–10).

Leaf cutter ants form long files as they bring cut leaf sections to the nest (Figure 16–11), then cut the sections into smaller pieces, and work the edges with their mandibles to make them wet and pulpy. They may deposit an anal droplet on the leaf before inserting the leaf into the garden. They place tufts of the fungal mycelium at intervals on the substrate, and the hyphae grow out in all directions from the tuft (Weber 1966). Cleaning operations help to keep foreign fungi out of the garden, but there is evidence that bacteriostatic and fungistatic factors are produced. By converting the indigestible cellulose of the leaves into sugars that are digestible by

FIGURE 16–11   Leaf-cutter ants en route to underground garden in the Amazonian rain forest
Source: Photo by Stephen H. Vessey.

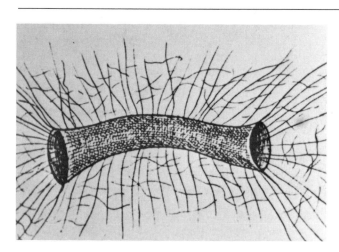

FIGURE 16–12   Lake-living caddisfly net
Larval insects of the genus *Polycentropus* manufacture nets with which to catch planktonic plants and animals.
Source: Drawing by Alice Loesch.

the ants, the fungus makes available to the ants the vast energy supply contained in the forest leaves. Other ant species rear aphids or other plant-feeding insects in or near their nests, protect them, and feed on their "honeydew" excretions (Wilson 1971).

TRAPPING AND DETECTING

Some invertebrates, such as ant lion (genus *Myrmeleon*) larvae, use entrapment to get their food. They make pits in the sand, bury themselves just below the pit, and knock ants into the pit with grains of sand that they hurl at them with tosses of their head. Other traps include the webs of spiders and the underwater nets of caddisfly larvae (order Tricoptera), as seen in Figure 16–12.

Unusual channels of communication may be used to capture food; for example, Kalmijn (1971) demonstrated that sharks can detect the weak electric fields of fish even when the latter are buried in sand (Figure 16–13). Electric fish (Gymnotidae, Mormyridae, and Gymnarchidae) generate their own electric fields (see Figure 6–4) and can identify the presence of potential prey (Lissmann 1958).

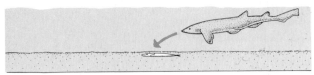

a. Shark detects fish under sand.

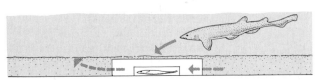

b. Shark detects fish in agar chamber.   Chamber doesn't block electric field.

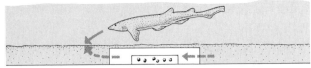

c. Shark is unable to detect pieces of fish in agar chamber.

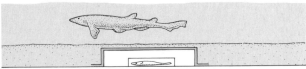

d. Fish in agar chamber is covered with metallic film.

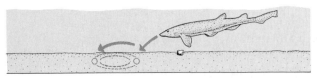

e. Electrodes produce dipole field.

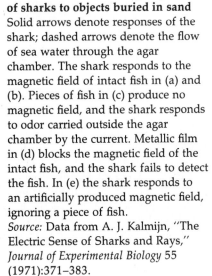

FIGURE 16–13   Feeding responses of sharks to objects buried in sand
Solid arrows denote responses of the shark; dashed arrows denote the flow of sea water through the agar chamber. The shark responds to the magnetic field of intact fish in (a) and (b). Pieces of fish in (c) produce no magnetic field, and the shark responds to odor carried outside the agar chamber by the current. Metallic film in (d) blocks the magnetic field of the intact fish, and the shark fails to detect the fish. In (e) the shark responds to an artificially produced magnetic field, ignoring a piece of fish.
Source: Data from A. J. Kalmijn, "The Electric Sense of Sharks and Rays," *Journal of Experimental Biology* 55 (1971):371–383.

AGGRESSIVE MIMICRY

A taxonomically widespread strategy for capturing prey is the use of lures, referred to as **aggressive mimicry.** In Chapter 13 we saw an example of sexual enticement by fireflies, in which the female mimics the mating signal of other species and eats the male that responds.

Fish of the order Lophiiformes have a modified first dorsal fin spine on the tip of the snout. At the end of the modified spine may be a fleshy appendage, a tuft of filaments, or, in deep-sea forms, an organ containing light-emitting bacteria. The bait often resembles worms or crustacea, and the rest of the fish resembles an inert object, such as an algae-encrusted rock or a sponge. The fish wriggles the bait and keeps the rest of its body still. See Figure 16–14 for a case in which the lure is a nearly exact replica of a small fish (Pietsch and Grobecker 1978).

FIGURE 16–14   Aggressive mimicry by fish
Angler fish (*Antennarius* spp.) and its lure are shown. In the bottom photo, a two-second time exposure shows the pattern of movement of the luring apparatus. Note the lure resembles a small fish.
Source: Photos by David B. Grobecker.

Siphonophores of the phylum Coelenterata capture prey with tentacles armed with stinging cells called nematocysts. Some species have tentacles with branches and clustered nematocysts that look like zooplankton such as copepods or fish larvae. When the siphonophore moves its tentacles, it lures predators of zooplankton into the web of nematocysts and the predators become food for the siphonophore instead (Purcell 1980).

USING TOOLS

Although the use of tools was once considered an exclusively human trait, it has evolved independently in several different species of animals. In most cases the tool is an unmodified inanimate object. The sea otter *(Enhydra lutris)*, for example, holds a rock on its chest and cracks shellfish, such as mussels, against it. Similarly, the Egyptian vulture *(Neophron percnopterus)* picks up rocks and drops them on ostrich eggs. The chimpanzee *(Pan troglodytes)*, however, sometimes makes a modified tool by stripping leaves from a twig, which it then inserts into an ant or termite nest (Figure 16–15); the insects cling to the stick, and the chimp eats those that hang on after it removes the stick. The woodpecker finch *(Cactospiza pallida)* of the Galápagos Islands uses sticks in a similar way to extract larvae from dead wood and may modify the stick by shortening it (reviewed in Lawick-Goodall 1970; Beck 1980).

FIGURE 16–15   Chimpanzee using tool to obtain ants
Chimps select sticks, modify them, and insert them into termite or ant mounds to extract the clinging insects. The infant observes the process from its mother's abdomen.
Source: Photo by James Moore from Anthro-Photo.

Tool use seems to have little relationship to central nervous system complexity. Rather, it enables generalized organisms, which have not evolved specialized appendages, to exploit a new resource. Morphologically specialized mammals, such as the anteaters (of the order Edentata) and pangolins (of the order Pholidota), have long snouts, sticky vermiform tongues, and large claws for digging up ant nests. These features restrict their ability to eat other types of food. In the absence of competition from woodpecker specialists, the woodpecker finch could evolve a flexible tool-using habit instead of a more efficient but more restricted morphological specialization during the adaptive radiation of finches on the Galápagos Islands.

Tool use was important in enabling humans to compete with specialized carnivores and scavengers millions of years ago in Africa. Humans could retain the flexibility of the generalist while enjoying the greater efficiency of the specialist. Tools also permitted the cultivation of plants more than ten thousand years ago and were the basis of the Industrial Revolution of 200 years ago. Each of these tool-using developments increased our energy-gathering capability and resulted in a substantial increase in population (Deevey 1960).

FEEDING AND SOCIAL BEHAVIOR

DEFENDING A TERRITORY

One way to increase net energy gain is to defend a food source against potential competitors. (For a discussion of territoriality see Chapter 13.) Since an animal must spend energy to advertise its presence and chase out intruders, only under certain circumstances will defending a territory be economical.

Pied wagtails (*Motacilla alba*) are insectivorous birds, some of which defend winter feeding territories along river banks, feeding on insects that are washed up on shore. The other wagtails feed in flocks in nearby pools. Territory holders, usually males, follow a circuit up one bank and down the other, a pattern that maximizes food intake as new insects are washed ashore (Davies and Houston 1983, 1984); intruding wagtails are chased away. When food on territories was very scarce, the owners fed elsewhere in flocks, but kept returning to the territory to evict intruders. When food was abundant, owners often shared the territory with a satellite, usually a juvenile or a female, which walked about one-half a circuit behind the owner (Figure 16–16). This meant that the food available to the owner was reduced by one-half, but the owner sometimes gained because the satellite helped chase away intruders. If food declined, the owner chased the satellite away. If food became extremely abundant, owners made no effort to defend their territories. Sometimes territory-holders fed on the territory even when they could have done better in the flock. Thus they were not maximizing energy intake in the short run. This system differs from the nectar-feeders, such as the golden-winged sunbirds (Gill and Wolf 1975), whose territory sizes vary as a function of food availability. The wagtails keep

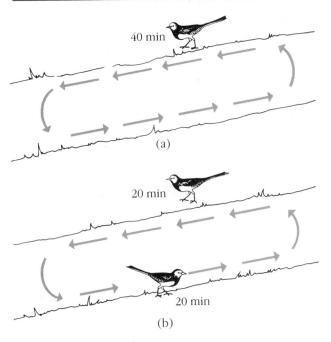

40 min

(a)

20 min

20 min

(b)

FIGURE 16–16 Pied wagtails feeding along a river bank
(a) Pied wagtails exploit their territories systematically. The circuit of the river bank takes on average 40 minutes to complete. (b) When a territory is shared between two birds, each walks, on average, half a circuit behind the other and so crops only 20 minutes worth of food renewal.
Source: N. B. Davies and A. I. Houston, 1984. Territory economics. In J. R. Krebs and N. B. Davies, (eds.), *Behavioural Ecology,* 2nd ed. Sunderland, Mass.: Sinauer Associates.

the same size area but vary their response to intruders as food supply fluctuates. Davies and Houston suggest that territory maintenance is a long-term investment to protect a more reliable food source than the pools where the flocks feed.

GROUP FEEDING

When food is spread widely and irregularly over the environment and cannot be defended, a species may group into flocks or herds, possibly associating with other species. Cody (1974) argued that by feeding in flocks, animals can recognize unexploited feeding areas more quickly. In addition, the flock can time its visits to an area so that food supplies have replenished sufficiently to make a return trip worthwhile. Assuming that the flock has exclusive use of the area, the flock could act as a "return-time regulator."

Within a foraging area, flocks move so as to maximize coverage. When encountering the edge of a food patch in a feeding area, a flock is expected to move ahead half to three-quarters of the time and have a strong right- or left-turn bias, according to computer simulations. Observations of finch flocks foraging in the Mojave desert have supported the predictions (Cody 1974).

Animals feeding in groups may forage more efficiently because each individual spends less time scanning the environment for predators. For example, downy woodpeckers *(Picoides pubescens)* wintering in mixed-species flocks scanned less and fed more than did solitary individuals (Sullivan 1984).

Group living also seems to increase the efficiency of food capture in some species of spiders. Members of a few genera build dense communal webs, which may be more than several meters across, as shown in Figure 16–17. Together the spiders attack trapped prey items, drag them back to a central retreat, and feed on them communally (Buskirk 1981).

Perhaps the most spectacular food-catching enterprise is the march of a colony of army ants (Figure 16–18). T. C. Schneirla spent most of his life trying to understand the complex life cycle of several New World species. Each night the colony, consisting of a queen, workers, larvae, pupae, and eggs, forms a bivouac (Schneirla and Piel 1948). Workers, hundreds of thousands strong, form a protective net by hooking their legs and bodies together. Each morning the bivouac dissolves

FIGURE 16-17   Communal spider web in Amazonian rain forest
Although most spiders are solitary, individuals of a few species cooperate to the extent that they build large, communal webs.
Source: Photo by S. H. Vessey.

FIGURE 16-18   Army ants emigrating
During the nomadic phase, when larvae are present, the entire colony of army ants moves to a new bivouac site at the end of each day's swarm raid.
Source: Photo by Arthur Christiansen from Frank W. Lane.

and the ants begin moving outward. A column emerges along the path of least resistance and heads away from the bivouac site. There are no true leaders; workers in the lead turn back into the swarm behind them every few centimeters. The ants lay down pheromone trails that guide those that follow and the column, as shown in Figure 16–19, may branch into a fan-shaped affair. Virtually any animal in the path is stung, cut into pieces, and transported to the rear. Army ants are thus able to attack and consume prey items as large as snakes, lizards, and small birds that they would be unable to handle as individuals. Arthropod life in these areas is temporarily depleted. When larvae are developing, the bivouac location changes each night (nomadic phase); during the egg laying and egg-development phase of the reproductive cycle, the bivouac remains in the same place each night (statary phase).

The relationship between feeding behavior and group size is evident in the great diversity of African antelope. Jarman (1974) has classified these herivores on the basis of five feeding styles, ranging from selective species that feed on only a few highly nutritious parts of localized plants to unselective species that feed primarily on grasses and browse of low nutritive value. The selective species are small, solitary, monogamous, and monomorphic and defend small territories. The least selective species are large, gregarious, sexually dimorphic, and polygamous and occupy large home ranges. Jarman argued that feeding style is an important determinant of group size; both group size and the pattern of movement over the home range affect reproductive strategies and social behavior.

FIGURE 16–19 Pattern of raiding employed by army ants
In this swarm raid of army ants *(Eciton burchelli),* which can be found on Barro Colorado Island in the Panama Canal Zone, the advancing front is made up of a large mass of workers. The swarm flushes a variety of prey, mostly invertebrates, but also snakes, lizards, and birds. The queen and immature forms remain at the bivouac site.
Source: Data from C. W. Rettenmeyer, "Behavioral Studies of Army Ants," *Kansas University Science Bulletin* 44 (1963):281–465.

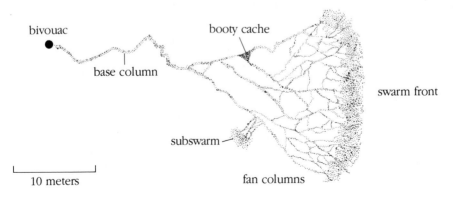

FIGURE 16–20 Lions with zebra
Male lions do relatively little killing of prey. Male on the right drags a zebra carcass that he probably scavenged from hyenas.
Source: Photo by George Schaller from Bruce Coleman.

SOCIAL CARNIVORES

Among the mammalian carnivores, one member of the family Felidae and several members of the family Canidae have evolved complex social behavior that is related to cooperative capture of prey. All of these species have been the subjects of intensive behavioral and ecological study in their natural habitats. The evolutionary advantages that would seem to accrue to individuals living in groups will be considered in more detail in Chapter 19; here we discuss only factors that relate to diet.

LIONS. The lion *(Panthera leo)* lives in closed social units and is most abundant in the grasslands and open woodlands of Africa. Schaller (1972) has studied the behavior and ecology of this species, its competitors, and its prey.

Lions gain from cooperation in several ways. First, by hunting in groups, they increase the resource spectrum; specifically, they add to their diet two species — buffalo *(Syncerus caffer)* and giraffe *(Giraffa camelopardalis)* — that an individual lion could never attack alone. Second, cooperative hunts are at least twice as successful as solitary ones. Third, a group can consume a captured prey more fully than can an individual. Fourth, the group can drive other predators and scavengers from the food. Schaller found that plains-dwelling prides kill less than half their food, relying on other predators, such as hyenas *(Crocuta crocuta)*, to make kills for them (Figure 16–20).

Although lions hunt in groups, the amount of cooperation is not extensive. When the lions encounter a herd of prey, such as zebras *(Equus burchelli)*, the pride females fan out, sometimes encircling the herd. Lions are not endurance runners and rely on a stalk-and-rush tactic. Although lions are more effective when hunting upwind, Schaller found no evidence that they do so more often than would be expected by chance.

Once a kill has been made, the males, which are larger than the females but which do little hunting, may drive the females from the kill. The cubs are the last to get food. Schaller believes that lion populations are regulated directly by food supply through starvation of cubs. The reproductive rate remains constant, but the mortality rate of the young can be very high, exceeding 80 percent when food is scarce. Not all cub mortality is the result of starvation; strange males that move into a group may kill cubs (Bertram 1975).

Caraco and Wolf (1975) used Schaller's data to relate ecological factors to lion foraging-group size (Figure 16–21). When hunting small prey, such as Thomson's gazelles *(Gazella thomsoni)*, two lions are more than twice as efficient as one, but three or more do no better than two. Since the available food for each lion per kill decreases with increasing group size, we would expect foraging groups of about two. Caraco and Wolf went on to show that only pairs of lions can take in the minimum daily amount of food if they feed exclusively on Thomson's gazelles; Schaller reported that the mean foraging-group size ranged from 1.5 to 2, close to the expected.

FIGURE 16–21   Capture efficiency, food availability, and estimated food intake as functions of lion group size for Thomson's gazelle prey
Capture efficiency (a) increases as the lion group size increases, showing the benefits of cooperative hunting. Little increase occurs beyond hunting parties of two, however, since Thomson's gazelle is a small prey-type for lions. In (b) as the lion group size increases, the food availability declines, since there are more mouths to feed. Given the relationships in (a) and (b), the projected food intake per lion as shown in (c) is maximized at a group size of two. If Thomson's gazelle were the only prey taken, two would be the only group size in which lions would be able to obtain enough food to survive. Prey that is larger and more difficult to capture would require larger lion group sizes.
Source: T. Caracao and L. L. Wolf, "Ecological Determinants of Group Sizes of Foraging Lions," *American Naturalist* 109 (1975):343–352. Reprinted with permission of the University of Chicago Press.

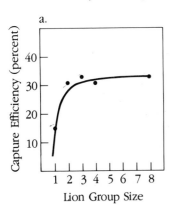
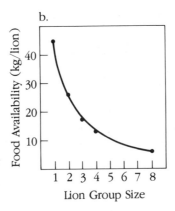
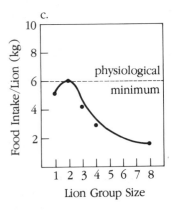

FIGURE 16–22   African wild dogs attacking zebra
Extensive cooperation is necessary among small social carnivores such as African wild dogs to enable them to capture much larger prey such as zebra. *Source:* Photo by George Schaller from Bruce Coleman.

In hunting larger prey, such as zebras and wildebeests (*Connochaetes taurinus*), a foraging-group size of 2 was again the most efficient, but groups of 1 to 4 could still attain their minimum daily requirement; however, Schaller found larger than expected foraging groups of about 4 to 7. Caraco and Wolf were forced to suggest that factors other than prey size, such as the higher reproductive success documented for larger prides (lionesses share in the feeding of cubs), could influence group size.

AFRICAN WILD DOGS.   Weighing only about 18 kg (40 lbs), the African wild dog (*Lycaon pictus*), shown in Figure 16–22, is an unlikely big game predator but typically catches prey weighing as much as 250 kg (Schaller 1972). The dogs live in mixed-sex packs that average ten adults and, when there are young in a den, range out from it to hunt. The pack is tightly organized. Of particular interest here is the prehunt ceremony, in which dogs draw back their lips to expose their teeth, nibble and lick each other's mouths, and run whining from pack member to pack member. The greeting seems to represent ritualized food begging. Setting out on a hunt, the dogs travel at a trot and fan loosely over the terrain; certain adults consistently lead some packs, but in other packs there seems to be no leader. Schaller (1972, p. 340) describes the hunt in his field notes:

A pack of 12 dogs becomes active at 1640; the animals greet and chase each other for 15 minutes before they trot off. About 8 km away is a large gazelle herd. The dogs race toward it at 45 km per hour. Several dogs run after one fleeing group, the rest toward another, but when a female gazelle breaks to one side, one dog pursues her immediately. She flees in a wide arc, the dog 10 m behind. The rest of the pack veers toward them and two dogs cut in front of the gazelle. She swerves to avoid them only to face two other dogs; she slows to a walk, unable to find an escape route, and a dog grabs her rump. The time is 1706.

The hunting success rate for African wild dogs is extremely high; about 90 percent of all hunts result in the capture of at least one prey item.

HYENAS.   The remaining social carnivore on the African plains is the spotted hyena (*Crocuta crocuta*), which was studied intensively by Kruuk (1972). The basic social unit is the clan, consisting of ten to sixty hyenas of both sexes and all ages. There is little exchange between clans; each defends a territory with a centrally located den. Because the prey, mainly wildebeests and zebras, migrate to the woodlands in the dry season, most hyena clans break up at that time, with some individuals becoming nomadic and following the prey. Thus food availability has a profound effect on social organization.

Although long thought to be scavengers, spotted hyenas at Serengeti kill more than two-thirds of their food, a higher percentage than the lion (Schaller 1972). The method of hunting varies with the prey. In searching for wildebeests, one or two hyenas may walk past a herd as close as 10 to 20 meters; suddenly one may dash randomly toward the herd or toward a lone wildebeest, causing the animals to run. The hyena may run, never at full speed, after the fleeing animals, then stop and watch, often joined by other clan members. The function of the random chase seems to be to make the prey run in order to spot any physically inferior individuals. Once an individual is selected, one or two hyenas begin to chase it at speeds of 40 to 50 kmph, much slower than top speed for wildebeests, which can run much faster than hyenas. Other hyenas may join in until the wildebeest stops and turns or runs into a lake or stream. After immobilizing the wildebeest with bites to the hind legs or loins, they quickly disembowel it.

When hunting zebras, hyenas always operate in packs. The group forms long before they get near the zebras, and it ignores other prey during the search, which clearly indicates a specific decision to hunt zebras. For such zebra hunts, hyenas come together from all over the range and often use a special meeting place, or "club," for several months. Hyenas of the same clan may hunt zebras exclusively for several days. The reproductive social unit of zebras consists of a stallion with one to six mares and their foals. Kruuk (1972, p. 176) gives the following description in his field notes:

One evening in December 1965, from 1850 to 1900 hr, hyenas belonging to the Lakeside clan had been gathering in the rapidly approaching darkness on a small area some 300 m from their den. They lay down there or sniffed around a bit, and then, exactly at 1900, eight hyenas began to slowly walk off together toward the nearest group of zebra, a family of twelve, less than a km away in the middle of wildebeest. The hyenas walked slowly up to the zebra, keeping very close together, and without showing any obvious interest in them. Five minutes later, when the hyenas were very close, the zebra stopped grazing and closed up together, their heads up. When the hyenas were about 4 m away, the stallion turned toward them and charged, head low and teeth bared. The hyenas scattered out of his way and the stallion immediately turned back to his family. The zebra bunched up and began running slowly (at a speed of 20 to 25 kmph) away from the hyenas. The stallion ran just behind his family; several times he charged the hyenas and tried to bite them, and once he kicked out with his hind legs. The eight hyenas galloped just behind the zebra, five of them close around the stallion and the other three just in front of him. Several times some of the zebra barked excitedly. The whole party still moved at the same fairly slow speed over the plains; the zebra stayed closely bunched together and the hyenas ran in a semicircle behind them.

When the stallion was a little farther behind than usual, the rest of the family made a 90° turn, bunched up till they were almost touching each other, and virtually stopped; again the stallion attacked a hyena, chasing him right around the zebra family. The zebras moved on again with the hyenas following. Now and then a hyena managed to get very near to the family or even in between the zebra, biting at their flanks. The speed still did not increase. By 1907 hr seven more hyenas had been attracted by the commotion and there were fifteen hyenas following the zebras, but otherwise the picture was unchanged. Suddenly one hyena managed to grab a young zebra while the stallion was chasing another member of the pack. This young zebra, which was between nine and twelve months old, fell back a little, and within seconds twelve hyenas converged on it; in 30 sec they had pulled it down while the rest of the family ran slowly on. More hyenas arrived, and the little zebra was completely covered by them. At 1917, 10 min after the victim had been caught, the last hyena carried off the head and nothing remained on the spot but a dark patch on the grass and some stomach contents. Twenty-five hyenas were involved, and the whole process of dismembering took exactly 7 min.*

WOLVES.   The social carnivore of northern temperate and arctic regions is the wolf *(Canis lupus),* which typically preys on deer *(Odocoileus* spp.), moose *(Alces americana),* buffalo *(Bison bison),* sheep *(Ovis* spp.), caribou *(Rangifer tarandus),* and elk *(Cervus canadensis).* Pack size varies greatly (Mech 1970), but most have fewer than eight members. Wolves hunt by traveling over their territory after consuming their previous kill. Most often they use scent; the lead animals stop when they detect the odor of prey. The group may stand nose to nose with tails wagging for a few seconds, and then all follow the leaders directly toward the prey. They may also use chance encounter and tracking. If they detect the prey at some distance, they proceed with restraint and stalk sometimes to within 10 meters of the prey. When the prey detects the wolves, it may stand its ground or flee. Once the prey is in flight, the wolves rush. Wolves must get close to their prey during the stalk and rush or they will quickly give up; if they cannot make an attack, they may chase the prey, usually for less than half a mile. Some earlier studies described wolves as coursers, but Mech has concluded that they are mainly stalkers and rushers (Figure 16–23).

One prey of the wolf, the musk ox *(Ovibos moschatus),* defends itself in an unusual way. Adults form a circle, facing outward and protecting the young inside. There is little evidence that wolves can break this defense; in areas where musk oxen are present, wolves eat mainly other prey or capture an occasional postreproductive individual.

When hunting moose, which are solitary, wolves rarely attack a prime individual (between one and six years old). Most studies of wolf predation have indicated that they are highly selective of young, old, and sick prey. Although earlier reports described the main killing tactic as hamstringing, wolves actually avoid the hooves and direct bites at the rump, flanks, shoulders, neck, and nose.

---

* H. Kruuk, *The Spotted Hyena: A Study of Predation and Social Behavior* (Chicago: University of Chicago Press, 1972). Copyright © 1972 University of Chicago Press. Reprinted with permission.

FIGURE 16–23 **Moose-wolf interactions**
Observations at Lake Superior's Isle Royale National Park show the results of 131 separate moose-wolf interactions. Circled numbers indicate moose actually encountered by wolves. Only 6 of the 131 moose detected by wolves were killed.
Source: Illustration from *The Wolf* by L. David Mech. Copyright © 1970 by L. David Mech. Reprinted by permission of Doubleday & Company, Inc.

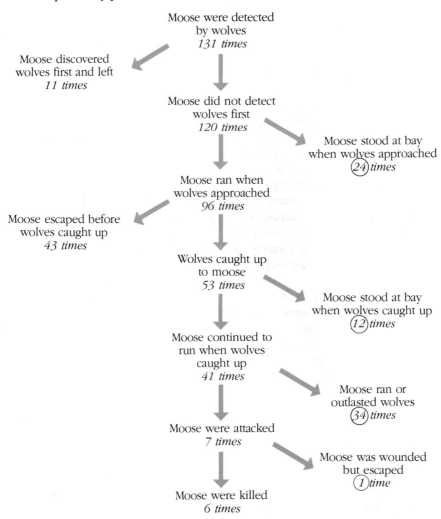

While the stalk-and-rush strategy is similar to that of the social cats, wolves sometimes chase for long distances and may attack weakened prey at intervals over several days. In the manner of prey selection, wolves resemble the hyena and the African wild dog. However, Mech believes that wolves catch certain individuals because these prey animals are less able than others to escape or defend themselves and not because the wolves deliberately pick out certain individuals to chase. Observers of hyenas and African wild dogs have implied that these predators, in contrast, deliberately evaluate a herd and select vulnerable individuals (Kruuk 1972; Schaller 1972).

One of the biggest contributions of modern studies of social carnivores has been the debunking of much of the folklore. Although social hunters may encircle prey, ambush prey, hunt upwind, or take turns chasing prey, the amount of cooperation in hunting seems to have been overstated. Lions hunt upwind no more often than they would be expected to do by chance, and they scavenge more than half of their food. Wolves do not hamstring their prey; hyenas may kill a higher percentage of their prey than do lions.

HUMANS.   Carnivorousness has been suggested as one of the key elements in the development of social behavior in humans. Schaller and Lowther (1969), among others, have suggested that those interested in the behavior of early humans might learn more from social carnivores than from the near relatives of humans, such as the gorilla. Humans were widely roaming scavengers and hunters for perhaps two million years. There is some doubt about how important meat was in the past; the bunodont (low-cusped) molars of humans, which resemble those of a pig, suggest an omnivorous diet in the past. However, some of the selective forces favoring sociality in carnivores also could have operated on humans. Cooperation in deciding where and what to hunt, encircling of prey, relay racing in pursuit of prey, capturing large prey, protecting food from competitors, bringing back food to the rest of the group —all could have been behaviors exhibited by early humans. Compared with most mammals, humans are exceptionally good endurance runners. Even today, African Bushmen capture small game by pursuing it over long distances at a medium pace, much as hyenas and African wild dogs do. The use of tools in capturing both large and small prey and in chasing other predators from their kills probably became important to humans more than a million years ago as well.

DEFENSE AGAINST PREDATORS

INDIVIDUAL STRATEGIES

Natural selection operates just as strongly on individuals with traits that reduce the likelihood of being eaten as it does on those with traits that increase prey-catching ability. We have seen that feeding habits such as meat eating may favor the evolution of social behavior, but so may avoiding predation. Most animals, except for a few top-trophic-level carnivores, are potential prey. Escape patterns are usually species-specific. Some social herbivores clump together when threatened, for example, zebras, while others scatter, for example, patas monkeys (*Erythrocebus patas*).

ESCAPING AND FREEZING. Many nonsocial animals keep close to a nest, burrow, or other refuge that is likely to be near the center of their home range. The way white-footed mice *(Peromyscus leucopus)* use the space away from their nests (usually holes in trees or logs) varies with distance; they are less likely to be trapped at increasing distances from the nest. The use pattern frequently follows a normal distribution, and practically all activity is close to the nest (Flemming 1977).

Another strategy in response to predators is to freeze. The presence of protective or cryptic coloration is often associated with this behavior, as in the spotted white-tailed deer fawn *(Odocoileus virginianus).* Some animals carry this strategy further and feign death. For example, the opossum *(Didelphis virginiana),* if harassed by a predator such as a dog, remains motionless on its back. The physiological correlates of this behavior are not completely understood (Francq 1969). Hog-nosed snakes *(Heterodon platyrhinos)* exhibit similar behavior, although they may precede it by threat behavior in which the hog-nosed snake resembles a cobra. Freezing and feigning death may work because many predators seem to respond only to moving prey. For example, wolves will not attack prey that stand motionless (Mech 1970).

Escape or defense reactions may be specific to the type of predator. Chickens *(Gallus gallus)* head for cover on the ground or crouch to escape aerial predators, but they fly up into trees to escape ground predators. Each response has its own warning call, and brain stimulation experiments have revealed that the responses have separate neural representation in the brain (Eibl-Eibesfeldt 1975). Vervet monkeys *(Cercopithecus aethiops)* also respond differentially to aerial and terrestrial predators and use different alarm calls (Struhsaker 1967), as do the Belding's ground squirrels *(Spermophilus beldingi)* discussed earlier.

DECEPTION. Another strategy used by vulnerable organisms is deception. The larvae of several moth species resemble snakes. When threatened, this type of larval insect inflates its head end to form an excellent representation of the head of a snake and it then waves this "head" (Figure 16–24a). Other larvae resemble inanimate objects in the environment, such as twigs, leaves, bark, or even bird droppings.

Many animals have behaviors or markings that may serve to misdirect or surprise predators. Some reef fish have conspicuous eye spots at the tail end that are much larger than the real eyes; these could frighten away potential predators or misdirect their attack to less vulnerable parts of the body. Moths (Figure 16–24b) may have similar spots. A large group of insects, the underwing moths (of the order Lepidoptera), are cryptically colored when they are at rest on tree trunks. If a predator comes near, the moth spreads its outer wings and reveals large and brightly colored spots on the underwings that seem to surprise the predator.

Disruptive color patterns consist of high-contrast markings that are thought to break up the outline of an organism or create a target that directs predator attack away from vital organs. For example, many species of butterflies and reef fish have disruptive stripe patterns. Indirect evidence that they provide protection against predation comes from the finding that chemically protected or unpalatable species usually lack the markings found in palatable species; however, few experimental field tests have been attempted. Silberglied, Aiello, and Windsor (1980) obliterated

FIGURE 16–24 Insect defense displays
The anterior end of an alarmed caterpillar (a) resembles the head of a snake and
frightens predators. A moth (b) exposes large eye spots on its underwing in order to
frighten away or deflect attacks by predators.
Source: Photos (a) by Lincoln Brower, (b) by Thomas Eisner.

a   b

the wing stripes on a species of tropical butterfly *(Anartia fatima)* with a felt-tipped
pen and compared their survival with that of unaltered controls (Figure 16–25). The
lack of a difference between experimentals and controls in survival rate or frequency
of wing damage (an indicator of predatory attack) suggests that the color patterns
have some other function.

TOXICITY. Many organisms are in some way harmful to predators. Plants may
contain toxins such as alkaloids; animals may have armor plates (as in the armadillo)
or spines (as in the porcupine, hedgehog, and three-spined stickleback). Many
invertebrates inject venom into attackers. Eisner (1966, 1970) has discovered a large
number of arthropod defense mechanisms. For instance, the bombardier beetle
(Brachinus spp.) has a plumbing system that resembles a liquid-fueled rocket. It
stores quinones and hydrogen peroxide in separate reservoirs. When the beetle is
disturbed, these two liquids are mixed in an outer vestibule; and, in the presence of
enzymes, the solution heats to the boiling point and is discharged as a noxious spray.

Noxious animals tend to be conspicuously colored, presumably so that predators can easily recognize and thus avoid them. Such warning, or **aposematic,** coloration occurs, for example, in skunks *(Mephitis mephitis)* and coral snakes *(Micrurus
fulvius)* (Wickler 1968).

Among arthropods such as butterflies, unpalatable species tend to resemble
each other and are referred to as *Mullerian mimics*—for example, the monarch
(Danaus plexippus) and queen *(Danaus gilippus)* butterflies, both of which may be
toxic because of the plants on which they feed. Mullerian mimics seem to have

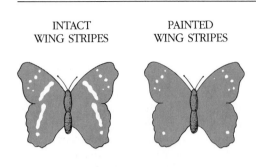

INTACT
WING STRIPES

PAINTED
WING STRIPES

FIGURE 16-25　Field experiment to test function of butterfly markings

Butterflies *(Anartia fatima)* with wing stripes intact, left, and covered with ink, right. If stripes function to reduce predation, butterflies with stripes obliterated should suffer higher mortality. Results of experiments, however, indicated no such function for the stripes. *Source:* Data from R. E. Silberglied et al., "Disruptive Coloration in Butterflies: Lack of Support in *Anartia fatima*," *Science* 209 (1980):617–619.

evolved because by looking alike and acting similarly, they provide fewer different prey types for predators to learn to avoid and thus are less likely to be eaten by mistake. Some palatable species, called *Batesian mimics*, may evolve morphologies and behaviors similar to those of unpalatable species. Batesian mimics enjoy protection because of the predators' learned avoidance of the toxic species. Brower et al. (1968) demonstrated "one trial conditioning" of blue jays *(Cyanocitta cristata)* fed monarch butterflies; consumption of one monarch butterfly often led to vomiting by the jay and avoidance of other monarchs or their mimics.

　　The toxicity of monarchs comes from the poisons (cardiac glycosides) they obtain from the milkweed plants they feed on as larvae. Adult females lay eggs on the toxic species of milkweed; the larvae have evolved a resistance to the poisons and a capacity to store them. We might question how such a system could evolve if a toxic adult must be consumed before a predator can learn to avoid it; any butterfly with mutant genes enabling it to store the toxins through metamorphosis would probably not survive. One possibility is that the predator need not consume the entire butterfly to reject it, and so the butterfly could live to reproduce. But in many cases the vomiting is delayed, and the butterfly does not survive. Since the predator will probably never again eat another butterfly resembling that one, we seem to have a case of altruism on the part of the consumed butterfly. The second possibility is that through the process of kin selection the toxic adult, by sacrificing its life to "educate" the predator, protects relatives that live in the area and share its genes (Hamilton 1963). From the latter theory one predicts that sedentary populations of monarchs, in which relatives are likely to be in close proximity, should be highly toxic, but migratory populations, in which relatives are likely to be scattered, should be nontoxic, as was found to be the case by Duffey (1970).

　　The tendency of predators to avoid harmless prey that resemble toxic prey may be put to practical use. In a field experiment with crows *(Corvus brachyrhynchos)* chicken eggs were placed in straw nests on the ground. Eggs that were injected with a bad tasting but nonlethal toxin were painted green, while untreated eggs remained white. The crows quickly learned to avoid the green eggs, selecting the white ones even when the green eggs were not injected (Nicolaus et al. 1983). If only green eggs were present, crows tended to shift to alternate foods rather than eat eggs.

　　Avoidance of dangerous or unpalatable prey may occur also in the absence of

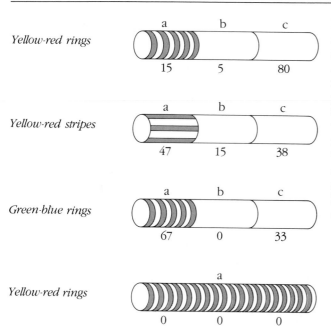

Yellow-red rings

Yellow-red stripes

Green-blue rings

Yellow-red rings

FIGURE 16–26 Wooden snake models used to test hand-reared motmots

Area (a) of each model was painted as indicated at left; areas (b) and (c) were plain wood. Numbers beneath each model show the percentage of pecks by the motmots. Motmots avoided those models or portions of models painted with yellow-red rings, the same colored rings that the coral snake has. *Source:* Data from S. M. Smith, "Innate Recognition of Coral Snake Pattern by a Possible Avian Predator," *Science* 187 (1975):759–760.

previous conditioning. For example, even laboratory-reared motmots *(Eumomota superciliosa)*, lizard- and snake-eating birds, avoid models painted with the color patterns of the coral snake (Smith 1975). (See Figure 16–26).

SOCIAL STRATEGIES

CONFUSION EFFECT. One of the main benefits of sociality is thought to be the protection gained, as stated in the adage, "safety in numbers." Living in flocks, herds, or schools increases the chance of spotting predators. When approached or attacked by predators, groups usually become more compact. Fish schools, however, usually form a vacuole around a predator. Tightly packed prey make a difficult target for the predator, which must select an individual toward which it can aim its attack. The predator tries to isolate an individual from the group; any individual that strays from the group or is in any way different from the rest is likely to be attacked. For example, when experimenters presented hawks with sets of ten mice, the hawks preferred the oddly colored mouse (Mueller 1971).

A primary means of protection is the **confusion effect,** described by Miller (1922). A variety of predators that attack swarming prey exhibit a lower prey capture rate, a longer period of hesitation, and more irrelevant behavior than when they attack solitary prey. Factors that enhance the confusion effect are swarm number, swarm density, and the uniformity in appearance of swarm members (Milinski 1979). The mechanism producing the confusion effect is not understood, but a predator seems to be more easily frightened when it feeds in a dense swarm of prey; perhaps the predator finds it harder to detect other animals that might prey on *it*. In laboratory experiments with three-spined sticklebacks *(Gasterosteus aculeatus)* feeding on water fleas *(Daphnia)*, fish with experience in feeding on dense swarms fed more efficiently when tested with dense swarms than did those experienced only in feeding on low-density populations (Milinski 1979).

DETECTION. We might expect to find groups of mixed species if the combination of different sensory capabilities is advantageous. Baboons and ungulates are frequently together on the African plains. The arboreal habits and keen vision of baboons and the well-developed olfactory systems of ungulates probably increase the distance at which prey can be detected. It is clear that these different species recognize and respond to each other's alarm calls.

Social living increases the likelihood that predators will be detected and allows group members to spend more time in other activities. Individuals in large, dense colonies of prairie dogs *(Cynomys* spp.) spend less time in alert postures than do those in small colonies (Hoogland 1979). Once predators are detected, alarm calls alert neighbors to danger. Marler (1955) has argued that in some birds the alarm call is difficult to locate because it is a high-pitched note in a narrow-frequency range. The alarm-calling ground squirrels studied by Sherman (1977) emitted calls that were easily located, and calling squirrels were more likely to be caught than non-callers. As we saw in Chapter 13, close relatives of the caller were likely to be close by and thus could benefit from this seemingly altruistic behavior.

GROUP DEFENSE: MOBBING. Some prey species are able to turn the tables and attack the predator. Mobbing of predators is common and may occur in species that usually live in widely spaced territories. If a crow *(Corvus brachyrhynchos)* flies over a field of nesting red-winged blackbirds *(Agelaius phoeniceus)*, the males and sometimes the females may rise up and pursue it en masse. Baboons *(Papio* spp.) sometimes mob lions; however, observers have disagreed about the response of baboon groups to predators. In some cases the dominant males move out to the periphery to fend off predators while the rest of the group flees to the safety of trees (DeVore 1965). Other accounts have suggested that the males are among the first to flee, leaving behind females with young to fend for themselves (Rowell 1966).

Hoogland and Sherman (1976), when considering the advantages and disadvantages of coloniality in bank swallows *(Riparia riparia)*, found that two possible advantages were increased foraging efficiency and reduced predation on adults, young, or eggs. The experimenters found that in larger colonies a stuffed weasel model was detected sooner and mobbed more intensely than was the case in smaller colonies. When they tethered young swallows at varying distances from the colony, they found that blue jay attacks on the young swallows caused intense mobbing by adult swallows when the young were close to the colony; but when the young were farther away, the mobbing was less intense and the blue jay attacks more often

TABLE 16-1   Field experiment to test the effectiveness of mobbing in deterring predators
A young bank swallow was tethered at varying distances from the main group of bank swallow burrows. On 17 occasions one of several different blue jays attempted to attack the tethered young. In all cases bank swallows mobbed the jays, but the mobbing was most effective as jay deterrence close to the main group of burrows.

	Distance from Burrows to Tethered Young Bank Swallow		
Behavior of Blue Jays	0–1 m (Bank Face)	9–11 m (Bank Base)	18–20 m (Center of Gravel Pit)
Times jays attack and kill the young bird *(N)*	0	2	7
Times jays attack but are unsuccessful in killing the young bird *(N)*	0	2	1
Times jays are deterred from attacking the young bird *(N)*	5	0	0
Successful attacks at each distance (%)	0	50	88

Source: J. L. Hoogland and P. W. Sherman, "Advantages and Disadvantages of Bank Swallow *(Riparia riparia)* Coloniality," *Ecological Monographs* 46 (1976):33–58. Copyright 1976, the Ecological Society of America. Reprinted with permission.

successful (Table 16–1). Emlen and Demong (1975) reported that increased foraging is a prime advantage of coloniality and synchronous breeding; the swallows follow each other to localized, ephemeral concentrations of food.

The extent to which individuals defend the group may be a function of the degree of relatedness to those group members. Mothers can be expected to defend offspring up to the point that the benefit to the offspring exceeds the cost to themselves in reduced reproductive output in the future. Similarly, males will defend their offspring and, to some extent, the mates for which they have competed. However, group-living species are mostly polygynous, and confidence of paternity may be low. Thus we may expect variability in the male role, with older, long-term residents being the most active in group defense. Among rhesus monkeys *(Macaca mulatta)* minor, extragroup threats are handled by low-ranking peripheral males. Serious challenges, such as attacks on infants, are met by the central, high-ranking males, who are more likely to be their fathers.

DISTRACTION DISPLAYS.   Individuals may use distraction displays—for example, the broken-wing displays of killdeers *(Charadrius vociferus)* and avocets *(Recurvirostra americana)*—to attract the attention of a predator and draw it away from group-mates or young (Figure 16–27). Many cursorial, or running, mammals have white rump or tail patches; in the presence of predators the hairs are erected and the tail waved, making a conspicuous display. Observers have suggested several functions for this behavior, such as: (1) distracting the predator from other members of the group, (2) warning other group members, (3) confusing the predator when many

FIGURE 16-27   American avocet performing broken-wing display This behavior pattern presumably functions to distract predators from the nest or young. Although considered by some to be altruistic behavior, it really is most easily understood as a form of parental investment.
Source: Photo by Leonard Lee Rue III from Animals Animals.

group members are displaying, (4) signaling the predator that it has been detected, and (5) eliciting premature pursuit. In the cases of (4) or (5), the predator may go off in search of less-alert prey or may be lured into an unsuccessful pursuit (Smythe 1977). Others have argued that rump patches have little to do with defense against predators and function in intraspecific social communication (Guthrie 1971).

SUMMARY

The populations that make up ecosystems can be classified according to trophic, or feeding, level. Because transfer of energy through the community is not 100 percent efficient, the number of trophic levels in the ecosystem is limited. Above the level of photosynthetic plants, or producers, we can classify organisms on a continuum of specialists to generalists. In maximizing the net rate of energy intake, competing organisms may become very efficient at harvesting a small number of different resources.

An organism makes decisions, consciously or otherwise, about whether it should include a newly encountered item in its diet, when it should move to a new resource area, where it should search next, and what pattern it should use to search an area. The potential energy available from a resource must be devalued by the energy needed to search for it, to pursue it, and to handle and eat it. Researchers have developed some simple models to test the hypothesis that natural selection operates on feeding behavior so as to maximize the net rate of energy intake. Although some tests have supported this hypothesis, during the nonbreeding season animals may minimize costs rather than maximize energy intake, since the emphasis is on survival rather than reproduction.

In general, species at low trophic levels are large and relatively sedentary, often simply filtering small organisms or detritus from the environment. Those at high trophic levels are likely to be few in number and active, with high metabolic rates.

Some adaptions have expanded the range of food sources available or have enhanced attainment of them. For example, closely related species may employ different feeding behaviors and thus reduce competition for food; herbivores may modify the environment so as to increase the food resource, as evidenced by some species of fungus-growing ants that tap a vast energy source unavailable to most animals; and animals may use a variety of aids — such as the lures used by anglerfish or the tools used by termite-feeding chimpanzees — to capture prey.

Social organization and food resources are often closely related; some form of territory or group living is common. A number of studies, mainly on insects, birds, and mammals, have demonstrated that group feeding, compared with feeding alone, is more efficient and increases the resource spectrum.

Adaptations that provide defense against predation are just as elaborate as those that increase feeding efficiency. Individual tactics include escaping, having cryptic coloration, feigning death, practicing deception, and having markings or behaviors that surprise the predator or that misdirect the attack of the predator. Many organisms produce or store toxic substances to gain protection, and nontoxic species may mimic the species with these toxic substances.

Living in groups may provide protection by making it difficult for a predator to single out a prey. Groups, particularly those made up of several species, may spot predators sooner or may mob a predator. Recent studies have explored the extent to which close relatives in a group warn or otherwise aid each other to escape predation.

Discussion Questions

1. Discuss the relationship between the distribution of food resources and social organization. What other factors affect social organization?

2. Discuss the similarities and differences between the evolution of alarm calls and the evolution of the ability to store substances toxic to predators.

3. Normal butterflies *(Nymphais io)* with large eyespots and altered butterflies with eyespots blotted out were presented equal numbers of times to six yellow buntings *(Emberiza citrinella)*, predatory birds. Data on the buntings' escape responses are presented here (from Blest 1957). Discuss these results in light of the findings of Silberglied, Aiello, and Windsor (1980) presented in the chapter.

Number of escape responses by yellow buntings

Yellow Bunting	Normal Butterfly (with large eyespots)	Altered Butterfly (with eyespots blotted out)
1	56	16
2	11	5
3	8	4
4	18	1
5	18	3
6	17	2
Total	128	31

Suggested Readings

Edmunds, M. 1974. *Defence in Animals.* Harlow, Essex, England: Longman.
Review of the means by which animals avoid being killed
and eaten by predators.

Kamil, A. C., and T. D. Sargent, eds. 1981. *Foraging Behavior — Ecological, Ethological, and Psychological Approaches.* New York: Garland STPM.
Proceedings of a symposium sponsored by the Animal
Behavior Society in 1978. Sections on optimal foraging
theory, field studies, learning, and behavior genetics.

Pyke, G. H., H. R. Pulliam, and E. L. Charnov. 1977. Optimal foraging: A selective review of theory and tests. *Quart. Rev. Biol.* 52:137–154.
One of the more intelligible treatments of a difficult subject,
which is currently receiving much attention from ecologists,
psychologists, and physiologists.

Schaller, G. B. 1972. *The Serengeti Lion: A Study of Predator-Prey Relations.* Chicago: University of Chicago Press.
Superb example of how a field study should be undertaken.
Part of a wildlife behavior and ecology series on individual
species.

References

Beck, B. B. 1980. *Animal Tool Behavior: The Use and Manufacture of Tools by Animals.* New York: Garland STPM.

Bernon, G. 1981. Species abundance and diversity of the Coleoptera of a south African cow dung community, and associated predators. Ph.D. dissertation, Bowling Green State University.

Bertram, B. P. 1975. The social system of lions. *Sci. Amer.* 232:54–65.

Blest, A. D. 1957. The function of eyespot patterns in the Lepidoptera. *Behaviour* 11:209–256.

Brower, L. P., W. N. Ryerson, L. L. Coppinger, and S. C. Glazier. 1968. Ecological chemistry and the palatability spectrum. *Science* 161:1349–1351.

Buskirk, R. E. 1981. Sociality in Arachnida. In H. R. Hermann (ed.), *Social Insects.* New York: Academic Press.

Caraco, T., and L. L. Wolf. 1975. Ecological determinants of group sizes of foraging lions. *Amer. Nat.* 109:343–352.

Carpenter, F. L., and R. E. MacMillen. 1976. Threshold model of feeding territoriality and test with a Hawaiian honeycreeper. *Science* 194:639–642.

Charnov, E. L. 1976. Optimal foraging: the marginal value theorem. *Theor. Pop. Biol.* 9:129–136.

Cody, M. L. 1974. Optimization in ecology. *Science* 183:1156–1164.

Colvin, B. A. 1984. Barn owl foraging behavior and secondary poisoning hazard from rodenticide use on farms. Ph.D. dissertation, Bowling Green State University.

Connor, V. M., and J. F. Quinn. 1984. Stimulation of food species growth by limpet mucus. *Science* 225:843–844.

Davies, N. B., and A. I. Houston. 1983. Time allocation between territories and flocks and owner-satellite conflict in foraging pied wagtails, *Motacilla alba. J. Anim. Ecol.* 52:621–634.

———. 1984. Territory economics. In J. R. Krebs and N. B. Davies (eds.), *Behavioural Ecology: An Evolutionary Approach,* 2nd ed. Sunderland, Mass.: Sinauer Associates.

Dawkins, R. 1976. *The Selfish Gene.* New York: Oxford University Press.

Deevey, E. 1960. The human population. *Sci. Amer.* 203:195–204.

DeVore, I., ed. 1965. *Primate Behavior: Field Studies of Monkeys and Apes.* New York: Holt, Rinehart and Winston.

Dix, M. W. 1968. Snake food preference: Innate intraspecific geographic variation. *Science* 159:1478–1479.

Duffey, S. S. 1970. Cardiac glycosides and distastefulness: Some observations on the palatability spectrum of butterflies. *Science* 169:78–79.

Eibl-Eibesfeldt, I. 1975. *Ethology: The Biology of Behavior,* 2nd ed. New York: Holt, Rinehart and Winston.

Eisner, T. E. 1966. Beetle spray discourages predators. *Natur. Hist.* 75:42–47.

———. 1970. Chemical defenses against predators in Arthropods. In E. Sondheimer and J. B. Simeone (eds.), *Chemical Ecology.* New York: Academic Press.

Emlen, S. T., and N. J. Demong. 1975. Adaptive significance of synchronized breeding in a colonial bird: A new hypothesis. *Science* 188:1029–1031.

Flemming, D. 1977. Home range pattern of the white-footed mouse. Master's thesis, Bowling Green State University.

Francq, E. N. 1969. Behavioral aspects of feigned death in the opossum *Didelphis marsupialis. Amer. Mid. Nat.* 81:556–568.

Gill, F. B., and L. L. Wolf. 1975. Economics of feeding territoriality in the golden-winged sunbird. *Ecology* 56:333–345.

Gilraldeau, L. A., and D. L. Kramer. 1982. The marginal value theorem: A quantitative test using load size variation in a central place forager, the eastern chipmunk *Tamias striatus. Anim. Behav.* 30:1036–1042.

Guthrie, R. D. 1971. A new theory of mammalian rump patch evolution. *Behaviour* 38:132–145.

Hamilton, W. D. 1963. The evolution of altruistic behavior. *Amer. Nat.* 97:354–356.

Harvey, P. H., and P. J. Greenwood. 1978. Antipredator defense strategies: Some evolutionary problems. In J. R. Krebs and N. B. Davies (eds.), *Behavioural Ecology: An Evolutionary Approach.* Sunderland, Mass.: Sinauer Associates.

Heinrich, B. 1983. Do bumblebees forage optimally, and does it matter? *Amer. Zool.* 23:273–281.

Hoogland, J. L. 1979. The effect of colony size on individual alertness of prairie dogs (Sciuridae: *Cynomys* spp.). *Anim. Behav.* 27:394–407.

Hoogland, J. L., and P. W. Sherman. 1976. Advantages and disadvantages of bank swallow (*Riparia riparia*) coloniality. *Ecol. Monogr.* 46: 33–58.

Jarman, P. J. 1974. The social organization of antelope in relation to their ecology. *Behaviour* 48:215–267.

Kalmijn, A. J. 1971. The electric sense of sharks and rays. *J. Exp. Biol.* 55:371–383.

Kamil, A. C. 1978. Systematic foraging for nectar by Amakihi (*Loxops virens*). *J. Comp. Physiol. Psychol.* 92:388–396.

Kormondy, E. G. 1976. *Concepts of Ecology.* Englewood Cliffs, N.J.: Prentice-Hall.

Krebs, J. R. 1978. Optimal foraging: Decision rules for predators. In J. R. Krebs and N. B. Davies (eds.), *Behavioral Ecology: An Evolutionary Approach.* Sunderland, Mass.: Sinauer Associates.

Krebs, J. R., A. Kacelnik, and P. Taylor. 1978. Test of optimal sampling by foraging great tits. *Nature* 275:27–31.

Kruuk, H. 1972. *The Spotted Hyena: A Study of Predation and Social Behavior.* Chicago: University of Chicago Press.

Lawick-Goodall, Jane van. 1970. Tool-using in primates and other vertebrates. *Adv. Study Behav.* 3:195–249.

Lissmann, H. W. 1958. On the function and evolution of electric organs in fish. *J. Exp. Biol.* 35:156–191.

MacArthur, R. H. 1958. Population ecology of some warblers of northeastern coniferous forests. *Ecology* 39:599–619.

———. 1972. *Geographical Ecology: Pattern in the Distribution of Species.* New York: Harper & Row.

MacArthur, R. H., and E. R. Pianka, 1966. On the optimal use of a patchy environment. *Amer. Nat.* 100:603–609.

McNair, J. N. 1982. Optimal giving-up times and the marginal value theorem. *Am. Nat.* 119:511–529.

Marler, P. 1955. Characteristics of some animal calls. *Nature* 176:6–8.

Mech, L. D. 1970. *The Wolf: The Ecology and Behavior of an Endangered Species.* Garden City, N.Y.: Natural History Press.

Milinski, M. 1979. Can an experienced predator overcome the confusion of swarming prey more easily? *Anim. Behav.* 27:1122–1126.

Miller, R. C. 1922. The significance of the gregarious habit. *Ecology* 3:122–126.

Mueller, H. C. 1971. Oddity and specific search-

ing image more important than conspicuousness in prey selection. *Nature* 233:345–346.

Nicolaus, L. K., J. F. Cassel, R. B. Carlson, and C. B. Gustavson. 1983. Taste-aversion conditioning of crows to control predation on eggs. *Science* 220:212–214.

Pietrewicz, A. T., and A. C. Kamil. 1979. Search image formation in the blue jay (*Cyanocitta cristata*). *Science* 204:1332–1333.

Pietsch, T. W., and D. B. Grobecker. 1978. The compleat angler: Aggressive mimicry in an Antennariid anglerfish. *Science* 201:369–370.

Pulliam, H. R. 1974. On the theory of optimal diets. *Amer. Nat.* 108:59–74.

Purcell, J. E. 1980. Influence of siphonophore behavior upon their natural diets: Evidence for aggressive mimicry. *Science* 209:1045–1047.

Pyke, G. H. 1979a. Optimal foraging in bumblebees: Rule of movement between flowers within inflorescences. *Anim. Behav.* 27:1167–1181.

———. 1979b. The economics of territory size and time budget in the golden-winged sunbird. *Am. Nat.* 114:131–145.

Pyke, G. H., H. R. Pulliam, and E. L. Charnov. 1977. Optimal foraging: A selective review of theory and tests. *Quart. Rev. Biol.* 52:137–154.

Rowell, T. E. 1966. Forest living baboons in Uganda. *J. Zool.* 149:344–364.

Schaller, G. B. 1972. *The Serengeti Lion: A Study of Predator-Prey Relations.* Chicago: University of Chicago Press.

Schaller, G. B., and G. Lowther. 1969. The relevance of carnivore behavior to the study of early hominids. *Southwest. J. Anthrop.* 25:307–341.

Schneirla, T. C., and G. Piel. 1948. The army ant. *Sci. Amer.* 178:16–23.

Schoener, T. W. 1971. Theory of feeding strategies. *Ann. Rev. Ecol. Syst.* 2:369–404.

Sherman, P. W. 1977. Nepotism and the evolution of alarm calls. *Science* 197:1246–1256.

Shettleworth, S. J. 1984. Learning and behavioural ecology. In J. R. Krebs and N. B. Davies (eds.), *Behavioural Ecology: An Evolutionary Approach,* 2nd ed. Sunderland, Mass.: Sinauer Associates.

Silberglied, R. E., A. Aiello, and D. M. Windsor. 1980. Disruptive coloration in butterflies: Lack of support in *Anartia fatima. Science* 209:617–619.

Smith, S. M. 1975. Innate recognition of coral

snake pattern by a possible avian predator. *Science* 187:759–760.

Smythe, N. 1977. The function of mammalian alarm advertising: Social signals or pursuit invitation? *Amer. Nat.* 111:191–194.

Struhsaker, T. T. 1967. Auditory communication among vervet monkeys. In S. A. Altmann (ed.), *Social Communication among Primates.* Chicago: University of Chicago Press.

Sullivan, K. A. 1984. The advantages of social foraging in downy woodpeckers. *Anim. Behav.* 32:16–22.

Weber, N. A. 1966. Fungus-growing ants. *Science* 153:587–604.

Wickler, W. 1968. *Mimicry in Plants and Animals.* London: World University Library.

Wilson, E. O. 1971. *Insect Societies.* Cambridge: Harvard University Press.

———. 1975. *Sociobiology: The New Synthesis.* Cambridge: Harvard University Press.

17

BEHAVIOR AND POPULATION REGULATION

The natural regulation of animal populations has long fascinated ecologists, and the role of behavior in this process has been the subject of much debate. A complex of abiotic substances (minerals and organic compounds), climate (mainly temperature, moisture, and wind), and biotic factors (other organisms of the same and other species) affects the growth and reproduction of any organism. A **population** may be defined as a group of organisms of the same species in a particular place. We know that all species have a reproductive potential that is seldom realized and that no population increases without limit, as was pointed out by Darwin and by Malthus before him. The number of individuals present in a population is determined by four population "forces": births, deaths, and movements out of (emigration) and into (immigration) an area. Thus the abundance of a species as well as its distribution is affected by behavior.

This chapter briefly reviews the extrinsic limits to population growth and then considers the evidence for these suggested intrinsic processes of population self-regulation: (1) purely behavioral mechanisms. (2) behavioral-physiological mechanisms, and (3) behavioral-genetic mechanisms. We then discuss how such systems could have evolved by natural selection.

LIMITING FACTORS

One axiom of ecology is that only a single factor limits the growth of a population at any one time. The idea originated with Liebig (1847), a plant physiologist who studied effects of nutrients, such as nitrogen, on plant growth and is referred to as Leibig's **law of the minimum.** Extending his idea to animal populations, we can test which factor limits a population's growth if we manipulate the amount of one factor while holding others constant and record changes in the population. If food is the limiting factor, an increase in the population will result if the food supply is augmented.

For example, Holling (1959) studied the population response of some small mammals to different amounts of a staple food—cocoons of the European pine sawfly *(Neodiprion sertifer).* Shrews of the genus *Sorex* increased more or less linearly with increasing amounts of the food supply up to a population of twenty-five shrews per acre; then the population leveled off (Figure 17–1). At that point some resource other than sawflies became limiting. Likewise, deermice *(Peromyscus)* increased in numbers up to about eight per acre, then remained constant despite the food supply. The abundance of sawflies seemed to be unrelated to the population density of another shrew of the genus *Blarina.*

While Liebig's law of the minimum can simplify our analysis of regulatory factors, it does not help us understand the level at which these factors may operate. Ultimately, food supply usually sets the limit to the number of organisms that can live in a place—that is, the **carrying capacity** of the environment. However, some behavioral mechanism involved with contest competition for food, such as defense of territory, may proximally limit the population below the carrying capacity if some individuals defend space in which to live and reproduce and thereby exclude others. Thus both territory and food supply may limit the population but operate at different levels, one proximate and the other ultimate.

The interaction of factors creates more complicated problems; for example, plants that are deficient in nitrogen are more susceptible to drought, and most

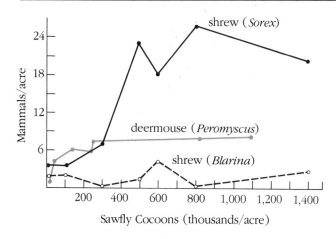

FIGURE 17-1   Number of mammals per acre as a function of food supply Populations of the shrew *(Sorex)* and deermouse *(Peromyscus)* increase in size, up to a point, as food supply increases. Their numbers, therefore, are limited by food supply. Shrews of the genus *Blarina,* however, show no response to an increased cocoon supply, so their numbers must be limited by some other resource. *Source:* C. S. Holling. "The Components of Predation as Revealed by a Study of Small Mammal Predation of the European Pine Sawfly," *Canadian Entomologist* 91 (1959):293–320. Reprinted with permission.

terrestrial insects can tolerate higher temperatures when humidity is high than when humidity is low. In addition, the impact of a limiting factor may occur during a very short period in the history of the population and thus be difficult to identify. A few days of severe winter weather may reduce a deer herd to the point that other regulatory factors never come into play. Therefore, it should not be surprising to find little agreement among population biologists about the mechanisms of population regulation.

DENSITY-INDEPENDENT AND DENSITY-DEPENDENT FACTORS

Regulatory factors are sometimes divided into two types: density independent and density dependent. **Density-independent** causes of mortality—such as climate, flood, and fires—are not themselves affected by population density; they kill organisms regardless of density. **Density-dependent** factors—which include competition, parasitism, disease, predation, and food supply—are related directly or inversely to density, so that as population density increases, death or emigration rates increase or birth rates decrease. Actually, this division is somewhat misleading, since some of the effects of so-called density-independent factors may be modified by population density. At high densities, organisms may gain protection from both wind and low temperature by grouping, or shelter from the elements may be limited. Thus the impact of climate may vary with population density. If we mean by population regulation the maintenance of numbers within certain limits, and not simply random fluctuations, we must look to density-dependent factors as the regulatory mechanisms. Only when a factor increases its negative effect on population growth as population increases can an equilibrium of numbers be maintained (Krebs 1985).

FOOD SUPPLY. Most ecologists agree that the supply of energy (food) most often sets the upper limit to the density of animal populations. Correlations between food abundance and population density, as shown in Figure 17–1, are one type of evidence. Another approach is to supplement the natural food supply and chart the response of the population in terms of density, reproduction, mortality, and movements. In one such study laboratory mouse-food pellets were distributed in a woodlot and the response of the resident white-footed mice *(Peromyscus leucopus)* was noted. By the end of the study the population was somewhat higher than was that in the control area, primarily due to higher survival of young in fall and winter (Hansen and Batzli 1978). When oats were supplemented to populations of deermice *(Peromyscus maniculatus)* breeding began within several weeks, even in the winter (Taitt 1981). Home ranges decreased in size and immigration tripled compared to untreated control plots, suggesting that spacing behavior may have been involved in limiting numbers (Figure 17–2).

 In spite of the presumed importance of food as a limiting resource for verte-

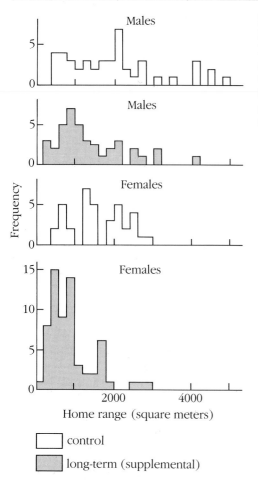

FIGURE 17-2   Effect of supplemental food on home range size in deermice
Home ranges were generally smaller where food was supplemented, as seen in these frequency distributions of home ranges of deermice.
Source: M. J. Taitt. 1981. The effect of extra food on small rodent populations: I. Deermice *(Peromyscus maniculatus). Journal of Animal Ecology* 50:111–124.

brate populations, naturalists have frequently noted that starving animals are rarely found in the wild, except where humans have altered the environment. These observations have contributed to the notion that behavioral mechanisms keep numbers below the point at which the food supply is exhausted.

SHELTER.   A few studies have demonstrated that the number of nesting or feeding sites limit population density. Filter feeders, such as sessile invertebrates, may compete actively for space. Barnacles *(Chthamalus stellatus* and *Balanus balanoides)* crowd each other out of attachment sites on rocks (Connell 1961).

   Some species of birds nest in tree holes. In a classic study by Haartman (1954) the numbers of pied flycatchers *(Ficedula hypoleuca)* increased significantly when nest boxes were added. The coqui *(Eleutherodactylus coqui),* a small tree frog of Puerto Rico, defends specific sites for retreats and nesting in the rain forest. When bamboo frog houses were added to some plots, they were quickly occupied by the frogs; one year later densities and reproductive rates were significantly higher than in control plots (Stewart and Pough 1983).

PREDATORS

Although most species of animals are preyed upon by some other species, there are few cases where predators have actually been shown to regulate the prey population. For regulation to occur, the predators must take an increasing percentage of the prey population as the number of prey increases. One approach has been to study islands, where predators may be absent. After censusing lizards and orb-weaving spiders on 93 islands in the Bahamas, Schoener and Toft (1983) found the numbers of spiders to be about 10 times lower on islands with lizard predators than on those without.

INTRASPECIFIC COMPETITION

Intraspecific competition has received much attention as a regulatory factor because it is the only one that is directly density-dependent. Other biotic regulatory factors involve other species of organisms that, in turn, are affected by other agents; thus they are not perfectly density-dependent. For example, predation involves a complex relationship between predator and prey. Although wolves eat caribou, they are not necessarily the regulators of the caribou population. Since the wolf population is affected by factors other than the numbers of caribou, such as diseases or the abundance of other prey species, it will not be able to parallel the caribou population exactly. An outbreak of disease might reduce the size of the wolf pack and allow the caribou to "escape" control and begin to destroy grazing land. Competition among the caribou, however, will vary directly with the population density and the limiting resource.

POPULATION SELF-REGULATION

The self-regulation school of thought argues that behavioral and physiological changes take place in animals because of competition with conspecifics. As a resource becomes scarce, the intensity of contest competition increases and causes changes within the organisms that increase the deaths and emigrations and/or decrease the birth rate. The adaptive value of this system is thought by some to be that the population is regulated below the limit set by the environment. If the limiting resource is food, overexploitation could permanently lower the carrying capacity; for example, overgrazing causes soil loss from erosion by wind or water. Contest competition in advance of such a shortage could adjust birth, death, and movement rates and thus avoid wasted energy in reproduction.

Although proponents of self-regulation have proposed many eloquent explanations, a big stumbling block is that most mechanisms seem to require individuals to sacrifice direct fitness by not breeding or by dying for the good of the rest of the population. The evolution of self-regulatory mechanisms of population regulation appears to involve selection at the level of the group. To put the problem in human terms, let us consider the so-called tragedy of the commons (Hardin 1968). Suppose that individuals graze their flocks of sheep in a communally owned pasture. Individual sheep owners will be tempted to increase the size of their own flocks, since by adding a sheep or two, they will get more lambs and wool (Figure 17–3). If the optimum number is already in the pasture, their action will reduce only slightly the

FIGURE 17–3   Sheep grazing in open pasture
If these sheep were owned by several individuals and were grazed on a communally owned pasture, individual sheep owners could increase their profits by adding a few sheep. But if each owner did that, the pasture would become overgrazed, and all the owners would suffer a loss, the tragedy of the commons. One solution would be for the owners to agree to limit the number of sheep each grazes on the pasture.
Source: Photo by Grant Heilman.

total yield and they will be ahead. But if all the other sheep owners behave selfishly and add a few sheep, the common pasture will soon be destroyed and all individuals will lose. The solution is for all the owners to come to a common agreement, or convention, to limit the size of their flocks and forgo any short-term gain in favor of the long-term benefit to the whole community. (We can also take the opposite view, expressed by Adam Smith (1776), that if individuals behave so as to maximize their own interests, they will in the long run promote conditions that are also optimal for the whole group. We examine this contrary view in more detail at the end of the chapter.)

BEHAVIORAL MECHANISMS

EPIDEICTIC DISPLAYS.   Wynne-Edwards (1962) argued that the aerial maneuvers performed by flocks of blackbirds or starlings in the evening are unrelated to feeding or defense against predators. He claimed that the main purpose of such social behavior is to communicate information about the supply of potentially limiting resources. The behaviors, called **epideictic displays,** act as conventions resulting in adjustments of birth, death, and movement rates, enabling populations to track closely and not overexploit the fluctuating external environment. Wynne-Edwards included among specifically timed communal displays the dancing of gnats and midges, the maneuvers of bats and birds at roosting time, and the choruses of birds, bats, frogs, fish, insects, and shrimps. He believed that displays are abstract conventions, highly symbolic, and produce a state of "excitation and tension closely reflecting the size and impressiveness of the display." A special time of day might be set aside for them, as well as a special place.

The mass feeding and roosting flights of birds are among the phenomena to which Wynne-Edwards applied his theory. Each night in the fall and winter, starlings *(Sturnus vulgaris)* gather by the millions (Figure 17–4). Birds in a roost use a common feeding area, and Wynne-Edwards considered it to be a communal territory. Most birds are constant in their choice of roost and feeding ground; the same roosts may be used for hundreds of years. Each bird tends to have its own place in the roost. As the birds go to roost in the evening, there is much communal singing and bickering over perching sites. As Wynne-Edwards (1962, p. 286) noted:

Moreover on fine evenings, especially early in the autumn, either a part or sometimes the whole of the noisy company not infrequently rises with a great roar of wings to engage in the most impressive aerial manoeuvres over the site: the massed flock may extend in a tight formation sometimes hundreds of yards in length, changing shape and direction like a giant amoeba silhouetted against the sky. On the grand scale these manoeuvres are by no means the least of the marvels of animal adaptation, so perfect is their co-ordination and so intense the urge to excel in their performance. It seems quite irrational to dismiss what is certainly the starling's most striking social accomplishment merely as a recreation devoid of purpose or survival value, and wiser to assume that a communal exercise so highly perfected is fulfilling an important function.

FIGURE 17-4 Aggregation of starlings
Wynne-Edwards argued that the roost serves primarily as a site for the morning and evening flight displays necessary for the regulation of population size. Other researchers offer explanations based on benefits to the individual, such as group protection in the roost or the improved foraging efficiency of groups.
Source: Photo by Eric Hosking.

Wynne-Edwards concluded that the primary function of the roost is to bring members of a population unit together for an epideictic display. The result is to stimulate the adjustment of population density through either emigration of individuals of poor quality or a reduction in the reproductive rate. The roost's function as a sleeping place is secondary to its function as the location of the morning and evening displays. In fact, the concentration in the roost makes the birds an easier target for predators than they would be if they slept individually.

Such synchronous mass behavior is not limited to colonial birds. The "evening rise," characteristic of brown trout (*Salmo trutta*) among others, occurs about twenty to thirty minutes after sundown and lasts about thirty minutes. The fish cruise in small circles in a particular area. Although the rise is influenced by weather and the abundance of flies (food), the relationship is not very close, and Wynne-Edwards considered the rise to be epideictic.

The Bogong moth (*Agrotis infusa*) of Australia forms large aggregations in caves in the summer. The moths become active for a half hour or so in the morning and evening. After vibrating their wings and crawling about, the moths fly around the cave and join others outside in a dense milling flight before they return to the same cave. There is no evidence of any feeding or mating associated with this behavior. The size of the population varies during the summer, and moths with depleted fat reserves disappear. Wynne-Edwards (1962) argued that the flights are epideictic displays that communicate information about population density and that function to reduce the population through the selective dispersal of individuals in poor condition. However appealing this hypothesis, it has been largely rejected by biologists for reasons we shall discuss shortly.

TERRITORY SIZE. Researchers recognized early the possible role of territory in population regulation in birds. Because populations of all species can potentially over-breed, exclusive defense of a resource may force some of the surplus population into marginal habitats, where birth rates are lower and death rates are higher than in optimal areas. One can readily demonstrate the presence of surplus animals by removing residents from optimal areas and then noting whether new animals come into the area from the marginal habitats. In his analysis of more than twenty years of breeding data for the great tit *(Parus major)* from Marley Wood in England, Krebs (1970) demonstrated that population is regulated by density-dependent decreases in both clutch size and hatching success. He studied the role of territoriality in limiting the number of breeding birds as follows. First Krebs (1971) demonstrated that defense of nest boxes causes the birds to be spaced out more than would be expected if nest box occupation were random, so that breeding density is reduced locally. Next he studied the effect of removing birds. The arrival of new pairs within hours after he had removed pairs of birds suggested that surplus birds were in the area (Figure 17–5). But where did these birds come from? Having color banded most of the pairs in the area, Krebs found that nearly all the newcomers were from the hedgerows surrounding the woods. The lower reproductive success of birds that breed in hedgerows compared with those that breed in the woods demonstrates again that territories limit breeding in the optimal habitat. However, Krebs found that although territory size varied considerably from year to year, it was not related to food supply. For this reason he concluded that territory is not very important in regulating the population as a whole, having only local effects.

    Several other lines of evidence suggest that territory size is related to limited resources and could therefore play a role in population regulation. The area defended by some species contains the minimum amount of resource to reproduce. For instance, in a study on red squirrels *(Tamiasciurus* spp.), Smith (1968) measured the

---

a.   b.

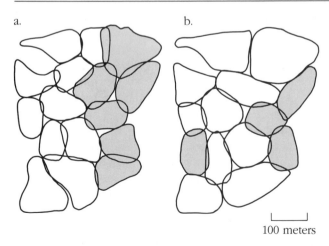

100 meters

FIGURE 17–5   Territorial replacement of great tits
Six pairs of great tits, which were defending territories as indicated by shaded areas on map (a), were shot. After three days, four new pairs had moved in from less suitable habitats, as indicated by shaded areas on map (b). The remaining residents also expanded their territories.
Source: J. R. Krebs, "Territory and Breeding Density in the Great Tit, *Parus major L,"* Ecology 52 (1971):2–22. Copyright 1971, the Ecological Society of America. Reprinted with permission of Duke University Press and the author.

food supply in each territory and concluded that each territory contained just enough food to provide the energy needs of its occupant. Studies of a number of species have shown that territories are larger in marginal habitats. Nevertheless, to prove that territory regulates population is difficult.

In polygynous species such as redwings male territory size is inversely related to habitat quality, but only other males are excluded and some feeding takes place off the territory. If the sex ratio is 50 : 50 and all females breed, it is not likely that male territory regulates the population as a whole.

In white-footed mice *(Peromyscus leucopus)*, male home ranges (which overlap to some extent and are therefore not strictly territories) decrease in size linearly with increasing density of males. It is therefore not likely that home ranges regulate population. Female home ranges are smaller, overlap little with those of other females, and are inelastic above a certain density; thus they possibly set an upper limit to the number of nesting females (Vessey, Dunsworth, and Flemming 1977).

The prevailing view of the function of territoriality is that it benefits individuals; any role in population regulation is likely to be localized. There is little experimental support for the idea that territoriality regulates populations or that it has evolved as a means of population regulation per se.

DOMINANCE HIERARCHIES.   Dominance hierarchies have also been implicated in the regulation of population. Lower-ranking individuals in groups with dominance hierarchies may have higher mortality and lower birth rates, as do surplus animals, those excluded from territories in prime habitats. In times of food shortage, high-ranking individuals survive while low-ranking individuals starve or disperse. Most group-living animals are polygynous, and the effects of competition on reproductive success are severe in the male. The elephant seal *(Mirounga angustirostris)* mating system demonstrates this effect spectacularly: the top-ranking male does as much as 80 percent of the mating (Le Boeuf 1974). However, a substantial turnover of males in the hierarchy takes place during the breeding season, so others do get a chance. Among redwings, nearly all females breed successfully each season; thus the male-based dominance system is ineffective in population regulation. Furthermore, in many other polygynous species with dominance hierarchies, the correlation between male rank and reproductive success is low.

Sometimes the reproductive differential occurs in the female hierarchy. In temperate North America, paper wasp *(Polistes fuscatus)* nests are usually started each spring by a single queen, but she is joined by several other females (West Eberhard 1969). The foundress is dominant, and she suppresses reproduction in the subordinates, whose role is to regurgitate food for the foundress (trophallaxis) and to care for the young. In packs of wolves *(Canis lupus)* the alpha female inhibits mating by lower-ranking females; she is usually the only female in the pack to mate (Mech 1970). Drickamer (1974a) and Sade and others (1976) found that among the rhesus monkeys *(Macaca mulatta)* on islands off the coast of Puerto Rico, matriarchies headed by high-ranking females are reproductively more successful than those headed by lower-ranking females. In fact, some of the latter actually lost numbers over the years. Although the preceding examples demonstrate density-dependent reproductive restraint, they do not prove that dominance hierarchies function to regulate populations.

SOCIAL PATHOLOGY OF OVERPOPULATION.   One research strategy for studying population regulation has been to create artificial populations in the laboratory. Such populations invariably cease to grow after following a roughly sigmoid (S-shaped) growth curve, even when food, water, and nesting space are kept in excess. Calhoun (1962, 1973) described the behavioral changes that take place in his mouse or rat "universes." Initially, males defend territories, and birth and juvenile survival rates are high. As density increases, aberrations begin. One is called a **behavioral sink:** animals become conditioned to eating and drinking in the presence of others and thus restrict their activities to a few places in the cage. In mouse universes, large numbers of "grouped withdrawn" spend nearly all their time in piles. Occasionally fights break out, and the jumping mice resemble popcorn. Those females that bear young build inadequate nests and may abandon litters.

FIGURE 17–6   Researcher John Calhoun in mouse universe
Although food, water, and nesting space are present in excess of needs at all times, populations in cages such as this one stop growing and eventually die out because of the behavioral and physiological changes brought about by crowding.
Source: Photo by Nilo Olin/National Institute of Mental Health.

Calhoun observed several aberrant kinds of behavior in male rats and mice in his "universes," as well as the territorial, aggressive behavior found in most natural populations (Figure 17–6). Pansexual males fail to discriminate between other males and females and make sexual advances to any creature that moves. Some males, called probers, are very active; although they do not compete for social rank or physical space, they show hypersexual, homosexual, and cannibalistic behavior. Normally an estrous female rat, after being pursued by a male, returns to her burrow while the male waits outside. After the male performs a courtship dance, the female emerges and they mate. The probers, however, follow the female inside the burrow and sometimes consume dead young that are there. Calhoun has referred to another group of males as "beautiful ones." These males have flawless coats with no scars and walk about the pen with impunity, being ignored by territorial males and even nesting females. At first, Calhoun thought these males were the highest ranking of all, but he later concluded conspecifics do not recognize them as adult rats or mice because they are in a state of physical and social immaturity even though of adult size. These males, when put in an uncrowded cage with receptive females, are unable to form dominance hierarchies, defend territories, or mate with females.

Females in these colonies show delayed maturation, and few breed at all. In fact, one of Calhoun's mouse universes went extinct because the females all became too old to reproduce without ever coming into estrus. Reproductive inhibition in response to density has also been demonstrated in confined populations of other species, such as deermice (Terman 1973).

Much attention has been paid to the role of agonistic behavior and interaction rates in laboratory populations of mice. Populations with particularly aggressive individuals tend to level off at lower densities than do those with less-aggressive members (Southwick 1955; Vessey 1967). When Vessey (1967), with the use of tranquilizers, lowered agonistic behavior of mice in populations that had reached the carrying capacity of the cage, litter survival improved and the population increased further (Figure 17–7). These experiments suffer from several problems that make extrapolation to natural populations risky: providing excess food, water, and nesting space creates an unrealistic environment, as does preventing emigration, which results in abnormally high densities.

Some researchers have tried to apply these findings to human populations. A number of studies have shown positive correlations between density and such variables as crime rate and mental illness, but these correlations tend to disappear when the level of income is controlled (Freedman 1980). People living in crowded, urban areas tend to have lower incomes than do those who live in low-density areas, and it is this poverty, rather than density itself, that is causally related to crime and mental illness. It also seems clear that sheer numbers per unit space are less important in social pathology than are the way space is utilized and the nature of the social interactions that take place. (See Chapter 12 for a discussion of the role of social disaggregation in aggression.)

Although humans respond physiologically to stress like many other animals, there is no firm evidence that birth rates are reduced in response to stress. However, a number of reproductive restraints are evident in hunter-gatherer populations.

FIGURE 17–7   Effects of tranquilizers on fighting, litter survival, and population growth of caged wild house mice

In (a) the control population of mice reached asymptote (upper limit), then slowly declined. Four-sided figures refer to survival of litters. In (b) the population of mice reached asymptote, then increased when chlorpromazine, a tranquilizer, was added to diet (downward pointing arrow). Note the increase in litter survival and the decrease in fights when the tranquilizer was added. When the tranquilizer was removed (upward arrow), litter survival decreased, fighting increased, and population declined. *Source:* From "Effects of Chlorpromazine on Aggression in Laboratory Populations of Wild House Mice" by S. Vessey, *Ecology*, 1967, 48, 367–376. Copyright © 1967 by the Ecological Society of America. Reprinted by permission.

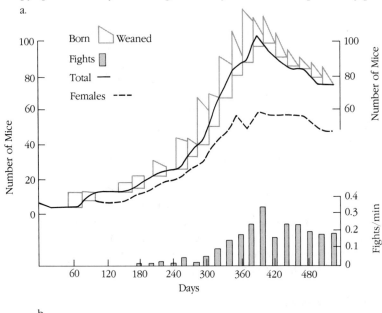

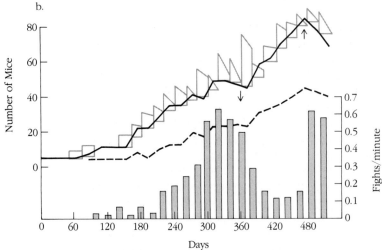

Delayed onset of puberty, prolonged lactation (which delays resumption of ovulatory cycles), and preferential female infanticide are the main ways family size is kept in check. Presumably these factors work in conjunction with mortality and emigration to adjust populations to the carrying capacity. Lee (1980) emphasized the importance of prolonged lactation; and Dickemann (1975) has argued the importance of infanticide.

BEHAVIORAL-PHYSIOLOGICAL MECHANISMS

Some animal species undergo seemingly regular nine-to-ten-year population cycles as in the case of the lynx *(Lynx canadensis)* and the snowshoe hare *(Lepus americanus)*, or three-to-four-year cycles as with the vole *(Microtus* spp.) and the lemming *(Dicrostonyx* spp. and *Lemmus* spp.). For example, during cyclical peaks, hares are under practically every bush; during these population peaks a disturbance as slight as a hand clap was enough to send them into convulsions and coma, followed by death due to hypoglycemic shock (Green, Larson, and Bell 1939). In 1950 Christian proposed that mammalian populations could be regulated by disease caused by exhaustion of the adrenal gland, following prolonged psychological stress from agonistic interactions at high population levels. This negative feedback loop, dependent on intraspecific competition, would be perfectly density dependent. The idea grew from the work of Selye (1950) on the **general adaptation syndrome,** in which nonspecific stressors, such as heat, cold, or defeat in a fight, produce a specific physiological response. ACTH released from the anterior pituitary, under control of the hypothalamus, stimulates production of glucocorticoids by the adrenal gland. These hormones, such as cortisone, function mainly to elevate blood glucose to prepare the body for fight or flight.

The phenomenon of death due to adrenal exhaustion turned out to be an extreme case, and Christian modified his hypothesis after a series of laboratory experiments (Christian 1978). There is, in fact, an increase in adrenocortical output in response to increasing population density. These hormones, along with ACTH and some others still being explored, seem to reduce direct fitness in numerous ways. The body's two main defense mechanisms, the immune and the inflammatory responses, are inhibited; these changes obviously increase the likelihood of morbidity or mortality. At the same time, growth and sexual maturation are inhibited, as are spermatogenesis, ovulation, and lactation. Some of these effects on reproduction persist even into subsequent generations, in spite of a reduction in population density (Figure 17–8). Field data supporting these findings have come from studies of a variety of mammals, such as deer, rats, mice, and woodchucks (Christian 1978). Probably an equal number of studies have failed to find a consistent relationship between adrenocortical output and population density. Part of the problem is that the stressor is not density, as seen earlier, but levels of agonistic behavior in competition for limiting resources. In the laboratory, as little as two minutes per day exposure to trained fighter mice produces a pronounced stress response as well as

FIGURE 17-8   Christian's model of population regulation in small mammals (simplified version)
As population increases (upper left), social contacts increase, activating the pituitary-adrenal gland system and inhibiting the production of sex hormones. The population then declines due to the decrease in the birth rate and the increase in the death rate. As the population declines, further social contacts are reduced, and hormone changes are reversed.
Source: Data from J. J. Christian, "Neurobehavioral Endocrine Regulation of Small Mammal Populations." In D. P. Snyder (ed.), Populations of small mammals under natural conditions. *The Pymatuni..g Symposia in Ecology* 5 (1978):143–158.

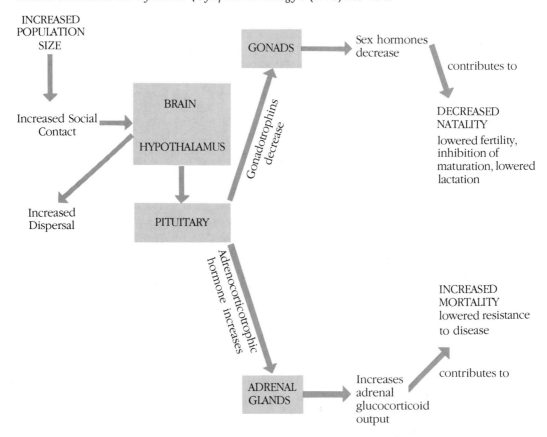

fifteenfold increase in the load of parasites in experimental mice compared with control mice (Patterson and Vessey 1973).

Clearly, this negative feedback loop works under some conditions, but its generality in natural systems remains to be demonstrated. Work with pheromones has augmented some of the above findings. Substances released in the urine produce effects on conspecifics without a physical encounter. In some species of mice, the smell of strange male urine causes a pregnancy block by preventing implantation

(Bruce 1966); in crowded or unstable populations, the birth rate could thus be lowered. Mice avoid the urine of a mouse recently defeated by another mouse, and urine from stressed mice produces an adrenocortical response in naive mice (reviewed by Bronson 1971, 1979). Females grouped together produce a substance that delays sexual maturation in other females (Drickamer 1974b); the same delay has been produced by using female urine from high-density populations of house mice in the field (Massey and Vandenbergh 1980). In contrast, odor from a mature male accelerates the sexual maturation of young females (Vandenbergh 1969). No single mechanism is likely to account for regulation of numbers in all species, and several mechanisms may operate within the same species.

BEHAVIORAL-GENETIC MECHANISMS

In general, the ideas examined so far are complementary; they are all based on the notion that increased interaction rates act to reduce population size. A competing hypothesis suggests that natural selection favors different genotypes at high and at low population levels. While working with voles (*Microtus* spp.) that have a three- to four-year population cycle, Chitty (1960) noticed that populations continued to decline even under seemingly favorable environmental conditions. Voles from declining populations were highly aggressive and intolerant and bred poorly; voles from increasing populations were mutually tolerant and were rapid breeders. He postulated that there was a change in the *quality* of animals in a declining population; through natural selection the proportion of aggressive individuals increased, and, even though they could compete in crowded conditions, their reproductive rates were low and the population declined (Figure 17–9). It seemed unlikely that a change in gene frequency could take place over a span of only a few years, but some confirming evidence has come forth that dispersal of animals of a certain genotype can produce rapid changes in the sedentary portion of the population. Myers and Krebs (1971) reported that the frequency of one allele in the blood serum of voles was significantly different in dispersers as compared to residents. However, the behavioral changes reported in vole populations have not been linked to specific genes. Evidence from the laboratory shows that the age of puberty can be shifted in mice *(Mus musculus)* after only a few generations of artificial selection (Drickamer 1981). The low birth rates observed in peak populations in the wild could be due in part to selection for late-maturing individuals.

Tamarin (1980) has emphasized the role of dispersal itself as a regulating mechanism. If dispersal is blocked by a fence or is blocked naturally, as on islands, regulation fails and high populations and depleted resources result (the "fence effect," Krebs et al. 1973). Dispersers are likely to be at a disadvantage reproductively since they are probably in a suboptimal habitat; also, predation is higher on transient individuals (Metzgar 1967). However, emigrating individuals have the potential of colonizing new areas and may be the founders of new species (Christian 1970).

FIGURE 17–9   Chitty's model of population regulation in small mammals (simplified version)
In response to increased population and social contact, there is selection for aggressive individuals with low reproductive rates leading to a decline in population. Modifications of the model emphasize the role of emigration of certain phenotypes in causing the decline. This model predicts genetic changes in the population as different genotypes are favored at high versus low densities.
Source: Data from D. Chitty, "The Natural Selection of Self-Regulatory Behavior in Animal Populations," *Proceedings of the Ecological Society of Australia* 2 (1967):51–78, and from C. Krebs, "The Lemming Cycle at Baker Lake, Northwest Territories, during 1959–1962." *Arctic Institute of North America Technical Paper No. 15, 1964.*

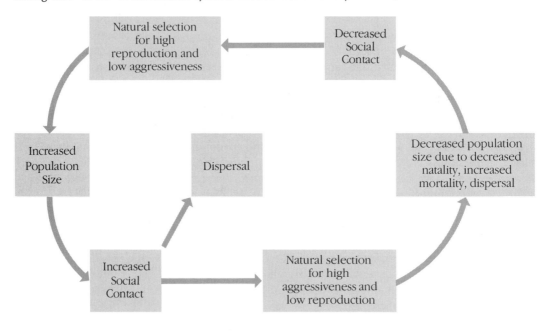

EVOLUTION OF POPULATION SELF-REGULATION

Proponents of self-regulation have not yet resolved two important problems. First, because most of the data are from laboratory populations, where densities are often unrealistically high and where emigration is usually prevented, more testing on natural populations is needed. However, in natural populations many extrinsic factors come into play, and no single mechanism can be implicated. Second, if we accept the idea that such self-regulatory mechanisms exist, do they evolve specifically for that purpose, and, if so, what is the unit of selection?

We can probably agree that self-regulation is adaptive for a population or species that has the potential for destroying or using up its resources, but many of the mechanisms we have discussed seem to be maladaptive for the individuals involved. An animal that is genetically programmed to respond to increased fighting by cutting back on reproduction will lose out, genetically speaking, to another individual that can keep on reproducing. If an animal that responds to aggression by reducing its defense mechanisms is, by definition, less fit, how could such a system evolve?

One answer is by group, or interdemic, selection, as Wynne-Edwards (1962) argued (Figure 17–10). Groups or populations that avoid overexploitation of the environment survive and may later colonize the habitats left vacant by imprudent groups that became extinct. While group selection is considered by most evolution-

FIGURE 17–10   Levels of selection, individual to interdemic
Small groups of four to six lions represent families of related individuals. On the left side of a barrier is deme A. This deme shows a case of individual selection in which the individual in the upper left dies, and a case of kin selection in which the family to the right is eliminated. On the right side of the barrier, group selection occurs when the entire deme B is wiped out.
Source: Data from E. O. Wilson, "Group Selection and Its Significance for Ecology," *BioScience* 23 (1973):631–638.

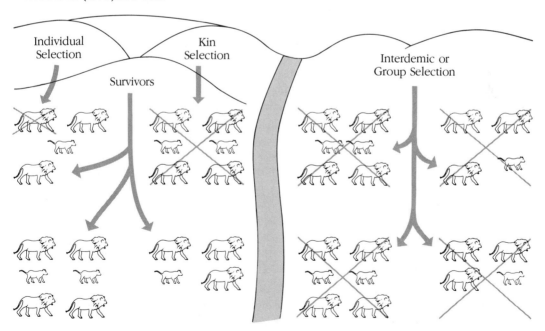

ary biologists to be possible, and a few examples of it have been documented (Lewontin and Dunn 1960), the rate of change in gene frequency would seem to be too slow, when compared with the rate of change through individual selection. To maintain a trait that is adaptive to the group but not to the individual, group extinction would have to be more rapid than individual extinction—and that does not seem to be the case. Therefore, most population biologists favor arguments based on individual selection and reject Wynne-Edward's hypothesis.

Arguments based on kin selection offer one way around the problem. We now know that many social groups are made up of close relatives. Thus behavioral and physiological traits that reduce direct fitness but increase indirect fitness, in this case by increasing the likelihood of survival of a kinship group, could be selected for. In other words, it might be worth your while, evolutionarily speaking, to sacrifice your reproductive output if the survival of the group as a whole, which contains relatives who share your genes through common descent, is ensured.

A second answer to the question of self-regulation is that the mechanisms did evolve through natural selection at the level of the individual. Reproductive restraint in the face of overcrowding may be the best individual strategy; later maturation or production of fewer young can save energy for a more optimum time. Ensuring the greatest genetic representation in the next generation is not synonymous with having the most offspring. Lack (1954) concluded that clutch size (the number of eggs laid) in birds evolved to an optimum number in terms of the number of offspring that will, in turn, survive and reproduce; laying too many eggs could result in survival of few or no young. We can understand many of the epideictic displays described by Wynne-Edwards (1962) simply as outcomes of intraspecific competition of individuals that are behaving so as to increase direct fitness.

Similarly, at times of high density, aggressive individuals that are intolerant of others would be selected for if limiting factors were operating (Chitty's hypothesis). The most successful individuals at times of high density might be those that reduce the fitness of others through competition. We should keep in mind that fitness is a relative term that refers to increasing one's genes in the next generation *relative* to others'.

The inhibition of the body's defense systems as a result of increased adreno-cortical output at high population is more difficult to explain (Christian's hypothesis). How could such a response be adaptive to the individual under any circumstances? The pituitary-adrenal response is part of the general adaptation syndrome; only animals that are being attacked and defeated show this response, whereas dominant animals, even in crowded situations, respond little and are like individuals at low densities. The general adaptation syndrome (GAS) enables an animal under stress to mobilize energy reserves just to survive. Other systems, such as the reproductive system, are temporarily shut down; producing offspring at such times would likely be a waste of energy. Low rank or exclusion from a territory suitable for breeding may be a temporary situation. In fact, many territory and dominance systems are age graded, and younger animals, usually males, attain prime territories or high rank if only they can live long enough.

The preceding arguments suggest that self-regulatory mechanisms do not evolve for the purpose of controlling populations. Most scientists who test hypotheses about the functions and origins of behaviors assume that the functions evolved at the lowest level consistent with the facts. Usually this level is no higher than that of parents and young (Williams 1966).

SUMMARY

The numbers of organisms present in a population equals the sum of births and immigrants minus the sum of deaths and emigrants. All populations have the capacity for rapid increase; understanding the abiotic and biotic factors in the environment that limit numbers is a central problem in ecology.

Density-independent factors, including floods, droughts, and fires, are unaffected by the individuals or the numbers of individuals in a population. Density-dependent factors are affected by the numbers of individuals in a population and cause increases in the death rate, decreases in the birth rate, or increases in emigration. These factors may include disease, predation, and competition.

Some evidence suggests that behavior involved in contest competition for limiting resources is important in population regulation. Most of the research on the role of behavior in population regulation has focused on self-regulatory mechanisms, in which increases in intraspecific competition produce behavioral or physiological changes. The result of these changes is that the population is regulated below the carrying capacity, and overexploitation of resources is avoided.

Communicating information about limiting resources might enable populations to adjust birth, death, and movement rates and to track environmental resources. Some observers argued that communal displays — such as dancing of gnats and midges, choruses of birds, as well as mass feeding and roosting flights — function as epideictic displays. Because this idea seems to involve sacrifice of individual fitness for the good of the group, attention has shifted more recently to regulatory mechanisms that can be understood in terms of the action of natural selection on the individual rather than on the group.

Population growth rates can be affected by territorial behavior, in which some individuals are excluded from breeding or are forced to occupy marginal habitats, or by dominance hierarchies, in which subordinate individuals are less successful in obtaining mates or rearing offspring. In laboratory populations, where emigration is usually restricted, manifestations of social pathology appear as the population increases. Birth rates are lowered by such things as interference in courtship or copulation, or by inadequate maternal care. Death rates also increase, and the population stabilizes or even declines. Researchers have shown that physiological changes associated with increasing population density and agonistic behavior increase death rates and decrease birth rates in laboratory and field populations. The most important changes are in adrenocortical hormones, which are secreted in response to defeats suffered in social fighting. Pheromones also may be important in population regulation because, for example, the odor of strange male mice blocks pregnancy in females, males avoid the odor of recently defeated mice, and group-housed females produce an odor that delays sexual maturation of other females.

Another possible self-regulatory mechanism is change in gene frequencies in a population. As numbers increase, emigration of certain genotypes may occur. Those that remain are genetically adapted to live under crowded conditions; they are highly aggressive but have low reproductive rates. As the population declines, natural selection favors individuals that are less competitive but that have high reproductive output.

Some self-regulatory mechanisms seem to involve individual sacrifice in reproduction or in survival for the good of the population. It is widely believed that group selection operates too slowly to lead to the evolution of such mechanisms; selfish individuals would increase in numbers at the expense of the altruists. However, if the populations are kin groups, then individuals sacrifice direct fitness for increased inclusive fitness by reducing the possibility of extinction of the kin group. Reproductive restraint is probably a temporary strategy that allows resources to be devoted to surviving under intense competition by postponing reproduction until resources might become more plentiful; selection here is operating at the level of the individual.

Discussion Questions

1. Understanding factors affecting the size of populations is a central problem in ecology. Why has social behavior been implicated in so many of the theories of population regulation?

2. Natural selection, acting at the level of the individual, would seem to favor those with high reproductive success. How can individuals that produce fewer young ever be favored?

3. Statements suggesting that many factors contribute to regulating animal populations seem to run counter to Liebig's law of the minimum. Is there any way to resolve this apparent contradiction?

4. The data below refer to mice that were exposed two minutes per day either to trained fighter mice, to trained nonfighter mice, or to no mice for Days 1 through 14. On Day 8 each mouse was fed equal numbers of tapeworm eggs. On Day 22 the mice were killed, their adrenal and thymus glands weighed, and the intestinal tapeworms counted and weighed. Adrenal gland weight is positively related to adrenocortical hormone secretion. The thymus is involved in the immune re-

sponse. Discuss these data as they may relate to population regulation and behavior.

Means of host organ weights and weights and numbers of tapeworms in the three experimental groups of mice

	Exposed to No Mice	Exposed to Trained Non-fighter Mice	Exposed to Trained Fighter Mice
Group size	10	10	7
Adrenals (mg)	4.39	4.42	5.88*
Adrenal weight/body weight (mg/gm $\times$ 100)	13.1	13.1	16.6*
Thymus weight (mg)	37.0[a]	35.0	21.9*
Body weight (gm)	33.7	34.0	35.8
Worm number/mouse	11.6[b]	15.4[b]	75.3**
Worm weight/mouse weight (mg)	6.37	4.04	82.28**

Source: Data from M. A. Patterson and S. H. Vessey, "Tapeworm *(Hymenolepis nana)* Infection in Male Albino Mice: Effect of Fighting among the Hosts," *Journal of Mammalogy* 54 (1973):784–786. Reprinted by permission.
* Significantly different from other two groups, $P < .01$.
** Significantly different from other two groups, $P < .001$.
[a] Mean based on nine weights.
[b] Three of these mice had no worms.

Suggested Readings

Cohen, M. N., R. S. Malpas, and H. G. Klein, eds. 1980. *Biosocial Mechanisms of Population Regulation.* New Haven: Yale University Press.
Reviews theories of self-regulation. Several papers deal with attempts to apply Calhoun's and Christian's ideas to population regulation in humans.

Krebs, C. J. 1985. *Ecology: The Experimental Analysis of Distribution and Abundance,* 3rd ed. New York: Harper & Row.
Part 3 covers population growth. Chapter 16 deals specifically with theories of population regulation.

Krebs, C. J. 1978. A review of the Chitty hypothesis of population regulation. *Can. J. Zool.* 56:2463–2480.
Reviews hypotheses of population regulation and the types of field data needed to test them.

McLaren, I. A., ed. 1971. *Natural Regulation of Animal Populations.* New York: Atherton Press.
A collection of ten review papers by different researchers presenting competing views of population regulation.

References

Bronson, F. H. 1971. Rodent pheromones. *Biol. Reprod.* 4:344–357.

———. 1979. The reproductive ecology of the house mouse. *Q. Rev. Biol.* 54:265–299.

Bruce, H. M. 1966. Smell as an exteroceptive factor. *J. Anim. Sci.*, suppl. 25:83–89.

Calhoun, J. B. 1962. Population density and social pathology. *Sci. Amer.* 206:139–148.

———. 1973. Death squared: The explosive growth and demise of a mouse population. *Proc. Roy. Soc. Med.* 66:80–88.

Chitty, D. 1960. Population processes in the vole and their relevance to general theory. *Can. J. Zool.* 38:99–113.

———. 1967. The natural selection of self-regulatory behavior in animal populations. *Proc. Ecol. Soc. Aust.* 2:51–78.

Christian, J. J. 1950. The adreno-pituitary system and population cycles in mammals. *J. Mammal.* 31:247–259.

———. 1970. Social subordination, population density and mammalian evolution. *Science* 168:84–90.

———. 1978. Neurobehavioral endocrine regulation of small mammal populations. In D. P. Snyder (ed.), *Populations of Small Mammals under Natural Conditions. The Pymatuning Symposia in Ecology* 5:143–158.

Christian, J. J., and D. E. Davis. 1964. Endocrines, behavior and population. *Science* 146:1550–1560.

Connell, J. H. 1961. The influence of interspecific competition and other factors on the distribution of the barnacle *Chthamalus stellatus*. *Ecology* 42:710–723.

Dickemann, M. 1975. Demographic consequences of infanticide in man. *Ann. Rev. Ecol. Syst.* 6:107–137.

Drickamer, L. C. 1974a. A ten-year summary of reproductive data for free-ranging *Macaca mulatta*. *Folia Primatologica* 21:61–80.

———. 1974b. Sexual maturation of female house mice: Social inhibition. *Dev. Psychobiol.* 7:257–265.

———. 1981. Selection for age of sexual maturation in mice and the consequences for population regulation. *Behav. Neur. Biol.* 31:82–89.

Freedman, J. L. 1980. Human reactions to population density. In M. N. Cohen, R. S. Malpass, and H. G. Klein (eds.), *Biosocial Mechanisms of Population Regulation.* New Haven: Yale University Press.

Green, R. G., C. L. Larson, and J. F. Bell. 1939. Shock disease as the cause of the periodic decimation of the snowshoe hare. *Amer. J. Hyg.* 30:83–102.

Haartman, L. von. 1954. Territory in the pied flycatcher *Muscicapa hypoleuca*. *Ibis* 98:460–475.

Hansen, L. P., and G. O. Batzli. 1978. The influence of food availability on the white-footed mouse: Populations in isolated woodlots. *Can. J. Zool.* 56:2530–2541.

Hardin, G. 1968. The tragedy of the commons. *Science* 162:1243–1248.

Holling, C. S. 1959. The components of predation as revealed by a study of small mammal predation of the European pine sawfly. *Can. Entomol.* 91:293–320.

Krebs, C. J. 1985. *Ecology: The Experimental Analysis of Distribution and Abundance,* 3rd ed. New York: Harper & Row.

Krebs, C. J., M. S. Gaines, B. L. Keller, J. H. Myers, and R. H. Tamarin. 1973. Population cycles in small rodents. *Science* 179:34–41.

Krebs, J. R. 1970. Regulation of numbers in the great tit (Aves: Passeriformes). *J. Zool. Lond.* 162:317–333.

———. 1971. Territory and breeding density in the great tit, *Parus major* L. *Ecology* 52:2–22.

Lack, D. 1954. *The Natural Regulation of Animal Numbers.* London: Oxford University Press.

Le Boeuf, B. J. 1974. Male-male competition and reproductive success in elephant seals. *Amer. Zool.* 14:163–176.

Lee, R. B. 1980. Lactation, ovulation, infanticide, and women's work: A study of hunter-gatherer population regulation. In M. N. Cohen, R. S. Malpass, and H. G. Klein (eds.), *Biosocial Mechanisms of Population Regulation.* New Haven: Yale University Press.

Lewontin, R. C., and L. C. Dunn. 1960. The evolutionary dynamics of a polymorphism in the house mouse. *Genetics* 45:705–722.

Liebig, J. 1847. *Chemistry Application to Agriculture and Physiology,* 4th ed. London: Taylor and Walton.

Massey, A., and J. G. Vandenbergh. 1980. Puberty delay by a urinary cue from female house mice in feral populations. *Science* 209:821–822.

Mech, D. 1970. *The Wolf: The Ecology and Behavior*

of an Endangered Species. Garden City, N.Y.: Natural History Press.

Metzgar, L. 1967. An experimental comparison of screech owl predation on resident and transient white-footed mice *(Peromyscus leucopus). J. Mammal.* 48:387–391.

Myers, J., and C. Krebs. 1971. Genetic, behavioral, and reproductive attributes of dispersing field voles *Microtus pennsylvanicus* and *Microtus ochrogaster. Ecol. Monogr.* 41:53–78.

Patterson, M. A., and S. H. Vessey. 1973. Tapeworm *(Hymenolepis nana)* infection in male albino mice: Effect of fighting among the hosts. *J. Mammal.* 54:784–786.

Sade, D. S., K. Cushing, P. Cushing, J. Dunaif, A. Figueroa, J. R. Kaplan, C. Laver, D. Rhodes, and J. Schneider. 1976. Population dynamics in relation to social structure on Cayo Santiago. *Yearbook Phys. Anthrop.* 20:253–262.

Schoener, T. W., and C. A. Toft. 1983. Spider populations: Extraordinarily high densities on islands without top predators. *Science* 219:1353–1355.

Selye, H. 1950. *Stress.* Montreal: Acta.

Smith, A. 1776. *Wealth of Nations.* New York: Modern Library. 1937.

Smith, C. C. 1968. The adaptive nature of social organization in the genus of tree squirrels *Tamiasciurus. Ecol. Monogr.* 38:31–63.

Southwick, C. H. 1955. The population dynamics of confined house mice supplied with unlimited food. *Ecology* 36:212–225.

Stewart, M. M., and F. H. Pough. 1983. Population density of tropical forest frogs: Relation to retreat sites. *Science* 221:570–572.

Taitt, M. J. 1981. The effect of extra food on small rodent populations: I. Deermice *(Peromyscus maniculatus). J. Anim. Ecol.* 50:111–124.

Tamarin, R. H. 1980. Dispersal and population regulation in rodents. In M. N. Cohen, R. S. Malpass, and H. G. Klein (eds.), *Biosocial Mechanisms of Population Regulation.* New Haven: Yale University Press.

Terman, C. R. 1973. Reproductive inhibition in asymptotic populations of prairie deermice. *J. Reprod. Fertil.*, suppl. 19:457–463.

Vandenbergh, J. G. 1969. Male odor accelerates female sexual maturation in mice. *Endocrinology* 84:658.

Vessey, S. H. 1967. Effects of chlorpromazine on aggression in laboratory populations of wild house mice. *Ecology* 48:367–376.

Vessey, S. H., T. S. Dunsworth, and D. P. Flemming. 1977. Relationship between home-range size and population density in the white-footed mouse. Paper presented at American Society of Mammalogists meeting.

West Eberhard, M. J. 1969. The social biology of Polistine wasps. *Misc. Pub. Mus. Zool.*, University of Michigan, Ann Arbor 140:1–101.

Williams, G. C. 1966. *Adaptation and Natural Selection: A Critique of Some Current Evolutionary Thought.* Princeton: Princeton University Press.

Wynne-Edwards, V. C. 1962. *Animal Dispersion in Relation to Social Behavior.* Edinburgh: Oliver and Boyd.

PART FIVE

BEHAVIOR AND EVOLUTION

18   Evolution of Behavior Patterns

19   Evolution of Societies

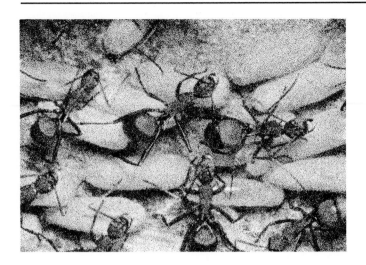

18

EVOLUTION OF BEHAVIOR PATTERNS

Throughout this book we have recognized that behavior has a genetic basis, even though the specific genes involved and their exact contribution to a behavior pattern are poorly understood. We have examined the diversity of behavior patterns, the physiological mechanisms that cause them, and the environmental factors, biotic and abiotic, that affect them. Part Five describes how the link between genes and behavior and the documentation of change in the frequencies of those genes can demonstrate the evolution of behavior.

This chapter considers the types of evidence used to document the evolution of behavior, with emphasis on the comparative method. After considering the role of behavior in the formation of new species we describe behavioral changes that occur through nongenetic means — tradition and culture. Chapter 19 looks specifically at the evolution of social behavior.

Wilson (1975) pointed out that behavior is the part of the phenotype that is most likely to change in response to environmental change. If selection pressure shifts, behavior will usually change first and physical structure second. Because behavior is the evolutionary pacemaker, studying change in behavior from an evolutionary perspective is important, but the lability of behavior — that is, its "distance from the genes" — makes the study difficult. The events that led to a change in behavior are, of course, rooted in the past. We can never duplicate those conditions exactly, so our evidence is largely indirect.

MICROEVOLUTION

Evolutionary biologists have generally assumed that natural selection produces large-scale, qualitative change through the accumulation of small, quantitative increments. These small changes, referred to as **microevolution,** may come about through changes in thresholds in the sensory, motivational, or motor systems that control behavior patterns (Manning 1971). Changes in thresholds lead to modifications in frequency or amplitude of displays. For example, the long calls of common and herring gulls (genus *Larus*), which function to advertise territory, are similar, but the two species emphasize different parts (Figure 18–1). The basic pattern involves jerking the head down, then throwing it back while calling. The herring gull has increased the amplitude of the downward jerk of the head; the common gull has increased the amplitude of the throwing back of the head (Tinbergen 1959). Such differences probably function in species recognition. Microevolution also seems important in enabling organisms to adjust to changing environments, but can it lead to the evolution of new species, as Darwin assumed? Gould and Eldredge (1977) think not, and have argued that new species have arisen in peripherally isolated populations through rapid change, rather than by slow transformation in the central ancestral area.

One of the cornerstones of research by ethologists, which began in the early 1900s, has been the idea that behavior patterns can be treated like anatomical features. They "dissected" behavior patterns to learn their "anatomy." They also demonstrated **homologies** in behavior, which are patterns shared by species

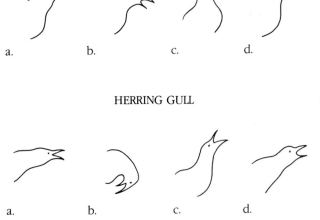

COMMON GULL

a.   b.   c.   d.

HERRING GULL

a.   b.   c.   d.

FIGURE 18–1   Oblique long call in common and herring gulls
The sequence of oblique long calls reads from left to right. The difference in the call between the two species is primarily the amplitude of the movement of the (b) and (c) components. The herring gull, for example, jerks its head further down (b) while the common gull throws its head back further (c).
Source: N. Tinbergen, "Comparative Studies of the Behavior of Gulls *(Laridae):* A Progress Report," *Behaviour* 15 (1959):1–70. Reprinted with permission.

through descent from a common ancestor. For instance, Van Tets (1965) studied birds of the order Pelecaniformes, scoring species for the presence or absence of behavioral traits and comparing the pattern with the phylogenetic tree as inferred from morphology (Figure 18–2). Some behavior patterns, such as the pre-landing call, are common to all members of the order; others, such as head wagging, as seen in gannets and boobies, are restricted to one family. The general conclusion from this and other comparative studies (for example, Lorenz 1972) is that the distribution of behavioral similarities and differences in a group of species tends to be correlated with phylogenetic relationships as disclosed by morphological or other means (for example, the fossil record or protein analysis).

There is a widespread tendency, particularly among nonbiologists, to talk about "progression" in evolution and to refer to some traits or species as "advanced" and others as "primitive." These terms are holdovers from orthogenetic interpretations that view evolution as proceeding in a predetermined direction independent of external factors. We should talk, however, of progress in evolution only in reference to either conformity to phyletic trends or to an approach to some *arbitrarily* designated final stage (Williams 1966); thus forms that share traits with that final stage could be said to be advanced. The evidence that groups "progress" in a more general sense or that adaptations become more perfect can usually be countered by examples of ancient, primitive forms that are highly successful (for example, sharks) and by the fact that adaptations in one direction usually mean trade-offs in some other area; improvements in terrestrial locomotion, as an example, are likely to be matched by decrements in swimming ability. Nevertheless, as will become evident from the examples that follow, many researchers speak of "improvements" in traits through evolutionary time.

EVIDENCE FOR THE EVOLUTION OF BEHAVIOR

In this section we examine the types of evidence used to demonstrate how behavior changes in a more or less permanent way through generations. The evidence is of three general types: (1) **phylogeny,** the tracings of the patterns of evolutionary change through time; (2) **adaptation,** the way natural selection actually works; and (3) **speciation,** the formation of new species (Alexander 1978).

PHYLOGENY

FOSSILS. Although the fossil record is one of the strongest lines of evidence supporting evolution, behavior patterns themselves do not leave fossil remains. However, we can infer much about behavior from bones, teeth, horns, tracks, and human and nonhuman artifacts such as tools.

FIGURE 18-2   Behavioral traits and phylogeny in the Order Pelecaniformes
The top half of this figure shows the distribution of behaviors among genera. Dots
indicate that the behavior is present in that genus. In the bottom half of the figure is
a phylogenetic tree based on morphological traits. Note that genera that are closely
related based on morphology also share the most behaviors, whereas those distantly
related share the fewest.
Source: Redrawn from G. F. Van Tets. "A Comparative Study of Some Social
Communication Patterns in the *Pelecaniformes.*" *Ornithology Monographs* 2 (1965):1–
88. Reprinted with permission.

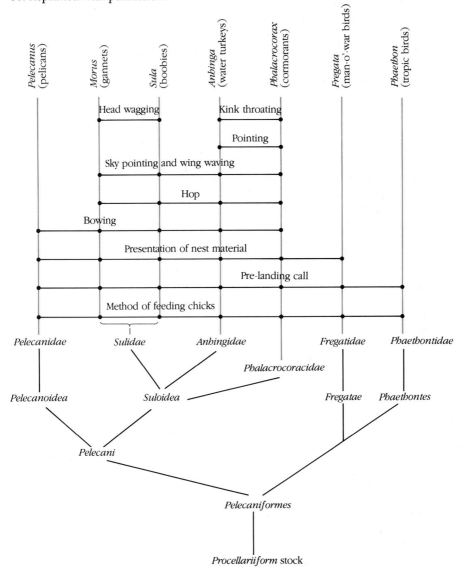

Feeding behavior has been inferred from trace fossils of the tunnels of apparently closely related worms of the Gaphoglypt family, dating from between 135 and 2 million years ago (Figure 18–3). The lines of descent show a gradual shift from rigid behavior patterns toward increasing flexibility, which enabled the worms to forage more efficiently (Seilacher 1967). There is also a trend toward specific changes in foraging patterns. The simplest foraging left tracks resembling the scribbling of children. The "program" for scribbling seems to have had one command: Keep heading to one side but don't stay in the same track. A more complex pattern was an outward spiral, the commands being: Keep circling in one direction and keep in touch with the spiral whorl made earlier. More complex, and more efficient in terms of covering the layer containing food, was a series of meanders. In this case the commands seem to have been: Move horizontally, keeping within a single layer of sediment; after advancing one unit of length, make a U-turn; never come closer to any other tunnel than some given distance.

Raup and Seilacher (1969) programmed a computer with these commands and generated tracks very similar to those found in the ancient sediments. These worm tunnels may have been more than deposit feeding trails or dwelling structures; they could have acted as traps for mobile microscopic organisms, much as the tunnels of moles are thought to entrap earthworms and arthropods (Ekdale 1980). Seilacher claims that scribbling appeared in Ordovician times (425 to 500 million years ago) and eventually disappeared. Complex meanders and dense double spirals did not appear until the Cretaceous period (63 to 135 million years ago).

Evidence of predation can be gained from drill holes in the fossilized shells of bivalve molluscs from the late Triassic (about 200 million years ago). These holes were made by carnivorous snails that penetrate the shells of their prey. Oddly, the habit seems to have disappeared, only to reappear some 120 million years later (Fürsich and Jablonski 1984). Thus, what we might consider a real breakthrough in predatory technique did not persist; the presumed selective advantage of such a behavior pattern may not be a good predictor of its long-term survival.

Evidence about the behavior of extinct mammalian carnivores was obtained

SCRIBBLES   MEANDERS

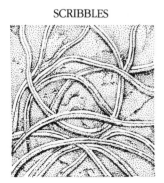 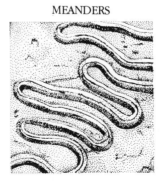

FIGURE 18–3 Fossilized worm tunnels from millions of years ago Worm trails in sediment provide a record of foraging behavior. Earlier types left scribbles (left), more recent types left meanders (right). *Source:* Data from A. Seilacher, "Fossil Behavior," *Scientific American* 217 (1967):72–80.

when Hunt, Xiang-Xu, and Kaufman (1983) excavated ancient burrow systems containing the skeletons of bear dogs from the early Miocene, 20 million years ago. These animals were about the size of a wolf or hyena and apparently used the dens much as do modern carnivores.

We can also glean much information about social behavior and mating systems by studying fossils. In mammals the amount of sexual dimorphism in body weight and length is a good predictor of the degree of polygyny (Alexander et al. 1979). Across a wide variety of species, the larger or heavier the male is relative to the female, the greater the number of females the male monopolizes. In many cases the amount of sexual dimorphism can be estimated from fossilized remains.

ADAPTIVE RADIATION.   Isolated areas such as oceanic islands offer the possibility of tracing phylogeny. Because of small target size and large distance from a colonizing source, such places are likely to be invaded by few species (see Chapter 15). In the relative absence of interspecific competition, new forms evolve rapidly. The best-studied case of such adaptive radiation is the Galápagos Islands, some 600 miles off the coast of Ecuador. Biologists believe that a single species of finch (family Fringilidae) colonized the Galápagos and radiated into fourteen species in four genera (Figure 18–4). (Darwin was the first to study these birds on his 1835 voyage on H.M.S. *Beagle*; D. Lack [1947] drew together much of the available information.) Although adaptive radiation is evident in such traits as beak morphology and plumage, feeding and breeding behaviors provide evidence of a common ancestry as well as mutability. The ancestral form was probably a heavy-beaked ground finch that fed on seeds, from which evolved: modern ground finches; a cactus-feeding form with a long, decurved bill; insectivorous forms, including the famous woodpecker finch, which holds a stick in its beak to probe underneath bark for insects; and slender-beaked warblerlike finches that feed on small insects.

DOMESTICATION.   Artificial selection by humans has changed the behavior of numerous animals with which humans have associated over thousands of years. Domestication involves more than just taming or socializing animals; humans also control the breeding, care, and feeding of domesticated animals. Some ethologists have argued that behavioral plasticity is greater in domestic animals than in their wild counterparts and that **neoteny,** the persistence of juvenile characteristics in adults, is typical in domestic animals. Others have emphasized the degeneracy of domesticants; they cite the laboratory rat. But few of these generalizations hold true when the entire spectrum of domesticated animals is considered (Ratner and Boice 1975; Price 1984). Certain species are preadapted to domestication. For instance, most carnivores maintain a nest, and urinate and defecate some distance away from it; they are thus suited to living in human habitations. Other factors facilitating the process of domestication include large social groups, the presence of males in the group year round, precocial young, and an omnivorous diet (Hale 1969).

The evolution of dog behavior provides the most striking example of selective breeding by *Homo sapiens* (Scott and Fuller 1965). The wolf *(Canis lupus),* ancestor of the dog, and the principal carnivore of the northern hemisphere prior to the arrival

FIGURE 18-4 Adaptive radiation of Darwin's finches on the Galápagos Islands
Small, isolated islands with low colonization rates serve as useful sites to study
evolution and adaptation. It is likely that a single species of finch gave rise to all
these types.
Source: Data from D. Lack. *Darwin's Finches: An Essay on the General Biological
Theory of Evolution* (New York: Harper & Brothers, 1961). First published in 1947 by
Cambridge University Press, Cambridge, England.

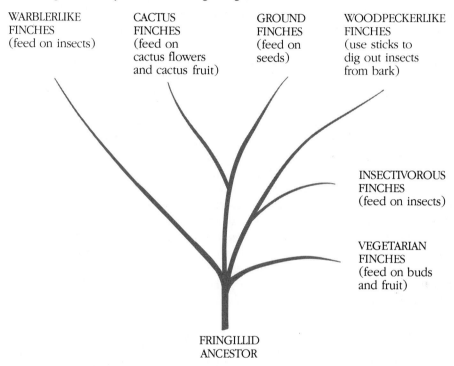

WARBLERLIKE
FINCHES
(feed on insects)

CACTUS
FINCHES
(feed on
cactus flowers
and cactus fruit)

GROUND
FINCHES
(feed on
seeds)

WOODPECKERLIKE
FINCHES
(use sticks to
dig out insects
from bark)

INSECTIVOROUS
FINCHES
(feed on insects)

VEGETARIAN
FINCHES
(feed on buds
and fruit)

FRINGILLID
ANCESTOR

of humans, has a complex, cooperative social system. As the wolf was domesticated
and transformed into the dog, it spread all over the world in conjunction with
humans and underwent adaptive radiation. Dogs *(Canis familiaris)* were used for
protection, for hunting, and for herding other domesticated animals. The relative
isolation of human tribes 8,000 to 10,000 years ago, with only occasional genetic
interchange, probably encouraged the evolution of distinct dog breeds. As human
contact increased with improved transportation, the different breeds of dogs mixed,
and many breeds, such as those in South Africa and North and South America, lost
their identity. Only in the last few hundred years have humans practiced rigid
artificial selection, which has resulted in increased separation of dog breeds. The
great diversity of form and behavior of current breeds illustrates the potential behavioral effects of selection, whether natural or artificial, of certain phenotypes.

Scott and Fuller (1965) concluded that the behavioral patterns of the dog and wolf are essentially the same. Selection has modified primarily the agonistic and investigatory systems. For example, the bulldog breed, in the interest of an English sport, was selected for its tendency to attack the nose of a bull and hang on, in contrast with the more typical slashing attack from the rear used by wolves. Terrier breeds have been selected for their tendency to attack prey relentlessly, regardless of any injury suffered; the usual wolf pattern is to snap and then withdraw to avoid injury. Other breeds have been selected for opposing reasons; scent hounds and bird dogs are examples of peaceable animals that can be kept in groups in a kennel.

Wolves and dogs search for prey rather than lie in wait and therefore have a well-developed investigatory repertoire. Some dogs, such as scent hounds, are bred for the ability to follow a trail. Others, such as bird dogs, use their visual and olfactory senses much more equally; after rapid quartering of ground, they locate prey by scent only when they are a few paces away, and then they freeze. Dogs must be trained for all of these tasks, but in each case artificial selection has produced a phenotype that is predisposed to specific behavioral traits.

COMPARATIVE SERIES.   Perhaps the best available evidence for the evolution of behavior has come from comparative studies of closely related forms. The presence of a series of intermediate forms between two extremes suggests what the evolutionary sequence was, but it may not represent the exact route because the living intermediates may have changed from the ancestral type. It is usually assumed that behaviors evolve from simple to complex, and such series are often referred to as progressions; but inferring goal-directed change in evolution is unwise, as we noted at the beginning of this chapter.

One comparative series is that of balloon flies, members of the family Empididae (order Diptera). According to Kessel (1955) the story began in 1875 when Baron Osten-Sacken was visiting in the Swiss Alps and noticed bright silvery flashes among the shadows of the fir forest. Believing that they were silvery insects he used his net to capture some, but ended up with dull-colored flies. Later he noticed that along with male flies, he had netted packets of filmy material. Osten-Sacken assumed that the males carried the material on their backs to attract females during the mating swarm. Another observer later discovered that the male flies carry the packets underneath their bodies and suggested that the somewhat flattened balloons serve as aeronautical surfboards on which the flies cavorted among the sunbeams. Another suggestion was that the devices serve as warning signals to birds and predaceous insects; however, no evidence exists that flies are distasteful to predators. The male flies seem to use the balloon for attracting females, stimulating mating, and reducing the likelihood that the female will try to eat the male once they have coupled. Kessel (1955) described the evolutionary stages that seem to have led to this situation.

Stage 1. In the majority of empidids, both sexes capture insects independently, and no presentation of prey is associated with mating. These flies sometimes prey on conspecifics; when the male attempts to copulate, it may be eaten by the female.

Stage 2. The male captures a prey item and presents it to the female. He copulates with her as she eats the prey instead of him. Many other insects follow this procedure, and natural selection theory predicts that intraspecific deception should occur if the deceiver gains a reproductive advantage over a conspecific. In an unrelated insect, the scorpionfly *Hylobittacus apicalis* (order Mecoptera), the male with a prey item advertises to females by means of a pheromone. But sometimes a male assumes the behavioral posture of a female; he approaches a male that has a prey item, lets the male try to copulate, then steals the prey item and either eats it or uses it to attract a female (Thornhill 1979).

Stage 3. Rather than taking the prey and searching for a female, some male empidids join other males, each with a prey item, in an aerial dance. The prey, Kessel posits, is now a stimulus for mating rather than a distraction to avoid mate cannibalism. The female enters the swarm and selects one of the males.

Stage 4. In many species of the genus *Hilara*, the male wraps the prey loosely with some silken threads, an action that seems to quiet the prey.

Stage 5. In several species of the genus *Empis*, from the western United States, the male applies elaborate silken wrappings to the prey, which then resembles a balloon. When male and female meet in midair, the male transfers the balloon to the female and climbs on her back. The pair alights on a plant, and the female rolls the balloon about, probes it, and eventually consumes the prey item while the male copulates with her.

Stage 6. The male catches a small prey and may consume its fluid so it is no longer edible; he then constructs a complex balloon. The female accepts the balloon and plays with it during copulation, but gets no meal from it.

Stage 7. The prey item is very small, of no food value to either sex, and pieces of it are plastered at the front end of the balloon. The balloon is now the sole stimulus for copulation.

Stage 8. The *Hilara sartor* male gives the female a balloon that has no prey at all (Figure 18–5). This behavior is thought to be the final stage in the series.

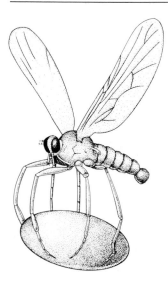

FIGURE 18–5   Male balloon fly carrying balloon
Males with balloons fly in a swarm from which a female selects a mate. That male gives the balloon to the female prior to copulation. In this species (*Hilara sartor*) the silken balloon has replaced the prey item that the male offers to the female in other closely related species.

Kessel argues that the nuptial feeding of the female initially functioned to reduce the chances of the male's being consumed by the female during mating. Thornhill (1976) suggested an alternative function based on parental investment. Male insects may invest in their offspring by providing nutrition to their mates in the form of glandular secretion, prey captured by the male, regurgitative food offering (see Figure 18–6), or the male himself. Females may select mates on the basis of the quality of food the males offer. In all cases, the substitution of a balloon for prey represents deception of the female by the male.

It is useful, if not critical, to have an independent assessment of phylogenetic relationship with which the behavioral series can be compared. For instance, many wasps (order Hymenoptera) capture prey to provide food on which they lay their eggs and on which the larvae subsequently feed. Some species carry the prey anteriorly with the mandibles; other species carry the prey posteriorly, sometimes on a terminal abdominal segment called an "ant clamp" (Evans 1962). The series, with many intermediate forms, correlates well with other evolutionary evidence, and we can be confident of the direction of change. The selection pressure directing the change from anterior to posterior carriage is thought to be cleptoparasitism. When carrying prey with the mandibles, the wasp must put the prey down to open or dig a burrow. Other flies or wasps can then lay eggs on the prey as well and thus diminish the food resource.

Among vertebrates, the evolution of pigmentation in the mouth-breeding cichlid fish (*Haplochromis* spp.) was studied by Wickler (1962). Presumably under the selection pressure of egg predation, some species of cichlid females pick up eggs after spawning (i.e., after the eggs have been laid and externally fertilized) and "incubate" them in the mouth. The faster the females pick up the eggs after laying,

FIGURE 18–6   Regurgitative food offering during copulation
Food offerings during copulation occur in many kinds of animals, such as in the stilt-legged fly (family Micropezidae) shown here. This behavior may be a form of parental investment by the male. By giving the female a protein-rich meal, the male may increase the survival of their offspring.
Source: Photo by E. S. Ross.

the less the chance of predation. In some species the females take the eggs before they are fertilized by the male. The male presents his anal fin, which bears yellow spots resembling eggs; the female snaps at these spots and takes in sperm, which fertilize the eggs in her mouth. Comparative studies have shown that these yellow spots evolved from less conspicuous pearly spots on the vertical fin that have no signaling function. Eibl-Eibesfeldt (1975) has pointed out that the following preadaptations existed:

1. Female readiness to pick up eggs,
2. Pearly spots on the vertical fins of the male,
3. A lateral display by the male during courtship and spawning.

The case in which intermediates are not available is more common and we must draw evidence from related species that have evolved more or less individually from some common ancestor. A weak inference can be made on the basis of the assumption that those behavior patterns common among the various living species represent ancestral traits.

Tinbergen (1959) compared the behaviors of different species of gulls (family Laridae); in particular he examined those patterns that are relatively constant throughout a species and of the **"fixed pattern"** type, that is, they are developmentally stable and resistant to environmental change. The basic similarities in displays throughout the family strengthen the conclusion, previously based on morphology, that the family is **monophyletic,** that is, it originated from a single ancestral type. If a species is judged to be closely related to other species because it shares many of their characteristics, any differences can be considered to have been recently acquired.

A variety of animals construct objects for prey capture, shelter, or mate attraction — for example, spider webs, caddisfly cases and nets, bee hives, bird nests, and bowers. It is relatively easy to examine these semipermanent structures and to compare closely related species. Collias and Collias (1963) studied the nests of African weaverbirds of the family Ploceidae, tracing the evolution of the nest from the ancestral cup nest to intricate domed structures with bottom entrances (Figure 18–7). Primitive members of the family begin the nest as a simple cup or platform; in some species a dome is added later.

In the true weavers (subfamily Ploceini) a ring is made from which the roof is built, followed by the floor. In other species a communal thatched roof is constructed, with brood chambers of many individual pairs underneath. The researchers developed a key to the nests of the true weavers and actually suggested some taxonomic revisions in the family based on nest construction.

The techniques used to weave the nest from grasses also vary among the weavers, and Collias and Collias noted trends toward more regularity in the pattern, tighter weaving, and finer stitches within the subfamily (Figure 18–8). In general, species making the more "advanced" nests used the more refined pattern of weaving. In many species the male builds the basic nest and, along with neighboring males, advertises it to the female, who makes the final choice. Males may make many nests before one is chosen for completion by the female, implicating sexual selection in the evolution of nest construction.

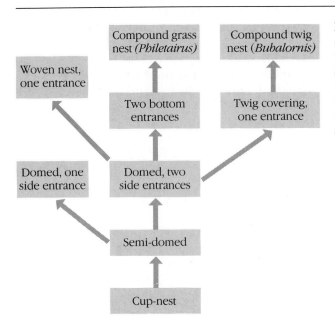

FIGURE 18-7   Main trends in evolution of nest building in the weaverbird family (Ploceidae)
Source: N. E. Collias and E. C. Collias. 1963. Evolutionary trends in nest building by the weaverbirds (Ploceidae). *Proc. XIII Intern. Ornith. Cong., Amer. Ornith. Union,* pp. 518–530.

EVOLUTION OF COMMUNICATIVE BEHAVIORS.   In his study of gulls Tinbergen (1959) concluded that displays are derived movements. For instance, herring gulls (*Larus argentatus*) sometimes peck the ground vigorously and pull grass—behavior that represents displacement collecting of nest material. The movement, however, is combined with attack postures; a bird pulls grass as hard as it would pull at the feathers of an opponent. Tinbergen's interpretation was that the initial movement is an attack movement redirected at the ground, while the grass-pulling behavior (displaced collecting of nest material) is superimposed. Tinbergen believed that finding the origin of this particular pattern requires no comparison with intermediate links because it is so similar to the patterns from which it is derived. However, many displays have undergone **ritualization,** an adaptive evolutionary change away from noncommunicative functions and towards increased efficiency as a signal. In our discussion of communication in Chapter 13, we found that the main sources of these derived movements are: (1) behavior patterns immediately evoked by the situation, such as attack movements triggered by the intrusion of another into one's territory, which are either performed incompletely (intention movements) or oriented at objects other than the evoking stimulus (redirected movements); or (2) movements from patterns other than those evoked by the stimulus, which are unexpected and functionally irrelevant (displacement activities).

BIOGENIC LAW.   Can the development of behavior during an organism's life cycle tell us anything about the evolutionary history of that behavior? If changes in a characteristic tend to be added on late in an individual's development, the earlier

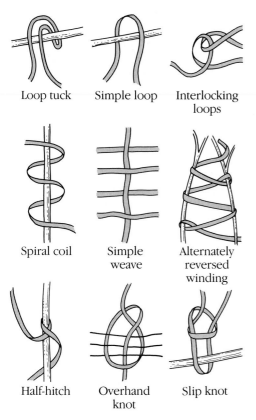

FIGURE 18-8 Types of stitches used by true weaverbirds *(Ploceini)*
Source: N. E. Collias and E. C. Collias. 1963. Evolutionary trends in nest building by the weaverbirds (Ploceidae). *Proc. XIII Intern. Ornith. Cong., Amer. Ornith. Union,* pp. 518–530.

stages could be considered primitive. Young peacocks, for example, use the primitive form of food enticement before developing the adult form (see Figure 13–18). This reasoning led Haekel to his biogenic law: ontogeny recapitulates phylogeny. But, as de Beer (1958) and, more recently, Gould (1977) have pointed out, no developing organism really goes through the actual endpoints of its ancestors. Although the mammal goes through a gill-slit stage during development, it is by no means a fish at that point. Haekel's law has been rejected by nearly all scientists, and its application to the study of the evolution of behavior is unwarranted.

STUDY OF ADAPTATION

Many studies have attempted to illustrate the adaptiveness of behavior by emphasizing the ways in which behaviors ensure an organism's survival and reproduction.

The courtship of ring doves (Chapter 7) is adaptive because it coordinates the physiological processes necessary for reproduction in the male and female. But does such evidence really tell us anything about evolution by natural selection? Natural selection leaves only the best adapted or optimal phenotypes for our inspection.

A common practice is to study an organism's traits or behavior patterns and to propose an adaptive story for each trait considered separately. Behavior patterns or structures found to be suboptimal are considered to be the results of trade-offs among competing demands within the same organism and thus also the result of adaptation. The possibility that present traits are not adaptive to present conditions is usually not considered. Gould and Lewontin (1979), among others, have criticized this approach; they argue that present traits are constrained by phyletic heritage, developmental pathways, and general architecture and are not just the result of current selective forces. As one example, they cite adaptive stories told about the utility of the tiny front legs of the dinosaur *Tyrannosaurus*, legs so small they didn't even reach its mouth. One purported function of such legs was to help the animal rise from a prone position. While the front legs were no doubt used for something, we know them to be homologous to the conventional, full-sized limbs of dinosaur ancestors. Rather than explaining the evolution of short limbs for a specific purpose, one can propose a developmental correlation between the apparent reduction in front leg size and the relative increase in head and hind limb size. The present function of a structure or behavior is not proof that it evolved for that purpose.

We can still test hypotheses critically, however, and make predictions about the way natural selection is supposed to work. A number of researchers are currently comparing behavior patterns with theoretical optima that might be achieved in the best of all possible worlds. The models of optimal foraging discussed in Chapter 16 were constructed to predict strategies based on maximizing some parameter, such as rate of energy intake. The degree to which such models fit data collected in the field demonstrates the importance of natural selection and selective processes in causing changes in gene frequency. The most useful models are those that are **falsifiable,** enabling us to reject hypotheses.

OBSERVING NATURAL SELECTION IN THE FIELD.   The best evidence for the evolution of behavior is documentation in a natural population. Perhaps the strongest case we have so far is the work of Kettlewell (1965) on morphological change in moths, where **melanism,** the development of dark pigmentation, was a result of the darkening of the substrate by soot. As the Industrial Revolution led to increased burning of coal, the bark of trees used as resting places by the moths became darker. Those moths whose darker color matched the background were subject to less predation than were the lighter-colored moths. A behavioral change in background preference must also have occurred so that the moths remained cryptic.

Intense natural selection has been demonstrated in one of the ground finches (*Geospiza fortis*) of the Galápagos (Boag and Grant 1981). On one of the islands, Daphne Major, regular rainfall throughout the early 1970s produced abundant seeds, and seed-eating finches thrived. In 1977 a drought occurred and the finches failed to breed, declining in numbers by 85 percent. The corresponding decline in seed supply was non-random, with small seeds becoming scarce and large, hard seeds remaining relatively common. The result was strong selection for birds with large beaks that could handle the seeds, and within a short period of time the average beak size increased dramatically. Boag (1983) went on to show that beak size is heritable and that offspring showed a strong phenotypic response to natural selection of their parents.

BEHAVIORAL ISOLATING MECHANISMS IN SPECIATION

During the process of speciation, behavioral changes take place that may restrict the exchange of genes between two segments of the population, should they come in contact (see Chapter 4). Such differences serve as isolating mechanisms, which prevent the production of less-fit hybrids that might result from matings between the two populations. Differences associated with mating signals have been studied extensively, and we look at a few examples here.

Lizards of the genus *Sceloporous* (Figure 18–9) are morphologically very similar and perform a species-specific head bob during courtship. Hunsaker (1962) developed a "head-bobbing machine" that moved the heads of models in various species-typical patterns. Females given a choice among models chose those that bobbed in the pattern of their own species (Figure 18–10).

Smith (1966) conducted experiments on the signal function in two natural populations of sympatric but noninterbreeding species of Arctic gulls. The glaucous gull *(Larus hyperboreus)* has a yellowish eye ring; Thayer's gull *(Larus thayeri)* has a reddish purple eye ring. When the eye rings of Thayer's males were painted to look like those of glaucous males, glaucous females paired with them. Once paired,

FIGURE 18-9   Spiny lizard of the genus *Sceloporus* used in head-bobbing experiments
Source: Photo courtesy of the Ohio Department of Natural Resources.

FIGURE 18-10 Head-bobbing patterns of spiny lizards

Males of each species have a distinctive head-bobbing pattern. The height of the bob is shown on the vertical axis and time is shown on the horizontal axis. The female is attracted to the male that bobs in the pattern of that female's species. *Source:* D. Hunsaker, "Ethological Isolating Mechanisms in the *Sceloporus torquatus* Group of Lizards," *Evolution* 16 (1962):62–74. Reprinted with permission.

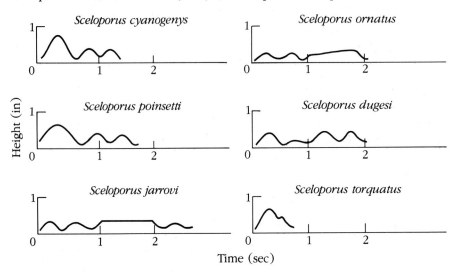

however, the painted Thayer's males refused to copulate with the glaucous females until the latter were painted to look like Thayer's females. Further experiments on other sympatric gull species demonstrated the critical importance of eye-ring color as an isolating mechanism. Such manipulations, which change signals to provide the "wrong" information, are a powerful means of demonstrating signal function.

Isolating mechanisms are important only when two species are sympatric and thus might interbreed. Natural selection leads to greater differences between two closely related species in areas where they overlap than in areas where they do not. Observers have noted this phenomenon, referred to as **character displacement,** primarily in morphological traits, particularly feeding structures, such as bill length in birds. The divergence in bill length is greatest in areas of sympatry because of intense competition between the closely related species: each specializes on a different resource. We can observe a similar effect for behavioral traits. Littlejohn (1965) recorded the songs of two species of Australian tree frogs, *Hyla ewingi* and *Hyla verreauxi.* In zones of sympatry the songs appear to be quite distinct; in zones of allopatry the songs appear to be similar (Figure 18–11). In playback experiments Littlejohn and Loftus-Hills (1968) gave gravid females a choice of two songs. They chose their own species' song when the choice was between the two species in sympatry, but showed no preference between songs of allopatric species. (See Chapter 4 for further examples.)

TRADITION

Although the evolution of persistent changes in form and behavior usually has a genetic basis, natural selection may act to produce both plasticity of expression and the tendency to transmit behavior to offspring through nongenetic means. For example, experiences of female rats can bias the behavior of offspring and even grandoffspring (Denenberg and Rosenberg 1967). The physiological mechanisms that cause these changes in behavior are poorly understood, but the effects of environment on the behavior of subsequent generations probably are more widespread and important in producing persistent changes in behavior than are generally recognized.

We have a better understanding of the role of **tradition,** the passing of specific forms of behavior from generation to generation by learning. Compared with genetically controlled behavior, tradition-transmitted behavior can spread through a population rapidly, sometimes in less than one generation. One common form is dialect in animal communication; birdsong has received the most attention. Males learn song during a critical period, when they imitate phrases heard in their neighborhood. Male immigrants into a local area may adopt the local dialect. The

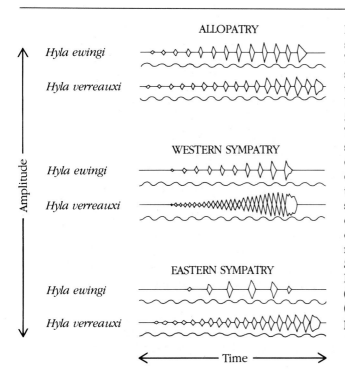

FIGURE 18-11 **Oscillograms of songs of two Australian frog species** The amplitude of each note of the song is shown on the vertical axis, and time is shown on the horizontal axis. Under each song recording is a 50-cycles-per-second reference line. The songs of the two species are similar when comparing individuals caught in areas where the two species do not overlap (allopatry). Where the two species do overlap (sympatry), song differences are pronounced. This divergence of behavior between closely related sympatric species is referred to as character displacement. *Source:* M. J. Littlejohn, "Premating Isolation in the *Hyla ewingi* Complex (Anura: Hylidae)," *Evolution* 19 (1965):234–243. Reprinted with permission.

adaptive significance of this phenomenon could be to enhance communication between male and female in pair bonding. It may also maintain the separateness of locally adapted populations (see Chapter 13).

Migratory animals tend to use the same routes and return to the same places as their ancestors. Presumably the young learn the routes from their elders. Waterfowl, for example, use traditional routes and stop at the same places each year during migration, and grazing mammals use the same trails for many generations. Breeding grounds may also be used year after year; in lek species the mating arenas may persist for centuries. Tradition is particularly important in species living in closed social groups with prolonged socialization of young.

Most examples of the social transmission of acquired behavior are from insects, birds, and mammals (reviewed by Galef 1976 and by Bonner 1980). Among coral reef fish, French grunts *(Haemulon flavolineatum)* exhibit social traditions of daytime schooling sites and twilight migration routes (Helfman and Schultz 1984). Individuals transplanted to new schooling sites and allowed to follow residents used the new migration routes and returned to the new sites even in the absence of resident fish. Naive, control grunts released at the same new sites failed to use the new routes or to return to the new sites.

Nonhuman primates can develop new behavior patterns through invention and incorporate them rapidly and rather permanently into their social systems. Chapter 10 discussed the techniques for washing sand from potatoes and wheat grains introduced by a Japanese macaque *(Macaca fuscata)* and copied by others in the group.

Another type of learned behavior that occasionally spreads through populations is tool use. Tools enable otherwise generalized species to specialize in particular tasks without evolving morphological adaptations; for example, animals may use sticks or similar objects, rather than specialized beaks, to extract insects from crevices in tree bark.

Although the mechanism of behavioral change involved in tradition and human culture usually is assumed to be quite different from the mechanism of genetic change brought about by natural selection and drift, learned responses can be highly stereotyped and may persist for many generations. The big difference between learned and genetically controlled traits is the speed with which learned traditions can spread through the population, most notably in human societies. An old controversy, which sociobiology has recently kindled anew, concerns the extent to which differences in human social behavior are influenced by genetics and by culture.

Durham (1979), among others, has attempted a synthesis of human cultural and genetic evolution and has argued for the coevolution of genes and culture. During the organic evolution of *Homo sapiens*, according to Durham, there was selection at the level of the individual for the capacity to modify and extend phenotypes on the basis of learning and experience. As the capacity for culture evolved, the developing culture in each population was adaptive for the members of that population and increased their inclusive fitnesses. The process was further accelerated by group-level cultural selection, which operates analogously to natural selection and independently of it. Competition among cultures results in selection of the "more fit" cultures, those that best exploit the environment to their own advantage (Dawkins 1976). Durham emphasizes, however, that cultural "fitness" is dependent on the reproductive success of the individuals that make up a culture.

SUMMARY

Behavior is that part of the phenotype most likely to change in response to environmental change; behavior is thus the evolutionary pacemaker. European ethologists were the first to treat behavior patterns like anatomical features, "dissecting" them to learn their structure and thus to infer their evolution. Homologous behaviors are those shared by species through descent from a common ancestor.

Evidence for the evolution of behavior comes from a variety of sources. Fossils occasionally tell us about behavior, though behavior must often be inferred from morphology. Adaptive radiation, the rapid appearance of new forms, as occurred on the remote islands of the Galápagos archipelago, permits comparison among closely related species and with the presumed ancestral type. Domestication shows the potential for modifying behavior and morphology, where humans, rather than nature, are the selective agents. Through comparisons among closely related species, we can sometimes trace the probable course of evolution from one extreme to another when the intermediate forms are still extant. Ethologists believe that communicative behaviors, or displays, have evolved from noncommunicative behaviors of several types, including intention movements, redirected movements, and displacement activities. The adaptiveness of behavior does not, by itself, demonstrate evolution, but observers can test predictive models of the way natural selection operates. Direct observation of evolution by natural selection is strong evidence, but it is rarely documented. Two examples are industrial melanism in moths and beak size in one species of Darwin's finches.

The evolution of behavior plays a crucial role in the restriction of genetic exchange between populations in the formation of new species. We are most familiar with the species-specific mating patterns that function as species-isolating mechanisms. Character displacement occurs in areas where closely related species overlap geographically. In such areas of sympatry, the species differ more than where they are allopatric (nonoverlapping).

Natural selection may act to produce plasticity of expression and the tendency to transmit behavior to others through nongenetic means, such as learning. Manipulating the maternal environment can produce behavioral changes in offspring and grandoffspring. Song dialects and knowledge of migration routes are examples of tradition, passed to others by learning. Tool use and novel feeding techniques, as seen in nonhuman primates, illustrate the speed with which these potentially permanent changes in behavior can spread through a population. The evolution of culture in humans initially involves natural selection for the ability to extend the phenotype on the basis of learning and experience. Cultures themselves may evolve in a manner analogous to natural selection.

Discussion Questions

1. Discuss some examples of homologous behavior patterns other than those already mentioned. Can you think of behaviors that are analogous—that is, behaviors that are similar between species but *not* because of common ancestry?

2. Discuss an example of what Gould and Lewontin (1979) have referred to as "adaptive stories" to explain the function of a behavior pattern. How could factors other than evolution by natural selection have led to such behavior?

3. Do you believe creation should be discussed as an alternative theory to evolution in public school science classes? For further discussion see Alexander (1978) and references contained therein.

4. The following table provides measurements (in mm) of skull length in closely related species of weasels and the culmen (the top ridge of a bird's beak) length in closely related species of nuthatches and finches. Both features reflect feeding habits. What conclusions can you draw about the role of competition in the evolution of these species?

	Sympatry			Allopatry		
	Male	Female	Both Sexes	Male	Female	Both Sexes
Weasels (skull length)						
Mustela nivalis	39.3	33.6		42.9	34.7	
Mustela ermina	50.4	45.0		46.0	41.9	
Nuthatches (culmen length)						
Sitta tephronota			29.0			25.5
Sitta neumayer			23.5			26.0
Finches (culmen length)						
Geospiza fortis			12.0			10.5
Geospiza fuliginosa			8.4			9.3

Source: Data from G. E. Hutchinson, "Homage to Santa Rosalia *or* Why Are There So Many Kinds of Animals?" *American Naturalist* 93 (1959):145–159.

Suggested Readings

Alexander, R. D. 1962. Evolutionary change in cricket acoustical communication. *Evolution* 16:443–467.
Using the comparative method Alexander reconstructs the evolution of cricket songs from courtship movements of the male.

Brown, J. L. 1975. *The Evolution of Behavior.* New York: Norton.
An advanced text that emphasizes the adaptedness of
behavior. Draws heavily on examples from the literature.

Gould, S. J., and R. C. Lewontin. 1979. The spandrels of San Marco and the Panglossian paradigm: A critique of the adaptationist programme. *Proc. Roy. Soc. Lond.*
B205:581–598.
Argues against the tendency of biologists to explain every
morphological and behavioral trait as an adaptation to
current environments.

References

Alexander, R. D. 1978. Evolution, creation, and biology teaching. *Am. Biol. Teach.* 40:91–107.

Alexander, R. D., J. L. Hoogland, R. D. Howard, K. M. Noonan, and P. W. Sherman. 1979. Sexual dimorphisms and breeding systems in pinnipeds, ungulates, primates and humans. In N. A. Chagnon and W. Irons (eds.), *Evolutionary Biology and Human Social Behavior: An Anthropological Perspective.* N. Scituate, Mass.: Duxbury.

Boag, P. T. 1983. The heritability of external morphology in Darwin's ground finches *(Geospiza)* on Isla Daphne Major, Galápagos. *Evolution* 37:377–384.

Boag, P. T., and P. R. Grant. 1981. Intense natural selection in a population of Darwin's finches *(Geospizinae)* in the Galápagos. *Science* 214:82–85.

Bonner, J. T. 1980. *Evolution of Culture in Animals.* Princeton: Princeton University Press.

Brown, J. L. 1975. *Evolution of Behavior.* New York: Norton.

Collias, N. E., and E. C. Collias. 1963. Evolutionary trends in nest building by the weaverbirds *(Ploceidae). Proc. XIII Intern. Ornith. Cong., Amer. Ornith. Union,* pp. 518–530.

Dawkins, R. 1976. *The Selfish Gene.* New York: Oxford University Press.

de Beer, G. 1958. *Embryos and Ancestors,* 3rd ed. London: Oxford University Press.

Denenberg, V. H., and K. M. Rosenberg. 1967. Nongenetic transmission of information. *Nature* 216:549–550.

Durham, W. H. 1979. Toward a coevolutionary theory of human biology and culture. In N. A. Chagnon and W. Irons (eds.), *Evolutionary Biology and Human Social Behavior: An Anthropological Perspective.* N. Scituate, Mass.: Duxbury.

Eibl-Eibesfeldt, I. 1975. *Ethology: The Biology of Behavior,* 2nd ed. New York: Holt, Rinehart and Winston.

Ekdale, A. A. 1980. Graphoglyptid burrows in modern deep-sea sediment. *Science* 207:304–306.

Evans, H. E. 1962. The evolution of prey-carrying mechanisms in wasps. *Evolution* 16:468–483.

Fürsich, F. T., and D. Jablonski. 1984. Late Triassic naticid drillholes: Carnivorous gastropods gain a major adaptation but fail to radiate. *Science* 224:78–80.

Galef, B. G. 1976. Social transmission of acquired behavior: A discussion of tradition and social learning in vertebrates. *Adv. Stud. Behav.* 6:77–100.

Gould, S. J. 1977. *Ontogeny and Phylogeny.* Cambridge, Mass.: Harvard University Press.

Gould, S. J., and N. Eldredge. 1977. Punctuated equilibria: The tempo and mode of evolution reconsidered. *Paleobiology* 3:115–151.

Gould, S. J., and R. C. Lewontin. 1979. The spandrels of San Marco and the Panglossian paradigm: A critique of the adaptionist programme. *Proc. Roy. Soc. Lond.* B205:581–598.

Hale, E. B. 1969. Domestication and the evolution of behavior. In E. S. E. Hafez (ed.), *Behavior of Domestic Animals,* 2nd ed. London: Balliere, Tindall, and Cassell.

Helfman, G. S., and E. T. Schultz. 1984. Social

transmission of behavioral traditions in a coral reef fish. *Anim. Behav.* 32:379–384.

Hunsaker, D. 1962. Ethological isolating mechanisms in the *Sceloporus torquatus* group of lizards. *Evolution* 16:62–74.

Hunt, R. M., X. Xiang-Xu, and J. Kaufman. 1983. Miocene burrows of extinct bear dogs: Indication of early denning behavior of large mammalian carnivores. *Science* 221:364–366.

Hutchinson, G. E. 1959. Homage to Santa Rosalia *or* why are there so many kinds of animals? *Amer. Nat.* 93:145–159.

Kawai, M. 1965. Newly acquired pre-cultural behavior of the natural troop of Japanese monkeys on Koshima Islet. *Primates* 6:1–30.

Kessel, E. L. 1955. The mating activities of balloon flies. *Syst. Zool.* 4:97–104.

Kettlewell, H. B. D. 1965. Insect survival and selection for pattern. *Science* 148:1290–1296.

Lack, D. 1961. *Darwin's Finches: An Essay on the General Biological Theory of Evolution.* New York: Harper and Brothers. First published in 1947 by Cambridge University Press, Cambridge, England.

Littlejohn, M. J. 1965. Premating isolation in the *Hyla ewingi* complex (Anura: Hylidae). *Evolution* 19:234–243.

Littlejohn, M. J., and J. J. Loftus-Hills. 1968. An experimental evaluation of premating isolation in the *Hyla ewingi* complex. *Evolution* 22:659–663.

Lorenz, K. Z. 1972. Comparative studies on the behavior of Anatinae. In P. H. Klopfer and J. P. Hailman (eds.), *Function and Evolution of Behavior: An Historical Sample from the Pens of Ethologists.* Reading, Mass.: Addison-Wesley.

Manning, A. 1971. Evolution of behavior. In J. L. McGaugh (ed.), *Psychobiology: Behavior from a Biological Perspective.* New York: Academic Press.

Price, E. O. 1984. Behavioral aspects of animal domestication. *Quart. Rev. Biol.* 59:1–32.

Ratner, S. C., and R. Boice. 1975. Effects of domestication on behaviour. In E. S. E. Hafez (ed.), *Behaviour of Domestic Animals,* 3rd ed. Baltimore, Md.: Williams & Wilkins.

Raup, D. M., and A. Seilacher. 1969. Fossil foraging behavior: Computer simulation. *Science* 166:994–995.

Scott, J. P., and J. L. Fuller. 1965. Genetics and the social behavior of the dog. Chicago: University of Chicago Press.

Seilacher, A. 1967. Fossil behavior. *Sci. Amer.* 217 (2):72–80.

Smith, N. G. 1966. Evolution of some arctic gulls (*Larus*): An experimental study of isolating mechanisms. *Ornith. Monogr.* 4:1–99.

Thornhill, R. 1976. Sexual selection and paternal investment in insects. *Amer. Nat.* 110:153–163.

———. 1979. Adaptive female-mimicking behavior in a scorpionfly. *Science* 205:412–414.

Tinbergen, N. 1959. Comparative studies of the behavior of gulls (Laridae): A progress report. *Behaviour* 15:1–70.

Van Tets, G. F. 1965. A comparative study of some social communication patterns in the Pelecaniformes. *Ornith. Monogr.* 2:1–88.

Wickler, W. 1962. Ei-Attrapen und Maulbrüten bei afrikanischen Cichliden. *Zeit. Tierpsychol.* 19:129–164.

Williams, G. C. 1966. *Adaptation and Natural Selection: A Critique of Some Current Evolutionary Thought.* Princeton: Princeton University Press.

Wilson, E. O. 1975. *Sociobiology: The New Synthesis.* Cambridge: Harvard University Press.

19

EVOLUTION OF SOCIETIES

Two of the most rapidly advancing areas of animal behavior studies are at opposite ends of the organizational spectrum. At one end is **psychobiology,** in which the main goal is understanding the nature of the central nervous system. At the other end is **sociobiology,** study of the biological basis of social behavior.

By the 1960s the study of animal societies increasingly emphasized field studies, particularly of insects, birds, and nonhuman primates, and focused on the relationship between social structure and ecology. Although animal behaviorists were always concerned with the apparent adaptiveness of particular social systems, they did not investigate to any great extent the population genetics of how sociality actually evolved. Many researchers sensed a potential conflict between Darwinian evolution, in which selection acts on individuals, and the existence of sophisticated cooperative behavior. They assumed that group, population, or species benefits likely outweighed individual sacrifice. More recently, observers consider society to be an outgrowth of the action of natural selection at the level of the individual or gene. (See chapters 4 and 17 for a further discussion of group selection.)

We can define a **society** as a group of individuals of the same species that is organized in a cooperative manner extending beyond sexual and parental behavior. In this chapter we briefly review examples of complex sociality in several taxonomic groups, discuss the evolutionary advantages and disadvantages of social behavior, consider in some detail the way natural selection has brought about the evolution of social systems, and finally mention some applications of sociobiology to our understanding of human social behavior.

EXAMPLES OF COMPLEX SOCIAL SYSTEMS

Sociality has evolved independently in many groups of animals, ranging from invertebrates to primates. We might expect an increase in complexity of social behavior as we move from simpler to more sophisticated organisms, yet by some criteria the opposite is true; some of the simplest invertebrates have more complex societies than do birds or mammals.

COLONIAL INVERTEBRATES

Cooperation, which is the aggregation of individuals for mutual benefit, and interdependence are so highly developed in many invertebrate forms that deciding what constitutes an individual and what constitutes a society presents a problem. Hydrozoans of the phylum Coelenterata (Cnidaria) are a striking example of a colonial system. Individuals are physically united in colonies and specialize as reproductive or nonreproductive types. The basic unit is the **polyp,** which consists of two cell layers that form a tubular body with a central mouth surrounded by soft tentacles. Although one genus *(Hydra)* consists of a single, free-living polyp, most of the genera are colonial, and the polyps (also called **zooids**) show a division of labor. In genera such as *Obelia,* some zooids feed and defend by means of stinging cells, called **nematocysts,** and others reproduce. Another form, which lives on the shells of hermit crabs, has three types of zooids: feeding, reproducing, and fighting. A fourth type of zooid, found in the Portuguese man-of-war *(Physalia)* (Figure 19–1), produces a gas-filled float.

    In deciding whether or not an organism like *Physalia* is a society or a single individual, we should note that the entire colony comes from a single fertilized egg (zygote). The different zooids form by **budding off** asexually. Since all the zooids are genetically identical, the man-of-war may be considered a single individual with several organ systems. However, in comparing other Hydrozoans showing varying degrees of coloniality, it appears that the formation of complex colonies originated from the grouping of many individual zooids; in some species the zooids forming the colony are unrelated (Wilson 1975).

    From the standpoint of the individual, why have some zooids evolved into feeding or fighting structures with no reproductive potential? Marine invertebrates would seem to gain several advantages from colonial life (Wilson 1975). First, many colonial invertebrates live in shallow-water coastal areas, where sediment and wave action would devastate a single hydroid polyp. Colonies such as corals have a calcareous base that anchors and elevates individuals above the ocean floor. Second, colonial forms such as *Physalia* are pelagic, or oceanic. Singly, the polyps are restricted largely to underwater surfaces, but, by teaming up with those that form the gas-filled float, the colony can travel on the high seas. Third, the large colonial invertebrates that are sessile can smother or "shade out" competing forms. Fourth,

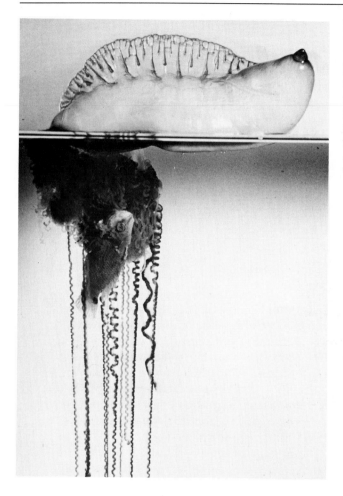

FIGURE 19-1 Portuguese man-of-war eating a fish
This colonial invertebrate of the genus *Physalia* is actually a group of individuals. Four types of polyps (zooids) make up the colony, with each type specializing in either flotation, feeding, defense, or reproduction.
Source: Photo © D. P. Wilson from Eric and David Hosking.

the concentration of zooids that bear nematocysts is vey large in some forms, such as *Physalia.*

SOCIAL INSECTS

Social complexity rivaling that of colonial coelenterates occurs in some groups of insects. Three traits characterize the truly social, or **eusocial,** insects: (1) cooperation in the care for young; (2) reproductive castes, with nonreproductives caring for reproductive nestmates; (3) overlap between generations such that offspring assist parents in raising siblings. Intermediate forms exist; thus quasisocial forms have only trait 1, and semisocial forms have traits 1 and 2. Eusociality among insects is limited to termites, ants, social wasps, and social bees (Wilson 1975).

FIGURE 19-2   Polyrhachis ant colony
Visible are workers, larvae, and pupae. In these eusocial insects, members of the worker caste are sterile females that care for their mother, the queen, and help to rear their siblings.
Source: Photo by Edward S. Ross.

Most of the eusocial insects are in the order Hymenoptera (wasps, bees, and ants). As an example we look briefly at ants, which, in terms of distribution and numbers, are the dominant social insects. (Chapter 16 discusses army ants and fungus-growing ants.) Ant colonies (Figure 19–2) range in size from a few hundred to thousands of individuals. The queens are winged when they emerge from the pupae; the workers, always female, are smaller and wingless. Often the winged males and queens engage in mass nuptial flights away from the nest; they swarm on prominent landmarks and the males attempt to copulate. Once the queen is inseminated, she sheds her wings, excavates a cell in the soil, and rears the first brood of workers. In the more advanced species of ants, only one queen is found in a nest. Workers typically come in several sizes: some forage, others care for the next brood, and soldiers defend the colony. Workers of many species lay **trophic** eggs, which are nonfertile eggs that nourish other workers and the queen. Males, which come from unfertilized eggs, disperse, mate with queens, and then die.

Communication is critical to the organization of such colonies. Wilson (1975) lists the following responses as typical of social insects:

1. Alarm;
2. Simple attraction (multiple attraction = "assembly");
3. Recruitment, as to a new food source or nest site;
4. Grooming, including assistance at molting;
5. Trophallaxis (the exchange of oral and anal liquid);
6. Exchange of solid food particles;
7. Group effect: either increasing a given activity (facilitation) or inhibiting it;
8. Recognition, of both nestmates and members of particular castes;
9. Caste determination, either by inhibition or by stimulation.

Eusociality is not restricted entirely to the Hymenoptera. Termites, of the order Isoptera, are eusocial insects from a totally different ancestry, although they have many similarities with eusocial ants (Figure 19–3). Most of the species are in the family Termitidae. Their energy source is wood cellulose, which is digested by flagellate protozoans that live in the intestinal tracts of the termites. Winged forms of termites disperse from the colony in a mass exodus. A male and female pair off and together begin construction of a new nest; they do not copulate until later. In early broods, workers of both sexes are produced; soldiers appear in later broods; and winged reproductives are the last to appear, sometimes many years after the founding of the colony. Although termites do not produce trophic eggs, they frequently exchange liquid oral and anal food, the latter being necessary for transmission of the protozoans. Like social hymenoptera, termites use pheromones in colony organization, territoriality, and the laying of trails.

VERTEBRATES

FISH. Nothing approaching coelenterate colonies or insect eusociality occurs in the cold-blooded vertebrates; schooling is perhaps the most conspicuous form of social behavior. Complex schooling behavior in fish that live in open water provides

---

FIGURE 19–3   Queen termite surrounded by workers
The reproductive male ("king") of the genus *Macrotermes* is in the upper left.
Termites are the only eusocial insects not in the order Hymenoptera.
Source: Photo by Edward S. Ross.

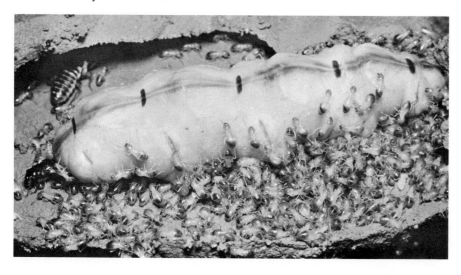

protection from predators, since single prey are picked off more easily than are grouped prey (Shaw 1978). Other advantages to schooling seem to be increased ability to locate patchy food resources, conservation of energy because heat is generated, creation of currents that facilitate locomotion, and increased likelihood of finding mates. Other types of social behavior in fish revolve around defense of nest sites and do not involve nonparents. Cooperation by nonparents may occur in one species of Antarctic fish, in which a male may take over the guarding of eggs if the female disappears (Daniels 1979). Daniels argues that such behavior is altruistic, since there is little likelihood that that particular male is the father.

AMPHIBIANS AND REPTILES. Amphibians and reptiles demonstrate many complex social interactions, but most are related to territory defense and obtaining mates. Large groupings are uncommon except in hibernacula (overwintering aggregations) and in mating choruses of some frogs and toads. Bullfrog *(Rana catesbeiana)* males establish and defend individual territories in chorusing areas in one part of a pond. A gravid female enters a territory and is grabbed by the resident male who fertilizes the eggs externally as the female lays them within that territory (Howard 1978). A small number of the larger males do most of the mating; and eggs laid in the territories of these males have higher hatching success, since the eggs are subject to less predation by leeches than are eggs laid in territories of smaller males. Some males use alternate mating strategies. For example, a parasitic male does not call but remains in the territory of a large male and grabs the female before she reaches the resident male (Figure 19–4). Others, referred to as opportunistic males, call from specific areas and act like territorial males; if challenged, however, they flee to another area and resume calling.

BIRDS. Many species of birds aggregate in feeding and roosting flocks; and well-organized breeding colonies are common. However, a majority of birds are monogamous and territorial breeders, and even those that breed in colonies do not approach the division of labor that is characteristic of eusocial insects. Perhaps the most complex social systems are those involving cooperative breeding, in which nonparents share in the rearing of young. Two types of cooperative breeding have been noted: communal nesting and helpers-at-the-nest.

Communal Nesting. Members of the cuckoo family (Cuculidae) are best known for their practice of laying eggs in the nests of other species and letting the host rear the cuckoo young, but a few species nest communally, with several females raising young in the same nest. Davis (1942), who studied some of the tropical New World cuckoos, reported a phylogenetic series within the subfamily. One species *(Guira guira)* nests colonially or in isolated pairs; another species, the greater ani *(Crotophaga major)*, always nests in colonies. In both species the flock is composed of mated pairs. The smooth-billed ani *(Crotophaga ani)* is promiscuous, and several females contribute to the same clutch. Territorial behavior in this species has evolved from defense of small areas by mated pairs to vigorous flock defense of the commu-

FIGURE 19-4   Territorial male bullfrog with parasitic male
The parasitic male is the smaller of the two and is facing forward. Females are attracted to the call of the territorial male; when a female enters the territory, the parasitic male attempts to mate with her before the territorial male has a chance.
Source: Photo by Richard D. Howard.

nal feeding and scarce nesting areas. Davis argued that flock defense evolved in a patchy environment, with clumps of trees scattered as islands in grasslands. Wilson (1975) more recently pointed out that small breeding populations and genetic isolation may be involved in the evolution of this type of cooperative living.

Vehrencamp (1977) studied groove-billed anis *(Crotophaga sulcirostris)* in Costa Rica and found that considerable competition takes place among adults. Dominant females tend to be the last to lay their eggs in the communal nest and throw out the eggs of subordinates. A subordinate female sometimes sneaks an egg into the nest later; the egg then has a better chance of surviving. The high-ranking females do relatively little incubating and let subordinate females and males do most of the work.

Helpers-at-the-Nest.   A more common type of cooperation in birds occurs when nonparents assist in rearing offspring of a single pair. Jays, New World members of the family Corvidae, have various grades of sociality (Figure 19–5). The California scrub jay *(Aphelocoma coerulescens)* exhibits the typical system, with a pair of birds defending a territory and rearing young without help from others. Another species, the piñon jay *(Gymnorhinus cyanocephalus),* has developed colonial nesting, a system associated with a reduction of individual territory, probably in response to unpredictable food supplies (Brown 1974). As many as two hundred adult pairs nest in clusters. Each pair defends only a small space around the nest. Colony members feed in dense flocks. Possible intermediate forms are the Steller's jays *(Cyanocitta stelleri)* and blue jays *(C. cristata),* which do not nest in colonies but have widely overlapping and undefended home ranges. A second departure from the typical system is toward the maintenance of pair territories, but with cooperative breeding,

FIGURE 19-5　Two pathways to advanced sociality in jays
According to Brown (1974) the original breeding system, as seen in California scrub jays, is one of territory defense by mated pairs. The upper route leads to colonial nesting, with mated pairs sharing a large home range and defending only a small area around the nest. The lower route leads to cooperative breeding, with nonparents helping to rear offspring.
Source: Data from J. L. Brown, "Alternate Routes to Sociality in Jays—With a Theory for the Evolution of Altruism and Communal Breeding," *American Zoologist* 14 (1974):63–80.

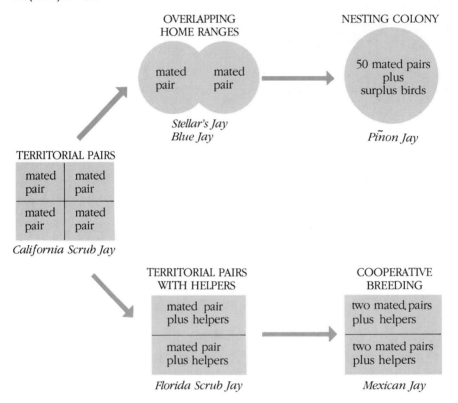

where birds other than the parents assist in feeding the young and defending the territory (Figure 19–6).

　　Florida scrub jay *(Aphelocoma coerulescens)* helpers really do help (Woolfenden 1975; Stallcup and Woolfenden 1978). They found that pairs with one or more helpers reared significantly more offspring than those without (Table 19–1). The total amount of food brought to the nest did not differ between the pairs with helpers and those without. Helpers, however, brought about 30 percent of the food, thereby reducing the reproductive effort of the parents and increasing the parents' chances

FIGURE 19-6 Scrub jays in nest with helper nearby
Helpers are usually older siblings of the nestlings; they help feed young and defend the nest against predators such as snakes. (Note that three adults are present.)
Source: Photo by Glen E. Woolfenden.

TABLE 19-1 Reproductive success of Florida scrub jays for each of five breeding seasons for pairs without and with helpers

Year	Number of Pairs		Fledglings per Pair ($\bar{x}$)		Independent Young per Pair ($\bar{x}$)	
	Without Helper	*With Helper*	*Without Helper*	*With Helper*	*Without Helper*	*With Helper*
1969	2	5	0	2.6	0	2.0
1970	8	8	2.0	3.5	1.3	1.8
1971	6	19	1.3	2.1	1.0	1.2
1972	13	17	0.3	1.6	0.1	1.1
1973	18	10	1.3	1.8	0.4	1.1

Source: G. E. Woolfenden, "Florida Scrub Jay Helpers at the Nest," *Auk* 92 (1975):1–15. Reprinted with permission.

of surviving to produce more offspring in the future. An additional advantage of helpers is the increased vigilance and mobbing protection against predatory snakes.

Mexican jays *(Aphelocoma ultramarina)* live in flocks of eight to twenty birds, with two or more breeding pairs. Nestlings are fed by all members of the group; parents do only about half the feeding. Through long-term banding studies Brown and Brown (1981) found that groups contained grandparents, uncles, aunts, and cousins in addition to parents and offspring. In one unit grandchildren helped rear their grandfather's offspring.

Helping has been reported in many other species of birds. Zahavi (1974) argued that the helpers in the groups of Arabian babblers *(Turdoides squamicaps)* he studied did more harm than good; they got in the way of the parents and attracted predators to the nest. However, in an experimental test of whether helpers really help, Brown and colleagues (1982) removed helpers from breeding groups of grey-crowned babblers, an Australian species *(Pomatostomus temporalis)*. Intact control groups with a full complement of helpers produced an average of 2.4 young, while

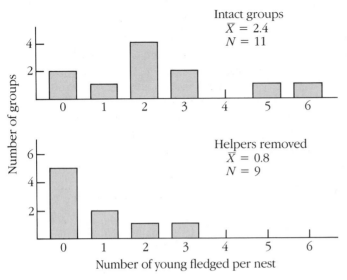

FIGURE 19-7 **The effect of artificial removal of helpers upon reproductive success in groups of grey-crowned babblers**
Source: Drawing from S. T. Emlen (1984), Cooperative breeding in birds and mammals. In J. R. Krebs and N. B. Davies (eds.), *Behavioural Ecology: An Evolutionary Approach.* Sunderland, Mass.: Sinauer Associates.

the depleted groups produced an average of 0.8 young (Figure 19–7). Evidence that helpers actually help does not, however, explain what benefit the helpers receive and why they don't go off and have young of their own. We will deal with this problem shortly.

MAMMALS. In contrast to birds, which are mostly monogamous, mammals tend to be either solitary, where the most complex social unit is a mother with her young, or polygynous, where males monopolize several females each and harem formation is common (Eisenberg 1981). Complex social organization has evolved in some species in nearly all mammalian orders, especially marsupials, carnivores, ungulates, and primates. Most of these systems are organized matrilineally; mothers and offspring may stay together, and groups are thus composed of mothers, daughters, sisters, aunts, and nieces. Because of the prevalence of polygyny and the associated tendency of males to disperse as they reach sexual maturity, researchers tend to ignore relationships among males.

The most complex societies have evolved among mammals that live in relatively open habitats; grouping seems to function largely as defense against predators and protection of resources through group territory. Although cooperative rearing of young — that is, nonmothers nursing or providing food for young — is not common, it does occur in social carnivores. For example, lionesses share the nursing of cubs in the pride, and subordinate wolves regurgitate food for the alpha female and her litter.

In the banded mongoose (*Mungos mungo*) of Africa several females breed synchronously, giving birth in a communal den and nursing each other's offspring. Although the dwarf mongoose (*Helogale parvula*) is monogamous, it lives in packs of about ten individuals. As in wolves, breeding is suppressed among subordinate females. These non-reproductives lactate and help nurse the young of the dominant female (Rood 1980).

Rodents such as house mice and white-footed mice may form communal nests with several litters of different ages present. Jarvis (1981) recently demonstrated the first case of mammalian eusociality in the naked mole rat *(Heterocephalus glaber)*. She captured forty members of a colony from their burrow system in Kenya and studied them for six years in an artificial burrow system in the laboratory. Only one female in the colony ever had young; mother and young were fed but not nursed by male and female adults of the worker caste; members of this caste were not seen to breed. Another caste of non-workers assisted in keeping the young warm; males of this caste bred with the female reproductive. This species fits our criteria for eusociality since there is cooperative care of young, there is a reproductive caste, with nonreproductives caring for reproductives, and there are offspring assisting parents in rearing siblings (Figure 19–8).

In the large, multimale groups found in many Old World monkeys and apes, cooperation in caring for the young does not extend to nursing or giving food to offspring other than one's own. We do not see the division of labor characteristic of colonial invertebrates or eusocial insects in any of the primates, but social roles are characteristic of some species. In rhesus monkeys *(Macaca mulatta)* the highest-ranking males act as control animals (Bernstein and Sharpe 1966). These males protect the group against serious extragroup challenges and reduce intragroup conflict by intervening in fights among group members (Vessey 1971). Since learning and early experience play a large part in determining social structure in primates, flexibility is perhaps the hallmark that sets primates apart from other groups of mammals.

FIGURE 19-8 The naked mole rat *(Heterocephalus glaber,* **Bathyergidae) of East Africa**
These fossorial rodents live in complex social groups approaching the eusociality of Hymenopteran insects.
Source: Photo by Graham C. Hickman, courtesy of the Mammal Slide Library.

EVOLUTIONARY ADVANTAGES AND DISADVANTAGES OF LIVING IN GROUPS

Although in many cases no strong evidence exists that links a particular selection pressure to the evolution of social behavior, here we list advantages that have been suggested over the years:

1. Protection from physical factors. For example, bobwhite quail *(Colinus virginianus)* survive low temperatures better when grouped than when isolated (Gerstell 1939).

2. Protection against predators. Detection of danger and alarm communication are faster and predator deterrence is enhanced by mobbing and group defense. Examples include prairie dogs *(Cynomys* spp.) (Hoogland 1979a), Belding's ground squirrels *(Spermophilus beldingi)* (Sherman 1977), and bank swallows *(Riparia riparia)* (Hoogland and Sherman 1976). (See also Chapter 16.)

3. Assembling members of sexual species together for location of mates. For example, mating swarms are common in insects and among some vertebrates (see Chapter 11).

4. Finding resources. Laboratory studies demonstrate that fish in larger schools find food faster than those in smaller schools (Pitcher, Magurran, and Winfield 1982). Communal roosts may serve as "information centers" where unsuccessful foragers learn the location of food sources by following successful foragers from the roost back to the source (Waltz 1982). It is not yet clear, however, how the unsuccessful foragers know which birds to follow; perhaps they observe the engorged crops of the successful foragers. A related advantage is created by tradition; knowledge about resource location can be transmitted to subsequent generations as in sheep *(Ovis canadensis)* (Geist 1971). (See also Chapter 16.)

5. Group defense of resources against conspecifics or competing species. Many examples of group territoriality are included here, such as the anis studied by Davis (1942) and by Vehrencamp (1977). Among invertebrates, large colonies may have a competitive advantage over smaller groupings. The bryozoan *Bugula turrita* occurs in dense colonies on pilings and rocks in shallow water along the coasts of North America. Another species, *Schizoporella errata*, frequently overgrows the more flexible colonies of *B. turrita*. Larvae of the latter species group together when forming a colony. The resulting dense colonies are less likely to be overgrown by *S. errata* (Buss 1981).

6. Division of labor among specialists. Roles in complex societies, most pronounced in colonial invertebrates and in the castes of eusocial insects, would seem to give one population a competitive advantage over another

lacking such divisions. Division of labor can include **mutualism,** in which two species cooperate.

7. Richer learning environment for young that develop slowly. This advantage is frequently suggested as important for mammals in general and primates in particular. Dependence on learning provides great plasticity but requires a long period of physiological and psychological dependence on others.

8. Population regulation. Wynne-Edwards (1962) hypothesized that the evolutionary significance of social behavior is that organisms are able to track resources in the environment more efficiently. Intraspecific competition became ritualized into contests whose intensity was proportional to the supply of the limiting resource. Because the result of such competition leads to reduced reproductive success of those participating, Wynne-Edwards believed that natural selection must be acting at the level of the group. Because of the slow pace of evolution via group selection compared with individual selection, his theory has been rejected by most behavioral ecologists. Nonetheless, social behavior may act to regulate populations incidentally to its other functions (see Chapter 17).

The following are several obvious disadvantages associated with living in groups:

1. Increased competition, possibility of intense aggression as social organisms crowd around a clumped resource;
2. Increased chance of spread of diseases and parasites;
3. Interference with reproduction, such as cheating in parental care or killing of progeny by nonparents;
4. Reduced fitness due to inbreeding.

One of the few studies that directly attempts to assess possible disadvantages of sociality is Hoogland's work (1979b) on prairie dogs. Ectoparasites such as fleas and lice were more numerous in larger and denser prairie dog colonies than in smaller ones. Fleas transmit bubonic plague, epidemics of which periodically decimate prairie dog colonies.

COOPERATION THROUGH GROUP ADVANTAGE

W. C. Allee, in *Cooperation among Animals* (1951), synthesized much of the thinking about animal societies through the first half of this century. The ideas expressed are best understood in relation to contemporary developments in the study of community ecology. Clements (1936) classified plant communities and started a school of "plant sociology," which emphasized the interrelationships of species and recognized the community as a "superorganism" with attributes transcending those of single-species populations. According to Clements living organisms reacted to the environment and so modified it that other species could not exist there. The various assemblages of plants were so interdependent that they were distributed as single units. The idea of communities and ecosystems as superorganisms with emergent properties reached a peak with the writings of Margalef (1963), who discussed the qualities of mature and immature ecosystems. He treated the ecosystem as the functional biological unit that organizes itself so as to conserve and manage information. Although not explicitly stated, such reasoning led to the notion that the whole ecosystem was the unit of selection. (For further discussion of these ideas see the ecology textbooks by Colinvaux 1973 or Krebs 1985). Among animal behaviorists, the popularity of group selection reached its peak with Wynne-Edwards's theory that social behavior evolved mainly for the function of population regulation (Wynne-Edwards 1962). (See Chapter 17 for further discussion.)

Allee (1951) recognized the importance of natural selection at the level of the individual and did much work on egoistic, or selfish, behaviors such as competition and dominance. However, he felt that altruistic or cooperative forces were stronger, and he devoted much study to what he called **natural cooperation,** a force supposedly separate from natural selection. Allee believed that the phenomena he studied could be represented as several grades of social behavior, listed here in generally increasing complexity:

1. *Invertebrate coloniality.* Allee thought the complexity and division of labor shown by such invertebrates as coelenterates is involuntary and, therefore, less advanced than the more voluntary cooperation seen in the vertebrates.
2. *Aggregation.* Organisms may be forced together by the actions of wind, tides, and other phenomena over which they have no control.
3. *Orientation to stimuli.* Animals may be brought together by their response to environmental gradients—for example, insects aggregate around lights. Such organisms must have at least a tolerance for each other.
4. *Locomotion to favorable locations.* When resources are in patches, animals may gather in taking advantage of them—for example, birds in a mulberry tree when the fruit is ripe.
5. *Clumping in the absence of substrate.* Mutual attraction and clumping occur in an attempt to find missing substrate. Brittle stars (Ophiuroidea) are typically dispersed in eel grass, but in an aquarium devoid of objects they cling to each other (Figure 19–9). If small glass rods are planted as artificial substrate, they disperse and attach to the rods.
6. *Sleeping group.* In many species, individuals that are more or less isolated come together to sleep. Such behavior may provide superior predator protection but may attract predators as well.
7. *Complex social life.* Allee considered highly developed social life an extension of sexual and family relations over a large portion of the life span. He posited an unconscious tendency to cooperate that predisposes such species to develop complex societies.

FIGURE 19-9   Aggregation of brittle stars in Allee's laboratory
On the left, immediately after having been placed in a container of seawater, brittle stars are dispersed. On the right, ten minutes later, they are aggregated. If glass rods are added as substrate, the brittle stars will cling to the rods instead of to each other, thereby remaining dispersed.
Source: Data from W. C. Allee (1951), *Cooperation among Animals.* Chicago: University of Chicago Press.

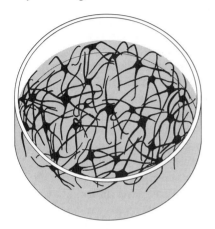

Brittle stars in
sea water

Aggregated brittle
stars ten minutes
later

Although he appreciated the negative effects of crowding, Allee also pointed out the unfavorable consequences of undercrowding. Typical of his experiments was the finding that goldfish (*Cyprinus* species) reared in a toxic colloidal suspension of silver survived longer in groups than they did alone; the slime produced by fish present in the "conditioned" water precipitated the silver to the bottom of the tank, thus protecting them. Other experiments showed that planarian flatworms (*Planaria* spp.) gain protection from ultraviolet rays by grouping and that the effect is not due simply to shading of one worm by another.

Allee (1951, p. 31), referring to the struggle for existence, best stated his philosophy about evolution in the following manner:

In the more poetic post-Darwinian days this struggle was thought of as so intense and so personal that an improved fork in a bristle or a sharper claw or an oilier feather might turn the balance toward the favored animal. Now we find that struggle for existence mainly a matter of populations, measured in the long run only, and then by slight shifts in the ratio of births to deaths.

COOPERATION THROUGH SELFISHNESS

Another, smaller group of biologists returned to a stricter interpretation of Darwin's theory of evolution. The initiator of this movement was R. A. Fisher, who wrote *The Genetical Theory of Natural Selection* in 1929. In his 1958 revision Fisher pointed out that his fundamental theorem referred strictly to "the progressive modification of structure or function only in so far as variations in these are of advantage to the individual." His theorem offered no explanation for the existence of traits that would be of use to the species to which an individual belongs. Darwin (1859) had stated that "if it could be proved that any part of the structure of any one species had been formed for the exclusive good of another species, it would annihilate my theory." But what about structures or functions of one individual that may have been formed for the exclusive good of other individuals of its own species? Understanding this problem is critical for any general theory of social behavior.

Williams (1966) has argued that, when considering any adaptation, we should assume that natural selection operates at that level necessary to explain the facts, and no higher—usually at the level of parents and their young. The argument is based on evidence that natural selection at the group or population level is so slow that it will almost always be outpaced by selection of individual phenotypes. Group selection is not impossible: if there are genes that decrease individual fitness but make it less likely that a group, population, or species will become extinct, then group selection will influence evolution (Maynard-Smith 1976). Most sociobiologists today follow Williams's rules of parsimony and argue that group selection is weak. (See also chapters 4 and 17.)

A parallel shift has been underway in ecology as researchers have collected new data on plant communities. Through the work of Whittaker (1975) and others, it has become apparent that the organization of a community is best understood as an aggregation of individual species, each responding to its own set of environmental tolerances. For example, Clement's theory (1936) predicts that species along some environmental gradient—say, up the side of a mountain—come and go in groups; in fact, the species come and go almost randomly. This is not to say that species in communities do not interact, but rather that the community is not really behaving as a superorganism.

THE SELFISH HERD

One way to explain gregarious behavior in animals is to suppose that it is a form of cover-seeking in which each individual tries to reduce its chances of being caught by a predator. Hamilton (1971) suggested how this might work in a hypothetical lily pond in which live some frogs and a predatory water snake. The snake sleeps on the bottom of the pond most of the time and feeds at a certain time of day. Because it usually catches frogs in the water, the frogs climb out on the edge before the snake wakes up. They don't move inland from the rim because of even more threatening terrestrial predators. The snake surfaces at some unpredictable place and grabs the nearest frog. What should each frog do to minimize the chances of its being eaten? Hamilton showed that it will jump around the rim moving into the nearest gap between two other frogs. The end result is an aggregation (Figure 19–10).

Such an example could be applied to herds of ungulates, flocks of birds, and schools of fish. When each individual behaves selfishly and minimizes its chances of being picked off, the group as a whole may be more vulnerable since the predator can go after the group rather than singling out dispersed individuals.

KIN SELECTION

To explain the evolution of altruistic behavior, Hamilton (1963, 1964) presented a theory incorporating both the gene and the individual as units of selection. In its simplest form this kin selection theory suggests that if a gene that causes some kind of altruistic behavior appears, the gene's success depends ultimately not on whether it benefits the individual carrying the gene but on the gene's benefit to itself. If the individual that benefited by the act is a relative of the altruist and therefore more likely than a nonrelative to be carrying that same gene, the frequency of that gene in the gene pool will increase. The more distant the relative, the less likely it will be to carry that gene, so the greater must be the ratio of benefit-to-recipient and cost-to-altruist. (See Chapter 4.)

Hamilton's idea about the evolution of cooperative behavior through kin selection has stimulated considerable research. Allee and others observed that most complex social groups are made up of relatives; it remained to be seen if cooperative acts are in some way distributed to relatives according to the degree of relatedness. Such studies, however, require information about paternal and maternal relationships that is usually not available to the observer of a natural population. Another

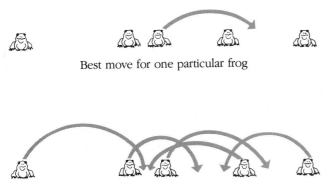

Best move for one particular frog

Series of moves that can be expected

FIGURE 19-10   Frog aggregation
Arrows denote the movement of frogs as they might move around the rim of a pond. The end result is an aggregation as each frog tries to decrease its distance from its neighbor and thereby reduce the probability that it will be eaten by a predator.
Source: W. D. Hamilton, "Geometry for the Selfish Herd," *Journal of Theoretical Biology* 31 (1971):295–311. Reprinted with permission.

problem is the recognition factor. Do animals have to be able to recognize kin in order to "correctly" distribute their altruistic acts? One primate species can do so: researchers found that pigtail monkeys *(Macaca nemestrina)* preferred to interact with related over nonrelated monkeys even after being reared apart from all relatives (Wu et al. 1980). A simpler, and perhaps more widespread, explanation is that a young animal forms social attachments to those in proximity. In a natural group proximate individuals probably will be close relatives; thus we can expect animals that are reared with nonrelatives to treat those nonrelatives as relatives.

Sherman's (1977) study of alarm calls in ground squirrels (see Chapter 13) and Massey's (1977) study of rhesus monkeys tested the kin selection model. Massey found that within an enclosed group, monkeys aid each other in fights in proportion to their degree of relatedness. In observations of the same species in a free-ranging situation, Meikle and Vessey (1981) found that when males leave the natal group, they usually join groups containing older brothers. In contrast to their behavior with nonbrothers, they associate with brothers in the new group, aid each other in fights, and avoid disrupting each other's sexual relationships.

Bertram (1976) found that males in a pride of lions are related on the average almost as closely as half siblings ($r = 0.22$), and females are related as closely as full cousins ($r = 0.15$). He also found that males show tolerance toward cubs and seldom compete for females in estrus and that cubs suckle communally from the pride females. Bertram has concluded that kin selection is at least in part responsible for these cooperative behaviors.

Black-backed jackals *(Canis mesomelas)* live in the brushland of Africa, where monogamous pairs defend territories, hunt cooperatively, and share food. Frequently, offspring from the previous year's litter help rear their siblings by regurgitating food for the lactating mother and for the pups themselves (Moehlman 1979). The number of pups surviving correlates highly with the number of helpers (Figure 19–11). One and a half extra pups survived, on the average, for each additional helper—a yield of one pup per adult involved. Only one-half pup per adult survived when just the parents were involved. Given the fact that helpers are as closely related to their siblings as they would be to their own offspring ($r = 0.5$), yearling jackals gain, genetically speaking, by aiding their parents. In addition to increasing their indirect fitness by aiding siblings, helpers may increase their direct fitness by gaining experience in rearing young, becoming familiar with the home territory, and possibly gaining a portion of the home territory, all of which would increase their own chances of breeding successfully at a later time. Woolfenden (1975) has also cited these advantages in his study of helpers in the Florida scrub jay.

A critical feature of a theory such as Hamilton's is that it is testable. While the theory cannot be proved simply by finding cases of nepotism, it can be discounted in specific cases by observing cooperation among nonrelatives in a natural population. For example, McCracken and Bradbury (1977) have demonstrated that the degree of relatedness of colonial bats *(Phyllostomus hastatus)* is too low to explain their coloniality on the basis of kin selection. Another instance of cooperation among nonrelatives may be exemplified by a species of Antarctic plunder fish *(Harpagifer bispinis)* studied by Daniels (1979). Females guard their eggs, which take four to five months

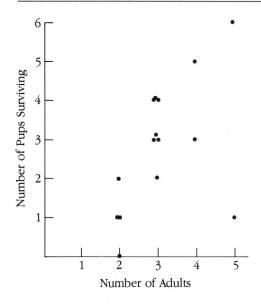

FIGURE 19-11   Black-backed jackal pup survival as a function of the number of adults in family Young from previous years may assist their parents to rear young. Each helper increases the number of surviving pups by a factor of about 1.5. Helpers may, therefore, increase their inclusive fitness more by rearing siblings than by attempting to breed on their own.
Source: Data from P. D. Moehlman, "Jackal Helpers and Pup Survival," *Nature* 277 (1979):382–383. Copyright © 1979 Macmillan Journals Limited. Reprinted with permission.

to hatch in the cold water. Unguarded eggs are quickly eaten. When the female is experimentally removed, males from outside the vicinity of the nest take over guard duty. Since the larval fish are planktonic, drifting long distances in the ocean, Daniels assumed that these replacement guards are not related to the developing eggs and thus that kin selection is not a likely explanation for this behavior. However, to consider such acts "truly" altruistic, we must also know that guarding the nest is a real cost to the guard and not just that it is a benefit to the supposedly unrelated eggs.

As a final example of cooperation among nonrelatives consider the green woodhoopoe *(Phoeniculus purpureus)* studied by Ligon and Ligon (1978a) in Africa. Only a single pair breeds, but there may be as many as fourteen helpers. Although some helpers are close relatives of the breeding pair, others are non-relatives. Clearly, factors other than kin selection must be operating. These other factors will be considered shortly.

At first glance, the ultimate in unselfishness appears to be the worker honeybee. She sacrifices her life when stinging intruders in the vicinity of the colony, and she sacrifices her own reproductive success for that of the colony. She produces no eggs and raises the young of others; her direct fitness is zero. Kin selection theory suggests that it would be significant if those others were close relatives. In fact, because of a peculiarity in the genetics of hymenopteran insects, the honeybee worker is more closely related to her sisters than she would be to her own daughters! Hamilton (1964) has argued that this peculiarity predisposes members of this order toward greater cooperation through kin selection. In fact, all eusocial insects, with the important exception of the termites, belong to the order Hymenoptera. According to Wilson (1975), eusociality has evolved independently in eleven different groups within the order Hymenoptera.

Hymenopteran males are haploid (1*n*), containing only one set of chromosomes which arise from an unfertilized egg; females are diploid (2*n*), containing two sets of chromosomes which arise from a fertilized egg. In this **haplodiploid** system, all the sperm from one male are genetically identical. If only one male mates with one queen to form the colony, the daughters will share identical genes from their father and will share half the genes from their mother. Among daughters, then, the coefficient of relationship (*r*) will be 0.75 instead of the usual 0.5. (See Chapter 4.) Table 19–2 shows that daughters are more closely related to each other than they would be to their own daughters. Thus any genes promoting care-giving behavior of sibs would increase faster than those promoting investment in offspring.

Trivers and Hare (1976) predicted from Hamilton's model that in species of eusocial insects in which the queen mates with a single male, workers should invest in female sibs over male sibs by a 3 : 1 margin, since they are related to sisters by 0.75 and to brothers by only 0.25. In twenty species of ants, the investment ratio was as predicted. In contrast, in slave-making ant species the queen's brood is reared not by her daughters but by slaves, workers of other species stolen from their own nests. Since the slaves are unrelated to the brood they rear, they should not favor one sex over another. Trivers and Hare found that the investment ratio in the slave species was 1 : 1, as predicted.

One criticism of the haplodiploid model is that in some social hymenoptera more than one male may fertilize the queen. In such cases the coefficients of relationship among daughters could be as low as 0.25, and it would be more profitable for daughters to produce their own offspring. Another problem is presented by the termites, which are diploid but have evolved eusociality and sterile worker castes. Termites are dependent on symbiotic protozoans that digest cellulose in the termite gut. Coloniality is necessary because the protozoans are lost at each molt, and reinoculation takes place through trophallaxis, or the exchange of liquid oral and anal food, among colony members (Wilson 1971); however, coloniality does not necessitate the evolution of sterile castes. One possible explanation is the recent finding that a substantial proportion of the genes in termites are linked to the X chromosome. Since the father has only one X chromosome, his daughters are identical with respect to these sex-linked traits and can share more than half their genes on the average. Thus termites may have a system analogous to the haplodiploidy of hymenoptera (Lacy 1980, 1984).

RECIPROCAL ALTRUISM

Kin selection may explain the evolution of much cooperative behavior, but there are other possible explanations. The idea of inclusive fitness necessitates a narrowing of the definition of altruism to include only behavior that benefits a nonrelative at a cost to the one performing the act (Trivers 1971). According to Trivers, natural

TABLE 19-2   The coefficients of relationship (r) among close kin in hymenopteran groups

	Mother	Father	Sister	Brother	Son	Daughter	Nephew or Niece
Female	0.50	0.50	0.75	0.25	0.50	0.50	0.38
Male	1.00	0.00	0.50	0.50	0.00	1.00	0.25

Source: Data from W. D. Hamilton, "The Genetical Evolution of Social Behavior I, II," *Journal of Theoretical Biology* 7 (1964):1–52.

selection acting at the level of the individual could produce altruistic behaviors if in the long run they benefit the organism performing them. Trivers first shows that if altruistic acts are dispensed randomly throughout a large population, genes promoting such behavior will disappear. However, if altruistic acts are dispensed nonrandomly among nonrelatives, genes promoting them could increase in the population if some sort of reciprocation occurs. The factors that affect that likelihood are: (1) length of lifetime: long-lived organisms will have a greater chance of meeting again to reciprocate; (2) dispersal rate: low dispersal rate will increase the chance that repeated interactions will occur; (3) mutual dependence: clumping of individuals, as in avoiding predation, will increase the chances for reciprocation. In any social system nonreciprocators (cheaters) can be expected, but if cheating reaches a high enough level, altruistic acts should become infrequent.

   Through the use of game theory, Axelrod and Hamilton (1981) developed a model for the evolution of cooperation between nonrelatives. The basis for their analysis was a game called Prisoner's Dilemma, in which two players have a choice of either cooperating with each other or defecting. The payoff matrix is shown in Table 19–3. No matter what the other player does, it always pays in the short run to

TABLE 19-3   The Prisoner's Dilemma game
The payoff to player A is shown with illustrative numerical values, where $T > R > P > S$.

	Player B	
Player A	C Cooperation	D Defection
C Cooperation	$R = 3$ Reward for mutual cooperation	$S = 0$ Sucker's payoff
D Defection	$T = 5$ Temptation to defect	$P = 1$ Punishment for mutual defection

Source: From R. Axelrod and W. D. Hamilton. 1981. The evolution of cooperation. *Science* 211:1390–1396.

FIGURE 19–12   Vampire bats
(Desmodus rotundus)
Source: Photo by William A. Wimsatt,
courtesy of the Mammal Slide Library.

defect, since *T*, the reward for defection when the opponent cooperates, is greater than *R*, the reward for mutual cooperation. However, if the game continues and if both defect, they do even worse than if both cooperated; hence the dilemma. How well you do depends on the behavior of your opponent. After conducting an international computer tournament for the best solution, the highest score was obtained by the simple strategy "tit for tat," in which one player cooperates on the first move and then does whatever the other player did on the preceding move. Axelrod and Hamilton argue that this strategy is evolutionarily stable and that it shows how cooperation based on reciprocity could get started in an asocial group.

    A few field studies have suggested the importance of reciprocal altruism. Working with olive baboons *(Papio anubis)* in Africa, Packer (1977) studied coalitions among males, in which two presumably unrelated males joined forces against a third male. If that third one was in consort with a female in estrus, one of the attackers might gain access to her. The pair of males tended to maintain the previously established coalition, and the next "stolen" female would be taken over by the other member of the male pair.

    A different sort of reciprocity has been demonstrated in vampire bats *(Desmodus rotundus)* by Wilkinson (1984), who studied them in Costa Rica. At night these bats feed on blood, primarily from cattle and horses, then return to a hollow tree to roost during the day (Figure 19–12). Wilkinson marked nearly 200 bats that roosted in fourteen trees and spent 400 hours observing them in their roosts. He recorded 110 cases of blood sharing, where one bat regurgitated blood that was then eaten by another bat. Not surprisingly, most of these exchanges were between mothers and offspring. In most of the other feedings he was able to determine both the coefficient of relationship between the pair and an index of association based on how often the pair had been together in the past. After statistical analysis Wilkinson concluded that both relatedness and association contributed significantly to the pattern of exchange (Figure 19–13).

FIGURE 19-13   Reciprocal food-sharing in vampire bats
Graphs A and B show the frequency of pairs observed, based on the degree of
association and relatedness. Graphs C and D show the frequency of food-sharing,
excluding mother-young pairs. Both degree of association and relationship independently predict food-sharing, implicating both reciprocal altruism and kin selection in
the evolution of this behavior.
Source: G. S. Wilkinson (1984), Reciprocal food sharing in the vampire bat. *Nature*
308:181–184.

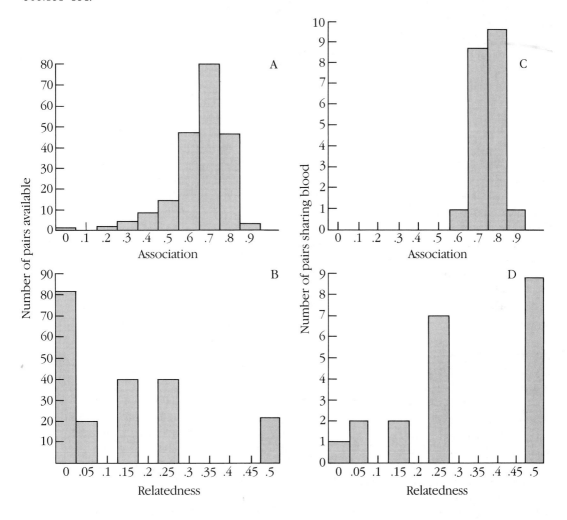

For reciprocity to persist: (1) the pairs must persist long enough to permit
reciprocation, (2) the benefit to the receiver must exceed the cost to the donor, and (3)
donors must recognize cheaters (those that don't reciprocate) and not feed them.
Through additional studies on captive animals, Wilkinson demonstrated that vam-

pire bats meet these conditions. It is important to note that both kin selection and reciprocal altruism are involved and that the mechanisms are not mutually exclusive.

Trivers (1971) also considered how cooperation between members of different species could have evolved through reciprocal altruism. One example he uses is that of cleaning symbioses in fish. Many fish clean ectoparasites off other species of fish (hosts), sometimes entering the host's mouth or gill chambers; both the cleaner and the host would seem to benefit from this mutualism (Figure 19–14). Fish that are cleaners have evolved distinct colors by which they can be identified. Some unrelated species mimic cleaners but, instead of cleaning the host, they nip a piece of fin or gill. Usually cleaners and hosts meet at stations; the host signals when it is ready to be cleaned and also when it has had enough.

Why does the host show restraint and not simply eat the cleaner after it is finished instead of signaling it to go away? Ectoparasitism is a serious problem for fish, and when cleaners are removed from a coral reef, the hosts succumb to a variety of diseases (reviewed by Feder 1966). More recent attempts to replicate this experiment have failed, however (Losey 1979). Cleaners may be in short supply, and hosts may be at some risk of predation when seeking a cleaner. If cleaners and hosts live a long time and hosts use the same cleaner repeatedly, both of which seem to be true, reciprocal altruism is possible. Thus, hosts that eat cleaners after having been cleaned by them may profit in the short run since they gain an easy meal, but would later be selected against as the supply of cleaners is reduced. Gorlick, Atkins, and Losey (1978) have reviewed the evidence and question whether hosts really incur a cost by not eating cleaners and whether hosts even benefit from cleaning. The relationship does not seem to be the clear case of reciprocity that Trivers thought.

FIGURE 19–14 Cleaner fish removing ectoparasites from around mouth of host fish
In this case of mutualism, both the cleaner and the host seem to benefit. The relationship may be an example of reciprocal altruism. If the host refrains from eating the cleaner after being cleaned, the cleaner can perform its services for the host on another occasion. However, recent evidence suggests that there is little, if any, cost to not eating cleaners and that the host may not even benefit from being cleaned.
Source: Photo from Douglas Faulkner/ Sally Faulkner Collection.

Lin and Michener (1972) have presented a mutualistic hypothesis for the evolution of eusociality that is based on reciprocity in social insects. They propose that groups of related or unrelated female bees cooperated in defense against parasites and predators. This mutualism gradually led to such a great degree of division of labor that some females gave up reproducing altogether. By cooperating, the individual gambles that the benefits it gives will be returned at a later time in a way that increases its fitness. While we can see how cooperation might evolve by means of this mechanism, kinship must be involved in the later stages when sterility appears.

PARENTAL MANIPULATION OF OFFSPRING

In the discussion of parent-offspring conflict in Chapter 11, we observed that offspring with only half of each parent's genes demand a greater investment from each parent than the parent should be willing to give. Our first reaction might be that offspring should win such conflicts since they are the ones that must survive and pass the genes on to future generations. But Alexander (1974) has argued that parents should win in the long run. Suppose that some offspring have genes that give the offspring a competitive advantage over sibs to the extent that they reduce the parents' lifetime reproductive success. For example, a highly competitive baby bird that pushed all of its siblings out of the nest would receive more food from its parents and probably increase its direct fitness. The parents' fitness would suffer, however, since they would only raise one young that year. When the young bird grew up and had young of its own, it would pass those competitive traits on and its fitness would be less than that of a bird with less competitive young that could coexist in the nest. Genes that will be favored are those that cause offspring to behave so as to maximize the lifetime reproductive success of the parent. Alexander points out that the parents thus "manipulate" the offspring to the parents' advantage. He lists the following types of behavior as examples of parental manipulation:

- Limiting the amount of parental care given to each offspring so that all have an equal chance to survive and reproduce;
- Restricting parental care or withholding it entirely from some offspring when resources become insufficient for an entire brood;
- Killing some offspring or feeding some offspring to others;
- Causing some offspring to be temporarily or facultatively sterile helpers at the nest;
- Causing some offspring to become permanent (obligately sterile) workers or soldiers.

The last type of behavior, the extreme form of parental manipulation, is a third hypothesis, in addition to haplodiploidy and reciprocity, to explain the evolution of eusociality in insects. The queen uses some of her offspring to help rear others, thereby forcing them to sacrifice their reproductive success in order to produce males and potential queens that have a competitive edge over offspring of other queens.

ECOLOGICAL FACTORS IN COOPERATION

In trying to understand the mechanisms involved in the evolution of cooperation, it is necessary to consider the type of environment in which the population lives. Emlen (1982, 1984) developed an "ecological constraints" model to explain the evolution of cooperative breeding in birds and mammals. He considers environmental factors that restrict the chances for individuals to breed independently (Table 19–4).

One condition that might favor staying at home and helping the parents or others rear offspring is a stable, predictable environment. In these cases unoccupied territories are absent or rare; young have little chance of dispersing and breeding on their own. Thus Stacey (1979) found that among communally breeding acorn woodpeckers (*Melanerpes formicivorous*) the fewer territories that were vacant, the more yearlings were retained as helpers. Presumably, the young remain at home, on familiar ground, to gain experience and social status and await an opportunity to obtain and defend a territory of their own. In the meantime they can increase their indirect fitness by aiding relatives. There is also a chance, as is the case in the Florida scrub jay, that all or part of their parents' territory will eventually become vacant.

Having explained helping on the basis of habitat saturation in stable environments, it is ironic to find that cooperative breeding is most common in arid regions of Africa and Australia, where rainfall is highly variable and unpredictable. In this

TABLE 19–4　Ecological constraints severely limit any possibility of personal reproduction

Type of constraint	*Cause of constraint*
1. Breeding openings are nonexistent	Species has specialized ecological requirements; suitable habitat is "saturated" and marginal habitat is rare (stable environments)
2. Cost of rearing young is prohibitive	Unpredictable season of extreme environmental harshness (fluctuating, erratic environments)

Result: Grown offspring postpone dispersal and are retained in the parental unit. The population becomes subdivided into stable, social, kin groups.

Source: S. T. Emlen. 1982. The evolution of helping. I. An ecological constraints model. *Am. Nat.* 119:29–39.

case, Emlen (1982, 1984) argues that the same behavior results for different reasons. Working with the white-fronted bee-eater *(Merops bullockoides)* Emlen found that in harsh years, when rain was low and insects were scarce, the number of helpers was high. In good years birds were more likely to breed independently, needing little or no help to get enough food for their young.

While the ecological constraints model may explain why some animals stay home and refrain from breeding, why should they bother to help? We have already mentioned the kin selection argument, whereby indirect fitness is enhanced. But often nonrelatives or very distant relatives are aided, as in the green woodhoopoe and the bee-eater. A number of purely selfish possibilities have been suggested, such as gaining experience and social status that might help later in an independent breeding effort. Another possibility is reciprocal altruism, as proposed for the wood-hoopoes by Ligon and Ligon (1978b). There is some evidence that birds that join breeding groups and help nonrelatives "inherit" that flock and are helped to breed themselves in a later effort. Larger flocks have a better chance of getting a breeding territory and providing for young. A similar argument has been used by Rood (1983) to explain why unrelated dwarf mongooses guard and feed young of the dominant pair. He observed one case where a female was later assisted by the very young she had helped rear.

IMPLICATIONS FOR UNDERSTANDING HUMAN SOCIAL BEHAVIOR

To explain cooperation among other animals on the basis of such ultimately selfish forces as kin selection, reciprocal altruism, and parental manipulation of offspring leads us to seek similar explanations in humans. The usual arguments made against such attempts are that human behavior is too heavily shaped by culture and that our behavior is so far removed from genetic control that natural selection can have little to do with our present societies. In addition, both social and natural scientists resist the idea that Darwinian evolution, with its emphasis on competition among individuals, is largely responsible for present-day human morphology and behavior. They emphasize the importance of random processes, such as genetic drift, in changing gene frequencies. Also, cultural evolution, although possibly influenced by natural selection at the group level, is not thought to be a result of selection of individuals. Unfortunately, much of the argument boils down to a new version of the old nature-nurture controversy. Social scientists generally attribute an overwhelming input to environment, and natural scientists tend to stress the underlying importance of genetic heritage.

In the last two decades a few anthropologists, biologists, and psychologists have tested the ability of the theories of Darwin, Fisher, Hamilton, Trivers, and others to explain some aspects of human social behavior. A few examples are considered here.

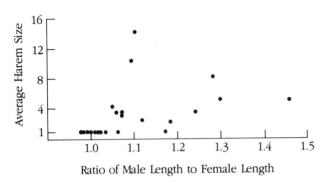

FIGURE 19-15   Relationship between harem size and degree of sexual dimorphism in primates
Each point represents a different species of primate. Males of species with the largest difference in male/female size have the largest harems.
Source: R. D. Alexander et al. in N. A. Chagnon and W. Irons (ed.), *Evolutionary Biology and Human Social Behavior: An Anthropological Perspective* (N. Scituate, Mass.: Duxbury Press, 1979). Reprinted with permission.

HUMAN MATING SYSTEMS

Polygyny is prevalent in mammals, and sexual dimorphism in body size and mating systems are related across a number of species in such a way that the average number of females a male can reproductively monopolize is positively correlated with the ratio of male to female size. The data on primates is shown in Figure 19–15. Human males are 5 to 12 percent taller than females, which should indicate that they are mildly polygynous. In fact, many societies do have harem polygyny (Figure 19–16) or promiscuity, which leads to greater variance in reproductive success in males than in females (Alexander et al. 1979). According to Murdock (1967) 708, or 83 percent, of 849 human societies sampled had at least occasional polygyny. Alexander and his

FIGURE 19-16   Polygyny in humans
A Bedouin man with his two wives (right), niece (left), and child. Polygyny is practiced in a large number of human societies.
Source: Photo by Lila Abu-Lughod from Anthro-Photo.

colleagues (1979) have further demonstrated that in societies in which monogamy is practiced (excluding those in which monogamy is institutionally imposed), males are less dimorphic than in societies in which polygyny is practiced. We find also that, like other polygynous species, human males take longer than females to mature, suffer a higher rate of mortality, and senesce more rapidly.

Trivers and Willard (1973) hypothesized that high-ranking females in good condition should invest more heavily in sons than in daughters (see Chapter 11). As predicted by Trivers and Willard, human males receive more parental care than females, and sons are favored in high-ranking families among stratified polygynous societies such as caste systems (Dickemann 1979). In another study of differential investment in humans, Smith, Kish, and Crawford (1984) analyzed one thousand probated wills. Wealthy decedents significantly favored male kin, and poor decedents favored female kin. Because reproductive success is more variable in males than in females, parental investment in males is more risky. However, wealthy families can increase the reproductive competitiveness of their sons through heavy material investment. Poor families take the safer option of investing in females. This argument assumes, as has been shown in some societies, that there is a positive correlation between reproductive success and material wealth.

KIN SELECTION

Testing the theory of kin selection in humans offers potential rewards because of the relative ease with which we can obtain data on genealogical relationships. Hamilton (1964, p. 19) stated the hypothesis as follows:

> The social behavior of a species evolves in such a way that in each distinct behavior-evoking situation the individual will seem to value his neighbor's fitness against his own according to the coefficients of relationship appropriate to that situation.

The Yanomamö, an Amerindian tribe in southern Venezuela, live in villages of several hundred individuals. Chagnon and Bugos (1979) recorded a crisis situation—an ax fight—precipitated when members of a recent splinter group visited their original village. One of the male visitors (Mohesiwä) was insulted by a resident woman, and he beat her. When she told her story, her kinsmen were angered, and the dispute escalated into an ax fight with several injuries. Chagnon and Bugos analyzed the fight in detail and compared the coefficients of relatedness. The supporters of Mohesiwä, one of the principals in the fight, were related to him on the average 7.8 times more than they were to the principal opponent, who was a half-brother of the beaten woman. Chagnon and Bugos argue that kinship behavior in this society is consistent with predictions based on Hamilton's theory.

Closer to home, Daly and Wilson (1982) studied homicides in Detroit among cohabitants. Out of 98 murders, 76 (77 percent) were by nonrelatives living in the home, usually the spouse. When corrected for their representation in the family, relatives were ten times less likely to kill each other than were nonrelatives. Kinship is also important in predicting patterns of child abuse. Parents are far more likely to abuse their stepchildren than their genetic offspring (Daly and Wilson 1981).

Kin selection may also explain some aspects of parental behavior. Human males are capable of great investment in offspring, compared to other mammals, but confidence of paternity must be high to make the investment profitable. In polygynous societies a male may not be sure that his wife's children are his own. In that case he might do better to invest in his sister's children, since they are likely to share at least one-eighth of his genes. As predicted, in highly polygynous societies the mother's brother often cares for his sister's children (Alexander 1974).

Returning to the analysis of wills by Smith, Kish, and Crawford (1984), close relatives received more wealth than did distant relatives, and younger relatives, with greater reproductive potential, received more wealth than did older relatives. In these cases humans could be acting to maximize their inclusive fitness.

Although some social systems can be understood in part as the result of kin selection, humans engage in reciprocity, which transcends close kinship. Wilson (1978) argues that reciprocity between distantly related or unrelated individuals is the key to human society. The constraints of rigid kin selection have been broken by our use of written and spoken language to fashion long-remembered agreements upon which civilizations can be built.

SUMMARY

A society is defined as a group of individuals of the same species, organized in a cooperative manner that extends beyond sexual behavior. An example of a complex social system is that fashioned by a colonial invertebrate such as the Portuguese man-of-war, in which individuals specialize in the functions of feeding, defense, locomotion, and reproduction. Other examples are the social systems of eusocial insects, cooperatively nesting birds, and primates.

The potential benefits of social behavior include protection from physical elements, predator detection and defense, group defense of resources, and division of labor. The costs are increased competition, spread of contagious diseases, and inbreeding.

Allee studied cooperation in a variety of animals and argued that altruistic forces are stronger than those promoting selfishness. He believed that natural cooperation is a force separate from natural selection, and he implied in his arguments that behavior and structure can evolve for the good of the group or species, even to the detriment of the individual.

Most research on the evolution of social behavior in the last decade has invoked Williams's rule of parsimony: When considering any adaptation we should assume natural selection to operate at the level necessary to explain the facts, and no higher. Usually that is the level of parents and their young; group selection is invoked only as a last resort. Organisms cooperate for genetically selfish reasons.

Aggregations may occur as individuals attempt to reduce their own chances of being picked off by a predator. Kin selection may explain cooperation among individuals that share genes by common descent. Individuals may behave so as to lower their own fitness but increase their inclusive fitness. For example, in many eusocial insects haplodiploidy results in female workers' being more closely related to their sisters than they would be to their daughters; this relationship may have led to the evolution of sterile workers.

Reciprocal altruism may be important in long-lived organisms that do not disperse extensively. Thus an individual stands a good chance of being "paid back" if he or she cooperates with others. Reciprocity is possibly involved in the evolution of symbiotic relationships between different species.

Parents may manipulate their own offspring to behave in ways beneficial to the parents but not to the offspring. Genes giving some offspring a competitive edge against sibs will be selected against if the reproductive success of the parent is lowered. In addition to kin selection, both reciprocal altruism and parental manipulation of offspring could explain the evolution of eusociality in insects.

The application of kin selection and reciprocity to the evolution of human social behavior has met with much resistance among social scientists, who downplay the role of natural selection of favored genotypes in determining human behavior. Nonetheless, some anthropologists, psychologists, and biologists are testing these ideas against the anthropological evidence and are coming up with suggestive results.

Discussion Questions

1. How might you argue that colonial invertebrates such as the Portuguese man-of-war have a more "perfect" society than does a group of chimpanzees?

2. What are the differences between kin selection and group selection? In what ways can they be considered similar?

3. Contrast Allee's and Wynne-Edwards's attitudes toward the evolution of social behavior with those of Hamilton, Trivers, and Alexander.

4. Distinguishing characteristics of human sexuality are the concealment of ovulation in women and the lack of pronounced monthly change in female sexual receptivity. Speculate on the possible evolutionary importance of these characteristics. See Alexander and Noonan (1979) and Burley (1979) for alternative explanations.

5. The data below represent lifetime reproductive success for the !Kung San, an undisturbed population of hunter-gatherers in Africa's Kalahari Desert. What conclusions can you reach about the social system in this population?

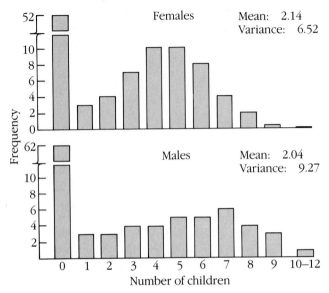

Source: Data from N. Howell, 1979. *Demography of the Dobe !King.* New York: Academic Press; as drawn in M. Daly and M. I. Wilson. 1983. *Sex, Evolution and Behavior,* 2nd ed. Boston: Willard Grant Press, p. 325.

Suggested Readings

Alexander, R. D. 1974. The evolution of social behavior. *Ann. Rev. Ecol. Syst.* 5:325–383.
One of the first comprehensive attempts to understand the evolution of sociality through Darwinian evolution. Thoroughly develops the idea of parental manipulation of offspring.

Alexander, R. D., and D. W. Tinkle, eds. 1981. *Natural Selection and Social Behavior.* New York: Chiron Press.
From a symposium held in 1978. Pioneering articles on theory and on field studies of organisms ranging from insects to humans.

Campbell, B., ed. 1972. *Sexual Selection and the Descent of Man 1871–1971.* Chicago: Aldine.

Assembled on the one hundredth anniversary of Darwin's treatment of the subject, this volume contains a number of important papers on sexual selection and human evolution.

Vehrencamp, S. L. 1979. The roles of individual, kin, and group selection in the evolution of sociality. In P. Marler and J. G. Vandenbergh (eds.), *Handbook of Behavioral Neurobiology*, vol. 3. *Social Behavior and Communication.* New York: Plenum.
An in-depth review of current sociobiological theory.

Wilson, E. O. 1975. *Sociobiology: The New Synthesis.* Cambridge: Harvard University Press.
Reviews the basic principles of social organization, then gives examples from different taxa. Clearly written and still the most comprehensive treatment available.

Wittenberger, J. F. 1981. *Animal Social Behavior.* Boston: Duxbury Press.
An in-depth text that exhaustively treats alternative hypotheses about the evolution of social behavior.

References

Alexander, R. D. 1974. The evolution of social behavior. *Ann. Rev. Ecol. Syst.* 5:325–383.

Alexander, R. D., and K. M. Noonan. 1979. Concealment of ovulation, parental care, and human social evolution. In N. A. Chagnon and W. Irons (eds.), *Evolutionary Biology and Human Social Behavior: An Anthropological Perspective.* N. Scituate, Mass.: Duxbury.

Alexander, R. D., J. L. Hoogland, R. D. Howard, K. M. Noonan, and P. W. Sherman. 1979. Sexual dimorphisms and breeding systems in pinnipeds, ungulates, primates, and humans. In N. A. Chagnon and W. Irons (eds.), *Evolutionary Biology and Human Social Behavior: An Anthropological Perspective.* N. Scituate, Mass.: Duxbury.

Allee, W. C. 1951. *Cooperation among Animals.* Chicago: University of Chicago Press.

Axelrod, R., and W. D. Hamilton. 1981. The evolution of cooperation. *Science* 211:1390–1396.

Bernstein, I. S., and L. G. Sharpe. 1966. Social roles in a rhesus monkey group. *Behaviour* 26:91–104.

Bertram, B. C. R. 1976. Kin selection in lions and in evolution. In P. P. G. Bateson and R. A. Hinde (eds.), *Growing Points in Ethology.* New York: Cambridge University Press.

Brown, J. L. 1974. Alternate routes to sociality in jays—with a theory for the evolution of altruism and communal breeding. *Amer. Zool.* 14:63–80.

Brown, J. L., and E. R. Brown. 1981. Extended family system in a communal bird. *Science* 211:959–960.

Brown, J. L., E. R. Brown, S. D. Brown, and D. D. Dow. 1982. Helpers: Effects of experimental removal on reproductive success. *Science* 215:421–422.

Burley, N. 1979. The evolution of concealed ovulation. *Amer. Nat.* 114:835–858.

Buss, L. W. 1981. Group living, competition, and the evolution of cooperation in a sessile invertebrate. *Science* 213:1012–1014.

Chagnon, N. A., and P. E. Bugos, Jr. 1979. Kin selection and conflict: An analysis of a Yanomamö ax fight. In N. A. Chagnon and W. Irons (ed.), *Evolutionary Biology and Human Social Behavior: An Anthropological Perspective.* N. Scituate, Mass.: Duxbury.

Clements, F. E. 1936. Nature and structure of the climax. *J. Ecol.* 24:252–284.

Colinvaux, P. 1973. *Introduction to Ecology.* New York: Wiley.

Daly, M., and M. I. Wilson. 1981. Abuse and neglect of children in evolutionary perspective. In R. D. Alexander and D. W. Tinkle (eds.), *Natu-*

ral Selection and Social Behavior. New York: Chiron Press.

———. 1982. Homicide and kinship. *Amer. Anthrop.* 84:372–378.

Daniels, R. A. 1979. Nest guard replacement in the antarctic fish *Harpigifer bispinus:* Possible altruistic behavior. *Science* 205:831–833.

Darwin, C. R. 1859. *Origin of Species.* London: Dent.

Davis, D. E. 1942. The phylogeny of social nesting habits in the Crotophaginae. *Quart. Rev. Biol.* 17:115–134.

Dickemann, M. 1979. Female infanticide, reproductive strategies, and social stratification: A preliminary model. In N. A. Chagnon and W. Irons (eds.), *Evolutionary Biology and Human Social Behavior: An Anthropological Perspective.* N. Scituate, Mass.: Duxbury Press.

Eisenberg, J. 1981. *The Mammalian Radiations.* Chicago: University of Chicago Press.

Emlen, S. T. 1978. The evolution of cooperative breeding in birds. In J. R. Krebs and N. B. Davies (eds.), *Behavioural Ecology: An Evolutionary Approach.* Sunderland, Mass.: Sinauer Associates.

Emlen, S. T. 1982. The evolution of helping. I. An ecological constraints model. *Am. Nat.* 119:29–39.

———. 1984. Cooperative breeding in birds and mammals. In J. R. Krebs and N. B. Davies (eds.), *Behavioural Ecology: An Evolutionary Approach,* 2nd ed. Sunderland, Mass.: Sinauer Associates.

Feder, H. M. 1966. Cleaning symbioses in the marine environment. In S. M. Henry (ed.), *Symbiosis,* vol. 1. New York: Academic Press.

Fisher, R. A. 1958. *The Genetical Theory of Natural Selection.* New York: Dover.

Geist, V. 1971. *Mountain Sheep: A Study in Behavior and Evolution.* Chicago: University of Chicago Press.

Gerstell, R. 1939. Certain mechanisms of quail losses revealed by laboratory experimentation, *Trans. Fourth N. Amer. Wildl. Inst.* 462–467.

Gorlick, D. L., P. D. Atkins, and G. S. Losey. 1978. Cleaning stations as water holes, garbage dumps, and sites for the evolution of reciprocal altruism? *Amer. Nat.* 112:314–353.

Hamilton, W. D. 1963. The evolution of altruistic behavior. *Amer. Nat.* 97:354–356.

———. 1964. The genetical evolution of social behavior I, II. *J. Theoret. Biol.* 7:1–52.

———. 1971. Geometry for the selfish herd. *J. Theoret. Biol.* 31:295–311.

Hoogland, J. L. 1979a. The effect of colony size on individual alertness of prairie dogs (Sciuridae: *Cynomys* spp.). *Anim. Behav.* 27:394–407.

———. 1979b. Aggression, ectoparasitism, and other possible costs of prairie dog (Sciuridae: *Cynomys* spp.) coloniality. *Behaviour* 69:1–35.

Hoogland, J. L., and P. W. Sherman. 1976. Advantages and disadvantages of bank swallow (*Riparia riparia*) coloniality. *Ecol. Monogr.* 46:33–58.

Howard, R. D. 1978. The evolution of mating strategies in bullfrogs, *Rana catesbeiana. Evolution* 32:850–871.

Jarvis, J. U. M. 1981. Eusociality in a mammal: Cooperative breeding in naked mole-rat colonies. *Science* 212:571–573.

Krebs, C. J. 1985. *Ecology: The Experimental Analysis of Distribution and Abundance,* 3rd ed. New York: Harper & Row.

Kurland, J. A. 1979. Paternity, mother's brother, and human sociality. In N. A. Chagnon and W. Irons (eds.), *Evolutionary Biology and Human Social Behavior: An Anthropological Perspective.* N. Scituate, Mass.: Duxbury Press.

Lacy, R. C. 1980. The evolution of eusociality in termites: A haplodiploid analogy? *Amer. Nat.* 116:449–451.

———. 1984. The evolution of termite eusociality: Reply to Leinaas. *Am. Nat.* 123:876–878.

Ligon, J. D., and S. H. Ligon. 1978a. The communal social system of the green woodhoopoe in Kenya. *Living Bird* 17:159–198.

———. 1978b. Communal breeding in green woodhoopoes as a case for reciprocity. *Nature* 276:496–498.

Lin, N. A., and C. D. Michener. 1972. Evolution of sociality in insects. *Quart. Rev. Biol.* 47:131–159.

Losey, G. S. 1979. The fish cleaning symbiosis: Proximate causes of host behaviour. *Anim. Behav.* 27:669–685.

Margalef, R. 1963. On certain unifying principles in ecology. *Amer. Nat.* 97:357–374.

Massey, A. 1977. Agonistic aids and kinship in a group of pigtail macaques. *Behav. Ecol. Sociobiol.* 2:31–40.

Maynard-Smith, J. 1976. Evolution and the theory of games. *Amer. Sci.* 64:41–45.

McCracken, G. F., and J. W. Bradbury. 1977. Paternity and genetic heterogeneity in the polygynous bat, *Phyllostomus hastatus. Science* 198:303–306.

Meikle, D. B., and S. H. Vessey. 1981. Nepotism among rhesus monkey brothers. *Nature* 294:160–161.

Moehlman, P. D. 1979. Jackal helpers and pup survival. *Nature* 277:382–383.

———. 1983. Socioecology of silverbacked and golden jackals, *Canis mesomelas* and *C. aureus.* In J. F. Eisenberg and D. G. Kleiman, (eds.), *Recent Advances in the Study of Mammalian Behavior.* Special Pub no. 7, Amer. Soc. Mammal.

Murdock, G. P. 1967. *Ethnographic Atlas.* Pittsburgh: University of Pittsburgh Press.

Orians, G. H. 1969. On the evolution of mating systems in birds and mammals. *Amer. Nat.* 103:589–603.

Packer, C. 1977. Reciprocal altruism in *Papio anubis. Nature* 265:441–443.

Pitcher, T. J., A. E. Magurran, and I. J. Winfield. 1982. Fish in larger shoals find food faster. *Behav. Ecol. Sociobiol.* 10:149–151.

Rood, J. P. 1980. Mating relationships and breeding suppression in the dwarf mongoose. *Anim. Behav.* 28:143–150.

———. 1983. The social system of the dwarf mongoose. In J. F. Eisenberg and D. G. Kleiman (eds.), *Recent Advances in the Study of Mammalian Behavior.* Special Pub. No. 7, Amer. Soc. Mammal.

Shaw, E. 1978. Schooling fishes. *Amer. Scientist* 66:166–175.

Sherman, P. W. 1977. Nepotism and the evolution of alarm calls. *Science* 197:1246–1253.

Smith, M. S., B. J. Kish, and C. B. Crawford. 1984. Inheritance of wealth as human kin investment. Paper presented to the Animal Behavior Society, Cheney, Washington.

Stacey, P. B. 1979. Habitat saturation and communal breeding in the acorn woodpecker. *Anim. Behav.* 27:1153–1166.

Stallcup, J. A., and G. E. Woolfenden. 1978. Family status and contributions to breeding by Florida scrub jays. *Anim. Behav.* 26:1144–1156.

Treisman, M. 1975. Predation and the evolution of gregariousness. I. Models for concealment and evasion. *Anim. Behav.* 23:779–800.

Trivers, R. L. 1971. The evolution of reciprocal altruism. *Quart. Rev. Biol.* 46:35–57.

Trivers, R. L., and H. Hare. 1976. Haplodiploidy and the evolution of the social insects. *Science* 191:249–263.

Trivers, R. L., and D. E. Willard. 1973. Natural selection of parental ability to vary the sex ratio of offspring. *Science* 179:90–92.

Vehrencamp, S. L. 1977. Relative fecundity and parental effort in communally nesting anis, *Crotophaga sulcirostris. Science* 197:403–405.

Vessey, S. H. 1971. Free-ranging rhesus monkeys: Behavioural effects of removal, separation and reintroduction of group members. *Behaviour* 40:216–227.

Waltz, E. C. 1982. Resource characteristics and the evolution of information centers. *Am. Nat.* 119:73–90.

Whittaker, R. H. 1975. *Communities and Ecosystems,* 2nd ed. New York: Macmillan.

Wilkinson, G. S. 1984. Reciprocal food sharing in the vampire bat. *Nature* 308:181–184.

Williams, G. C. 1966. *Adaptation and Natural Selection. A Critique of Some Current Evolutionary Thought.* Princeton: Princeton University Press.

Wilson, E. O. 1971. *Insect Societies.* Cambridge: Harvard University Press.

———. 1975. *Sociobiology: The New Synthesis.* Cambridge: Harvard University Press.

———. 1978. *On Human Nature.* Cambridge: Harvard University Press.

Woolfenden, G. E. 1975. Florida scrub jay helpers at the nest. *Auk* 92:1–15.

Wynne-Edwards, V. C. 1962. *Animal Dispersion in Relation to Social Behavior.* Edinburgh: Oliver and Boyd.

Wu, H. M. H., W. G. Holmes, S. R. Medina, and G. P. Sackett. 1980. Kin preference in infant *Macaca nemestrina. Nature* 285:225–227.

Zahavi, A. 1974. Communal nesting by the Arabian babbler: A case of individual selection. *Ibis* 116:84–87.

AUTHOR INDEX

Abbott, I., 70
Able, K. P., 420, 421, 422, 426, 427, 431
Abrams, T. W., 282
Ader, R., 224
Adkins-Regan, E., 180
Adler, K., 431
Aiello, A., 496, 498, 504
Albon, S. D., 308
Albrecht, H., 363
Alexander, B. K., 368
Alexander, M., 360
Alexander, R. D., 308, 323, 537, 540, 581, 584, 586
Alkon, D. L., 282, 283
Allee, W. C., 569, 570, 571
Altmann, J., 48, 358
Anand, B. K., 138
Anderson, B., 294
Anderson, L. T., 460, 461
Antony, C., 309
Arbib, M. A., 136, 290
Archer, J., 224
Ardrey, R., 347
Arendash, G. W., 143
Arendt, J., 202
Arey, L. B., 327
Aristotle, 14–15, 28
Arling, G. L., 234
Arnold, A. P., 169

Aronson, L. R., 330
Aschoff, J., 192, 193, 196, 202
Askenmo, C. E. H., 26
Asmundson, S. J., 205
Atkins, P. D., 580
Attneave, F., 381
Austad, S. N., 309
Axelrod, R., 577, 578
Azrin, N. H., 362

Bachus, S. E., 134–135
Bailey, E. D., 238
Baker, M. C., 384, 385, 392, 454
Baker, R. R., 304, 436
Bandler, R. J., 359
Banfield, A. W., 417
Banks, E. M., 98, 99
Barash, D. P., 10–11, 24
Barbour, R. W., 430
Barfield, R. J., 142, 162
Barlow, G. W., 255
Barnett, S. A., 230, 286, 288
Barraclough, C. A., 165
Barron, A., 278
Barron, D. H., 222
Barth, R. H., 163, 164
Bass, A. H., 395
Bastock, M., 99
Bateman, A. J., 316
Bateson, P. O. G., 229, 455

Batzli, G. O., 510
Beach, F. A., 24, 37, 153, 160, 177, 256, 257
Beck, B. B., 484
Beck, S. D., 203
Bekoff, M., 220, 221, 246, 247, 248, 377
Belanger, P. L., 38
Bell, J. F., 521
Bellrose, F. C., 421, 426
Benamar, O., 309
Bennett, M. A., 162
Bentley, D., 21, 125
Benzer, S., 103
Berkowitz, L., 363, 367
Bernon, G., 468
Bernstein, I. S., 358, 360, 567
Beroza, M., 117
Bertram, B., 490, 574
Berven, K. A., 313, 457
Bhanvani, R., 249, 250
Bielfelt, S. W., 95
Bingman, V. P., 426
Birch, H. C., 275
Bitterman, M. E., 23, 284, 286, 287
Blair, W. F., 74
Blest, A. D., 504
Block, G. D., 201
Block, R. A., 280
Boag, P. T., 548, 549

Bodenstein, D., 132–133
Bodmer, W. F., 101, 102
Boice, R., 38, 540
Bolles, R., 288
Bonner, J. T., 552
Bonsall, R. W., 334
Borgia, G., 307
Bossert, W. H., 68, 381
Boudreau, J. C., 94, 95
Bouissou, M. F., 167
Boulos, Z., 201
Bovjberg, R. V., 378
Bower, G. H., 269
Bowne, G., 360
Boyce, M. S., 320
Boycott, B. B., 282
Boysem, S., 403
Bradbury, J. W., 574
Bradtke, J., 202
Brady, J., 196, 197, 198
Brandt, E. M., 337
Brazier, M. A. B., 133
Breed, M. D., 244
Broadhurst, P. L., 98
Brobeck, J. R., 138
Brody, P. N., 172
Bronson, F. H., 168, 390, 523
Brookman, M. A., 427
Brooks, R. J., 383
Brookshire, K. H., 286
Brower, J. V., 275
Brower, L. P., 275, 419, 498
Brown, A. I., 426
Brown, E. R., 565
Brown, F. A., 190
Brown, J. H., 307
Brown, J. L., 563, 564, 565
Brown, S., 202
Brown, S. D., 565
Brown, W. L., 74
Bruce, H. M., 176, 310, 523
Bruderer, B., 426
Bruner, J. S., 249
Buckle, G. R., 384
Bugos, P. E., 585
Bull, J. J., 305
Bullock, T. H., 395
Burg, R. von, 279
Burghardt, G. M., 133, 221, 231
Burnet, B., 241
Busch, D. E., 162
Buskirk, R. E., 487
Buss, L. W., 568
Byers, J. A., 247

Caldorola, P. C., 199
Calhoun, J. B., 518, 519
Calvert, W. H., 419
Campbell, B., 323

Campbell, C. B. G., 24, 37, 257, 289
Caracao, T., 490, 491
Carew, T. J., 282
Carey, R. J., 24
Carlisle, D. B., 164
Carlisle, T. R., 318
Carlson, A. D., 396
Carlson, R. B., 498
Carpenter, F. L., 352
Carr, A., 433, 434, 435
Cassel, J. F., 498
Castellucci, V., 282
Catalano, S. M., 275
Cavalli-Sforza, L. L., 101, 102
Chadab, R., 388
Chagnon, N. A., 585
Charniaux-Cotton, H., 164
Charnov, E. L., 476
Chase, J., 430
Cheney, D. L., 386, 453, 454
Cheng, M. F., 172
Chevalier-Skolnikoff, S., 335
Chiszar, D., 363
Chitty, D., 106, 523, 524, 526
Chitty, H., 106
Christenson, T. E., 26
Christian, J. J., 358, 521, 522, 523, 526
Chu-Wang, W., 220, 255
Clark, A. L., 247
Clark, L. H., 361
Clark, M. M., 231
Clarke, C. A., 71
Clemens, L. G., 168
Clements, F. E., 569, 570, 572
Cloudsley-Thompson, J. L., 202, 203
Clutton-Brock, T. H., 26, 308
Cody, M. L., 486
Colby, D. R., 162, 240
Cole, J. E., 244
Cole, L. J., 167
Colinvaux, P., 570
Collias, E. C., 545, 546, 547
Collias, N. E., 545, 546, 547
Collier, G., 452
Colvin, B. A., 472, 473
Comer, C., 130
Coniglio, L. P., 168
Conklin, P. M., 224
Connell, J. H., 511
Connolly, K., 241
Connor, V. M., 479
Cook, A., 282
Cook, R., 241
Cooper, M. L., 330
Cooper, R., 91, 92
Copeland, J., 396, 397
Corbet, P. S., 190

Corning, W. C., 279, 280, 281, 283
Coulson, J. C., 311
Cowan, W. M., 201
Cox, C. R., 309
Cox, V. C., 134
Craig, W., 20, 229
Crawford, C. B., 585, 586
Crews, D., 24, 176, 177
Cross, B. A., 178
Crow, J. F., 78
Crowcroft, P., 238
Cullen, E., 255

Daan, S., 202
da Cunha, A. B., 241
Daly, M., 303, 585, 586
Daniels, R. A., 562, 574, 575
Darchen, R., 244
Darwin, C., 16, 17, 18, 28, 63, 64, 65, 70, 305, 306, 307, 364, 365, 386, 508, 536, 540, 571, 572, 583
Datta, L. G., 284
Davidson, J. M., 142, 165
Davies, N. B., 458, 485, 486, 584
Davis, D. E., 26, 358, 361, 562, 563, 568
Davis, S. F., 284
Davis, W. H., 417
Dawe, A. R., 202
Dawkins, R., 78, 318, 405, 467, 553
de Beer, G., 547
DeCoursey, P. J., 192, 196
Deevey, E. S., 485
De Fries, C., 91
Delage, B., 244
Delgado, J. M. R., 135, 359
DeLong, K. T., 238
Demong, N. J., 501
Denenberg, V. H., 138, 224, 551
de Souza, H. M. L., 241
de Roth, S., 379, 380
Derscheid, J. M., 249
Dethier, V. G., 132–133
Deutsch, J. A., 290, 293, 294
DeVoogd, T., 140
DeVore, I., 349, 500
Dewey, J., 22
deWilde, J., 169
Dewsbury, D. A., 10, 24, 25, 37, 257, 286, 287, 316, 327, 328, 329
Diamond, J. M., 449, 451, 452, 453
Diamond, M., 305
Dickemann, M., 521, 585
Dickinson, J. L., 310
Dillon, L. S., 75
Dingle, H., 418, 420
Dix, M. W., 474
Domjan, M., 285

Dörner, G., 165, 169
dos Santos, E. P., 241
Doty, R. L., 334
Dow, D. D., 565
Dowling, J. E., 130
Drickamer, L. C., 162, 163, 236, 240, 250, 251, 333, 338, 356, 428, 517, 523
Drury, W. H., 421, 426
Drummond, H., 40
Duffey, S. S., 498
Dunbar, R. I. M., 48
Dunn, L. C., 526
Dunsworth, T. S., 517
Durham, W. H., 552, 553
Dyal, J. A., 280, 283
Dyer, R. G., 178

Ealey, E. H. M., 324
Ebert, P. D., 361
Ebling, F. J., 451
Edwards, D. A., 166
Ehrenfeld, D. W., 435
Ehrhardt, A. A., 167
Ehrman, L., 101, 102
Eibl-Eibesfeldt, I., 21, 403, 496, 545
Eichler, V. B., 201
Eik-Nes, K. B., 175
Eisenberg, J., 325, 327, 566
Eisner, T. E., 497
Ekdale, A. A., 539
Eldredge, N., 536
Ellins, S. R., 275
Ellis, P. E., 164
Elmes, G. W., 446, 447
Emlen, S. T., 26, 311, 382, 383, 420, 422, 424, 425, 426, 427, 429, 501, 566, 582, 583
Enright, J. T., 202
Erickson, C. J., 172
Erlenmeyer-Kimling, L., 101
Ervin, F. R., 288
Erwin, J., 236
Eskin, A., 196, 201
Evans, H. E., 544
Ewert, J. P., 130

Fagen, R., 246, 247, 249, 389, 408
Falconer, D. S., 97
Fallon, A. M., 22
Falls, J. B., 383
Farner, D. S., 324, 414, 415, 422
Fechner, G., 22
Feder, H. H., 165, 173
Feder, H. M., 580
Ferguson, D. E., 432
Ferreira, A. J., 224

Feshbach, S., 367
Fingerman, M., 165
Fisher, R. A., 19, 89, 303, 306, 307, 572, 583
Flemming, D. P., 496, 517
Fletcher, H. J., 285
Fletcher, T. J., 167
Flourens, P., 22
Flynn, J. P., 359
Ford, E. B., 103
Fouts, R., 288
Fox, M. W., 247, 330
Fraenkel, G. S., 169, 420, 455
Francq, E. N., 496
Fraser, D. F., 74
Fredericson, E., 362
Free, J. B., 244
Freedle, R., 237
Freedman, J. L., 519
Freeman, F. N., 101, 102
Frisch, K. von, 21, 284, 398, 399, 431
Frumin, N., 133
Fuller, C. A., 195, 201
Fuller, J. L., 19, 96, 97, 233, 234, 247, 540, 541
Fürsich, F. T., 539
Futuyma, D. J., 75

Galambos, R., 429, 430
Galef, B. G., 229, 230, 231, 278, 552
Gandelman, R., 138, 168
Garcia, J., 288
Gardner, B. T., 288, 401, 402
Gardner, R. A., 288, 401, 402
Gaston, S., 201
Gauthreaux, S. A., 421
Geist, V., 21, 462, 568
Gelber, B., 279
Gelperin, A., 132–133
Gerall, A. A., 139
Gergltts, W., 426
Gerhardt, C., 22
Geronimo, J., 199
Gerstell, R., 568
Giantonio, G. W., 139
Gilbert, D. G., 309
Gilbert, L. I., 169
Gilhousen, H. C., 284
Gill, T. W., 377, 401, 402, 472, 485
Gilliard, E. T., 307
Gilpin, A. R., 281
Ginsberg, B. E., 95
Giraldeau, L. A., 476
Girgis, M., 136
Gladue, L. G., 168
Goldfoot, D. A., 334, 335

Golding, D. W., 164
Goldman, B., 202
Gonzalez, R. C., 285
Gordon, T. P., 360
Gorlick, D. L., 580
Gorski, R. A., 143, 165, 169
Gottlieb, G., 23, 222, 223, 228
Gould, J. L., 22, 398, 399, 400, 427, 431
Gould, S. J., 536, 547, 548
Goy, R. W., 166
Graber, R. R., 416
Grant, B. R., 70
Grant, E. C., 41
Grant, P. R., 70, 548
Green, R. G., 521
Greenberg, L., 384
Greenberg, N., 176
Greenwood, P. J., 314, 452, 453
Greer, M. A., 294
Griffin, D. A., 413, 417, 420, 423, 429, 430
Griffin, D. R., 50
Grobecker, D. B., 483
Grobstein, P., 130
Gromko, M. H., 309
Grossfield, J., 241
Grossman, M. I., 293, 294
Grota, L. J., 175
Guadioso, V., 167
Guhl, A. M., 162
Guiness, F. E., 308
Gulick, W. L., 118
Gunderson, H. L., 417
Gunn, D. L., 420, 455
Gustavson, C. B., 498
Gustavson, C. R., 275
Guthrie, R. D., 502
Gwynne, D. T., 318, 319

Haartman, L. von, 511
Haekel, 547
Hafez, E. S. E., 331
Hailman, J. P., 231, 232, 233, 457
Hale, E. B., 229, 329, 331, 540
Hall, E. R., 464
Hall, E. T., 350
Halpern, M., 133
Hamburger, V., 221, 222
Hamilton, W. D., 27, 80, 305, 498, 572, 573, 574, 575, 576, 577, 578, 583, 585
Hankins, W. G., 288
Hansen, E. W., 322
Hansen, L. P., 510
Hansen, R. M., 452
Hardin, G., 513
Harding, C. F., 361
Hare, H., 576
Harker, J. E., 197

Harlow, H. F., 23, 233, 234, 235, 236, 278, 285, 337
Harlow, M. K., 23
Harnly, M. H., 258
Harris, G., 165
Harris, V. T., 459
Harrison, J. L., 324
Hart, B. L., 139
Hartwick, P., 426
Hashimoto, H., 431
Hasler, A. D., 431, 461
Hassell, M. D., 430
Hastings, J. W., 188
Hausfater, G., 347
Hawkins, R. D., 282
Hayes, J. H., 401
Haynes, C. M., 24
Hazlett, B. A., 381
Hebb, D. O., 138, 291
Hegmann, J. P., 93
Heiber, L., 231
Heiligenberg, W., 125
Heinrich, B., 472
Helfman, G. S., 552
Helmholtz, H. von, 22
Henderson, P. W., 230
Hendrickson, A. E., 201
Hermayer, K. L., 426
Herndon, J. G., 135, 360
Hess, E. H., 227, 231
Hilgard, E. R., 269
Hinde, R. A., 22, 48, 225, 287
Hirsch, J., 19, 24, 90, 94, 95, 102
Hitchcock, H. B., 417
Hock, R. J., 202
Hodgkin, A. L., 114
Hodos, W., 24, 37, 257, 289
Hoffman, K., 193
Hölldobler, B., 244
Holling, C. S., 509
Holmes, W. G., 384
Holz-Tucker, A. M., 153
Holzinger, K. J., 101, 102
Hoogland, J. L., 453, 500, 501, 540, 568, 569
Hoover, J. E., 240
Hopkins, C. D., 116, 395
Hotta, Y., 103
Hotten, N., 284
Houston, A. I., 485, 486
Howard, H. E., 350
Howard, R. D., 352, 457, 540, 562
Howell, N., 588
Hoy, R. R., 21, 125
Hrdy, S. B., 310, 346, 347
Hubel, D. H., 24, 139, 140
Huck, U. W., 38
Hudson, D. J., 200
Hughes, J., 368
Hulse, F., 68

Hunsaker, D., 549, 550
Hunt, R. M., 540
Hutchinson, G. E., 71, 554
Hutchinson, J. B., 162
Hutchinson, R. R., 362
Hutt, C., 249, 250
Huxley, J., 62, 75, 375
Hyde, J. S., 361

Ibuki, T., 178
Immelmann, K., 229
Ingle, D., 130
Inglis, I. R., 387, 388
Isaacson, A. J., 387, 388
Itani, J., 278

Jablonski, D., 539
Jacklet, J. W., 199, 200
Jackson, J. R., 249
Jacobson, M., 220, 255, 281
Jallon, J. M., 309
Janowitz, H. D., 293, 294
Jarman, P. J., 488
Jarvik, L. F., 101
Jarvis, J. U. M., 567
Jenner, C. E., 191
Jenni, D. A., 314
Jensen, A. R., 102
Jensen, D. D., 279
Jerison, H. J., 237
Johansson, B., 202
John, E. R., 290
Johnson, C. G., 419
Johnson, R. N., 363
Jolly, A., 249
Jones, F. G. W., 255
Joslin, J. K., 428, 429
Jung-Hoffman, I., 244
Jussiaux, C., 167

Kacelnik, R. A., 477
Kakolewski, J., 134
Kalmijn, A. J., 395, 482
Kamil, A. C., 475
Kandel, E. R., 282
Kaufman, I. C., 336, 362
Kawai, M., 278
Kawakami, M., 165, 169, 178
Kayser, C., 205
Kearney, M., 241
Keesey, R. E., 138
Keeton, W. T., 421, 422, 423, 426, 427
Keiner, M., 178
Keith, J. A., 426
Keller, B. L., 523
Keller, R., 249
Kelley, D. B., 178

Kelly, R., 24
Kelly, S., 280, 281
Kerlinger, P., 426
Kessel, E. L., 542, 544
Kettlewell, H. B. D., 548
Ketterson, E. D., 456
Kety, S. S., 291
Kiepenheuer, J., 426
Kimble, D. P., 281
Kimura, M., 78
King, J. A., 26, 93, 103, 105, 162, 218, 219, 255
King, J. R., 422
King, R. G., 201
Kirschvink, J. L., 116, 427
Kish, B. J., 585, 586
Kitching, J. A., 451
Kleiman, D. G., 325, 327
Kleinholz, L. H., 164, 165
Klopfer, M. S., 228
Klopfer, P., 228, 277, 457, 458, 459
Knipling, E. F., 117
Kodric-Brown, A., 307
Koelling, R. A., 288
Kogure, M., 191
Köhler, W., 275
Komisaruk, B. I., 178
Kramer, D. L., 476
Kramer, G., 422, 423
Krasne, F. B., 284
Krebs, C. J., 78, 106, 447, 448, 510, 523, 570
Krebs, J. R., 26, 405, 458, 477, 516, 566,
Kreithen, M. L., 426, 427
Krieger, D. T., 201
Kroodsma, D. E., 237
Kruuk, H., 388, 391, 492, 493, 495
Kruyt, W., 276
Kummer, H., 22
Kupfermann, I., 282
Kuo, Z. Y., 223

Lack, D., 447, 450, 526, 540, 541
Lacy, R. C., 576
Landreth, H. F., 432
Lank, D. B., 314
Larkin, T. S., 427
Larson, C. L., 521
Lashley, K. S., 136, 291
Lawick, H. van, 388
Lawick-Goodall, J. van, 288, 388, 484
Leakey, L. S. B., 14
Le Boeuf, B. J., 309, 317, 318, 359, 517
Lee, R. B., 521
Lehner, P. N., 48, 49

Lehrman, D. S., 24, 171, 172, 174, 176, 254, 255, 325, 326, 340
Lenn, N. J., 201
Lent, P. C., 417
Leon, M., 138
LePage, A., 367
Leshner, A. I., 162
Lettvin, J. Y., 129–130
Levine, S., 165, 169
Lewis, M., 237
Lewontin, R. C., 526, 548
Ley, W., 14
Licht, P., 176
Lickey, M. E., 200
Liebig, J., 508, 509
Ligon, J. D., 575, 583
Ligon, S. H., 575, 583
Lillie, F. R., 167
Lin, N. A., 581
Lincoln, F. C., 416
Lindauer, M., 431
Lissmann, H. W., 116, 482
Littlejohn, M. J., 550, 551
Lloyd, J. E., 396, 397
Lockard, R. B., 24, 37
Loeb, J., 420
Loehlin, J. C., 91
Loftus-Hills, J. J., 550
Lohrer, W., 164
Loizos, C., 247
Lombardi, J. R., 390
Longhurst, A. R., 206, 207
Lorden, J. F., 141
Lorenz, K., 21, 225, 228, 292, 294, 344, 365, 366, 367, 537
Losey, G. S., 305, 580
Lott, D. F., 174, 175, 340
Low, B. S., 325
Lowther, G., 495
Lubin, M., 174
Lund, N. L., 139
Luttge, W. G., 165
Lutz, P. E., 191
Lyell, C., 16
Lyman, C. P., 205
Lynch, C. B., 93

Maas, D., 93
MacArthur, R. H., 80, 448, 449, 468, 472, 474, 478, 479, 480
McCann, S. M., 294
McCleery, R. H., 477
McConnell, J. V., 280, 281
McCoy, M., 135, 360
McCracken, G. F., 574
McEwen, B. S., 24, 178
McGill, T. E., 21, 24, 166
McGuire, T. R., 102

Machin, K. E., 116
McIntosh, T. K., 142
McKeown, J. P., 432
Mackintosh, J. H., 41
McLaren, I. A., 207
MacMillen, R. E., 352
McNab, B. K., 351
McNair, J. N., 476
McVay, S., 393
Maderut, J. L., 220, 255
Magurran, A. E., 568
Malan, A., 205
Malthus, T., 16, 63, 508
Manning, A., 22, 99, 166, 241, 536
Maple, T., 236
Margalef, R., 570
Marler, P., 237, 238, 239, 368, 374, 375, 382, 386, 392, 500
Marsden, H. M., 333, 336, 356
Marshall, A. J., 324
Marx, J. L., 402
Mason, W. A., 234
Massey, A., 240, 523, 574
Masson, G. M., 175
Mast, S. O., 420
Masterton, R. B., 284
Mathews, G. V. T., 420, 423
Maturana, H. R., 129
Maxim, P. E., 135
Mayer, A. D., 174
Mayer, D. J., 136
Maynard Smith, J., 80, 365, 572
Mayr, E., 58, 62, 63, 64, 75
Mazur, A., 361
Meany, M. J., 167
Mech, L. D., 248, 493, 494, 496, 517
Meier, A. H., 413, 414
Meier, G. W., 337
Meikle, D. B., 320, 336, 356
Meites, J., 176
Menaker, M., 196, 201
Mendel, G., 19, 28
Menzel, E. W., 388
Merle, J., 241
Merrell, D. J., 103, 105
Metzgar, L. H., 275, 276, 523
Mewaldt, L. R., 414
Meyer, C. C., 225
Meyer, D. R., 285
Meyer, P. M., 285
Meyer, R. K., 165
Michael, R. P., 334, 335
Michel, R., 164
Michener, C. D., 242, 243, 244, 581
Milinski, M., 500
Miller, D. B., 228
Miller, J. W., 447
Miller, K. R., 281
Miller, N. E., 293
Miller, R. C., 499

Milstein, S., 284
Mitchell, G., 234, 236, 337
Mock, D. W., 347
Moehlman, P. D., 247, 574, 575
Moller, P., 395
Moltz, H., 138, 175, 254, 255
Money, J., 167
Moore, J. A., 43
Moore-Ede, M. C., 195
Moore, R. Y., 195, 201
Morgan, C. L., 19
Morrell, J. I., 178
Morris, D., 403
Morton, M. L., 414
Moyer, K. E., 345
Moynihan, M., 375
Mueller, H. C., 429, 430, 499
Muller, H. J., 303
Mullins, R. F., 165
Mundinger, P., 237
Murdock, G. P., 584
Murie, O. J., 417
Murphy, R. X., 240
Myer, J. S., 367
Myers, J. H., 106, 523

Neumann, D., 190
Newman, H. H., 101, 102
Nice, M. M., 324
Nicolaus, L. K., 498
Nicoll, C. S., 176
Niedeffer, J. D., 284
Nisbet, I. C. T., 421
Noakes, D. L. G., 244
Noirot, E., 50
Noonan, K. M., 540
Nordenskiöld, E., 15
Nottebohm, F., 22, 140, 178, 255, 291
Nottebohm, M., 140
Novak, M. A., 236
Numan, M., 138–139

O'Connell, M. E., 173
Olds, J., 135
Olgivie, D. W., 105
Oliverio, A., 24, 103
Oltmans, G. A., 141
Oppenheim, R. W., 23, 220, 223, 255
Orians, G. H., 312, 313, 314, 315, 324, 352, 452
Oring, L. W., 311, 314
Orozco, E., 242
Orr, R. T., 418
Osten-Sachen, Baron, 542

Packer, C., 453, 578
Page, T. L., 199, 201
Panksepp, J., 134
Papi, F., 426
Parker, G. A., 81, 304, 309, 365
Parks, M., 175
Parsons, P. A., 101, 102, 456
Partridge, L., 456, 458
Paton, J. A., 140, 291
Patterson, D. J., 279
Patterson, M. A., 522, 529
Pavlov, I., 270, 271, 274
Payne, R. S., 393
Pener, M. P., 164
Pengelley, E. T., 205
Perachio, A. A., 135, 360
Perdeck, A. C., 231
Peters, S., 237, 238, 239, 392
Pfaff, D. W., 178
Pfaffenberger, C. J., 95
Phillips, J. B., 426
Phoenix, C. H., 166, 333, 360
Pianka, E. R., 449
Pickert, R., 237
Piel, G., 487
Pietrewicz, A. T., 475
Pietsch, T. W., 483
Pinsker, H., 282
Pitcher, T. J., 568
Pittman, R., 220
Pittendrigh, C. S., 199, 202
Pliny, 15, 28
Plomin, R. J., 91
Popolow, H. B., 139
Pough, F. H., 512
Poulsen, H. R., 363
Powley, T. L., 138
Pratt, C. L., 261
Premack, D., 401
Price, E. O., 38, 248, 540
Price, G. R., 365
Prince, G. J., 456
Pruitt, C. H., 133
Pugesek, B., 322
Purcell, J. E., 484
Pyke, G. H., 471, 472

Quadagno, D. M., 98, 99
Quartermus, C., 244
Quinn, J. F., 479

Rainey, R. C., 431
Ralls, K., 407
Rao, T. R., 450
Ratner, S. C., 281, 540
Raup, D. M., 539
Rawson, K. S., 188

Reboulleau, C., 173
Reinisch, J. M., 168, 361
Reiter, R. J., 157, 202
Renner, M., 193, 194
Rettenmeyer, C. W., 388, 488
Ribbands, C. R., 243, 244
Richards, M. P., 174
Richardson, W. J., 421, 426
Richmond, R. C., 309
Richter, C. P., 195
Riddiford, L. M., 163
Riedel, I. B. M., 244
Ringo, J. M., 241, 242
Riopelle, A. J., 278
Robbins, D., 175
Roberts, E. P., Jr., 456
Roberts, R. C., 97
Roberts, S. K., 197, 199
Robinson, B. W., 360
Roeder, K. D., 120–121, 393
Romanes, G. J., 18–19, 28
Rood, J. P., 567, 583
Rose, R. N., 360
Rosenberg, K. M., 224, 551
Rosenblatt, J. S., 138, 174, 175, 176, 330
Rosenblum, L. A., 336
Rosenwasser, A. M., 201
Ross, D. M., 279, 280
Ross, R. M., 305
Rowe, F., 239
Rowell, T. E., 354, 355, 500
Rozin, P., 288
Rumbaugh, D. M., 377, 401, 402, 403
Rusak, B., 201
Rushforth, R. D., 279
Rusiniak, K. W., 288
Russell, P. F., 450
Rust, C. C., 165

Sackett, G. P., 261
Sade, D. S., 335, 337, 356, 394, 517
Sakri, B., 241
Sale, P. F., 456, 457
Salzen, E. A., 225
Sanders, G. D., 282
Sauer, E. G. F., 416, 423, 424
Saunders, D. S., 191
Savage-Rumbaugh, E. S., 403
Sawyer, C. H., 178
Schaller, G., 489, 490, 491, 492, 495
Schechinger, S. A., 275
Schein, M. W., 229, 329, 331
Schenkel, R., 405
Schjelderup-Ebbe, T., 354
Schleidt, W. M., 255, 256, 329, 331

Schlichte, H. J., 422
Schmidt-Koenig, K., 421, 422, 426
Schneider, D., 390
Schneiderman, H. A., 241
Schnierla, T. C., 47, 487
Schoener, T. W., 512
Scholz, S. D., 340
Schröder, J. H., 100, 101
Schultz, E. T., 552
Schultz, P., 229
Scott, J. P., 247, 345, 354, 362, 363, 365, 367, 368
Searcy, W. A., 237
Seilacher, A., 539
Seligman, M. E. P., 23, 287
Selye, H., 521
Seyfarth, R. M., 386, 453, 454
Shannon, C. E., 380, 381
Shapiro, D. Y., 305
Sharpe, L. G., 567
Shaw, E., 562
Shell, W. F., 278
Shepherd, J. G., 310
Sheppard, P. M., 71
Sherman, P. W., 27, 384, 386, 387, 454, 500, 501, 540, 568, 574
Shettleworth, S. J., 288, 477
Shields, J., 101, 102
Shields, W. M., 302, 454
Shockley, W. B., 102
Shumake, S. A., 275
Siegel, H. I., 174
Silberglied, R. E., 310, 496, 498, 504
Silver, R., 172, 173
Simonds, P. E., 336
Simons, E. L., 204
Simpson, J. I., 244
Singer, R. D., 367
Sisk, C. L., 201
Skinner, B. F., 23, 257, 271
Slater, P. J. B., 374, 381
Sluckin, W., 225
Smetzer, B., 324
Smith, A., 514
Smith, C. C., 516
Smith, D. G., 356
Smith, M. S., 585, 586
Smith, N. G., 549
Smith, R. E., 202
Smith, R. T., 102
Smith, S. M., 499
Smith, V. G. F., 304
Smith, W. J., 377, 381
Smythe, N., 502
Snodgrass, R. E., 244
Sokolove, P. G., 199
Sontag, L. W., 224
Southern, W. E., 427
Southwick, C. H., 361, 363, 519
Spalding, D. A., 225

Sperry, R., 136, 138, 140
Spieth, H. T., 241
Spurrier, W. A., 202
Stacey, P. B., 582
Stallcup, J. A., 564
Stanley, S. M., 18
Stebbins, G. L., 62
Steidinger, P., 426
Stein, A., 221
Stephan, F. K., 201
Sterner, R. T., 275
Stevenson-Hinde, J., 287
Stewart, J., 167
Stewart, M. M., 512
Stinson, R. H., 105
Stuart, J., 428
Struhsaker, T. T., 386, 496
Sullivan, K. A., 487
Sulzman, F. M., 195
Sund, M., 100, 101
Suomi, S. J., 23, 236
Suthers, R. A., 430
Swann, J. M., 201
Sweeney, B. M., 188
Sylva, A., 249

Taitt, M. J., 510, 511
Takahashi, J. S., 196, 201
Tamarin, R. H., 523
Tamarkin, L., 202
Tamura, M., 392
Taylor, D. H., 431
Taylor, P., 477
Terasawa, E., 178
Terkel, J., 138, 175, 176
Terman, C. R., 519
Terman, M., 201
Terrace, H. S., 402
Thompson, R., 280
Thompson, W. R., 20, 96, 97, 224
Thorndike, E. L., 23–24, 256
Thornhill, R., 543, 544
Thorpe, W. H., 255, 269, 277,
    392, 459
Tilford, B. L., 320
Tinbergen, N., 21, 33, 34, 35, 36,
    121, 122, 231, 253, 276, 365,
    388, 536, 545, 546
Toft, C. A., 512
Tollman, J., 162
Toran-Allerand, C. D., 165
Travis-Niedeffer, M. N., 284
Treat, A. E., 120, 393
Treichler, F. R., 285
Trillaud, M., 167
Trivers, R. L., 27, 310, 316, 318,
    320, 321, 322, 323, 576, 577,
    580, 583, 585
Truman, J. W., 22, 163
Tryon, R., 89

Turnispeed, M. R., 432
Turpin, B., 37–38
Twiggs, D. G., 139

Uexküll, J. von, 20–21
Ulrich, R. E., 362

Valenstein, E. S., 134–135
Valenstein, P., 177
Vandenbergh, J. G., 162, 163,
    223, 240, 333, 390, 523
Van Hemel, P. E., 367
Van Tets, G. F., 537, 538
Van Valen, L., 304
Vehrencamp, S. L., 563, 568
Vernikos-Danellis, J., 195
Vessey, S. H., 320, 333, 336, 338,
    356, 358, 368, 517, 519, 520,
    522, 529, 567
Vetta, A., 102
Vince, M. A., 222, 223, 389
Vinogradova, Y. B., 191
Vivien-Roels, B., 202
vom Saal, F. S., 168, 360

Waage, J., 27, 310
Wagoner, N., 201
Walcott, C., 116, 284, 422, 427
Waldman, B., 384
Walker, B. W., 190
Wallace, A. R., 17–18, 28
Wallace, R. F., 224
Wallen, K., 162
Waloff, Z., 431
Walsh, J., 367
Waltz, E. C., 568
Wang, L. C. H., 205
Ward, J. A., 244, 245
Ward, P., 315
Warren, J. M., 278, 285
Waterman, T. H., 431
Watkins, L. R., 136
Watson, J. B., 23, 256, 257
Wayner, M. J., 24
Weaver, N., 244
Weaver, W., 380
Weber, N. A., 479, 480
Wecker, S. C., 459, 560
Weigl, P. D., 456
Weisman, R. G., 93, 105
Wells, J., 118–119, 164
Wells, M. J., 118–119, 164
Wells, P. H., 284
Wenner, A. M., 398, 431
Werblin, F. S., 130
Werren, J., 27
West Eberhard, M. J., 517
Whalen, R. E., 165
Whimbey, A. E., 224
Whitman, C. O., 20

Whittaker, R. H., 572
Wickler, W., 22, 397, 497, 544
Wiesel, N., 24, 139–140
Wilcox, R. S., 394, 399
Wilding, N., 244
Wiley, R. H., 353
Wilkinson, G. S., 578, 579
Willard, D. E., 320, 321, 585
Wille, A., 242
Williams, G. C., 27, 78, 302, 304,
    321, 527, 537, 572, 586
Williams, J. M., 430
Williams, T. C., 430
Willis, J. S., 205
Willows, A. O. D., 133, 283
Wilson, E. O., 26, 68, 74, 242,
    345, 374, 376, 381, 382, 384,
    390, 448, 449, 481, 525, 535,
    558, 559, 560, 563, 576, 586
Wilson, M. I., 585, 586
Wiltschko, R., 427
Wiltschko, W., 427
Windsor, D. M., 496, 498, 504
Winfield, I. J., 568
Winget, C. D., 195
Witham, T. G., 455
Wittenberger, J. F., 302
Wodinsky, J., 330
Wolf, L. L., 472, 485, 490, 491
Wolff, J. R., 220
Wolfson, A., 416
Woody, C. D., 291
Woolfenden, G. E., 564, 565, 574
Wortis, R. P., 172
Wright, S., 89
Wu, H. M., 574
Wyatt, G. R., 22
Wyman, R. L., 244, 245
Wynne-Edwards, V. C., 79, 80,
    514, 515, 525, 526, 569, 570

Xiang-Xu, X., 540

Yerkes, R. M., 256
Young, W. C., 165

Zahavi, A., 306, 307, 565
Zahl, P. A., 419
Zarrow, M. X., 138, 174
Zatz, M., 196
Zeveloff, S. I., 320
Zigmond, R. E., 178
Zimen, E., 247
Zimmerman, N., 201
Zimmerman, R. R., 234
Zolman, J. F., 23
Zubek, J., 91, 92
Zucker, I., 201
Zumpe, D., 335
Zwick, H., 118

SUBJECT INDEX

Acanthurus triostegus. See
Hawaiian manini
Acheta domesticus. See House
cricket
Acorn woodpecker *(Melanerpes
formicivorous)*, cooperation
among, 582
Acrididae. See Locust
Acris gryllus and *A. crepitans. See*
Cricket frogs
Acromyrmex myrmex. See Leaf
cutter ant
Action-potential, 113, 114
Actitis macularia. See Spotted
sandpiper
Acquisition, 272
Activational effects, of hormones
on behavior, 161–165
Adaptation
behavior and, 537, 547–549
biological evolution and, 63
evolution and, 71–74, 83
genetics and, 105–106
survival and, 71
Adaptive behavior, 23
Adaptive radiation, behavior
pattern evolution and, 540,
541
Adrenal exhaustion, death due to,
521–523, 526
Adrenaline, 157, 158
aggression and, 360
secretions, 157, 158, 169

Adrenocorticotrophic hormone
(ACTH), 157, 160, 169, 179,
521
Aedes spp. *See* Mosquito
African antelope (Bovidae)
feeding styles and, 488
intrasexual selection in, 308
African blue monkey *(Cercopithecus mitis)*, alarm call of, 386
African green monkey *(Cercopithecus sabaeus)*, dominance
hierarchy and, 354
African hammerheaded bat
(Hypsignathus monstrosus),
lek in, 353
African weaver finch *(Quelea)*
evolution of nest and, 545
nomadism of, 315
African wild dog *(Lycaon pictus)*
cooperative behavior of, 36–37
feeding relationship of,
387–388, 491, 495
Agelaius phoeniceus. See Red-
winged blackbird
Aggregation, 570
of frogs, 572, 573
Aggression
as agonistic behavior, 344–347,
368
defined, 344–345
display and, 385–386
dominance, 345, 353–359,
367–368

external factors in, 361–363, 369
extreme forms of, 346–347
forms of, 345
games theory model and,
365–366
humans and, 361, 363
internal factors in, 359–361
langur monkeys and, 346
mouse urination and, 390
natural selection and, 523–524,
526
population interaction and,
518–520
restraint on, 364–368
in rhesus monkeys, 135
in rodents, 40, 41
sexual behavior and, 161–163
social control of, 367
social disorganization and,
367–368
social factors and, 363–364
space use, 347, 348–353
testosterone and, 39–46
in three-spined stickleback,
253–254
See also Competition; Predation
Aggressive mimicry, 483–484
Agonistic behavior, 344–347. *See
also* Aggression
Agrotis infusa. See Bogong moth
Aix sponsa. See Wood duck
Alarm, communication for, 386–
387

Alces alces. See Moose
Algae, reproduction in, 302
Alleles, 60, 83
Allopatry, 550
Alouatta spp. *See* Howler monkey
Alpha male, among rhesus
    monkeys, 357–358
Altricial young, 225
Altruistic behavior, 80, 83
    kin selection and, 573–576
    reciprocal, 576–581
Ambystoma triginum. See Tiger
    salamander
American jacana *(Jacana spinosa)*
    parental investment in, 318
    polyandry in, 314
American toad *(Bufo americanus)*,
    and kin recognition, 384
Amphibians
    orientation in, 432–435
    ovulation in, 325
    social behavior in, 562
    See also specific amphibians
Amplitude, of biological rhythm,
    188
Anabrus simplex. See Mormon
    cricket
Anartia fatima. See Tropical
    butterfly
Anas discors. See Teal
Anas platyrhynchos. See Mallard
    duck; Peking duck
Androgens, 157, 158, 162, 166,
    167, 168, 169
Anemone. See Sea anemone
Angler fish *(Antennarius* spp.),
    aggressive mimicry in,
    483–484
Anguilla rostrata. See Eel
Ani, 568
    greater, 563
    groove-billed, 563
    smooth-billed, 563
Anisogamy, 304, 310, 319
Annelids. *See* Earthworms
Anolis carolinensis. See Green
    anole lizard
Anopheles spp. *See* Mosquito
Anopheles culifacies, egg laying in,
    450
Ant
    army, 388, 487–488
    chemicals emitted by, 382
    communication among, 560
    entrapment by ant lion,
    481–482
    as eusocial insects, 559–561
    fungus growing, 479–480
    haplodiploidy in, 305
    heathland, 446, 447

leaf cutter, 479–480, 481
    navigation and orientation and,
    431
    polyrhachis colony, 560
    as prey, 485
    sun compass orientation and,
    431
Antagonism, in endocrine system,
    160
Antarctic plunder fish *(Harpagifer
    bispinis)*, nest guarding in, 574
Anteater (Edentata), feeding
    technique, 485
Antelope
    African, 308
    feeding style, 488
Antennarius spp. *See* Angler fish
Anthias squamipinnis. See Sea bass
Anthropomorphism, classical
    Romans and, 15
Anthus pratensis. See Meadow pipit
Anthus trivialis. See Tree pipit
Antipredatory aggression, 345
Ant lion *(Myrmeleon)*, entrapment
    used by, 481–482
Apes, great, 322
Aphagia. *See* Hypophagia
Aphelocoma coerulescens. See
    Scrub jay
Aphelocoma ultramarina. See
    Mexican jay
Aphid *(Phemphigus betae)*,
    reproduction success and, 455
Apis mellifera. See Honeybee
Aplysia dactylomela. See Sea hare
Aposematic, 497
Arabian babbler *(Turdoides
    squamiceps)*, as nest
    "helpers," 565
Ardea herodias. See Great blue
    heron
Arenicola marina. See Lugworm
Armadillo, defense against
    predators, 497
Army ant *(Eciton burchelli)*,
    feeding relationship of, 388,
    487–488
Artificial selection, 76, 94–95
    aggression and, 361
    domestication and, 540–542
Arthropods, learning and,
    283–284
Aschoff's Rule, 192–193
Asexual reproduction, 302
Asteroidea. See Starfish
Attachment, imprinting and, 225
Attack behavior, 361–362. *See
    also* Aggression
Auditory imprinting, 228
Aurelia. See Jellyfish

Australian tree frog *(Hyla ewingi;
    Hyla verreauxi)*, character
    displacement in, 550, 551
Autoradiography, for endocrine
    system study, 161, 178
Avocet *(Charadrius vociferus)*, bro-
    ken-wing display of, 500–501
Avoidance learning, 275
Axon, 112, 144

Baboon *(Papio* spp.)
    core area of, 349
    defense against predators, 500
    dispersal and, 453–454
    dominance in, 335, 368
    grooming in, 394
    hamadryas, 204
    home range of, 349
    olive, 578
Bacteria, transformation in, 302
Baiomys taylori. See Pygmy mouse
Balanus balanoides. See Barnacle
Baleen whale *(Mysticeti)*, song of,
    393
Balloon fly *(Empis; Hilara sartor)*,
    comparative behavior series
    in, 542–544
Banded mongoose *(Mungos
    mungo)*, breeding habits of,
    566
Barn owl *(Tyto alba)*, choice of
    food and, 472, 473, 475
Bank swallow *(Riparia riparia)*
    coloniality of, 500–501
    protection against predators,
    568
Barnacle *(Balanus balanoides;
    Chthamalus stellatus)*, and
    population regulation, 511
Bat
    African hammerheaded, 353
    colonial, 574
    echolocation in, 50, 51, 393,
    429, 430
    hoary, 417
    little brown, 417, 430
    migration, 417
    See also Vampire bat
Bateman effect, 316, 320
Batesian mimics, 498
Beachhoppers *(Talitrus)*, lunar
    orientation and, 431
Bee
    antennae of, 394
    bumblebee, 471–472
    chemicals emitted by, 382
    discrimination of color and, 284
    haplodiploidy in, 305
    honeybee, 193, 194, 242–244,
    394, 398–400, 405

sweat, 384
Beetle
  bombardier, 497
  dung, 470
  speciation in, 76
  sun compass orientation and,
    431
Behavior
  environment-endocrine
    interaction with, 169–178
  hormones and, *see* Endocrine
    system
  population self-regulation and,
    514–524
  See also Learning; Motivation
Behavioral ecology, 8–9, 20,
    25–27
Behavioral genetic mechanisms,
    in population regulation,
    523–524
Behavioral isolation, in speciation,
    77
Behavioral mechanisms in
    population regulation, 514–
    521
Behavioral physiological
    mechanisms in population
    regulation, 521–523
Behavioral sink, 518
Behavior development
  in adult life, 250–252
  early post natal events and,
    225–232
  embryology, 219–225
  epigenesis and, 63, 90, 102–104
  experimental design and,
    218–219
  food preference, 229–232
  generalized prenatal stimula-
    tion, 219–225
  genetic influence, 252–256
  juvenile events and, 232–246
  imprinting, 225–229
  modal action patterns, 255–256
  motor system, 221–224
  nervous system, 220–221
  in rhesus monkeys, 336–337
  sensory system, 221–224
Behavior genetics, 90, 107
  evolution and, 103, 105–106
  genes and behavior, 102–104
  genetic determination models,
    90–99
  inheritance modes, 100
Behaviorism, 23, 256, 257
Behavior patterns, evolution of
  adaptation, 537, 547–549, 553
  adaptive radiation and, 540, 541
  communicative behavior, 546
  comparative series of, 542–546

domestication, 540–542
fossils and, 537, 539, 540
homologies in, 536–537
isolating mechanisms, 549–551
microevolution and, 536–537
phylogeny and, 537, 538, 540,
    544
progression in, 537, 539
speciation, 537
tradition and culture and,
    551–553
See also Social behavior;
    specific behaviors
Belding's ground squirrel
    (Spermophilus beldingi), alarm
    call in, 386–387, 454, 496,
    568
Belted kingfisher, bill adaptation
    of, 72
Betta splendens. See Siamese
    fighting fish
Bilaterally symmetrical nervous
    system, 125–126, 145
Bioassays for endocrine system
    study, 161
Biogenic law, 546
Biological clocks, 187, 189, 413
Biological communication, 374
Biological evolution, 63–65
Biological pacemaker, location
    and physiology for,
    196–202, 207
Biological rhythms, 187, 188–195
  pacemaker for, 196–202
  model system of, 196, 197
Biological timekeeping, signifi-
    cance of, 202–207
Bird of paradise, display of, 18
Birds
  aggression in, 347, 361
  alarm call of, 500
  bill adaptation in, 72
  bill length of, 550
  biological rhythms in, 201
  birdsong, 382–383, 392
  butterfly consumption by, 275
  clutch size, 526
  communication, 389, 552
  courtship displays in, 306, 403
  fertilization, 328, 329
  imprinting, 225–227, 228, 229
  individual distance and, 348,
    349
  migration in, 190, 206, 411–414
  monogamy in, 311, 312
  nectar-feeding, 352
  orientation and navigation in,
    420–427
  parental investment, 318, 321,
    322

polyandry in, 314
prenatal behavior develop-
    ment, 222–223
prenatal sensory and motor
    development, 222
reproduction in, 324
search image of, 475
social behavior in, 562–566
territory and, 350, 351
territory and population
    regulation, 515–517
See also specific birds
Birdsong, 236–238, 382–383
Birth. *See* Reproduction
Bison *(Bison bison)* copulation, 328
Black-backed jackal *(Canis
    mesomelas)*, pup survival and
    helpers in, 574, 575
Blackbirds
  epideictic display of, 514–515
  migration in, 414
  territoriality and, 352
  tricolored, 314–315, 352
  See also Red-winged blackbird
Black-headed gull *(Larus
    ridibundus)*, eggshell removal
    in, 33–36
Black-tailed deer *(Odocoileus
    hemionus)* copulation, 328
Black-tailed prairie dogs *(Cynomys
    ludovicianus)*, dispersal and,
    453
Blarina. See Shrew
Blood transfusions, for endocrine
    system study, 161
Blowfly *(Phormia regina)*, feeding
    behavior of, 132–133
Bluegill sunfish *(Lepomis
    macrochirus)*, feeding and
    aggression in, 363–364
Blue jay *(Cyanocitta cristata)*
  bank swallow attacked by, 500–
    501
  monarch butterfly eaten by, 497
  search image of, 475, 476
  sociality in, 563, 564
Blue tit *(Parus caeruleus)*, habitat
    selection, 456–457, 458
Blue whale feeding technique, 478
Bobwhite quail *(Colinus virgin-
    ianus)*
  prenatal vocalization and,
    222–223
  sociality of, 568
Bogong moth *(Agrotis infusa)*,
    epideictic display of, 515
Bombardier beetle *(Brachinus
    spp.)*, defense against
    predators, 497
Bombykol, 116, 117, 390

Bombyx mori. See Silkworm moth
Bonnet macaque (*Macaca radiata*)
　copulation, 328
　infancy in, 336
　hierarchy in, 362
Bovidae. See African antelope
Bowl and doily spider (*Frontinella
　pyramitela*), intrasexual
　competition and, 309
Brachinus spp. *See* Bombardier
　beetle
Breeding
　choice of sites, 452–458
　inbreeding, 91–92, 95
　selective, 93–97
　See also Reproduction
Brent goose, alarm posture of, 388
Brittle stars (*Ophiuroidea*),
　aggregation in, 570, 571
Brown-headed cowbird (*Mo-
　lothrus ater*), courtship in, 308
Brown trout (*Salmo trutta*),
　epideictic display of, 515
Bruce effect, 310
Bryozoan (*Bugulla turrita;
　Schizoporella errata*), 568
Buffalo (*Bison bison*), as prey of
　wolf, 493
Buffalo, African (*Syncerus caffer*),
　as prey of lion, 489
Bufo americanus. See American
　toad
Bugulla turrita. See Bryozoan
Bullfrog (*Rana catesbeiana*)
　dominance in, 352–353
　reproduction, 562, 563
　territoriality and, 352
Bumblebee, foraging, 471–472
Butterfly
　avoidance conditioning and,
　275
　disruptive selection and, 71
　intrasexual competition and,
　310
　markings, function of, 497–498
　monarch, 275, 418, 419
　Papilio dardanus, 71
　queen, 496
　tropical, 497
　viceroy, 275

Cactospiza pallida. See Wood-
　pecker finch
Caddisfly (*Polycentropus*),
　underwater nets of, 481
Calidris canutus. See Knots
California gulls (*Larus californicus*)
　monogamy in, 312
　reproductive effort in, 322
　territoriality in, 351

Callicebus. See Titi monkey
Calopteryx maculata. See Damselfly
Camouflage, 156
　weasel and, 165
Canary (*Serinus canarius*),
　relationship between brain
　organization and songs,
　140–141
Canids, dependency and parental
　care and, 322
　play behavior and, 247–249
　See also Dog; Wolf
Canis familiaris. See Dog
Canis latrans. See Coyote
Canis lupus. See Wolf
Canis mesomelas. See Black-backed
　jackal
Cannibalism, 346–347
Cannulation, nervous system
　investigation with, 142–143
Cape May warbler (*Dendroica
　tigrina*), resource partitioning
　of, 478, 479, 480
Care, communication for giving
　and soliciting, 388–389. *See
　also* Parental care
Caribou (*Rangifer tarandus*)
　migration of, 417
　wolf as predator on, 493, 512
Carnivores
　extinct, 539–540
　feeding relationships of,
　489–495
　See also specific carnivores
Carrying capacity, 509
Castration, aggression and, 360
Cat (*Felis domesticus*)
　aggression in, 359
　light stimuli processing in
　brain, 139–140
　sexual behavior of, 161, 325,
　330, 363
Cattle, 167
Caterpillar, defense display of,
　496, 497
Catocala. See Noctuid moth
Causation, 51, 52
Cave paintings, depicting interest
　in animal behavior, 14–15
Cebus monkey, communication
　among, 382
Central filtering processes,
　121–122
Central place foragers, 476
Cercopithecus aethiops. See Vervet
　monkey
Cercopithecus ascanius. See
　Red-tailed monkey
Cercopithecus mitis. See African
　blue monkey

Cercopithecus sabaeus. See African
　green monkey
Cervidae. See Deer
Cervus canadensis. See Elk
Chaffinch (*Fringilla coelebs*)
　autoradiography and, 178
　song in, 392
Character displacement, 74, 550
Charadrius vociferus. See Avocet
*Charmosyna placentis; C.
　ribrigularis. See* Lorikeet
Chelonia mydas. See Green turtle
Chelydra serpentina. See Snapping
　turtle
Chemical signals. *See* Pheromones
Chicken (*Gallus gallus*)
　courtship display in, 403–404,
　405
　defense against predators, 496
　dominance in, 354, 360
　prenatal behavior develop-
　ment, 220, 221, 223
　prenatal motor system develop-
　ment, 223
Chimpanzee (*Pan troglodytes*)
　food-hunting communication,
　388
　insight learning in, 275, 276
　language in, 288, 289, 377, 401–
　403, 406
　tool use in feeding, 484
Chipmunk (*Tamias striatus*)
　foraging and, 476
　hibernation, 205
　natural cycle and, 196
Chipping sparrow (*Spizella
　passerina*), habitat selection,
　458–459
Chthamalus stellatus. See Barnacle
Cichlasoma biosallatum. See
　Cichlid fish
Cichlid fish
　aggression in, 367
　mouth-breeding, 544–545
Circadian rhythm, 189, 190
　of birds, 201
　of cockroach, 197–199
　of monkeys, 201
　of plankton, 206–207
　of rats, 201
　of sea hare, 199–201
　of vertebrates, 201–202
　of woodlouse, 203
Circannual rhythm, 189, 190, 191
　of birds, 206
　of mammals, 202
　of marine organisms, 206–207
　See also Hibernation
Circular arena, imprinting and,
　226

Cistothorus palustris. See Marsh wren
Classical conditioning, 269–270
 in planaria, 280
Classical world, interest in animal behavior of, 14–15
Class recognition, 384
Climate, reproduction and, 323–325
Clumping, 570
Clunio marinus. See Midge
Coal tit *(Parus ater)*, habitat selection, 458
Cockroach *(Periplaneta* spp.)
 antennae of, 394
 pacemakers for biological rhythms and, 197–199
 sexual attraction in, 163
Coefficient of relationship, 78, 79
 parent-offspring conflict and, 322–323
Coelenterates
 learning and, 279–280
 nerve net in, 122–123
Colinus virginianus. See Bobwhite quail
Collistella scabra. See Limpet
Colonial bat *(Phyllostomus hastatus)*, relatedness in, 574
Coloration
 as defense against predators, 496–497
 in fiddler crabs, 194
 in fish, 397
 for group coordination, 382
 in noxious animals, 498–499
 in vertebrates, 156
Columba livia. See Homing pigeon
Common gull, oblique long call in, 536
Communication, 374, 406
 channels of, 389–398, 406
 dialect in, 551
 evolution of behavior of, 546
 functions, 381–389, 406
 honesty in, 404–406
 language, 398–403
 measurement of, 378–381
 metacommunication, 377–378
 signals conveying information, 375–378, 382–383
 social insects and, 560
 See also Displays
Comparative behavior series, 542–546
Comparative learning, 286–287
Comparative method, 18–19
Comparative physiology, 20
Comparative psychology, 6, 20, 25, 53

advantages and disadvantages, 38–39, 53
 definition, 37
 study using, 37–38
Competition, 71
 adaptation and, 73–74, 83
 aggression and, 345–346
 behavior and, 16–17
 contest, 345–346, 351, 512–514
 diet and, 474
 habitat selection and, 450–452
 intraspecific, 512
 natural selection and, 63–65
 reproductive success and, 517
 social behavior increasing, 568–569
 territorial defense and, 351
Composite signals, 376–377
Compound eyes, of insects, 125, 126
Compression hypothesis, 474
Conditioning
 avoidance, 275
 classical, 269–270
 operant, 270–271, 272
 to noxious animals, 498
 pseudoconditioning, 280
Confusion effect, as defense against predator, 499–500
Connochaetes taurinus. See Wildebeest
Conspecifics, 26
Contest competition, 345–346
Control groups, 43–44
Cooperation, 569–571
 definition, 558
 See also Altruistic behavior; Social behavior
Copulation, 325–329
 regurgitative food offering during, 544
 tactile communication in, 394
 See also Reproduction
Coqui *(Eleutherodactylus coqui)*, shelter and, 511–512
Coral reef fish
 color patterns in, 382
 sex change in, 303
Coral snake *(Micrurus fulvius)*, defense against predator, 497
Core area, 348
Correlation, 51, 52
Corvus brachyrhynchos. See Crow
Corvus corax. See Crow
Cost of meiosis, 302
Coturnix coturnix. See Japanese quail
Courtship behavior, variations in, 103, 105. *See also* Reproduction

Cowbird, communication and, 377
Coyote *(Canis latrans)*
 play behavior in, 247–248
 sheep predation by, 275
Crab, fiddler, 194, 269
Crabs, mazes and, 284
Crateroscelis murina, C. robusta, 451, 452
Crayfish
 communication measured in, 378–381
 mazes and, 284
Cricket
 field, 308
 house, 125
Cricket frog *(Acris crepitans; Acris gryllus)*
 character displacement in, 74, 75
 orientation in, 432–433
Critical period
 of hormonal injections, 168–169
 for imprinting, 227
Crocuta crocuta. See Hyena
Cross-fostering, 97–99, 107
Cross-sectional design, 219
Crotophaga ani. See Smooth-billed ani
Crotophaga major. See Greater ani
Crotophaga sulcirostris. See Groove-billed ani
Crow *(Corvus brachyrhynchos; C. corax)*, feeding behavior, 34, 35, 498, 500
Crowding, aggression and, 363
Crustaceans
 color change in, 165
 molting in, 164–165
 See also specific crustaceans
Cryptic moth, blue jay detection of, 475–476
Cuckoo, social behavior of, 562.
 See also Ani
Cues, habitat selection and, 455–457
Culture, behavior learned via, 552–553
Cyanocitta cristata. See Blue jay
Cyanocitta stelleri. See Steller's jay
Cycle, of biological rhythm, 188
Cyclical selection, 78
Cynomys spp. *See* Prairie dog
Cynomys ludovicianus. See Black-tailed prairie dog

Damselfly *(Calopteryx maculata)*
 intrasexual competition and, 310

Danaus plexippus. See Monarch
　　butterfly
Danaus gilippus. See Queen
　　butterfly
Danger, communication for,
　　386–387
Daphnia. See Water flea
Dark-eyed junco (*Junco hyemalis*),
　　habitat selection and, 456
Data points, independent, 44–46
Deception, as defense against
　　predators, 496–497
Deer (*Odocoileus* spp.)
　　antlers of, 307, 308, 376
　　black-tailed, 328
　　intrasexual selection in, 308
　　population densities of, 462
　　as prey of wolf, 493
　　red, 167
　　white-tailed, 496
Deermouse (*Peromyscus* spp.)
　　food supply and, 510–511
　　grassland, 105, 459–460
　　habitat selection, 459–460
　　latent learning, 275
　　nest size, 93
　　orientation and navigation in,
　　　428–429
　　population regulation and, 509
　　woodland, 105, 459–460
Deme, 79–80, 83
　　recognition, 384
Dendrites, 112, 144
Dendroica coronata. See Yellow-
　　rumped warbler
Dendroica tigrina. See Cape May
　　warbler
Density-dependent factors,
　　population regulation and,
　　510
Density-independent factors,
　　population regulation and,
　　510
Dependency, parental care and,
　　322
Dependent variables, 43
Deprivation, behavior develop-
　　ment and, 233–236
Dermasterias imbicata. See Starfish
Desert locust (*Schistocera gregoria*),
　　development, 164
Desert tortoise, migration in, 420
Desmodus rotundus. See Vampire
　　bat
Detection
　　as defense against predators,
　　　500
　　as feeding technique, 481–482
Detritus feeders, 478

Development of behavior. *See*
　　Behavior development
Dewlap display patterns, 177
Diapause phase, 191
Dicrostonyx spp. *See* Lemming
Didelphis virginiana. See Opossum
Digger wasp (*Philanthus
　　triangulum*), burrow location,
　　276
Dimorphic, 305
Diploid, 59
Directional selection, 70, 83
Discrete signals, 375
Dispersal
　　habitat use and, 447–450, 463
　　population regulation and, 453,
　　　523, 524
Displacement activities, displays
　　and, 403
Displays, 375
　　aggression and, 364, 385–386
　　definition, 20
　　as derived movements, 546
　　distance of, 375–376
　　distraction, 501–502
　　epideictic, 514–515
　　as insect defense, 497
　　origin of, 403–404, 406
　　reproduction and, 306
　　ritualization of, 546
　　sequence of, 376–377
　　social status and, 385–386
　　tactile, 394
　　visual, 396–397
Disruptive selection, 70, 71, 83
Distraction display, as defense
　　against predators, 501–502
Distribution of a species, 445. *See
　　also* Habitat selection
Disuse hypothesis, 234
Diurnality, 203–205
Diversifying selection. *See*
　　Disruptive selection
Division of labor, social behavior
　　benefiting, 568–569
Dizygotic, 101
Dog
　　African wild, 36–37, 387–388,
　　　491, 495
　　copulation in, 328
　　deprivation and, 233–234
　　domestication, 540–542
　　drinking behavior, 294
　　feeding behavior, 291, 294
　　guide dogs, 95
　　metacommunication with play
　　　bow in, 377
　　play behavior in, 247–249
　　salivary reflex, 270, 271, 274

threatening and submissive
　　postures in, 364
　　vocalizations of, 393
Dominance, 345, 353–354
　　advantages of, 358–359
　　maintenance, 368
　　in rhesus monkey, 356–358
Dominance hierarchies, 354–358,
　　369
　　population regulation and,
　　　517–518
Dominant alleles, 60, 61
Dominant mode, of inheritance,
　　100, 107
Dove, ring, 161–162, 170–174,
　　325, 326, 548
Downy woodpecker (*Picoides
　　pubescens*), group feeding,
　　486–487
Dragonfly (*Tetragoneura cynosura*)
　　circadian rhythm and, 190
　　circannual rhythm of, 191
Drive, 268
　　as homeostatic mechanism, 291
　　See also Motivation
Drosophila grimshawi. See
　　Hawaiian fruitfly
Drosophila melanogaster. See Fruit
　　fly
Duck
　　communication synchronizing
　　　hatching in, 389
　　habituation in, 269
　　imprinting, 226, 227, 228
　　mallard, 226, 227, 228–229,
　　　277, 412
　　mandarin, 403, 404
　　migration, 412–413
　　Peking, 222
　　prenatal sensory and motor
　　　development, 223
　　sexual imprinting in, 229
　　teal, 412–413
　　wing preening in, 403
　　wood, 229
Dugesia trigrina. See Planaria
Dung beetle, robber fly feeding
　　on, 470
Dwarf mongoose (*Helogale
　　parvula*), breeding habits of,
　　566–567
Dynamic hypothesis, of memory,
　　290

Eagle, golden, 72
Early experience, 458–461
　　habitat selection and, 458–461
Early humans, interest in animal
　　behavior and, 13–14

Earthworm *(Lumbricus terrestris)*
  chemical alarm substances of,
    386
  learning in, 281
  nervous system in, 125–126
  reproduction in, 302
Eastern bluebird *(Sialia sialis)*,
    448
Ecdysone, 151, 169
Echinoderms. *See* Starfish
Echolocation
  in bats, 50, 51, 116, 393, 429,
    430
  in marine mammals, 393
  in mice, 50
Eciton burchelli. See Army ant
Eclosion, 163–164
Ecological niche, 71
Ecology
  mating systems and, 314–315,
    338
  parental care and, 321–322
Ecosystems, feeding relationships
    and, 468–470, 502–503
Ectotherms, biological rhythms
    of, 195
Edentata. *See* Anteater
Eel *(Anguilla rostrata)*, migration,
    420
Egyptian vulture *(Neophron
    percnopterus)*, feeding tech-
    nique, 484
Electric field, communication by,
    395
Electric fish *(Gymnarchus niloticus)*
  electroreceptors of, 395, 482
  food detection, 482
  Mormyridae, 395
  sensation in, 116–117
Electrode stimulation, nervous
    system investigation
    through, 129–131, 133–136
Electrophoresis, 106
Elephant *(Loxodonta africana)*,
    feeding technique, 478
Elephant seal *(Mirounga angusti-
    rostris)*, reproduction in, 309,
    317, 318, 359, 517
Elephant tortoise, 17
Eleutherodactylus coqui. See Coqui
Elk *(Cervus canadensis)*, as prey of
    wolf, 493
Embryology, of behavior,
    219–225, 258
Emotionality, open-field test
    arena for, 226
Empis. See Balloon fly
Encephalization, in vertebrates,
    127, 128

Endocrine system
  activational effects on
    behavior, 161–165
  antagonism and, 160
  endocrine-environment
    behavior interactions, 169–
    178
  feedback loops in, 158–160,
    171, 179
  invertebrate, 153–154
  organizational effects on
    behavior, 161, 165–169
  research techniques for,
    160–161
  synergism and, 160
  vertebrate, 154–160
  See also Hormones
Endogenous pacemaker, 191–195
Endorphins, 136
Endotherms, biological clocks of,
    195. *See also* specific
    endotherms
Energy
  definition, 468
  feeding relationships and,
    468–470
  net energy gain, 485
  net rate of energy intake, 470
Engram, 291
Enhydra lutris. See Sea otter
Entrainment, 195
Environment, 71
  adaptation and, 73
  behavior-endocrine interactions
    with, 169–178
  genes and, *see* Epigenesis
  territoriality and, 281–282
  See also Experience
Environmental cues, 455–457
Epicycles, 189, 190
Equation, 286
Epideictic displays, 514–515
Epigamic selection, 351–352
Epigenesis, 63, 90, 102, 104, 219,
    252–258
Equipotentiality, 136
Equus burchelli. See Zebra
Erithacus rubecula. See European
    robin
Erythrocebus patas. See Patas
    monkey
Escape, as defense against
    predation, 496
Eschrichtius gibbosus. See Gray
    whale
Estrogen, 157, 158, 160,
    165–166, 168–169
  aggression and, 360–361
  parturition and, 175

  reproductive behavior and,
    171, 177–178
Estrous cycle, 325, 327, 333, 334
Estrus, 325, 332
Ethogram, 20, 33, 34, 53
Ethology, 6, 20–22
  advantages and disadvantages,
    36–37, 53
  compared with comparative
    psychology, 24–25
  definition, 20, 33
  genetic control of behavior
    and, 252–256
  study using, 33–37
Etroplus maculatus. See Orange
    chromide
Eumomota superceliosa. See
    Motmot
European robin *(Erithacus
    rubecula)*, migratory restless-
    ness in, 425
European starling *(Sturnus
    vulgaris)*
  dispersal, 447–448
  epideictic display of, 514–515
  individual distance and,
    349–350
  orientation and navigation, 422
Eusocial insects, 559–561
  kin selection and, 575–576
  See also Ant; Bee; Wasp
Evolution, 16–18
  adaptation, 71–74
  biological, 63–65
  causes of, 67–69
  evaluation, 82, 83
  evolutionary stable strategy
    (ESS), 80–81, 83, 365
  genetics and, 103, 105–106
  migration and, 414–416
  nervous system and, 122–131
  ritualization, 403
  selection, 69–71
  self-regulation and, 524–527
  speciation, 74–75
  synthetic theory of, 62–63
  See also Behavior patterns, evo-
    lution of; Genetics; Natural
    selection; Population
    genetics; Social behavior
Evolutionary processes, 9–10
Evolutionary stable strategy
    (ESS), 80–81, 83, 365
Experience
  aggression and, 361–363
  behavior and, 254–255
  definition, 254
  epigenesis, 63, 90, 102–103,
    224–228

Experimental methodology,
    53–54
  control groups and, 43–44
  definitions and records and, 39–
    41
  designation of variables and, 43
  development of independent
    data points and, 44–46
  hypothesis formulation and,
    41–43
  sample size and, 46
Exteroreceptors, 115–116
Extinction, 273
Extirpation, endocrine system
    and, 160

Fatigue, 269
Feedback loops, 171, 179
  in endocrine system, 158–160
  negative, 521–523
Feeding relationships, 467
  aggression and, 363
  begging and offering communi-
    cation, 388
  courtship display and, 403
  defense against predators, 495–
    502
  ecosystems and trophic levels
    and, 468–470
  evolution of, 539
  feeding technique, 477–485
  food choice, 472–476
  foraging strategies, 470–477
  group feeding, 486–488
  hunting communication, 387–
    388
  relocation and, 476
  search image and, 475
  social behavior and, 485–495,
    568
  social carnivores, 489–495
Felids, dependency and parental
    care, and, 322. *See also*
    specific felids
Felis domesticus. See Cat
Female access polyandry, 311
Female defense polygyny, 311
"Fence effect," 523
Fertilization, 325–329
  mammals and, 327–329
  See also Reproduction
Ficedula hypoleuca. See Pied
    flycatcher
Fiddler crab (*Uca pugnax*)
  burrowing of, 269
  color change in, 194
Field cricket (*Gryllidae*), chirp of,
    308
Field methods, 47–48
Filial imprinting, 225

Filter feeding, 478
Finch
  adaptive radiation in, 540, 541
  African weaver, 315
  cactus feeding, 76
  foraging flocks, 486
  ground, 76, 548–549
  woodpecker, 484, 485
  zebra, 229
Firefly (*Photuris versicolor*)
  aggressive mimicry in, 483
  flash codes of, 396–397, 405
  light emitted by, 375
Fish
  agonistic behavior and,
    363–364
  cleaning symbioses in, 580
  habitat selection in, 456–457
  homing in, 431–432
  polarized light and, 431
  reproduction in, 301, 303, 328
  social behavior in, 244–246,
    561–562
  See also specific fish
Fitness
  direct, 80
  inclusive, 80
  indirect, 80
Fixed action pattern (FAP), 253,
    255, 256
  of communication, 374–375
Flatworms. *See* Planaria
Flesh fly (*Sarcophaga argyrostoma*),
    circannual rhythm of, 191
Flower pecker (*Melanocharis niger;
    M. longicanda; M. versteri*)
    distribution, 452, 453
Fly
  balloon, 542–544
  blowfly, 132–133
  flesh, 191
  pine sawfly, 509–510
  robber, 468–470
  scorpionfly, 543
  stilt-legged, 544
  yellow dung, 81
  See also Fruit fly
Fixed interval, 273
Fixed pattern, 545
Fixed ratio, 273
Flying squirrel (*Glaucomus volans*)
  entrainment and, 195
  free-running rhythms,
    191–192, 193
Foci, study of mosaics and, 103
Follicle-stimulating hormone
    (FSH), reproductive behavior
    and, 157, 158, 159, 160, 171,
    176, 179
Food. *See* Feeding relationships

Food chains, 468
Food supply, population
    regulation and, 510–511
Foraging strategies, 470–477. *See
    also* Feeding relationships
Fossils, evolution of behavior pat-
    terns studied from, 537, 539,
    540
Founder effect, 69
Fox (*Vulpes fulva*), reproductive
    cycle and, 325
Freemartin, 167
Free-running rhythms, 191–193
Freeze, as escape against
    predators, 496
French grunts (*Haemulon
    flavolineatum*), social tradi-
    tions and, 552
Fringilla coelebs. See Chaffinch
Frog
  aggregation, 572–573
  Australian tree, 550, 551
  autoradiography and, 178
  bullfrog, 562, 563
  coqui, 511
  cricket, 74, 75, 432–433
  fertilization in, 329
  migration in, 420
  nervous system of, 129–130
  orientation in, 432–433
  variation in natural activity
    periods, 194–195
  wood, 457
Frontinella pyramitela. See Bowl
    and doily spider
Fruit fly (*Drosophila melanogaster*)
  artificial selection and, 94–95
  "Bateman effect" in, 316–317
  circadian rhythm and, 189
  gene frequency in, 103, 105
  habitat selection, 455–460
  Hawaiian, 241
  juvenile events and, 241–244
  mating behavior, 99, 241–242
  mosaics and, 103
  wing development, 257–258
Frustration, aggression caused by,
    365
Functional neuroanatomy,
    investigation of nervous
    system through, 139–141
Functionalism, 22–23
Fundamental niche, 446

Gallus gallus. See Chicken
Gametes, 59
Ganglia, 117
Gaphoglypt, foraging behavior
    from tunnels of, 539

Garter snake (*Thamnophis sirtalis*), food preference of, 133, 474–475

Gasterosteus aculeatus. *See* Three-spined stickleback

Gazella thomsoni. *See* Thomson's gazelle

Gene flow, 68, 69, 83

Gene pools, 58, 83

General adaptation syndrome (GAS), 521, 526

Generalists, feeding behavior and, 468

Generalization, 274

Generalized vertebrate nervous system, 126–129, 145

Generation time, 240

Genes, 58–59, 82–83
environment and, *see* Epigenesis
frequencies, 68–69, 70, 83, 103, 105
See also Genetics

Genetic determination, models, 90–99

Genetic disharmony, 77

Genetic drift, 68–69, 83

Genetics
adaptations and, 105–106
behavior influenced by, 252–256, 361
evolution and, 103, 105–106
group selection, 79–80
kin selection, 79–80
mutations, 99
polymorphism, 77–78
principles of, 59–61
species and gene pools, 58–59
See also Behavior genetics;
Evolution; Population genetics

Genetic relationship, measurement of, 78, 79. *See also*
Coefficient of relationship

Genotype, 58, 83, 107

Geographic isolation, speciation by, 75–76

Geological change, evidence for, 16

Geomagnetic cue, in bird navigation and orientation, 426–427

Geospiza fortis. *See* Ground finch

Gerbil. *See* Mongolian gerbil

Gerris remigis. *See* Water strider

Giant silkworm moth (*Hyalophora euryalis*), sensation in, 116

Giraffe (*Giraffa camelopardalis*), as prey of lion, 489

Glaucomys volans. *See* Flying squirrel

Glaucous gull (*Larus hyperboreus*), signal function of eye ring in, 549–550

Goat
drinking behavior in, 294
imprinting and, 227

Golden-crowned sparrow (*Zonotrichia atricapilla*), migration in, 413–414

Golden eagle, bill adaptation, 72

Golden mouse (*Ochrotomys nuttalli*), copulation in, 328

Golden plover (*Pluviahs domenica*) migration habits and, 414

Golden-winged sunbird (*Nectarinia reichenowi*)
defense of feeding territory and, 472
territorial defense and, 485

Goldfish, grouping and, 571

Gonads, 155
reproduction and, 177, 178
secretions, 157, 158, 165–169

Goose
alarm posture of, 387, 388
brent, 388
female feeding rate in flocks, 46
migration, 411

Gopher, pocket, 452

Gorilla (*Gorilla gorilla*), feeding technique, 478

Graded signals, 375

Grasshopper (*Melanopus*)
brain neurosecretory cells and glands of, 153–154
development, 151–152

Grassland deermouse (*Peromyscus maniculatus bairdi*)
early experience and, 459–460
food choices, 250–252
habitat adaptation in, 105

Gray whale (*Eschrichtius gibbosus*), migration, 418

Grazing, 478

Great ape, dependency and, 322

Great blue heron (*Ardea herodias*), food choice of, 474

Greater ani (*Crotophaga*), communal nest of, 562

Great tit (*Parus major*), population regulation in, 516
reward ratio and, 477

Green anole lizard (*Anolis carolinensis*), reproduction in, 176–178

Green heron, individual distance and, 349

Green turtle (*Chelonia mydas*), orientation in, 433–435

Green woodhoopoe (*Phoeniculus purpureus*), cooperation among, 575

Grey-crowned babblers (*Pomatostomus temporalis*), breeding habits and, 565–566

Grooming, 394

Groove-billed ani (*Crotophaga sulcirostris*), communal nesting, 563

Ground finch (*Geospiza fortis*), natural selection and, 548–549

Ground squirrel (*Spermophilus* spp.)
hibernation, 202, 205
predator avoidance, 269, 496, 574

Group defense, as defense against predator, 500–501

Group feeding, 486–488
social carnivores and, 489–495

Group selection, 79–80, 525, 572

Group spacing, communication for, 382

Grunion (*Leuresthes* spp.), lunar cycle and, 189, 190

Gryllidae. *See* Field cricket

Guinea pigs, sexual behavior, hormones and, 166–167

Guira cuckoo (*Guira guira*), 562

Gull
behavior patterns in, 545
black-headed, 33–36
California, 312, 322, 351
common, 536
eggshell removal by, 33–36
glaucous, 549–550
herring, 34, 35, 121–122, 388, 536, 546
kittiwake, 35
laughing, 231–232, 233, 457
monogamy in, 311, 312
oblique long call of, 536
ring-billed, 427
sea, 311, 350
Thayer's 549–550

Gymnarchus niloticus. *See* Electric fish

Gymnorhinus cyanocephalus. *See* Piñon jay

Gymnotidae. *See* Electric fish

Habitat, compression hypothesis and, 474

Habitat imprinting, 459–460

Habitat isolation, 77

Habitat selection, 445, 463
behavioral restriction on habitat use, 450

Habitat selection *(cont.)*
  environment cues, 455–457
  definition, 445
  determinants of, 458–461
  physical and chemical factors
    and, 452
  restrictions on habitat use,
    446–452
  See also Orientation and
    navigation
Habituation, 269, 279
Haemulon flavolineatum. See
    French grunts
Hamadryas baboon, evolution of,
    204
Hamster
  aggression and sexual
    behavior, 162–163
  effects of melatonin on, 202
  heritability, 96–97
  hibernation and, 205
  Syrian golden, 162–163
Haplochromis spp. *See* Mouth-
    breeding cichlid fish
Haplodiploidy, 305, 576
Haploid, 59
Harbor seal *(Phoca vitulina)*,
    migration, 418
Hardy-Weinberg formula, 67–68
Hare, snowshoe, 521
Harpagifer bispinis. See Antarctic
    plunder fish
Hatching, communication
    synchronizing, 389
Hawaiian fruit flies, lek in, 353
Hawaiian honeycreeper *(Vestiaria
    coccinea)*, territoriality and,
    352
Hawaiian manini *(Acanthurus
    triostegus)*, habitat selection
    in, 456
Hawfinch, bill adaptation, 72
Hawks, confusion effect and, 499
Heathland ant *(Lasius niger)*,
    distribution, 446–447
Hedgehog, defense against pred-
    ators, 497
Heifers, and postnatal treatment
    with androgen, 167
Helogale parvula. See Dwarf
    mongoose
Heredity
  habitat selection and, 458
  population regulation and, 523
  See also Genetics
Heritability, measurement,
    95–97, 102, 107
Hermaphrodites, 302, 303
Hermissenda crassicornis. See
    Marine snail

Heron
  great blue, 474
  green, 349
Herring gull *(Larus argentatus)*, 34
  display in, 546
  food solicitation in chicks, 388
  oblique long call in, 536
  retrieval behavior in, 121–122
Heterocephalus glaber. See Mole rat
Heterodon platyrhinos. See
    Hog-nosed snake
Heterozygous condition, 61
Hibernation, 205–206
  as circannual rhythm, 190–191
Hierarchies. *See* Dominance
    hierarchies
Hilara sartor. See Balloon fly
Hill kangaroo *(Macropus robustus)*,
    reproduction in, 324–325
Hoary bat *(Lasiturus cinereus)*, mi-
    gration in, 417
Hog-nosed snake *(Heterodon
    platyrhinos)*, defense against
    predator, 496
Homeostatic mechanism, drive
    as, 268, 291–292
Home range, 348–349
  pheromones used for staking
    out, 389–391
  of rodents, 428–430
  signals regulating, 382
  size of, 351
Homing
  definition, 411, 431
  in fish, 431–432
  in rodents, 428–430
  See also Navigation
Homing pigeon *(Columba livia)*
  orientation and navigation and,
    420–421, 426–427
  sexual behavior, 160
Homo erectus, interest in animal
    behavior and, 13–14
Homozygous condition, 61
Homologies, in behavior patterns,
    536–537
Homo sapiens, interest in animal
    behavior and, 14
Homozygosity, 61
  inbreeding and, 91–92
Honeybee *(Apis mellifera)*
  development of caste and,
    242–244
  eusociality and, 575–576
  food location communication,
    398–400
  visiting frequency of, 193, 194
  waggle dance of, 375, 394,
    398–400, 405, 406, 431, 575
Honeycreeper, Hawaiian, 352

Hoplistomerus nobilis. See Robber
    fly
Hormone-brain relationship, 178
Hormone replacement therapy,
    161
Hormones
  aggression and, 360–361
  population regulation and, 521–
    523
  reproduction and, 323–332
  See also Endocrine system; spe-
    cific hormones
House cricket *(Acheta domesticus)*,
    sound production in, 125
House mouse. *See* Mouse
Howler monkey *(Alouatta* spp.),
    distance-maintaining signals
    of, 382, 391–393
Humans *(Homo sapiens)*
  control of aggression in,
    366–368
  cultural and genetic evolution
    and, 552–553
  entrainment and, 196
  heritability of intelligence in,
    101–102
  hormonal treatments and, 167,
    361
  links to aggression and, 361, 363
  personal space and, 350
  play behavior and, 249–250
  social behavior of, 495,
    583–586
  stress and pregnancy, 224
Humpback whale *(Megaptera no-
    vaeangliae)*, song of, 393
Hyalophora euryalis. See Giant
    silkworm moth
Hybrid vigor, 100, 107
Hydra *(Obelia)*
  coloniality in, 558
  nerve net in, 122–123
Hydrozoans, as colonial system,
    558
Hyena *(Crocuta crocuta)*
  feeding relationship of, 388,
    492–493, 495
  kills for lion made by, 489
  Mungi, 390–391
  territoriality in, 390–391
Hyla ewingi. See Australian tree
    frog
Hyla verreauxi. See Australian tree
    frog
Hylobittacus apicalis. See
    Scorpionfly
Hymenoptera, kin selection and,
    575–576. *See also* Ant; Bee;
    Wasp
Hyperphagia, in rats, 138

Hypophagia, in rats, 138
Hypothalamus, 155, 156, 158, 169, 178, 359, 413
Hypothesis formation, 41–43
Hypsignathus monstrosus. See African hammerheaded bat

Imitation, 277–278
Impeyan pheasant (*Lophorus impejanus*), courtship display in, 404, 405
Imprinting, 225–229, 258–259
    habitat, 459–460
Inbreeding, 91–92, 95, 107
Inclusive fitness, 80
Independent data points, 44–46
Independent variables, 43
Indigo bunting (*Passerina cyanea*)
    individual recognition among, 382–383
    orientation and navigation in, 424–425
    song of, 382–383
Individual distance, 349, 350
Individual recognition, communication for, 382–383
Induced ovulators, 325
Information analysis, 380–381
Inheritance, 19
    modes of, 100, 107
    See also Genetics; Heredity
Innate releasing mechanism (IRM), 253
Insects
    biological clock of, 195
    chemicals emitted by, 382
    circannual rhythm of, 191
    class recognition in, 384
    defense displays of, 496, 497
    eclosion, 163–164
    egg-laying behavior, 255
    eusociality in, 559–561, 575
    growth and metamorphosis in, 169
    hormones and behavior in, 153
    migration, 418–419
    orientation and navigation in, 431
    pheromones, 389–390
    regurgitative food offerings of, 544
    sexual attraction, 163
    visual receptors in, 125–126
    See also specific insects
Insight learning, 275, 276
Instinct model of behavior, 252–256
Instrumental learning. *See* Operant learning
Intention movement, 403

Interference competition, 346
Interdemic selection, 79–80, 525
Intermediate mode of inheritance, 100
Interobserver reliability tests, 40–41
Intersexual selection, 305, 306–308
Intrasexual selection, 305, 308–310, 338
Intraspecific competition, population regulation and, 512
Invertebrates
    coloniality, 558–559, 570
    defense against predators, 497–498
    as detritus feeder, 478
    endocrine systems of, 153–154
    migration of, 418–419
    tactile communication of, 394
    See also specific invertebrates
Irritability, 122
Island
    phylogeny from study of, 540
    species distribution on, 449–450
Isolating mechanisms, in speciation, 549–550
Isolation experiments
    for endogenous control of biological clocks, 193
    evaluation of, 255
Isolation-induced aggression, 363
Isopod (*Porcellio*), habitat selection, 455
Iteroparity, 321, 460–461

Jacana spinosa. See American jacana
Jackal, black-backed, 574, 575
Japanese macaque (*Macaca fuscata*), food habits of, 277–278, 368, 552
Japanese quail (*Coturnix coturnix*), 455
Jay
    blue, 475, 476, 497, 500–501, 563, 564
    Mexican, 564, 565
    piñon, 563, 564
    scrub, 563–565, 574
    sociality in, 562–566
    Steller's, 563
Jellyfish (*Aurelia*), nerve net in, 122–123
Jumping mouse, hibernation, 205
Junco hyemalis. See Dark-eyed junco

Juvenile hormone (JH), insect growth and metamorphosis, 169
Juvenile stage, 232

Kangaroo, hill, 324–325, 346–347
Karyotype, 59
Kea (*Nestor notabilis*), play behavior in, 249
Killdeer, 500–501
Kineses, 420
Kingfisher, belted, 72
Kin recognition, 384
Kin selection, 79–80, 526, 573–576
    in humans, 585–586
    insect eusociality and, 559–561, 575
Kittiwake (*Rissa tridactyla*), 35
Knot (*Calidris canutus*), migration, 412

Labile phase, of memory, 290
Lagomorph, reproduction in, 346, 365
Lampyridae. *See* Firefly
Language
    chimpanzees and, 288, 289, 401–403
    honeybees and, 398–400
Langur monkey (*Presbytis entellus*), infancy and, 310, 336, 346, 347
Larus argentatus. See Herring gull
Larus atricilla. See Laughing gull
Larus californicus. See California gull
Larus delawarensis. See Ring-billed gull
Larus hyperboreus. See Glaucous gull
Larus ridibundus. See Black-headed gull
Larus thayeri. See Thayer's gull
Lasioglossum zephyrum. See Sweat bee
Lasiurus cinereus. See Hoary bat
Lasius niger. See Heathland ant
Latent learning, 275–276
Laughing gull (*Larus atricilla*)
    feeding behavior of chick, 231–232, 233
    habitat selection, 457
Law of the minimum, 509
Leaf-cutter ant (*Acromyrmex myrmex*), food supply of, 479–480, 481
Learning
    aggression and, 361–363
    avoidance, 275

Learning (cont.)
    behavior development and,
        256–257
    classical conditioning,
        269–270, 272, 295
    comparative, 286–287
    constraints on, 287–290
    definition, 268
    habituation, 269, 295
    imitation, 277–278
    insight, 275, 276
    latent, 275–276
    memory, 290–291
    observational, 277–278
    operant, 270–271, 295
    phylogeny of, 278–287
    preparedness, 287–288
    sets, 278
    social behavior benefiting, 569
    See also Motivation
Learning curve, 272, 273
Learning sets, 278
Lek, 312
Lemming (Dicrostonyx spp.;
    Lemming spp.), population
    cycle of, 521
Leodice viridis. See Palolo worm
Leopard frog (Rana pipiens),
    129–130
Lepomis macrochirus. See Bluegill
    sunfish
Lepus americanus. See Snowshoe
    hare
Lesions, nervous system
    investigation by, 136–139
Lesser yellowleg (Tringa flavipes),
    migration, 413
Leuresthes spp. See Grunion
Limbic system
    aggression and, 359–360
    in primates, 128, 129
Limenitis archippus. See Viceroy
    butterfly
Limpet (Lottia gigantea; Collistella
    scabra), stimulation of food
    supply, 479
Linear peck order, rhesus monkey
    and, 356
Lion (Panthera leo)
    communication and, 377
    cooperative rearing of young
        in, 566
    feeding relationship of,
        489–491
    kin selection in, 574
    social behavior and, 325
Little brown bat (Myotis lucifugus)
    echolocation in, 430
    migration in, 417

Lizard
    biological clock in, 193, 195
    of Galápagos Islands, 16
    as predators, 512
    reproduction in, 176–178
    spiny, 549–550
Location/fear dichotomy,
    227–228
Loci, 60
Locust (Acrididae)
    desert, 164
    grasshopper, 151–152, 164
    migration in, 419
    migratory, 58
    orientation and navigation in,
        431
Longitudinal design, of test, 219
Long-term memory, 290
Lophiiformes, aggressive mimicry
    in, 483–484
Lophorus impejanus. See Impeyan
    pheasant
Lorikeet (Charmosyna placentis; C.
    rubrigularis), distribution, 453
Loris, evolution of, 204
Loss-of-information hypothesis,
    234
Lottia gigantea. See Limpet
Loxodonta africana. See Elephant
Lugworm (Arenicola marina),
    epicycles of, 189, 190
Lumbricus terrestris. See Earth-
    worm
Lunar cycle, 189, 190
    reproduction and, 325
Luteinizing hormone (LH), 157,
    158, 159, 160, 171, 179
    aggression and, 361
    reproductive behavior and, 171
Lycaon pictus. See African wild dog
Lymnaea. See Snails
Lynx (Lynx canadensis), population
    cycle of, 521

Macaca arctoides. See Stump-tailed
    macaque
Macaca fuscata. See Japanese
    macaque
Macaca mulatta. See Rhesus
    monkey
Macaca nemestrina. See Pigtailed
    macaque
Macaca radiata. See Bonnet
    macaque
Macaque (Macaca spp.)
    bonnet, 328, 336, 362
    delayed response problem,
        and, 285
    grooming in, 394

    hierarchy and, 367–368
    Japanese, 277, 278, 552
    pigtailed, 336, 362
    stump-tailed, 335
    territoriality and, 367–368
    See also Rhesus monkey
Macropus robustus. See Hill
    kangaroo
Macrotermes. See Termite
Malayan forest rat, reproductive
    cycle in, 324
Male dominance polygyny, 311
Mallard duck (Anas platyrhynchos)
    auditory imprinting, 229
    imprinting, 226, 227, 228
    migration, 412
    observational learning in, 277
    sexual imprinting, 229
Malthusian population growth, 16
Mammals
    effects of melatonin and, 202
    entrainment and, 195–196
    migration in, 417–420
    orientation and navigation in,
        427–430
    parental investment and,
        319–320
    pheromones in, 389–391
    play behavior in, 246
    reproduction in, 322, 329, 363
    social behavior among,
        566–568
    sounds of marine, 393
    territorial size and boundary of,
        351
Mandarin duck, wing preening
    in, 403, 404
Marginal value theorem, 476, 477
Marine mammals
    migration of, 418
    sounds of, 393
Marine mollusc (Tritonia),
    electrode stimulation of,
    133–134
Marine snail (Hermissenda
    crassicornis), conditioning
    and, 282–283
Marmota flaviventris. See
    Yellow-bellied marmot
Marmota monax. See Woodchuck
Marmota olympus. See Olympic
    marmot
Marsh wren (Cistothorus
    palustris), song learning and,
    237
Mass action, 136
Mating behavior. See Reproduc-
    tion
Mating systems, 310–311

lek, 312
parental investment and, 315–321
resource distribution and, 314–315
sexual selection and, 305–310
species recognition for, 382
See also Reproduction
Meadow pipit (Anthus pratensis), habitat, 450
Meadow jumping mouse, as prey of owl, 473
Meadow vole (Microtus pennsylvanicus)
copulation, 328
cyclical selection in, 78
epicycle of, 189–190
population cycles of, 106
as prey of owl, 472, 473
Mechanical isolation, 77
Melanism, in moths, 548
Melanocharis niger; M. longicauda; M. versteri. See Flower pecker
Melanophore-stimulating hormone (MSH), 156, 157, 179
color change and, 165
Melanoplus. See Grasshopper
Melanerpes formicivorous. See Acorn woodpecker
Melatonin, 157, 202
Meleagris gallopavo. See Turkey
Membrane permeability, 113
Memory, 290–291, 296
Menses, 327
Menstrual cycle, 327
Mephitis mephitis. See Skunk
Meriones unguiculatus. See Mongolian gerbil
Merops bullockoides. See White-fronted bee-eater
Mesocricetus auratus. See Syrian golden hamster
Metacommunication, 377–378, 406
Metamorphosis, in insects, 169
Meteorological cues, in bird orientation and navigation, 426
Mexican jay (Aphelocoma ultramarina), sociality in, 563, 564
Microevolution, behavior patterns and, 536–537
Microhabitat choice and reproductive success, 455
Micropezidae. See Stilt-legged fly
Microtus montanus. See Montane vole

Microtus pennsylvanicus. See Meadow vole
Micrurus fulvius. See Coral snake
Midge (Clunio marinus), lunar cycles and, 189, 190
Migration, 206–207, 411–420, 435
birds and, 411–414
definition, 411
invertebrates, 431
mammals, 427–430
tradition, 552
vertical, 206–207
See also Orientation and navigation
Migratory locust (Schistocera gregoria), classification, 58
Migratory restlessness, 413, 414
in birds, 424–426
geomagnetic cue and, 427
Minimum, law of the, 509
Mink, ovulation in, 325
Mirounga angustirostris. See Elephant seal
Mobbing, as defense against predators, 500–501
Modal action pattern (MAP), behavior development and, 255–256
Mole rat (Heterocephalus glaber), social behavior of, 567
Mollusks, marine, 133–134. See also Octopodes
Molothrus ater. See Brown-headed cowbird
Molting, 164–165
Molting hormone (MH), insect growth and metamorphosis and, 169
Monarch butterfly (Danaus plexippus)
migration, 418–419
poisons in, 275, 497
Mongolian gerbil (Meriones unguiculatus)
copulation, 328
testosterone and aggression in, 39–46
Monkey
African blue, 386
African green, 354
aggression in, 356–357
alarm calls in, 386–387
Cebus, 352
howler, 352
langur, 310, 336, 346–347
limbic system of, 128
metacommunication in, 377
Old World, 222

patas, 495
red-tailed, 386
squirrel, 44, 391
titi, 394
vervet, 354, 386, 453, 454, 496
See also Rhesus monkey
Monogamy, 310, 311
humans and, 585
Monophyletic, 545
Monozygotic, 101
Montane vole (Microtus montanus), copulation, 328
Moose (Alces alces)
as prey of wolf, 493, 494
population densities, 462
Morgan's canon, 19
Mormon cricket (Anabrus simplex), mating behavior and, 318, 319
Mormyridae. See Electric fish
Mosaics, 103
Mosquito (Aedes spp.; Anopheles spp.)
circannual rhythm of, 191
egg-laying location, 450
Motacilla alba. See Pied wagtails
Moth
antennae of, 126
Bogong, 515
cryptic, 475–476
display behavior of, 496, 497
eclosion, 163–164
intrasexual competition and, 310
melanism in, 548
noctuid, 120–121, 393
sexual attraction, 163
See also Silkworm moth
Motivation, 291–294, 296
definition, 268
models of, 292–294
Motmot (Eumomota superciliosa), coral snake recognition by, 499
Motor system, prenatal development of, 221–224
Mountain sheep (Ovis canadensis), habitat selection, 462
Mouse
aggression in, 361, 362, 518, 519
bedding preference of, 460, 461
Bruce effect in, 310
chemical alarm substance of, 386
communal nest of, 567
copulation, 328
distress calls of young, 388
dominance in, 353–354
echolocation and, 50–51

Mouse *(cont.)*
  effects of melatonin and, 202
  environmental effects on
    behavior of, 98–99
  genetic combinations, 60, 61
  golden, 328
  habitat selection, 456
  heritability in, 96
  homing in, 428–430
  jumping, 205
  life cycle chromosome changes,
    60, 61
  meadow jumping, 473
  nest-building activity, 92–93
  overpopulation and, 518–521
  pheromones in, 390–391, 522
  position and sexual behavior
    and, 168
  prey of owl as, 472, 473
  puberty age and artificial selec-
    tion, 238–240, 523
  sexual maturation, 238, 240
  survival strategies, 205–206
  urination in, 390, 523
  water escape behavior, 100–101
  wheel-running activity, 96, 100
  wild titmouse, 458
  See also Deermouse; White-
    footed mouse
Mouth-breeding cichlid fish
    (Haplochromis spp.)
  change in, 397
  pigmentation in, 544–545
Mudpuppy *(Necturus maculosus)*,
    retina response and, 130–131
Muller's ratchet, 303, 304
Mullerian mimics, 497–498
Mungos mungo. See Banded mon-
    goose
Musk ox *(Ovibos moschatus)*, as
    prey of wolf, 493
Mus musculus. See Mouse
Mussel *(Mytilus edulis)*, habitat
    restriction, 451
Mustela erminea. See Short-tailed
    weasel
Mustela frenata. See Long-tailed
    weasel
Mustelids, ovulation, 325. *See also*
    specific mustelids
Mutation, 65, 99
Mutation pressure, 68, 83
Myotis lucifugus. See Little brown
    bat
Myrmeleon. See Ant lion
Mysticeti. See Baleen whale
Mystilus edulis. See Mussel

Nasonia vitripennis. See Parasitic
    wasp

Natural cooperation, 570, 571
Natural selection, 63–65, 68, 69–
    70, 83
  aggression and, 523–524, 526
  agonistic behavior and, 361
  altruistic behavior and, 577
  behavior transmission and, 551
  competition and, 569
  defense against predators and,
    495
  evolution by, 17–18, 548
  evolutionary stable strategies,
    80–81
  feeding techniques and,
    477–485
  migration and, 416
  observation in the field and,
    548–549
  parent-offspring conflict and,
    322
  reproduction and, 330, 338
  sociality and, 572–581
  speciation and, 550
Nature-nurture controversy,
    252–258, 260
Navigation, 411. *See also*
    Orientation and navigation
Nectarinia reichenowi. See Golden-
    winged sunbird
Necturus maculosus. See Mud-
    puppy
Needs, 268. *See also* Motivation
Negative feedback loop,
    population regulation and,
    521–523
Nematocyst, 558
Neodiprion sertifer. See Pine sawfly
Neophron percnopterus. See
    Egyptian vulture
Neoteny, 540
Nerve net, 122–123, 144
Nerve tracts, 123
Nervous system, 111
  bilaterally symmetrical,
    125–126
  development of, 220–221
  endocrine system compared
    with, 152
  evolution of, 122–131
  generalized vertebrate, 126–
    129
  impulse communication,
    112–114, 115
  investigation methods;
    129–139, 145
  nerve cell properties, 114–115
  perception, 117–118
  radially symmetrical, 123–124
  sensation, 115–117
  sensory filters, 120–122

  simple, 122–123
  units, 111–112
Nestor notabilis. See Kea
Net energy gain, 485
Net rate of energy intake, 470–471
Neural stimulation, nervous
    system investigation
    through, 133–136
Neuroendocrine glands, 151,
    153–154
Neuromodulator, 114
Neuron, 111, 112, 144
Neurosecretions, 151. *See also*
    Endocrine system
Neurotransmitters, 114, 115
Niche
  ecological, 73–74
  fundamental, 446
  realized, 446
  See also Habitat selection
Noctuid moth *(Catocala)*, 120–
    121
  bat sounds received by, 393
Noradrenaline, 157, 158
  aggression and, 360
Norway rat *(Rattus norvegicus)*, 58
  attack behavior in, 362, 367
  autoradiography and, 178
  behavior transmission and, 551
  biological clock and, 201
  brain lesion in, 136
  classification, 58
  copulation in, 141–142, 328
  defecation rates, 98
  drinking behavior of, 134–135,
    294
  effects of handling and,
    224–225
  effects of melatonin and, 202
  estrus cycle and, 178
  food habits of, 138, 141, 142,
    229–231, 288
  inbreeding, 91–92
  intrauterine position and sexual
    behavior, 168
  maternal behavior, 138–139
  maze learning in, 272
  operant conditioning and, 284
  overpopulation, 518–519
  parturition and maternal be-
    havior of, 174–176
  playfighting, 167
  poison bait and, 286
  prenatal stimulation, 222, 224
  prey of owl as, 473
  self-stimulation and, 135
  sensory development in em-
    bryo, 222
  sexual behavior, 139, 151–153,
    160, 165–166, 168, 291

in Skinner box, 271
social dominance and mating at garbage dump, 52–53
social experience and environmental preference of, 37–39
stress and, 169
transplantation of preoptic tissue, 143
uses, 24
wild versus domestic, 38

Obelia. See Hydra
Observational learning, 277–278
Observational method, 14, 19
identifying subjects and, 49
natural variation vs. manipulation and, 49–50
observing vs. watching, 48–49
Ochrotomys nuttalli. See Golden mouse
Octopodes
learning and, 281–282
polarized light and, 431
tactile discrimination in, 118, 119, 282, 283
visual receptors in, 125
Octopus (*Octopus* spp.)
color waves and mood of, 397
sexual behavior of, 330
Odocoileus spp. *See* Deer
Odocoileus hemionus. See Black-tailed deer
Odocoileus virginianus. See Deer, spotted white-tailed
Odors
bird navigation and orientation and, 426
for communication, 389–391
See also Pheromones
Old World monkeys, dependency and, 322
Olfactory cues, in bird orientation and navigation, 426
Olive baboon (*Papio anubis*), reciprocal altruism in, 578
Olympic marmots, social behavior of, 10, 11
Omatidia, 125, 126
Oncorhynchus spp. *See* Pacific salmon
Ontogeny, recapitulating phylogeny, 547
Operant learning, 270–271, 272
Operational sex ratio, 304
Ophiuroidea. See Brittle star
Opossum (*Didelphis virginiana*), defense against predator, 496
Optimality theory, 80
Orange chromide
courtship pattern in, 34

development of behavioral patterns and, 244–246
Orconectes rusticus. See Crayfish
Organizational effects, of hormones on behavior, 161, 165–169
Orientation and navigation, 420, 570
in amphibians and reptiles, 432–435
birds and, 420–427
definition, 411
invertebrates and, 431
fish and, 431–432
kineses, 420
mammals and, 427–430
in taxes, 420, 421
Ostreidae. See Oyster
Ovary, secretions, 157, 158
Overdominant mode, 100, 107
Overpopulation, social pathology of, 518–521
Ovibos moschatus. See Musk ox
Ovis spp. *See* Sheep
Ovis canadensis. See Mountain sheep
Ovulation, vaginal stimulation for, 394. *See also* Reproduction
Ox, musk, 493
Oxytocin, 156, 157, 179
milk secretion and, 174–175
Oyster (*Ostreidae*)
shell valve-opening time in, 190
tidal rhythms and, 189, 190, 191

Pacific salmon (*Oncorhynchus* spp.), migration, 201, 321, 418, 420, 431–432, 461
Pain, aggression and, 362–363
Palolo worm (*Leodice viridis*), reproductive cycle in, 324
Pancreas, 155
Pangolin (*Pholidota*), feeding technique, 485
Panthera leo. See Lion
Pan troglodytes. See Chimpanzee
Paper wasp (*Polistes fuscatus*), dominance hierarchy in, 517
Papilio dardanus, 71
Papio spp. *See* Baboon
Paramecium aurelia. See Protozoans
Parasitic wasp (*Nasonia vitripennis*), circannual rhythm, 191
Parental care
dependency and, 322
ecological factors and, 321–322
monogamy and, 311

parental investment, 315–321, 322
polyandry and, 313–314
polygyny and, 312–313
reproductive effort and, 321–322
Parental investment, 315–321, 339
humans, 583–586
parent-offspring conflict and, 322–323
regurgitative food offerings as, 544
Parent-offspring relationship
aggression in, 345
communicating care, 388–389
conflict, 322–323
offspring manipulation, 581–582
Parsimony, law of, 19
Partial dominance, 60
Parulidae. See Warbler
Parus ater. See Coal tit
Parus caeruleus. See Blue tit
Parus major. See Great tit
Passerina cyanea. See Indigo bunting
Patas monkey (*Erythrocebus patas*), defense against predators, 495
Pavo. See Peacock
Peacock (*Pavo*),
courtship display in, 306, 404, 405
food enticement and, 547
Peacock pheasant (*Polyplectron bicalcaratum*), courtship display of, 306, 404, 405
Peck dominance hierarchies, 354
Peck right hierarchies, 354
Peking duck (*Anas platyrhynchos*), prenatal auditory development, 222
Perception, 117–118
Perceptual psychology, 22
Perceptual worlds, understanding, 50, 51
Period, of biological rhythm, 188
Peripheral filters, 120–121
Periplaneta. See Cockroach
Peromyscus leucopus. See White-footed mouse
Peromyscus maniculatus bairdi. See Grassland deermouse
Phalaropes, parental investment and, 318
Phase, of biological rhythm, 188–189
Phasianus colchicus. See Ring-necked pheasant

Pheasants
  communication synchronizing
    hatching in, 389
  courtship display in, 403–404
  impeyan, 404, 405
  peacock, 404, 405
  ring-necked, 404, 405
Phemphigus betae. See Aphid
Phenotype, 58, 63, 69–70, 83, 107
Phenotype matching, 384
Pheromones, 163, 179
  for communication, 389–391
  population regulation and,
    522–523
Philanthus triangulum. See Digger
  wasp
Philodota. See Pangolin
Philopatric, 314, 452
Phoca vitulina. See Harbor seal
Phoeniculus purpureus. See Green
  woodhoopoe
Phormia regina. See Blowfly
Phototaxis, planaria and, 420, 421
Photuris versicolor. See Firefly
Phylostomus hastatus. See Colonial
  bat
Phylogeny, 537–547
  adaptive radiation and, 540
  fossils and, 537–540
  learning and, 278–287
  ontogeny recapitulating, 547
Physalia. See Portuguese
  man-of-war
Physical contact, communication
  through, 394
Physiological processes, 6, 8
Physiological psychology, 22
Picoides pubescens. See Downy
  woodpecker
Pied flycatcher *(Ficedula
  hypoleuca),* population
  regulation, 511
Pied wagtails *(Motacilla alba),*
  territorial defense and,
  485–486
Pigeon
  homing, 160, 421–422, 426, 427
  sensation in, 116
  sun for orientation and
    navigation, 423
Pigtailed macaque *(Macaca
  nemestrina)*
  hierarchy in, 362
  infancy, 336
  kin selection in, 574
Piloting
  definition, 411
  in rodents, 428
Pineal gland, 155, 157, 165,
  201–202

Pine sawfly *(Neodiprion sertifer),*
  population regulation,
  509–510
Piñon jay *(Gymnorhinus cyanocephalus),* sociality in, 563, 564
Pipefish, parental investment
  and, 318
Pipit
  habitat, 450
  meadow, 450
  tree, 450
Pituitary gland, 155, 157, 413
  secretions, 156, 157
Placenta, secretions, 157, 158
Planaria *(Dugesia trigrina)*
  habituation and, 280–281
  negative phototaxis in, 420, 421
  ultraviolet rays and, 571
Plankton, vertical migration,
  206–207
Platyhelminthes. See Planaria
Play behavior, 246–250
Play, communication soliciting,
  389
Pleiotropism, 99
Plethodon hoffmani. See Pletho-
  dont salamander
Plethodon punctatus. See Pletho-
  dont salamander
Plethodont salamander *(Plethodon
  hoffmani; P. punctatus),*
  competition, 73–74
Plexis, in starfish, 123–124
Ploceini. See Weavers
Pluviahs domenica. See Golden
  plover
Pocket gopher *(Thomomys bottae;
  T. talpoides),* distribution, 452
Poison, as defense against
  predators, 497–499
Polecats, reproduction in, 363
Polistes fuscatus. See Paper wasp
Polyandry, 310, 311, 313–314
Polycentropus. See Caddisfly
Polygamy, 310–311
Polygyny, 310, 311, 312–313, 447
  humans and, 584–585
  threshold, 312
Polymorphism, 77–78, 83
Polyplectron bicalcaratum. See
  Peacock pheasant
Pomatostomus temporalis. See
  Grey-crowned babbler
Pond slider turtle *(Pseudemys
  scripta),* 118
Population genetics, 65, 83
  and change, 68, 69
  Hardy-Weinberg formula,
    67–68, 83
  mutation, 65

  recombination, 65–66
  relationship measurement, 78,
    79
Population regulation, 508, 527
  density factors and, 510–512
  food supply and, 510–511
  limiting factors, 508–510
  overpopulation pathology,
    518–521
  self-regulation, 512–524
  self-regulation, evolution of,
    524–527
  social behavior benefiting, 569
  territory and, 350
Porcellio. See Isopod
Porcupine, defense against
  predators, 497
Portuguese man-of-war *(Physalia),*
  coloniality of, 558, 559
Posture, as alarm call, 388
Postzygotic isolating mechanisms,
  77, 83
Prairie deermouse. *See* Grassland
  deermouse
Prairie dog *(Cynomys* spp.)
  defense against predators, 500,
    568
  epidemics in, 569
Predation
  as aggression, 345
  avoidance conditioning and,
    275
  cannibalism, 346–347
  defense against, 275, 495–502,
    568
  population regulators as, 512
  prey habitat restricted by, 451
  See also Feeding relationships
Prenatal development, of
  behavior, 219–225
Preparedness, learning and,
  287–288
Presbytis entellus. See Langur
  monkey
Prezygotic isolating mechanisms,
  77, 83
Primates
  and diurnality, 203–205
  grooming in, 394
  limbic system in, 128, 129
  menstrual cycle in, 327
  reciprocal altruism in, 578
  sexual dimorphism in, 584
  social behavior among, 568
  spacing signals used by, 382
  tradition and behavior
    transmission and, 552
  visual displays of, 396–397
  See also specific primates
Priming pheromones, 390

Prisoner's Dilemma, 577–578
Progesterone, reproductive
    behavior and, 171, 174, 177
Progestogens, 157, 158, 160
Prolactin, 157, 179
    milk secretion and, 175
    reproductive behavior and, 171
Promiscuity, 310
Protandry, 302
Protogyny, 302
Protozoan (Paramecium aurelia;
    Vorticella convallaria),
    learning and, 279
    turning rates, 455
Proximate factors, 10
Pseudemys scripta. See Pond slider
    turtle
Pseudoconditioning, 280
Psychobiology, 557
Psychopharmacology, nervous
    system investigation with,
    141–142
Ptilonorhynchus violaceus. See
    Satin bowerbird
Pygmy mouse (Baiomys taylori),
    environmental effects of
    behavior of, 98–99

Quail, bobwhite, 222–223
Queen butterfly (Danaus gilippus),
    toxicity of, 497
Quelea. See African weaver finch
Quinault Indians, 68

Rabbit
    behavior and habitat selection
    of, 26
    ovulation, 325
Racoon, survival strategies, 206
Radially symmetrical nervous
    system, 123–124, 145
Radioimmunoassay, for endocrine
    system study, 161
Rana catesbeiana. See Bullfrog
Rana pipiens. See Leopard frog
Rana sylvatica. See Wood frog
Rangifer tarandus. See Caribou
Rat. See Norway rat
Rattus norvegicus. See Norway rat
Realized niche, 446
Recessive allele, 60, 61
Reciprocal altruism, 576–581
Recognition, communication for,
    382–384
Recombination, 65–66
Red squirrel (Tamiascirus spp.),
    population regulation,
    516–517
Red-tailed monkey (Cercopithecus
    ascanius), alarm call of, 386

Red-winged blackbird (Agelaius
    phoeniceus)
    annual cycle of, 7–10
    classification, 58
    competitive exclusion and, 452
    defense against predator, 500
    dominance hierarchy in, 517
    epaulets of, 376
    mating system of, 314–315
    migration in, 414
    polygyny in, 314
    population regulation in, 517
    territory and, 32–33, 47, 352
Reef fish, defense against
    predators, 496
Refractory period, 113
Relative genetic fitness, and
    evolution, 62
Replication, of experiment, 41
Reproduction
    asexual, 302–303
    in cat, 330
    climatic-hormonal interaction
    and, 323–325
    competition and, 517–518
    copulation, 325, 327, 328
    costs of, 302–304, 338
    displays in, 384
    diversity of, 302
    embryo resorption, 346
    fertilization, 325, 327, 328
    humans and, 583–586
    mating systems, 310–311
    migration and, 416
    in octopus, 330
    ovulation, 325, 327, 328
    in rhesus monkey, 332–338
    rodents and, 346
    sex ratios, 304–305
    sexual selection, 305–310
    social-hormonal interaction,
    325–332
    in turkey, 329–330, 331
    See also Parental care;
    Population regulation
Reproductive effort, parental care
    and, 321–322
Reproductive isolation, speciation
    by, 77
Reptiles
    orientation in, 432–435
    social behavior among, 562
    See also specific reptiles
Resource defense polyandry, 311
Resource defense polygyny, 311,
    312–313
Resource partitioning, 478
Response to selection, 96
Rhesus macaque (Macaca mulatta)
    aggression in, 135, 360, 363, 368

brain lesions in, 136
circadian rhythm and, 201
classification, 58
defense against predators, 501
deprivation in, 234–236
dominance in, 356–358, 378,
    517, 567
fear and tension responses by,
    135–136
grooming in, 395
kin selection in, 574
metacommunication in, 378
parent-offspring conflict in, 322
perianal region as signal, 376
reproduction in, 332–338
sex ratio and, 320–321
sexual behavior, 166–167, 328
Rhinencephalon. See Limbic
    system
Ring-billed gull (Larus delawaren-
    sis), navigation and orienta-
    tion in, 427
Ring dove (Streptopelia risoria),
    reproduction in, 161–162,
    170–174, 325, 326, 548
Ring-necked pheasant (Phasianus
    colchicus), courtship display
    in, 404, 405
Riparia riparia. See Bank swallow
Rissa tridactyla. See Kittiwake
Ritualization, 403
    of display, 546
Robber fly (Hoplistomerus nobilis),
    as dung beetle specialist,
    468, 470
Robin (Turdus migratorius)
    migration in, 414
    red feather reaction, 291
    See also European robin
Rodent
    aggression in, 39–41
    cannibalism among, 346–347
    copulation in, 327–328
    diurnality and, 204–205
    home range of, 428–430
    nest-building activity, 271
    orientation and navigation in,
    427–430
    piloting in, 428
    See also specific rodents
Rooster, cocks comb in castrated,
    161

Saddleback wrasse (Thalassoma
    duperrey), sex change in, 305
Sage grouse, lek, 353
Saimiri sciureus. See Squirrel
    monkey
Salamander
    competition and, 73–74

Salamander *(cont.)*
  fertilization in, 329
  plethodont, 73–74
  polarized light and, 431
Salmon. *See* Pacific salmon
Salmo trutta. See Brown trout
Sample size, 46
Sandpiper, migration, 413
  parental investment in, 318
Sarcophaga argyrostoma. See Flesh
  fly
Satin bowerbird *(Pitilonorhynchus
  violaceus)*, intrasexual
  selection in, 307–308
Sawfly, pine *(Neodiprion sertifer)*,
  509–510
Scatophaga stercoraria. See Yellow
  dung fly
Sceloporous. See Spiny lizard
Schistocera gregoria. See Desert
  locust
Schizoporella errata. See Bryozoan
Scorpionfly *(Hylobittacus apicalis)*,
  precopulatory food offering
  in, 543
Scramble competition, 313, 345
Scrub jay *(Aphelocoma coerulescens)*, sociality in, 563–565,
  574
Scyliorhinus caniculus. See Shark
Sea anemone
  chemostimulation and,
  279–280
  nerve net in, 122–123
Sea bass, 303, 305
Sea gull
  aggression in, 363
  individual distance and, 350
  monogamy in, 311
  reproduction in, 363
  See also specific gulls
Sea hare *(Aplysia* spp.)
  biological rhythm of, 199–201
  habituation and, 232
Sea horse, parental investment
  and, 318
Seal
  elephant, 309, 317, 318, 359,
  517
  harbor, 418
Sea otter *(Enhydra lutris)*, feeding
  technique of, 484
Search image, foraging strategies
  and, 475
Sea urchins, nervous system in,
  123
Secondary sex characteristics
  hormones affecting, 158, 161
  intrasexual competition and,
  305

Selection
  artificial, 94–95
  cyclical, 78
  group, 79–80, 525, 572
  kin, 79–80, 526, 573–576
  population self-regulation and,
  526
  response to, 96–97
  See also Natural selection
Selection differential, 96
Selective breeding, 93–97
  dog domestication and, 540–
  542
Selfishness, cooperation through,
  572–583
Self-regulation, population
  regulation and, 512–514
  evolution of, 524–527
Self-stimulation, nervous system
  investigation through, 135–136
Semelparity, 321, 460–461
Sensation, 115–117, 144
Sensitive period, 227
Sensitization, 280
Sensory adaptation, 269
Sensory filters, 120–122
Sensory system, prenatal devel-
  opment of, 221–224
Serinius canarius. See Canary
Sex change, 302, 305
Sex ratio, 304–305
Sexual aggression, 345
Sexual attraction, hormones and,
  163
Sexual behavior
  aggression and, 161–163
  hormones and, 165–169
  See also Reproduction
Sexual dimorphism
  in humans, 584
  in primates, 584
Sexual imprinting, 229
Sexual selection, 305–310
Shannon-Weiner index, 380–381
Shark *(Scyliorhinus caniculus)*,
  electroreceptors in, 395, 482
Sheep *(Ovis* spp.)
  coyote predation on, 275
  feeding relationship in, 568
  tradition in mountain, 462
  tragedy of the common and,
  513–514
  wolf predation on, 493
Shelter, limiting population factor
  as, 511–512
Short-tailed shrew, prey of owl
  as, 473
Short-tailed weasel *(Mustela
  ermina)*, feeding behavior of,
  271

Short-term memory, 290
Shrew *(Blarina; Sorex)*, population
  regulation, 509–510
Sialia sialis. See Eastern bluebird
Siamese fighting fish *(Betta
  splendens)*, aggression and,
  361
Signal function, 549–550
Signaling pheromones, 390
Signals
  chemical; see Pheromones
  as information conveyors, 375–
  378
  spacing, 382
Sign stimulus, 21, 252
Silkworm moth *(Bombyx mori)*
  circannual rhythm and, 191
  giant, 116
  sensation in, 116–117
  sexual attractant in, 390
Siphonophores, aggressive
  mimicry in, 484
Skinner box, 271
Skunk *(Mephitis mephitis)*,
  defense against predators,
  497
Sleeping behavior, cooperation
  and, 570
Smooth-billed ani *(Crotophaga
  ani)*, communal nesting, 563
Snails *(Lymnaea)*
  habituation and, 281–282
  reproduction in, 302
Snake
  coral, 499
  hog-nosed, 496
Snapping turtle *(Chelydra
  serpentina)*, food preferences
  of, 231
Snowshoe hare *(Lepus americanus)*, population cycle,
  521
Social behavior, 558, 586
  advantages of, 568–569
  colonial invertebrates and,
  558–559
  complex social systems and,
  558–568
  cooperation, 569–571
  cooperation through selfishness, 572–583
  disadvantages of, 568–569
  ecological factors in, 582–583
  evolution of, 540
  feeding and, 485–495
  humans and, 583–586
  parental manipulation of
  offspring and, 581–582
  social insects, 559–561
  vertebrates, 561–568

Social cues, choice of habitat and, 456
Social deprivation, behavior development and, 223–226
Social disorganization, aggression and, 367–368
Social-hormonal interactions, reproduction and, 325–332
Social insects. See Eusocial insects
Social status, display and, 385–386. See also Dominance
Social strategies, as defense against predators, 499–502
Society, definition of, 557. See also Social behavior
Sociobiology, 6, 20, 26–28, 462, 557
    definition of, 557
Song birds, mating behavior, 51, 52
Song sparrow (Melospiza)
    reproductive cycle in, 324
    song in, 392
Sorex. See Shrew
Sound, communication by, 391–393
Space, social use of, 348–353, 369
    definitions of, 348–350
    territorial size and boundary, 351–353
Spacing signals, 382
Sparrow
    chipping, 458–459
    golden-crowned, 413–414
    song, 324, 392
    swamp, 237–238, 239, 392
    white-crowned, 324, 374, 384, 385, 415, 454
    white-throated, 383–384, 413, 414
Specialists, feeding behavior and, 463
Speciation, 63, 74–76, 83, 537
    isolating mechanisms in, 549–551
Species, 58–59, 82, 97–99
    distribution, 445–446
    natural selection and, 63–65
    See also Habitat selection
Species-isolating mechanisms, 77
Species-typical behaviors, 25
Sperm competition, 309
Spermophilus beldingi. See Belding's ground squirrel
Spermophilus tridecemlineatus. See Ground squirrel
Spider
    communal web of, 487
    intrasexual competition and, 309

prey as, 512
sounds and, 284
sun compass orientation and, 431
web communication among, 394
See also Bowl spider
Spiny lizard (Sceloporous), head-bobbing patterns of, 549–550
Spizella passerina. See Chipping sparrow
Split-brain techniques, nervous system investigation by, 136–139
Spontaneous ovulators, 325, 327
Spontaneous recovery, 273, 274
Spoonbill, bill adaptation, 72
Spotted hyena. See Hyena
Spotted sandpiper (Actitis macularia), polyandry and, 314
Squirrel
    Belding's ground, 386–387, 454, 496, 500, 471
    flying, 191–192, 193, 195
    ground, 202, 205, 269, 496, 574
    red, 516–517
Squirrel monkey (Saimiri sciureus), 44
    sound as communication among, 391
Stabilizing selection, 70, 83
Starfish (Dermasterias imbiceta)
    chemical alarm substances of, 386
    nervous system in, 123–124
    sea anemones and, 279–280
Starling. See European starling
Stellar cues, in bird navigation, 423–426
Steller's jay (Cyanocitta stelleri), sociality in, 563, 564
Stephanauge. See Sea anemone
Stickleback. See Three-spined stickleback
Stilt-legged fly (Micropezidae), regurgitative food offering during copulation, 544
Stimulus and response, learning and, 269
Stomphia. See Sea anemone
Strain differences, 92–93
Strangers, aggression and, 363, 369–370
Streptopelia risoria. See Ring dove
Stress
    adrenal secretions and, 169
    prenatal, 224
    response to, 521, 523

Stress-of-emergence hypothesis, 234
Stump-tailed macaque (Macaca arctoides), sexual behavior, 335
Sturnus vulgaris. See European starling
Sun, bird orientation and navigation and, 422–423
Surface vibration, communication by, 394
Survival, evolution and, 62–63
Swallow, bank, 500–501, 568
Swamp sparrow (Melospiza), song in, 237–238, 239, 392
Sweat bee (Lasioglossum zephyrum), kin recognition and, 384
Symbolization, 403
Sympatric speciation, 76, 314, 550
Synapse, 114, 115
Synaptic processes, 112
Syncerus caffer. See Buffalo, water
Synergism, in endocrine system, 160
Synthetic theory, of evolution, 62–63
Syrian golden hamster (Mesocricetus auratus), aggressive and sexual behavior in, 162–163
Systematic variation, 286

Tactile communication, 394
Talitrus. See Beachhoppers
Tamias striatus. See Chipmunk
Tamiasciurus spp. See Red squirrel
Taxes, 420, 421
Teal (Anas discors), migration, 412
Telemetry, nervous system stimulation by, 135
Termite (Macrotermes), 561
    chemicals emitted by, 382
    coloniality in, 576
Territory
    defense of, 351, 485–486
    definition, 350
    lek, 353
    maintenance, 368
    model, 352–353
    population regulation and, 350, 515–517
    size and boundaries, 351–353
    social behavior and, 569–571
    species recognition and, 382
    See also Home range
Testes, secretions, 157, 158
Testosterone, 158, 159, 173
    aggression and, 161–163, 360, 363
Testudo hermanni. See Turtle

Tetragoneura cynosura. See
Dragonfly
Thalassoma duperrey. See
Saddleback wrasse
Thamnophis sirtalis. See Garter
snake
Thayer's gull *(Larus thayeri)*,
signal function in, 549–550
Thomomys bottae; T. talpoides. See
Pocket gopher
Thomson's gazelle *(Gazella
thomsoni)*, as prey of lion, 490
Thorndike puzzle box, 24
Three-spined stickleback *(Gasterosteus aculeatus)*, 21
aggression in, 253–254
classification, 58
color changes in, 156
courtship dance, 255
defense against predators, 497
feeding behavior, 500
Thyroid, 155
secretions, 169
Tidal rhythms, 189, 190, 191
Tinamous, parental investment,
318
Tit
blue, 458
coal, 458
great, 477, 516
Titi monkey *(Callicebus)*, tactile
communication in, 394
Toads, nervous system of, 130
Tools, 552
as feeding technique, 484–485,
552
Topographic features, bird
orientation and navigation
and, 421–422
Tortoise, elephant, 17
Touch, communication through,
394
Toxicity, as defense against
predators, 497–499
Tradition
behavior transmission and,
551–553
habitat selection and, 462, 464
Tragedy of the commons,
513–514
Transection, nervous system
investigation by, 132, 133
Transformation, 301
Translocation experiments, for
endogenous control of
biological rhythms, 193–194
Transplantation, nervous system
investigation with, 143
Trapping, as feeding technique,
481–482

Tree pipit *(Anthus trivialis)*
habitat, 450
song in, 392
Trial-and-error learning. *See*
Instrumental learning
Tricolored blackbird *(Agelaius
tricolor)*
competitive exclusion and, 452
mating system of, 314–315
territoriality and, 352
Tringa flavipes. See Lesser
yellowleg
Tritonia. See Marine mollusk
Trophic hormones, 154, 155, 157,
158
Trophic levels, feeding relationships and, 468–470
Trophic neurosecretions, 154, 155
Tropical butterfly *(Anartia fatima)*,
function of wing-marks in,
497
Trout, brown, 515
True language, 401
Turdoides squamiceps. See Arabian
babbler
Turdus migratorius. See Robin
Turkey *(Meleagris gallopavo)*
inter-gobble intervals of, 255,
256
reproductive pattern in,
329–330, 331
sexual imprinting, 229
Turtle
auditory sensitivity of, 118
green, 433–438
hibernation, 202
snapping, 231
Tyto alba. See Barn owl

Uca pugnax. See Fiddler crab
Underwater sound, 393

Vaginal cornification, 166
Vampire bats *(Desmodus rotundus)*, reciprocal altruism
in, 578, 579
Variable interval, 273
Variable ratio, 273
Variables, designation of, 43
Variance, 47
Variation
discrete, 46
continuous, 46
Vasopressin, 156, 157, 179
Vertebrates
biological rhythms, 201–202
circadian rhythm, 201–202
circannual rhythms, 202
endocrine systems of, 154–160
learning in, 284–286

nervous system of, 126–130
social behavior of, 561–568
Vertical migration, 206–207
Vervet monkey *(Cercopithecus
aethiops)*
alarm call in, 386
defense against predators, 496
dispersal and, 453, 454
linear hierarchy in, 354
Vestiaria coccinea. See Hawaiian
honeycreeper
Viceroy butterfly *(Limenitis
archippus)*, coloration pattern,
275
Vision, communication by,
396–397
Vole *(Microtus* spp.*)*
population cycle, 521, 523
See also Meadow vole
Vorticella convallaria. See Protozoans
Vulpes fulva. See Fox
Vulture, Egyptian, 378

Warbler
Cape May, 478, 479, 480
migration in, 412, 413
orientation and navigation, 424
resource partitioning of, 478,
479
yellow-rumped, 478, 479, 480
Wasp
digger, 276
paper, 517
parasitic, 191
prey of, 544
Water buffalo. *See* Buffalo
Water flea *(Daphnia)*, three-spined
stickleback feeding on, 500
Water strider *(Gerris remigis)*,
surface vibration communication by, 394
Weasel
color change in, 165
feeding behavior, 271
short-tailed, 165
Weavers *(Ploceini)*, development
of nesting habits and, 545,
546, 547
Whale
baleen, 393
blue, 478
gray, 418
humpback, 393
migration, 418
song of, 393
toothed (dolphin), 393
White-crowned sparrow *(Zonotrichia leucophrys)*
deme recognition and, 384, 385

meteorological migratory cues in, 426

migratory restlessness in, 415

outbreeding depression and, 454

reproductive cycle in, 324

singing of, 374

White-footed mouse (Peromyscus leucopus)

communal nest, 567

escape as defense, 496

food choices, 250–252, 510

home range, 348, 349, 517

population regulation and, 63, 64, 517

prey of owl as, 472, 473

prey of weasel as, 271

territory of, 350, 351

White-fronted bee-eater (Merops bullockoides), cooperation among, 582–583

White-throated sparrow (Zonotrichia albicollis)

neighbor recognition in, 383–384

migration in, 413, 414

song of, 383–384

species recognition in, 383–384

White rat. See Norway rat

Wildebeest (Connochaetes taurinus)

as prey of hyena, 492–493

as prey of lion, 491

Wisconsin General Test Apparatus (WGTA), 284–285

Wolf (Canis lupus)

cooperative rearing of young by, 567

copulation in, 328

domestication, 540–542 (see also Dog)

dominance hierarchy in, 517

feeding relationship of, 493–495, 512

metacommunication in, 377

play behavior in, 247–249

vocalization of, 393

Woodchuck (Marmota monax)

circadian rhythm of, 189, 190

hibernation and, 202, 205

social behavior of, 10

Wood duck (Aix sponsa), auditory communication and, 228

Wood frog (Rana sylvatica), egg laying location, 457

mating behavior and, 313

Woodland deermouse

early experience and, 459–460

habitat adaptation in, 100

Woodlouse, circadian rhythm of, 203

Woodpecker finch (Cactospiza pallida), tool use in feeding technique, 484, 485

Wood rat, reproductive behavior and, 328

Worm

foraging behavior evolution and, 539

palolo, 324

See also Planaria

Wren, marsh, 237

Xenophobia, aggression and, 363

Yellow-bellied marmot, social behavior of, 11

Yellow dung fly (Scatophaga stercoraria), captures, 81

intrasexual selection in, 309, 310

Yellow-rumped warbler (Dendroica coronata), resource partitioning of, 478, 479, 480

Zebra (Equus burchelli)

composite signals in, 376

defense against predator, 495

as prey of African wild dog, 491

as prey of hyena, 492–493

as prey of lion, 489, 491

Zebra finch (Columba spp.), sexual imprinting, 229, 361

Zeitgebers, 187, 195, 196, 206, 207

Zonotrichia albicollis. See White-throated sparrow

Zonotrichia atricapilla. See Golden-crowned sparrow

Zonotrichia leucophrys. See White-crowned sparrow

Zooid, 558, 559

Zugunruhe. See Migratory restlessness